APPLIED FINITE MATHEMATICS

RICHARD J. COPPINS
Virginia Commonwealth University

PAUL M. UMBERGER
Virginia Commonwealth University

ADDISON-WESLEY PUBLISHING COMPANY
Reading, Massachusetts • Menlo Park, California
Don Mills, Ontario • Wokingham, England • Amsterdam
Sydney • Singapore • Tokyo • Mexico City • Bogotà
Santiago • San Juan

Sponsoring Editor: Jeffrey M. Pepper
Production Supervisor: Marion E. Howe
Copy Editor: Marret McCorkle
Text Designer: Marie McAdam
Geri Davis, Quadrata, Inc.
Illustrator: Textbook Art Associates
Photo Researcher: Darlene Bordwell
Manufacturing Supervisor: Ann DeLacey
Cover Designer: Marshall Henricks

Photo credits: Page 2, Hazel Hankin/Stock, Boston. Pages 36, 74, 274, 302, 422, 482, Darlene Bordwell. Page 4, Everett C. Johnson/Leo DeWys. Page 186, Burt Glinn/Magnum. Page 220, SharonBazarian/The Picture Cube. Page 348, Jim Domke/Black Star. Page 372, Courtesy Oldsmobile Division, General Motors Corporation. Page 536, Brude Davidson/Magnum.

Library of Congress Cataloging in Publication Data

Coppins, Richard.
Applied finite mathematics.

Includes index.
1. Mathematics—1961– . I. Umberger, Paul M.
II. Title.
QA37.2.C674 1985 510 85-1228
ISBN 0-201-10312-5

ABCDEFGHIJ-HA-89876

To Our Wives

Rosemarie Coppins
and
JoAnn Umberger

PREFACE

Our purpose in *Applied Finite Mathematics* is to introduce the mathematical tools most needed by students of business, economics, biology, and social science. We have tried to demonstrate the use of these tools to increase students' motivation. The material in this book is more than sufficient to provide a one-semester course in applied finite mathematics.

The book is intended for students who have had between one and two years of high school algebra. We have begun the book with two chapters of primarily review material. Depending upon the preparation of the students in the class, the instructor may include both of these chapters or refer to them as needed.

Important features of the book are as follows:

Modeling Approach A modeling approach is used throughout the book. This teaches students to think about how to attack a problem, rather than simply "plunging in." As a result, the modeling approach provides a better way to understand the mathematics in the book.

Writing Style The book is written so that students can read it. Through the use of a conversational, casual style, we neither talk down to students nor do we bore them.

Examples All examples in the book are completely worked out. Thus we ensure that students will not miss the purpose of an example because they can't supply a missing algebraic (or arithmetic) manipulation. Moreover, each chapter contains many examples.

Practice Problems Most examples are followed by a practice problem (and answer) to reinforce the concept demonstrated by the example.

Problem From Life Each chapter begins with a "Problem From Life" to help answer the questions "What good is this stuff?" and "Why do I need mathematics?" Each of these problems is worked as an example in the chapter.

Chapter Reviews Each chapter ends with a review and a set of review problems. The chapter review provides a summary, in alphabetical order, of the important terms and procedures the student should know. This provides an excellent starting point for study and a ready reference for review.

Exercise Sets Each chapter contains several exercise sets with numerous problems. Problems range in difficulty from basic drill to more-advanced exercises that are designed to test basic comprehension. Answers to the odd-numbered problems are provided in the back of the book.

Calculator The use of a calculator is integrated into the book, with examples that demonstrate explicitly the steps required to use a calculator.

Supplements An instructor's manual, which includes detailed answers to the even-numbered problems, is available. In addition, a student solutions manual, which provides detailed answers to the odd-numbered problems in the book, is also available.

Acknowledgments

We are extremely grateful for the excellent suggestions provided by the reviewers of the manuscript:

Wilson P. Banks, Illinois State University
Ron Barnes, University of Houston/Downtown
Raymond Coughlin, Temple University
Sam Councilman, California State University, Long Beach
Duane E. Deal, Ball State University
Frank Deane, Berkshire Community College
Murray Eisenberg, University of Massachusetts, Amherst
Curtis C. McKnight, University of Oklahoma
Tom Rodgers, Tyler Junior College
Peter Tannenbaum, California State University, Fresno
Carrol G. Wells, Western Kentucky University

Any errors that remain are solely our responsibility.

We also wish to thank Beverly Bowlas for typing the problem solutions and the staff at Addison-Wesley who transformed a rather messy manuscript into this book. We especially wish to thank Pat Mallion, Director of Development, for her faith in us and Jeff Pepper, Mathematics Editor, for his understanding, suggestions, encouragement, and great patience.

Richmond, Virginia

R.J.C.
P.M.U.

CONTENTS

8

9

10

11

APPLIED FINITE MATHEMATICS

1 NUMBERS, CALCULATORS, AND SETS

Gary and Georgia Vanning ate dinner with another couple, the Coopers. The Vannings paid $14.50 for the drinks in the bar before dinner. The bill for dinner (including sales tax of 5%) was $107.50. The Coopers and the Vannings decide to split the entire cost fifty–fifty (each couple will pay one half the total). The tip should be 15% of the cost of the dinner *without* tax. How much should each couple pay? How much more do the Vannings need to pay? (To be solved in Example 1, Section 1.1.)

PROBLEM FROM LIFE

INTRODUCTION

The world today, especially in the area of business, is far more complex than it was 30 years ago. The problems you encounter today are complicated because the world around you is not only complex but also changing rapidly. Approaches that might have been acceptable 30 years ago are likely to be inadequate today. Thus the more tools that you can use to make decisions, the better your decisions will be. Some of the tools that are currently used in business include accounting, economics, finance, statistics, operations research, and computers. To understand and utilize these tools properly, you **must** have the appropriate mathematical skills. We intend to help you learn these skills.

In this chapter, we'll begin a review of the basics that you'll need to be able to understand the material in this book. The chapter includes a review of the real number system and of sets and also a section that will help you use your calculator more effectively. In addition, we'll begin the chapter with a brief discussion of problem solving and modeling and how they relate to the mathematics you'll learn with this book.

1.1 PROBLEM SOLVING AND MODELING

As you go through this book, we not only want you to learn a set of skills, but we also want you to develop the ability to use these skills to solve problems. Here's a simple description of a quantitative approach to problem solving.

A Quantitative Approach to Solving Problems

1. Define the problem. This requires you to wade through the detail and determine what needs to be done.
2. Build a mathematical model of the problem. A **model** is an abstraction of the real world. Thus a mathematical model is a set of symbols and relationships that gives a reasonable portrayal of the real problem.
3. Analyze the model and determine the solution.
4. Implement the solution in the real world. This requires you to translate your paper solution into something that is meaningful in actual practice.

Most people believe that step 3, analyzing the model (performing the calculations), is the hardest part of problem solving. Actually, it's one of the easier parts. Steps 1 and 2, defining the problem and then converting it into symbols, are more difficult. As you work your way through the book, try to learn not only how to solve a problem, but how to recognize it and how to model it. We'll help you achieve this by providing practice in modeling and problem recognition.

You should have noticed the Problem from Life that began the chapter. We'll begin every chapter with one, which we'll analyze after you've learned the necessary skills. Now, let's take a look at the dinner problem presented to the Vannings and the Coopers. Although we won't solve it now, we do want to take you through the first two steps of our problem-solving approach.

EXAMPLE 1

Problem from Life Garry and Georgia Vanning ate dinner with another couple, the Coopers. The Vannings paid \$14.50 for the drinks in the bar before dinner. The bill for dinner (including sales tax of 5%) was \$107.50. The Coopers and the Vannings decide to split the entire cost fifty-fifty (each couple will pay one half the total). The tip should be 15% of the cost of the dinner *without* tax. How much should each couple pay? How much more should the Vannings pay?

SOLUTION

How can we solve this problem? It certainly doesn't appear to require any advanced mathematics. In fact, it is quite easy to solve once you devote some thought to it. The first step of our four-step approach is to define the problem. In this case it means that you should begin by *reading the problem completely and carefully*. Determine what value(s) you must know in order to solve the problem. Make note of the facts you do know.

Let's try that. The next-to-last sentence tells us that we want to determine how much each couple should pay. This means that we need to compute the total cost of the dinner (including drinks), because each couple should pay one half the total. Once we know the total cost of the dinner, it will be easy to compute how much more the Vannings owe. We know the following facts;

1. The cost of the drinks before dinner was \$14.50 (paid by the Vannings).
2. The cost of the dinner including tax was \$107.50.
3. The tax on the meal is 5%.
4. The tip should be 15% of the cost of the meal (not including tax).

The second step in the four-step approach is to develop a mathematical model. A good way to begin is to develop a simple *verbal* description of how to combine the different values to arrive at the solution. Once you have done this, you can decide which facts are relevant and which values are missing. It will then be much easier to develop the mathematical model. Don't be discouraged if your first model turns out to be incorrect. Sometimes you have to try several models. If a model does turn out to be incorrect (or inadequate), think about what's wrong with it before you revise it. In this case, we might develop the model shown in Fig. 1.1. Study it for a moment. Is it correct?

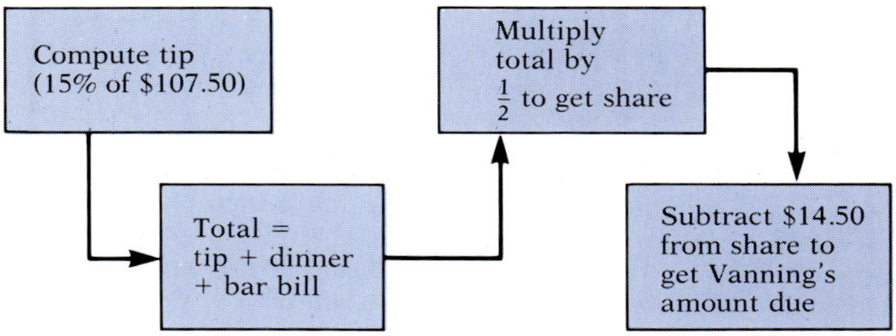

Figure 1.1 **First model, Vanning-Cooper dinner (Example 1).**

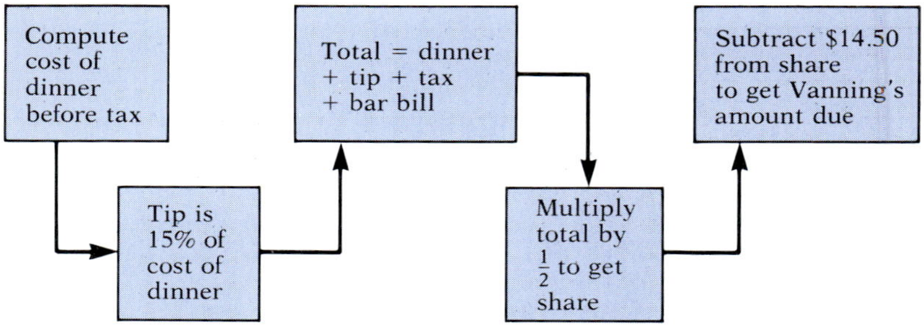

Figure 1.2 **Second model, Vanning-Cooper dinner (Example 1).**

Congratulate yourself if you said no. In Fig. 1.1 the tip is computed with the tax included. The correct method is to determine the tax, subtract it from the given cost, and then compute the tip. (See Fig. 1.2.)

For this example, there is no need to introduce a formal model with symbols and mathematical notation. You could proceed to calculate the answer directly from the model shown in Fig. 1.2 (step 3). (We'll do so in Section 1.3 of this chapter when we discuss calculators.) Moreover, because the situation described is so simple, there would be no difficulty in deciding how to arrive at a practical solution based on your mathematical model. □

1.2 THE REAL NUMBER SYSTEM

Numbers are a very important part of the language of mathematics. In order to be able to use numbers properly in both modeling and solving problems, you must be familiar with the "grammar" of numbers: what is allowed and what isn't allowed. You've seen various types of numbers such as 2, -3, $\frac{3}{8}$, 1.85, $\sqrt{2}$, and π (pi). Let's see which ones we need and how they relate to one another.

Types of Numbers and the Number Line: Rounding Off

As children, we all learned to count with the numbers 1, 2, 3, ... This group of numbers is called the **natural numbers.** We also learned that the natural numbers can't give us answers to certain questions. For instance,

1. Suppose your checking account has $100 in it and you write a check for $150. How much does it have in it now?
2. How can you divide two apples evenly among three friends?

To answer the first question we must expand the natural numbers to include the number 0 and the negative numbers such as -1, -2, -3. . . . This expanded group is called the **integers.** Integers can be positive (the natural numbers), 0, or negative. To answer the second question, we must introduce numbers such as $\frac{2}{3}$. The **rational numbers** are ratios (quotients) of the form m/n, where both m and n are integers and $n \neq 0$ (n is not equal to 0). Thus numbers such as $\frac{1}{2}$, $\frac{5}{3}$, $-\frac{6}{7}$, and $-10/-3$ are all rational numbers. *We cannot divide by zero.* Quotients of the form $m/0$ are often called **undefined.**

WARNING

BE CAREFUL OF QUOTIENTS THAT INCLUDE 0

$$\frac{0}{n} = 0 \qquad \text{whenever } n \neq 0,$$

$$\frac{m}{0} \qquad \text{is undefined (\textit{not} 1).}$$

We will find it very useful to represent numbers on a line called the **number line.** As Fig. 1.3 shows, we can associate each of the types of numbers we discussed with a point on the number line. The natural numbers lie to the right of 0. The inclusion of 0 and the negative integers extends the line to the left. The rational numbers are represented by points in between the integers. Do you think each of the points on the number line corresponds to one of the numbers we've discussed? Are there any "gaps" in the number line?

Figure 1.3 **The number line.**

In many cases we convert noninteger rational numbers to their **decimal equivalent.** This conversion is done by simply dividing the **denominator** (bottom) of the ratio into the **numerator** (top). Because a decimal number is just another way of expressing a rational number, it is graphed on the number line in the same place as its rational equivalent. (See Fig. 1.4.)

Some decimal numbers can be written with only one or two places to the right of the decimal point: For example, $\frac{1}{4}$ is 0.25, and

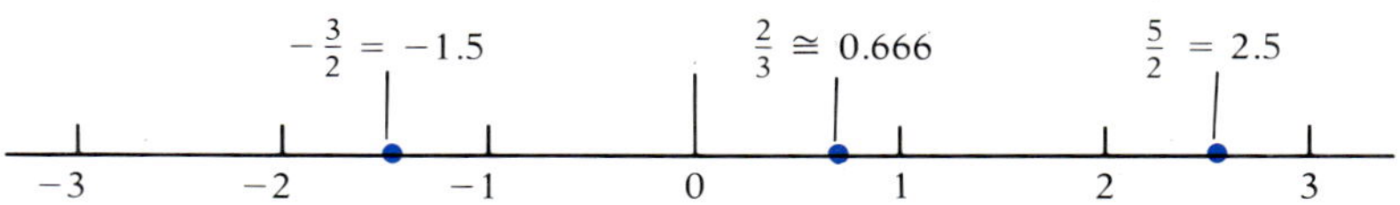

Figure 1.4 **Decimal numbers are equivalent to rational numbers.**

$\frac{375}{1000}$ is 0.375. Other rational numbers require an infinite number of places to the right of the decimal place to be represented properly: For example, $\frac{1}{3} = 0.33\underline{3}$, where the underscore indicates that the digit 3 repeats endlessly. Sometimes a group of digits will repeat endlessly: For example, $\frac{1}{7} = 0.142857\underline{142857}$, where the group of digits 142857 repeats endlessly. Is it possible to have a decimal number that requires an endless number of places to the right of the decimal point and yet does *not* repeat? The answer to this question is yes. It is indeed possible for a decimal number with an endless number of places not to be repeating. Such numbers are called **irrational** because they cannot be expressed as the ratio of two integers. Moreover, the irrational numbers are needed to complete the number line because without them there would be gaps between the rational numbers.

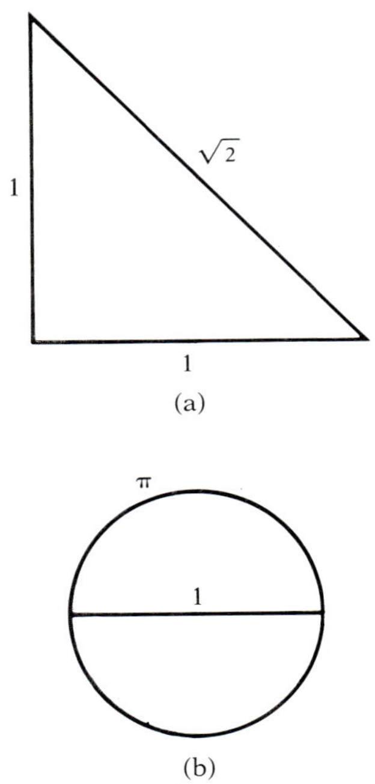

Figure 1.5 Irrational numbers can occur in simple geometrical figures.

You are probably familiar with several irrational numbers. Both $\sqrt{2}$ and π are irrational numbers that arise quite naturally from simple geometry. For example, suppose we construct a right triangle with both legs of length 1. By the Pythagorean theorem, the length of the hypotenuse (the side opposite the right angle) is the square root of the sum of the squares of the lengths of the other two sides. Thus we have a hypotenuse with length $\sqrt{1 + 1} = \sqrt{2}$. (See Fig. 1.5a.) As another example, consider a circle with a diameter of 1. The circumference of a circle is always π times the diameter, so our circle has a circumference of π. (See Fig. 1.5b.)

All the different groups of numbers we've discussed are referred to collectively as the **real numbers.** A representation of the real number system is shown in Fig. 1.6. The group of all decimal numbers is exactly the real number system if we include whole numbers (integers), decimal numbers that terminate or repeat endlessly (ra-

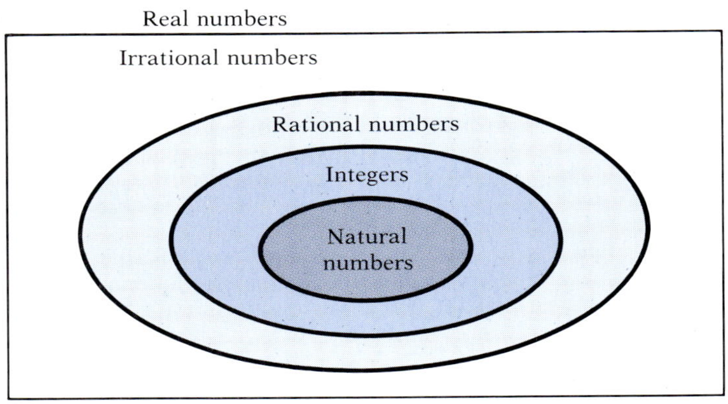

Figure 1.6 The real number system and its members.

tional numbers), and decimal numbers that go on endlessly without repeating (irrationals). Furthermore, each real number corresponds to a point on the number line, and each point on the number line corresponds to a real number.

It is not unusual for a decimal number to have too many decimal places for practical purposes. In this case we **round off** the number to the desired number of places. This is done by the following procedure:

Rounding Off

1. Determine the number of places desired. The remaining digits will be discarded.
2. Look at the *first* digit to be discarded.
   i. If it is 4 or less, simply drop it and all subsequent digits.
   ii. If it is 6 or more, add 1 to the *last digit to be kept,* carry out the addition if necessary and drop the subsequent digits.
   iii. If it is 5, transform the *last digit to be kept* to an even number by either adding 1 or leaving it unchanged. Drop all subsequent digits.

EXAMPLE 2

Round off the following numbers to the number of places shown.

a) 123456.7898764 to two decimal places: 123456.78(98764) DISCARD
The first digit to be discarded is 9, so the result is 123456.79.

b) 3.1415926535 to five decimal places: 3.14159(26535) DISCARD
The first digit to be discarded is 2, so the result is 3.14159.

c) 175.865 to two decimal places: 175.86(5) DISCARD
The first digit to be discarded is 5, so the result is 175.86.

d) 203.4375 to three decimal places: 203.437(5) DISCARD.

The first digit to be discarded is 5, so the result is 203.438 □

Practice Problem

Round off the following decimal numbers to the appropriate number of places.

a) 348.1298 to two places **b)** 8421.1666667 to four places
c) 2345.674359 to two places **d)** 751.83565 to four places

Answer: **a)** 348.13; **b)** 8421.1667; **c)** 2345.67; **d)** 751.8356

Properties of Real Numbers

Let a, b, and c be any real numbers. The following properties are true (except as noted).

1. Transitive Property

$$\text{If } a = b \text{ and } b = c, \text{ then } a = c.$$

2. Commutative Properties

$$a + b = b + a \quad \text{and} \quad ab = ba$$

3. Associative Properties

$$a + (b + c) = (a + b) + c \quad \text{and} \quad a(bc) = (ab)c$$

4. Distributive Properties

$$a(b + c) = ab + ac \quad \text{and} \quad (b + c)a = ba + ca$$

5. Inverse Properties

i. For each real number a, there is a unique real number denoted by $-a$ (called the **additive inverse of *a*** or the **negative of *a***) such that

$$a + (-a) = 0.$$

ii. For each real number *a except 0*, there is a unique real number denoted by a^{-1} $\left(\text{or } \dfrac{1}{a}\right)$ (called the **multiplicative inverse of *a***) such that

$$a(a^{-1}) = 1.$$

6. Order Property

For any two real numbers a and b, one of the following three conditions must hold:

i. $a < b$ (a is less than b, or a is to the left of b on the number line)

ii. $a = b$

iii. $a > b$ (a is greater than b, or a is to the right of b on the number line)

The additive inverse property allows us to define **subtraction:** $a - b$ means $a + (-b)$, where $-b$ is the negative of b. For example, $4 - 2$ means $4 + (-2)$.

Similarly, the multiplicative inverse property allows us to define **division:** for $b \neq 0$, $a \div b$, or a/b, we have

$$\frac{a}{b} = a(b^{-1}) = a\left(\frac{1}{b}\right).$$

We can use these properties to derive a number of important facts concerning operations with real numbers. In the following discussion, a, b, c, and d are real numbers, and all denominators are different from zero.

Operations with Real Numbers

Fact	Example
1. $-(-a) = a$	$-(-3) = 3$
2. $a - (-b) = a + b$	$3 - (-4) = 3 + 4 = 7$
3. $-(a + b) = -a - b$	$-(3 + 4) = -3 - 4 = -7$
4. $-(a - b) = -a + b$	$-(3 - 4) = -3 + 4 = 1$
5. $a \cdot 0 = (-a)0 = 0$	$3 \cdot 0 = (-3)0 = 0$
6. $(-a)b = -(ab) = a(-b)$	$(-3)4 = -(3 \cdot 4) = 3(-4)$
7. $\frac{a}{-b} = -\frac{a}{b} = \frac{-a}{b}$	$\frac{3}{-4} = -\frac{3}{4} = \frac{-3}{4}$
8. $\frac{-a}{-b} = \frac{a}{b}$	$\frac{-3}{-4} = \frac{3}{4}$
9. $\frac{0}{a} = 0$	$\frac{0}{3} = 0$
10. $a\left(\frac{b}{a}\right) = b$	$3\left(\frac{4}{3}\right) = 4$
11. $\frac{1}{a} \cdot \frac{1}{b} = \frac{1}{ab}$	$\frac{1}{3} \cdot \frac{1}{4} = \frac{1}{3 \cdot 4} = \frac{1}{12}$
12. $\frac{ab}{c} = \left(\frac{a}{c}\right)b = a\left(\frac{b}{c}\right)$	$\frac{3 \cdot 5}{4} = \left(\frac{3}{4}\right) \cdot 5 = 3\left(\frac{5}{4}\right)$
13. $\frac{a}{bc} = \left(\frac{a}{b}\right)\left(\frac{1}{c}\right) = \left(\frac{1}{b}\right)\left(\frac{a}{c}\right)$	$\frac{4}{3 \cdot 5} = \left(\frac{4}{3}\right)\left(\frac{1}{5}\right) = \left(\frac{1}{3}\right)\left(\frac{4}{5}\right)$
14. $\frac{a}{b} = \left(\frac{a}{b}\right)\left(\frac{c}{c}\right) = \frac{ac}{bc}$	$\frac{3}{5} = \left(\frac{3}{5}\right)\left(\frac{4}{4}\right) = \frac{3 \cdot 4}{5 \cdot 4}$
15. $\frac{a + b}{c} = \frac{a}{c} + \frac{b}{c}$	$\frac{3 + 5}{4} = \frac{3}{4} + \frac{5}{4}$
16. $\frac{a}{b} + \frac{c}{d} = \frac{ad + bc}{bd}$	$\frac{3}{4} + \frac{7}{5} = \frac{3 \cdot 5 + 7 \cdot 4}{4 \cdot 5} = \frac{43}{20}$
17. $\frac{a}{b} \cdot \frac{c}{d} = \frac{ac}{bd}$	$\frac{3}{4} \cdot \frac{7}{5} = \frac{3 \cdot 7}{4 \cdot 5} = \frac{21}{20}$
18. $\frac{a}{b/c} = a \div \frac{b}{c} = \frac{ac}{b}$	$\frac{3}{4/7} = 3 \div \frac{4}{7} = \frac{3 \cdot 7}{4} = \frac{21}{4}$

19. $\dfrac{a/b}{c} = \dfrac{a}{b} \div c = \dfrac{a}{bc}$ $\qquad \dfrac{3/4}{7} = \dfrac{3}{4} \div 7 = \dfrac{3}{4 \cdot 7} = \dfrac{3}{28}$

20. $\dfrac{a/b}{c/d} = \dfrac{a}{b} \div \dfrac{c}{d} = \dfrac{ad}{cb}$ $\qquad \dfrac{3/4}{7/5} = \dfrac{3}{4} \div \dfrac{7}{5} = \dfrac{3}{4} \cdot \dfrac{5}{7} = \dfrac{15}{28}$

Properties 10–20 deal with fractions and are especially important for you to know. We find that many students have difficulty with algebra because they don't know how to work with fractions.

WARNING

BE CAREFUL WHEN DEALING WITH FRACTIONS

Property 15 tells us (for example) that

$$\frac{3+5}{3} = \frac{3}{3} + \frac{5}{3} = \frac{8}{3}.$$

It is *not* valid to cancel the 3's as follows:

$$\frac{3+5}{3} \text{ is not equal to } \frac{\cancel{3}+5}{\cancel{3}} = 5.$$

EXERCISE SET 1.2

Locate the following numbers on the number line.

1. 1 **2.** -4 **3.** $\dfrac{2}{3}$ **4.** $\dfrac{-3}{-2}$

Simplify the following if possible.

5. $-5 + (-3)$
6. $-9 + 3$
7. $-4 - (-5)$
8. $(-3) \cdot 8$
9. $7(-2)$
10. $(-4) \cdot (-11)$
11. $-(-5)$
12. $-3 - 4$
13. $-5 - (-2)$
14. $8 - (-4)$
15. $3(-4 + 2)$
16. $7[-3(2) + 4(5)]$
17. $(-2)(-3)(-4)$
18. $-[-7 + (-3)]$
19. $(3 + 7)0$
20. $9\left(\dfrac{1}{6}\right)$
21. $\dfrac{3}{2} \cdot \dfrac{6}{5}$
22. $\dfrac{1}{4} + \dfrac{1}{3}$
23. $\dfrac{7}{9} - \dfrac{3}{5}$
24. $\dfrac{1}{2} \div \dfrac{1}{3}$
25. $\dfrac{6}{5/2}$
26. $\dfrac{7/8}{3}$
27. $\dfrac{9/2}{4/3}$
28. $\dfrac{0}{3}$
29. $\dfrac{1/3}{2/5} \div \dfrac{3/2}{4/3}$
30. $\dfrac{1/3}{0}$
31. $\dfrac{5+7}{7}$

1.3 CALCULATORS

The calculator has become an extremely valuable aid. As we go through this book, we will demonstrate how your calculator can be used to help you solve problems. We strongly urge that you become thoroughly familiar with its use *now*.

Before we begin, a word to the wise. Our discussion will provide a general description of how to use your calculator. However, your owner's manual will always provide an accurate description of how *your* calculator operates. When in doubt, refer to it.

We will assume your calculator uses **algebraic notation.** If you have one that employs **reverse Polish notation** (RPN), consult your owner's manual. If you're not sure, examine the keyboard of your calculator. Calculators that make use of algebraic notation have an [=] key, while RPN calculators do not.

A word of caution! Calculators with algebraic notation can operate in two different modes. Some calculators simply perform the operations sequentially as they are entered on the keys. Others, however, are designed to follow a hierarchy of operations. For such calculators, certain operations (for example, [×] and [÷]) are more important than other operations (for example, [+] and [−]). To see how your calculator operates, do the following calculation. Enter 3 [+] 2 [×] 4 [=]. If the display shows 20, your calculator performed the operations sequentially as you entered them. In other words, it performed the calculation $(3 + 2) \cdot 4$. Alternatively, if the display shows 11, your calculator accepted the entire string of operations and then performed the multiplication *before* the addition. In this case we say that addition is a *pending* operation. It will be performed only after the entire sequence of operations is entered and the calculator responds according to its priorities. To avoid errors, we urge you to *think* about calculations before performing them. If your calculator has parentheses, use them to establish the order in which you wish the calculations to be performed.

Using the [0]–[9], [.], and [+/−] Keys; Accuracy

Most calculators have a *floating decimal point:* As you enter numbers, the decimal point remains to the right of the numbers until you hit the [.] key. At this point the fractional portion of the number is entered, with the decimal point "floating" to the left in front of the fractional digits.

To enter a negative number, first enter the number and then use the [+/−] key. To change the sign of a number in the display, simply use the [+/−] key. *In either case, you need not use the* [−] *key.* It is used solely for subtraction. (Note: on calculators without a [+/−] key, the [−] key will be used for subtraction and to change signs.)

A calculator typically displays eight digits. It will work internally with eleven digits and then round off the results to eight places for display. To enter an eleven-digit number, you could break it into two parts, each with eight or fewer digits, and then add the two parts together. However, it is unlikely that you will ever have to perform calculations with a number with eleven digits of accuracy.

Using the [+], [−], [×], [÷], and [=] Keys

Study the following example.

EXAMPLE 3

Calculate $4 + 3 - 2$.

SOLUTION

Simply press the keys as follows, watching the display as you do so. Enter 4 [+] 3 [−] 2 [=]. □

EXAMPLE 4

Calculate $4 + 3 - 2 \cdot 2$.

SOLUTION

The intent of this sequence of numbers and operations is $4 + 3 - (2 \cdot 2)$. In other words, we assume the multiplication is to be performed before the subtraction. If your calculator has parentheses, you can simply enter the operations in the same sequence we used above. However, if your calculator lacks parentheses and you can't remember whether it has built-in priorities, you can always do the multiplication first and then the other operations.

With parentheses: Enter

[(] 4 [+] 3 [)] [−] [(] 2 [×] 2 [)] [=].

(Observe what happens as you close the parentheses and enter the [=] key.)

Without parentheses: Enter

2 [×] 2 [=] [+/−] [+] 4 [+] 3 [=]. □

EXAMPLE 5

Calculate $(4 + 3 - 2) \div 3$.

SOLUTION

Here we can enter

4 [+] 3 [−] 2 [=] [÷] 3 [=],

which allows us to proceed without parentheses. □

Practice Problem

Perform the following computations.

a) $150 - 10 - 25.50 - 45.05 + 55.65$
b) $(3 + 10) - (2 \cdot 7) \div 14$

Answer: **a)** 125.1; **b)** 12

Using the [STO], [RCL], [EXC], and [SUM] Keys

These keys enable you to use the **memory** of your calculator, a special part of the calculator that can hold a number without being affected by your operations on the display. The [STO] key simply takes the number shown on the display and "stores" it in memory (replacing whatever was there previously). (*Note:* Some calculators have more than one memory, in which case you enter [STO] and then a number or letter to designate which memory.) The [RCL] key displays the number stored in memory (without affecting memory), which can then be used in calculations. Memory is particularly useful when you must use an unusual (lengthy) number in many different calculations. Store the number in memory and simply use [RCL] each time you need it (rather than reentering it).

The [SUM] (or [M+]) key enables you to add the contents of the display to the number stored in memory. This would enable you, for instance, to keep a running total in a grocery store and leave the display free to compute other items.

The [EXC] (or [X→M] or similar) key swaps the contents of memory and the number currently displayed without affecting either. This can be useful in certain situations.

EXAMPLE 6

Calculate $\dfrac{4 + 5}{-2 + (4 \cdot 2)}$ without using parentheses.

SOLUTION

We can do this by first calculating the denominator and then storing it in memory. We then can compute the numerator and recall the denominator when we need it. Enter

4 (×) 2 (−) 2 (=) (STO) 4 (+) 5 (÷) (RCL) (=).

(If your calculator has parentheses, you can solve this problem without memory. Try it both ways.) □

EXAMPLE 7

Compute $144 - 9\left[\frac{3}{8}(44 - 16)\right]$ and $157 - 2\left[\frac{3}{8}(44 - 16)\right]$.

SOLUTION

Note that the quantity $\frac{3}{8}(44 - 16)$ appears in both calculations. This suggests computing it first and storing it in memory, and then recalling it as necessary. Enter

44 (−) 16 (=) (×) 3 (÷) 8 (=) (STO).

Next, enter

9 (+/−) (×) (RCL) (=) (+) 144 (=)

to do the first calculation and enter

2 (+/−) (×) (RCL) (=) (+) 157 (=)

to do the second. □

Practice Problem

Calculate $\dfrac{4 - 5(-3 + \frac{2}{3})}{3 \cdot (4 - 2 \cdot 9)}$.

Answer: -0.373, to three places

Clearing the Calculator: the (ON/C) Key

When you press the (=) key, all operations have been performed and the calculator is ready for a new problem (or to continue the current one if you so choose) without having to clear it. Some calculators have a (CE) key, which enables you to clear only the number in the display. Others use the (ON/C) key to perform a variety of functions. First, it turns the calculator on. During computations, when pressed once, it will clear *only* the number currently in the display. However, when pressed twice, it will clear the display, clear the memory, and cancel any operations.

Using the [1/x], [π], and [K] Keys

These keys enable you to do calculations more easily and accurately. The [π] key simply enters the value of π for you. The [1/x] (or [x^{-1}]) key enables you to compute the multiplicative inverse of the number in the display. Thus, to compute the value of $\frac{1}{3}$, enter 3 [1/x] (which means you do not need to divide). Finally, the [K] key enables you to store a number *and* an operation, which can save you considerable effort.

EXAMPLE 8

Add 343.33 to the following numbers: 75.22, 89.65, and 123.33.

SOLUTION

First, enter 343.33 [+] [K]. Next,

enter	75.22 [=]	giving 418.55,
enter	89.65 [=]	giving 432.98,
enter	123.33 [=]	giving 466.66. □

EXAMPLE 9

Divide the numbers 75.22, 89.65, and 123.33 by 25.42.

SOLUTION

First, enter 25.42 [÷] [K]. Next,

enter	75.22 [=]	giving 2.95909 to five places,
enter	89.65 [=]	giving 3.52675,
enter	123.33 [=]	giving 4.85169.

(You could also do this by entering 25.42 [1/x] [×] [K] and then entering the other numbers. Try it.) □

Practice Problem

Compute the following (round off to three places).

a) Multiply 56.98, 134.52, and 1234.11 by $\frac{2}{3}$.

b) Divide 448.75, 349.92, and −6744 by 32.25.

Answer: **a)** 37.987, 89.68, and 822.74; **b)** 13.915, 10.850, and −209.116

In addition to the material we've presented, we'll show you how to use other keys on your calculator as they become relevant.

WARNING

CALCULATORS CAN GIVE INCORRECT ANSWERS

Your calculator is a powerful tool that enables you to perform a multitude of calculations. However, your results will be accurate only if you enter the numerical values and the operations accurately. If you make a mistake, your calculator will quickly give you an *incorrect* answer. Put another way, GIGO ("garbage in = garbage out": a phrase commonly used by people who work with computers). One way to help prevent errors is to "ballpark" the answer. Try to get a quick mental estimate of the correct answer. For example, suppose we have to add \$16.16, \$17.45, \$18.92, and \$60.15. We can round the numbers mentally to \$16, \$17, \$19, and \$60, which add up to \$112. Or, we could round them to \$20, \$20, \$20, and \$60, which add to \$120. The correct answer is \$112.68. The point here is that we should be able to detect that an answer such as \$58.68 (which we could get by incorrectly entering \$6.15 instead of \$60.15) is clearly wrong. Many mistakes can be prevented if you think about what your calculations should produce rather than simply accepting the result. If the calculation is important enough, *do it twice.*

Now let's go back to our Problem from Life and finish it.

EXAMPLE 10

Problem from Life, revisited In Example 1 we determined a scheme for solving the problem of the Vannings and the Coopers. The steps involved and the computations are as follows. (See Fig. 1.2.)

1. Determine the cost of the meal before tax. *Be careful here.* The tax is 5% of the meal, which does *not* mean it is 5% of \$107.50. Actually, we have meal + tax = \$107.50 and since tax = 5%(meal) this is

$$\text{meal} + 5\%(\text{meal}) = \$107.50 \quad \text{or} \quad 105\%(\text{meal}) = \$107.50,$$

so

$$\text{meal} = \$107.50 \div 1.05 = \$102.38.$$

2. The amount of the tip is then 15% of the meal, so

$$\text{tip} = 15\%(\$102.38) = \$15.36.$$

3. The total bill is the cost of the meal + tax + tip + bar bill, so we have

$$\text{total} = \$107.50 + \$15.36 + \$14.50 = \$137.36.$$

(Notice that we already know the cost of the meal + tax!)

4. Each couple's share is one-half the total, so

$$\text{share} = .5(137.36) = \$68.68.$$

5. The Vannings still owe their share less the bar bill, so Vannings owe = $68.68 − $14.50 = $54.18. □

EXERCISE SET 1.3

Solve Problems 1–20 on your calculator to three places of accuracy.

1. $3.62 + 0.72$
2. $72.5642 + 6.25338$
3. $10.625 - 3.721$
4. $-426.85 + 32.851 - 25.432$
5. $0.0032 \cdot 962.41$
6. $0.725 \cdot 8.72$
7. $-8.562 \cdot 743$
8. $-107.3 \cdot 0.0525$
9. $-98.72 \cdot -25.431$
10. $-95.321 \cdot 7.663 \cdot -0.0543$
11. $32.4 \div 16.4$
12. $0.876 \div 0.625$
13. $(-0.035 \div 0.00633) \div 0.0754$
14. $5 \cdot (6 + 4)$
15. $2 \cdot (8 + 16) + 5 \cdot (7 + 4)$
16. $8 \cdot [3 \cdot (4 + 7) + 6 \cdot (3 + 1)]$
17. $25 \cdot \{[7 + 2 \cdot (3 + 9) - 6 \cdot (4 + 3)] + 6\}$
18. $-3 \cdot [5 \cdot (2.5 + 8 - 4.2) - 6 \cdot (5 - 8.8)]$
19. $2.8 \cdot \{[(6 + 2.9) + 16 - 3.2] + \frac{2}{5}\}$
20. $125 \cdot [(2.9 + 6.2 + 15.92) \div 3 \cdot (2.7 + 15)]$

1.4 SETS

In this section we will review some of the basic concepts associated with set theory. You'll find this material of benefit throughout the book, especially in the discussion of probability.

Basic Definitions

The theory of sets is rather new compared to that of most areas of mathematics. George Boole, an English mathematician, established the basic concepts of symbolic logic (a way of using mathematical symbols and operations to solve problems in logic) in the last century. George Cantor, a German mathematician, applied some of the concepts of symbolic logic to sets of numbers. The theory he developed is called *set theory*.

A **set** is a well-defined collection of items. The term **well-defined** means that we can tell precisely whether or not a particular item belongs to a set. For example, the set of students registered for your mathematics class and the set of presidents of the United States are well defined. The "set" of weeds is not well defined, because what to some people is a weed is to other people a wildflower.

We generally represent sets by capital letters such as A, B, C, and so on. The individual items in a set are called the **members** or **elements** of the set; they are generally represented by small letters. We indicate membership in a set as follows:

$a \in A$ means that a is a member of the set A,

$a \notin A$ means that a is not a member of the set A.

A set without *any* members is called the **empty** or **null** set. We indicate the null set by the symbol $\varnothing$. For example, the set of all green people with wings who are less than 1-foot tall is an empty set.

There are two ways to describe a set. The first method is simply to list all of the members of the set. The other method is to give a rule (usually in mathematical form) that enables us to determine whether an item is a member of the set. In either case, we enclose the exhaustive listing or the rule in braces { }.

EXAMPLE 11

a) Consider the set, D, that consists of the days of the week. We can denote this set formally by either listing its members:

$$D = \{\text{Sunday, Monday, Tuesday, Wednesday, Thursday, Friday, Saturday}\},$$

or we could specify D with a rule:

$$D = \{d \mid d \text{ is a day of the week}\}.$$

(We are assuming, of course, that anyone likely to encounter this set is familiar with the days of the week.) The vertical line after the first d is read "such that."

b) Consider the set of odd natural numbers N. We can define it by listing

$$N = \{1, 3, 5, 7, 9, \ldots\}$$

or by using the rule

$$N = \{n \mid n \text{ is an odd natural number}\}.$$

c) Consider the set of numbers S that when squared yield 4, 9, or 16. We can define it by listing

$$S = \{2, -2, 3, -3, 4, -4\}$$

or by the rule

$$S = \{s \mid s^2 = 4 \text{ or } s^2 = 9 \text{ or } s^2 = 16\}. \quad \square$$

Practice Problem

Consider the set P of prime numbers between 5 and 15. Describe the set by both listing its members and by using a rule.

Answer: $P = \{5, 7, 11, 13\}$ or $P = \{p \mid 5 \leq p \leq 15 \text{ and } p \text{ is prime}\}$

Suppose A and B are two sets. We say that B is a **subset** of A, represented by $B \subseteq A$, if *every* member of B is also a member of A. (Thus every set is a subset of itself.) For example, the set of students in your mathematics class is a subset of the set of students in your school. We say that B is a **proper subset** of A, represented by $B \subset A$, when B is a subset of A and, moreover, A contains at least one additional member that isn't in B. For example, the set of U.S. senators is a proper subset of the set of elected officials. Two sets are **equal,** represented by $A = B$, if they have exactly the same members. Although we won't prove it, it can be shown that the empty set $\varnothing$ is a subset of every set.

EXAMPLE 12

Consider the sets $A = \{1, 2, 3, 4, 5, 6, 7, 8, 9\}$, $B = \{1, 3, 5, 7, 9\}$, $C = \{2, 4, 6, 8\}$, and $D = \{1, 2, 4, 6, 8, 10\}$. Find all sets that are proper subsets of other sets.

SOLUTION

Both B and C are proper subsets of A, which we indicate by $B \subset A$ and $C \subset A$. Moreover, C is a proper subset of D, that is, we have $C \subset D$. However, D is not a subset of A nor is A a subset of D. We can indicate these relationships by $D \not\subset A$ and $A \not\subset D$. $\square$

Practice Problem

Given the sets V = {Alice, Bob, Carol, Don}, W = (Alice, Carol, Don}, X = {Alice, Bob, Carol, Ed}, Y = {Alice, Bob}, and Z = {Bob}. Which sets are subsets of other sets? Which sets are proper subsets of other sets? (Use set notation to describe your answer. You need not include relationships such as $W \subseteq W$.)

Answer: $W \subset V$, $Y \subset V$, $Z \subset V$, $Z \subset Y$, $Y \subset X$, and $Z \subset X$

There always exists a set consisting of all the elements under consideration, called the **universal set.** We denote the universal set by U. For example, suppose we're discussing sets that consist of the students registered for various courses at your school. The universal set in this case would be the set of all students registered for one or more courses at your school.

There are situations in which we would like to be able to choose elements that belong to one of two sets or elements that belong to both sets. For example, if an employer wanted someone knowledgeable about computer programming, she could hire either an information systems major or a computer science major. Another employer might want someone knowledgeable about both accounting and marketing, so he could hire a student with a major in both marketing and accounting. We can deal with such problems by defining **set operations.**

The **union** of two sets A and B is made up of all objects that belong to either A or to B (or to both). In symbols we have

$$\text{Union:} \quad A \cup B = \{x \mid x \in A \text{ or } x \in B\}.$$

In this case, the word *or* means that x may be a member of A or of B or of both.

The **intersection** of two sets A and B is made up of the objects that belong to *both* sets. Symbolically, this is

$$\text{Intersection:} \quad A \cap B = \{x \mid x \in A \text{ and } x \in B\}.$$

It is possible that two sets have no elements in common, in which case their intersection is the empty set. For example, the set of people and the set of cats have no elements in their intersection. Sets whose intersection is empty are said to be **disjoint.**

EXAMPLE 13

Let $A = \{1, 3, 5\}$, $B = \{2, 4, 6, 7\}$, and $C = \{1, 2, 3\}$. Find $A \cup B$, $A \cap B$, $B \cup C$, and $B \cap C$.

SOLUTION

$A \cup B = \{1, 2, 3, 4, 5, 6, 7\}$, $\quad A \cap B = \varnothing$ (A and B are disjoint),
$B \cup C = \{1, 2, 3, 4, 6, 7\}$, $\quad B \cap C = \{2\}$. □

Practice Problem

Let A = {Alice, Bob, Carol}, B = {John, Carol, Bob, Ed}, and C = {John, Mary, Sue}.

a) What is the union of A and B?

b) Which pair(s) of sets are disjoint?

Answer: **a)** $A \cup B$ = {Alice, Bob, Carol, Ed, John}; **b)** only A and C

We can extend the ideas of union and intersection to more than two sets. Thus the **union of a number of sets** is made up of the elements that belong to at least one of the sets. Similarly, the **intersection of a number of sets** is made up of the elements that belong to *all* of the sets.

EXAMPLE 14

A university uses the following scheme to classify the teaching faculty:

U = set of all teaching faculty
M = set of male faculty
F = set of female faculty
I = set of faculty with rank of instructor
A = set of faculty with rank of assistant professor
S = set of faculty with rank of associate professor
P = set of faculty with rank of professor
T = set of faculty with tenure
L = set of faculty who have been at the university for at least 10 years
B = set of faculty who have published at least one book

Describe the following sets:

a) $F \cap P \cap T$ **b)** $P \cap L \cap M$ **c)** $(S \cup A) \cap B \cap T$

SOLUTION

a) $F \cap P \cap T$ is the set of tenured female professors.
b) $P \cap L \cap M$ is the set of male professors who have been at the university at least 10 years.
c) $(S \cup A) \cap B \cap T$ is the set of assistant or associate professors with tenure who have published at least one book. □

Practice Problem

Referring to Example 14, describe the following sets.

a) $(M \cup I) \cap B$ **b)** $(I \cup P) \cap M$ **c)** $F \cap L \cap B$

Answer: **a)** The set of males or instructors who have published a book; **b)** the set of instructors or professors who are male;

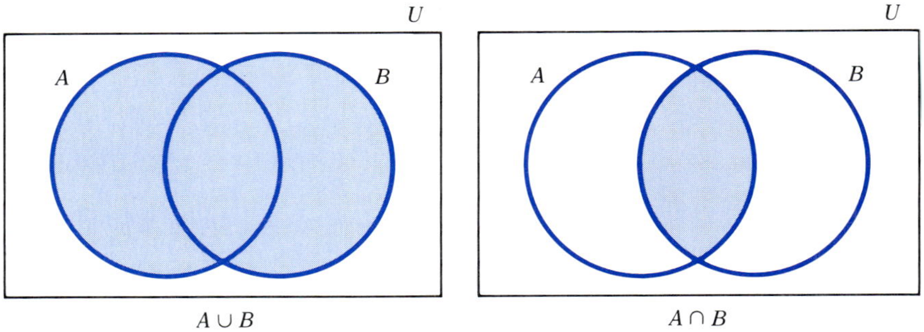

Figure 1.7 **Venn diagrams of the union and intersection of two arbitrary sets.**

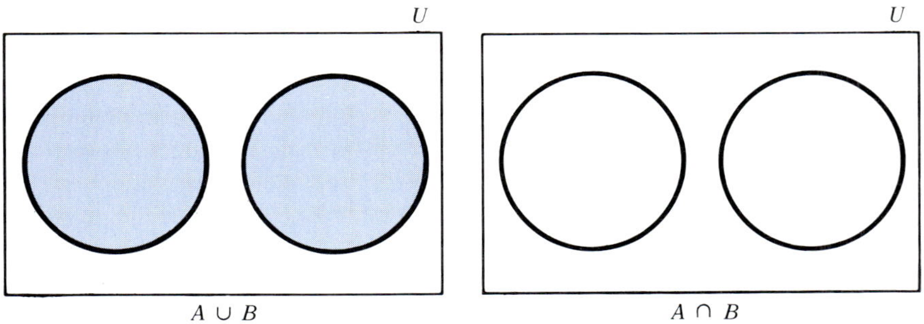

Figure 1.8 **Venn diagrams of the union and intersections of two disjoint sets.**

c) the set of females who have been at the university at least 10 years and who have published at least one book.

We can use **Venn diagrams** to help us represent and analyze many relationships among sets. A Venn diagram consists of a rectangle representing the universal set with the other sets being represented by circles within the rectangle. Figure 1.7 shows Venn diagrams in which the shaded portions represent the union and intersection of two arbitrary sets. Figure 1.8 shows Venn diagrams showing the union and intersection of two disjoint sets.

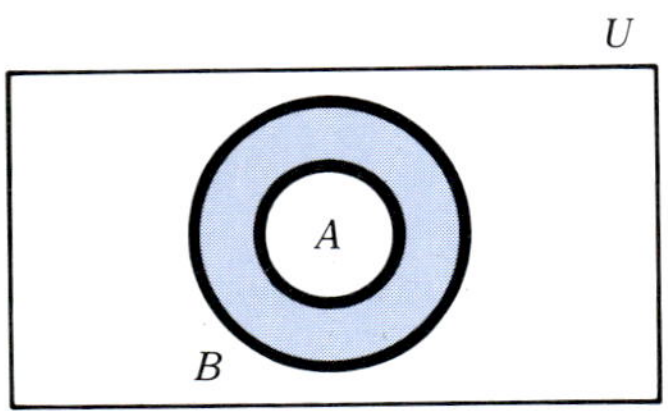

Figure 1.9 **The complement of *A* relative to *B*.**

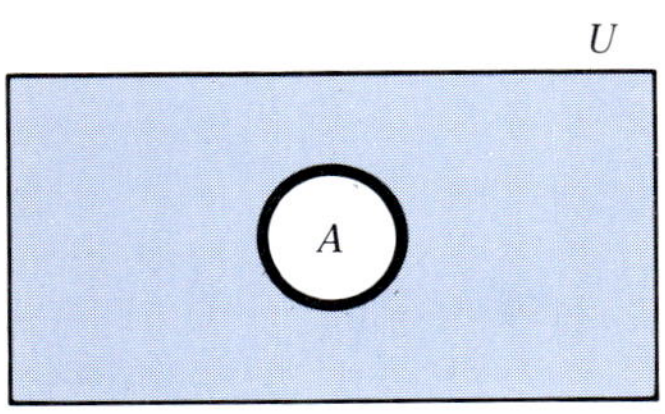

Figure 1.10
The complement of *A*.

Suppose A is a subset of another set, B. The **complement of A relative to B,** represented by $B - A$, is the set of all members of B that are not members of A. (See Fig. 1.9.) If A is *not* a subset of C, then the complement of A relative to C *is not defined*. The **complement of A,** denoted by A', is the set of all members of U that are not members of A. (See Fig. 1.10.) A very useful fact is $A \cup A' = U$ (study Fig. 1.10 to see this).

EXAMPLE 15

Let $U = \{1, 2, 3, 4, 5, 6, 7, 8, 9\}$, $A = \{1, 3\}$, $B = \{1, 3, 5, 7, 9\}$, and $C = \{2, 4, 6, 8\}$. Find A', $B - A$, and $C - A$.

SOLUTION

$$A' = \{2, 4, 5, 6, 7, 8, 9\},$$
$$B - A = \{5, 7, 9\},$$
$C - A$ is not defined since $A \not\subseteq C$. □

Practice Problem

Let $U = \{m, n, o, p, q\}$, $A = \{m, o, q\}$, $B = \{n, p\}$, and $C = \{m, o, p, q\}$. Find

a) B' **b)** $B - A$

c) $C - A$ **d)** C'

Answer: **a)** $\{m, o, q\}$; **b)** not defined; **c)** $\{p\}$; **d)** $\{n\}$.

Counting with Sets

In a variety of problems we will be concerned only with describing intersections, unions, and complements of sets. In others, however, we will find it necessary to be able to *count* the number of elements in certain unions and intersections. Consider the following example.

EXAMPLE 16

A survey of 1000 people in a small town showed that 450 read the morning paper, 300 read the evening paper, and 150 read both. The company that owns the newspapers would like to know how many people read at least one paper.

SOLUTION

We can analyze this situation by drawing a Venn diagram. First, we must define some sets: Let

M = set of people who read the morning paper
E = set of people who read the evening paper

Study the Venn diagram shown in Fig. 1.11. To begin with, we know that 150 people read both papers. But the set of people who read both papers is $M \cap E$, so we can place the number 150 in the intersection of the two sets. The rest is easy. A total of 450 read the morning paper, and we know that 150 of them read both papers.

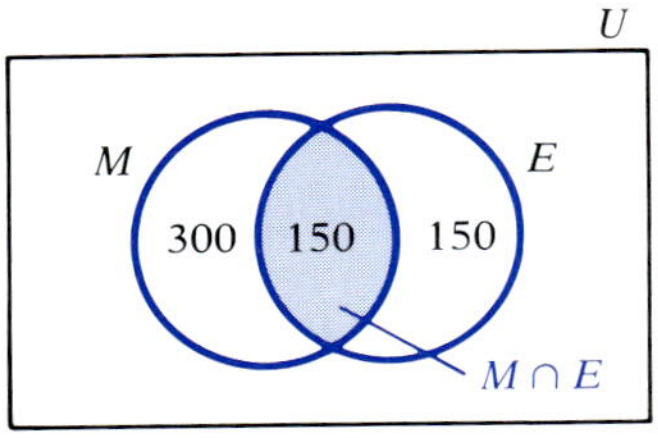

Figure 1.11 Using a Venn diagram to count elements in sets.

Therefore, $450 - 150 = 300$ must read only the morning paper (which are the people in the set $M \cap E'$). Finally, a total of 300 people read the evening paper, and 150 of them read both papers, so $300 - 150 = 150$ people read only the evening paper (which are the people in the set $M' \cap E$).

Now we can answer the question of how many people read at least one paper. We want the total number of people within the circles M and E: in other words, the total number in the union of M and E. By looking at Fig. 1.11, we see there are 300 people who read only the morning paper, 150 who read both papers, and 150 who read only the evening paper. Therefore the total number of people who read at least one paper is $300 + 150 + 150 = 600$. □

Another approach to the problem posed in Example 16 would be to simply add up the numbers of people who read each paper: $450 + 300 = 750$. Obviously, this is incorrect (because it doesn't agree with our answer). Can you see why it's wrong?

The difficulty is that we would *double count* the people who read both papers. Look at the Venn diagram in Fig. 1.11 again. Notice that adding up the 450 people in M to the 300 people in E means including the people in $M \cap E$ twice.

Let's generalize this experience by giving some rules for counting. For any set S, let $n(S)$ denote the number of elements of S.

Counting Rules for Sets

Let A and B be two arbitrary sets. Then

$$n(A \cup B) = n(A) + n(B) - n(A \cap B).$$

If A and B are disjoint, then we have the special case

$$n(A \cup B) = n(A) + n(B) \qquad \text{when } A \cap B = \varnothing.$$

EXAMPLE 16

Revisited Referring to the data in Example 16, we have

$$n(M) = 450, \qquad n(E) = 300, \qquad n(M \cap E) = 150.$$

Using our counting rule,

$$\begin{aligned} n(M \cup E) &= n(M) + n(E) - n(M \cap E) \\ &= 450 + 300 - 150 \\ &= 600 \qquad \text{(which we obtained previously)} \end{aligned}$$ □

EXAMPLE 17

The manufacturing process for a certain item involves three stages, labeled 1, 2, and 3. At the end of each stage, an inspector examines the item. A record is kept of all defective items and at which stage the defect was detected. On one particular day when 2000 items were manufactured, the following data were collected:

Number Defective Items	Stages
200	1
150	2
160	3
40	1 and 2
60	1 and 3
30	2 and 3
10	1, 2, and 3

The quality manager would like to have answers to the following questions:

a) How many items had at least one defect?
b) How many items had no defect?
c) At each stage of the process, how many items had exactly one defect?
d) How many items had defects at exactly two of the three stages?

SOLUTION

Let's see how we can solve the quality manager's problem. Our first step (recall our four-step approach to solving problems) is to define the problem, which we have already done quite clearly. Our next step is to build a mathematical model. (Since we've been discussing sets, you might guess that we'll use sets to model this problem.) This problem is similar to the previous one, except that we have more possibilities. We hope a similar approach will work. Let's define some sets:

U = the universe of 2000 items
A = the set of items with a defect from stage 1
B = the set of items with a defect from stage 2
C = the set of items with a defect from stage 3

Let's restate the data given to us using set notation.

$$n(U) = 2000 \qquad n(A \cap B) = 40$$
$$n(A) = 200 \qquad n(A \cap C) = 60$$
$$n(B) = 150 \qquad n(B \cap C) = 30$$
$$n(C) = 160 \qquad n(A \cap B \cap C) = 10$$

Figure 1.12 is a Venn diagram showing how three sets intersect. Note especially the labels on the sets and the intersections.

The third step in our four-step approach is to perform the analysis. Note that all three sets intersect to form the set $(A \cap B \cap C)$. In addition, we know that $n(A \cap B \cap C) = 10$. Now we can determine the number of items in $A \cap B \cap C'$, because we know $n(A \cap B) = 40$, which must include the 10 items in $A \cap B \cap C$. Hence

$$n(A \cap B \cap C') = 40 - 10 = 30.$$

Similarly,

$$n(A \cap B' \cap C) = 60 - 10 = 50$$

and

$$n(A' \cap B \cap C) = 30 - 10 = 20.$$

(See Fig. 1.13.)

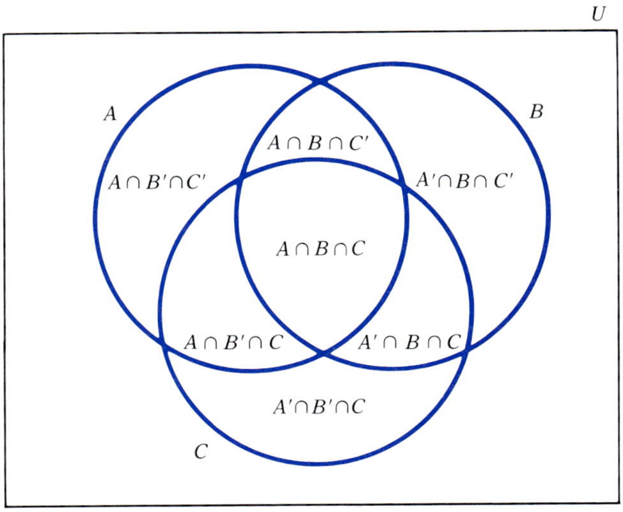

Figure 1.12 **A Venn diagram for the intersection of three sets.**

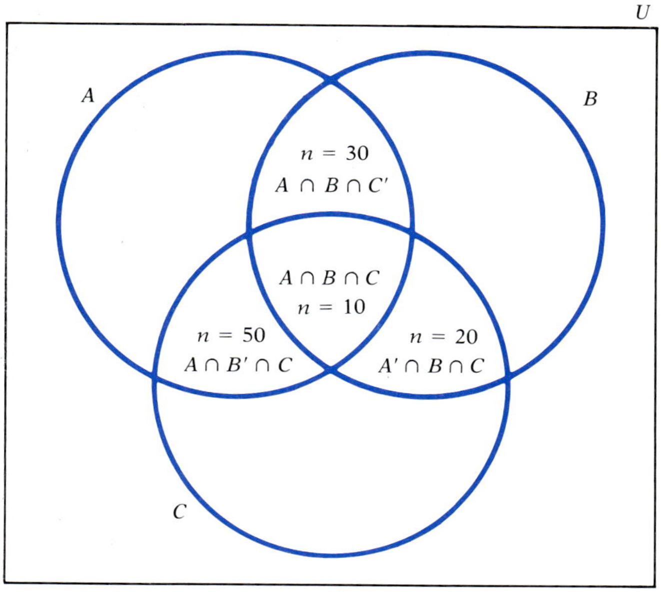

Figure 1.13 **Data for intersections of only two stages, Example 17.**

Finally, we can finish our analysis by considering the remaining sets. We have

$$n(A \cap B' \cap C') = 200 - (50 + 10 + 30) = 110,$$
$$n(A' \cap B \cap C') = 150 - (30 + 10 + 20) = 90,$$
$$n(A' \cap B' \cap C) = 160 - (50 + 10 + 20) = 80.$$

(See Fig. 1.14.)

Now we can finally answer the quality manager's questions. To do so, we must translate the questions into set notation.

a) How many items had at least one defect? This is equivalent to asking for $n(A \cup B \cup C)$. By addition, we have $110 + 30 + 90 + 50 + 10 + 20 + 80 = 390$.

b) How many items had no defect? This is equivalent to asking for $n(A \cup B \cup C)'$. Since

$$(A \cup B \cup C) \cup (A \cup B \cup C)' = U,$$

we have

$$n(A \cup B \cup C) + n(A \cup B \cup C)' = 2000.$$

Hence, $n(A \cup B \cup C)'$ is $2000 - 390 = 1610$.

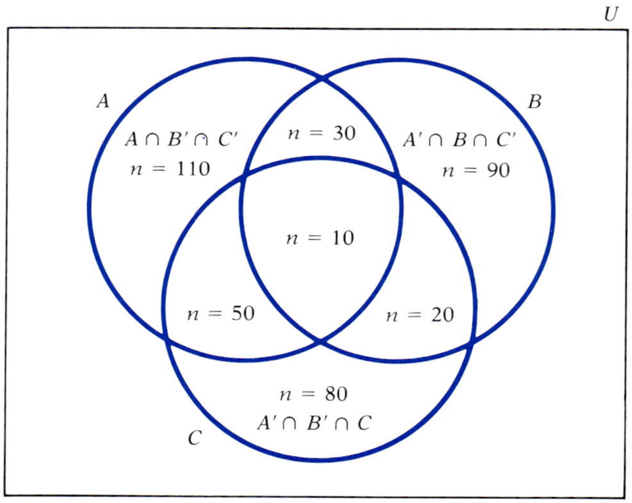

Figure 1.14 **Data for all intersections, Example 17.**

c) At each stage of the process, how many items had exactly one defect? This is equivalent to asking for $n(A \cap B' \cap C')$, $n(A' \cap B \cap C')$, and $n(A' \cap B' \cap C)$, which are 110, 90, and 80, respectively.

d) How many items had exactly two of the three defects? This is equivalent to asking for $n(A \cap B \cap C') + n(A \cap B' \cap C) + n(A' \cap B \cap C) = 30 + 50 + 20 = 100$. □

Practice Problem

Suppose it was discovered that a mistake had been made in recording the data in Example 17 and that $n(A \cap B \cap C) = 20$, not 10. Recompute the number of units in each of the parts of the Venn diagram shown in Fig. 1.14.

Answer: $n(A \cap B \cap C) = 20$, $n(A \cap B \cap C') = 20$, $n(A \cap B' \cap C) = 40$, $n(A' \cap B \cap C) = 10$, $n(A \cap B' \cap C') = 120$, $n(A' \cap B \cap C') = 100$ and $n(A' \cap B' \cap C) = 90$

EXERCISE SET 1.4

Label each of the following true or false.

1. A set is any group of objects or ideas.
2. $3 \in \{7, 8, 9\}$

3. A set is shown by enclosing its members in braces.
4. The empty set is also referred to as the null set.
5. $\emptyset \subseteq \{7, 9, 11\}$
6. $a \in \{a, b, c, d\}$
7. $\{5\} \not\subset \{4, 5, 6\}$
8. $\{7, 8\} \subseteq \{7, 9, 11\}$
9. The universal set consists of all objects being considered at any one time.
10. Two sets that are equal have the same elements.
11. $\{7, 8, 9\} = \{9, 8, 7\}$
12. $c \notin \{a, b, d, e\}$
13. $\{7, 8\} \not\subset \{6, 7, 8, 9\}$
14. $\{\emptyset\} = \emptyset$
15. $\{0\} = \{\ \}$

Refer to the Venn diagram in Fig. 1.15 and indicate the number of elements in each of the sets in Problems 16–34, given $n(A \cup B \cup C)'$ is 40.

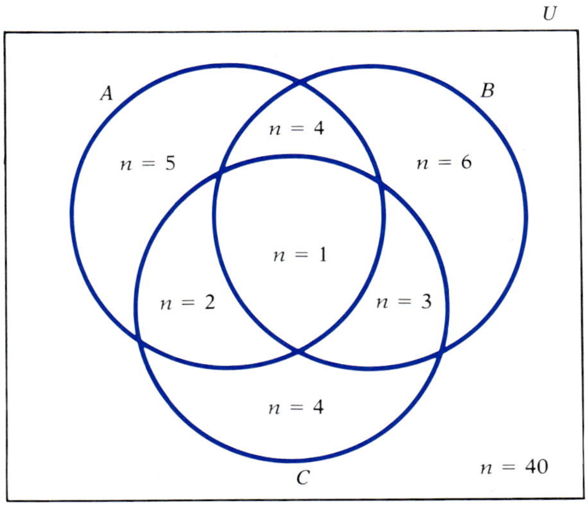

Figure 1.15 **Venn diagram for Exercise Set 1.4, Problems 16–34.**

16. A
17. $A \cup B$
18. U
19. $A \cap B$
20. $A \cap C$
21. $B \cup C$
22. $B \cap C$
23. A'
24. B
25. C'
26. $(A \cup B)'$
27. $(A \cap B)'$
28. C
29. B'
30. $(A \cup C)'$
31. $A \cup (B \cup C)$
32. $A \cap (B \cup C)$
33. $A \cap (B \cap C)$
34. $A \cup (B \cap C)$

Refer to the following sets to answer Problems 35–40.

$$U = \{5, 6, 7, 8, 9, 10\}$$
$$X = \{5, 6, 7\}$$
$$Y = \{6, 10\}$$

35. Draw a Venn diagram representing the sets U, X, and Y.

36. Using set notation, describe the set made up of members of both X and Y.

37. Using set notation, describe the set made up of members of X or Y.

38. Using set notation, describe the set made up of elements that are not members of both X and Y.

39. Using set notation, describe the set of elements that are members of X but not members of Y.

40. Using set notation, describe the set of elements that are not members of X or of Y.

CHAPTER REVIEW

IMPORTANT TERMS AND SYMBOLS

1.1 Problem Solving and Modeling

problem solving Four-step approach:

1. Define the problem.
2. Build a mathematical model.
3. Analyze the model.
4. Implement the solution in the real world.

model Abstraction of the real world.

1.2 The Real Number System

decimal numbers Real numbers that are written with a decimal point.
integers $\ldots, -3, -2, -1, 0, 1, 2, \ldots$
Negative integers: $\ldots, -3, -2, -1$
Positive integers: $1, 2, 3, \ldots$
irrational numbers Numbers that *cannot* be expressed as m/n, for integers m and n, equivalent to having an endless number of decimal places without repeating.
natural numbers $1, 2, 3, \ldots$
number line A graphical representation of numbers.

properties of real numbers (for a, b, c real numbers)

1. Transitive: if $a = b$ and $b = c$ then $a = c$.
2. Commutative: $a + b = b + a$ and $ab = ba$.
3. Associative: $a + (b + c) = (a + b) + c$ and $a(bc) = (ab)c$.
4. Distributive: $a(b + c) = ab + ac$ and $(b + c)a = ba + ca$.
5. Inverse: $a + (-a) = 0$ and $a(a^{-1}) = 1$, $a \neq 0$.
6. Order: Either $a < b$, $a = b$, or $a > b$ for any pair a and b.

rational numbers A ratio of two integers, m/n, where $n \neq 0$.
real numbers All rational and irrational numbers.
rounding off Procedure for reducing the number of decimal places in a number to a predetermined number.

1. Determine the number of places desired. The remaining digits will be discarded
2. Look at the first digit to be discarded.
   - **i.** If it is 4 or less, simply drop it and all subsequent digits.
   - **ii.** If it is 6 or more, add 1 to the *last digit to be kept* and drop the subsequent digits.
   - **iii.** If it is 5, *transform the last digit to be kept* to an even number by either adding 1 or leaving it unchanged. Drop all subsequent digits.

1.3 Calculators

algebraic notation Calculator notation that follows normal rules of algebra.
reverse Polish notation (RPN) Calculator notation that some calculators use.

1.4 Sets

complement of *A* All elements of universal set not in A, written A'.
complement of *A* relative to *B* (for $A \subset B$) All elements of B not in A; written $B - A$.
counting formula Used to give number of elements in set [for set A, written $n(A)$]:

$$n(A \cup B) = n(A) + n(B) - n(A \cap B)$$
$$n(A \cup B) = n(A) + n(B) \qquad \text{when } A \text{ and } B \text{ are disjoint}$$

disjoint sets Two sets with no elements in common.
element Member of a set; written with small letters such as a, b, c.
equal sets Have exactly the same members; written $A = B$.
proper subset First set is subset of second set and second set has other elements that don't belong to first set; written $A \subset B$.
set Well-defined collection of items (either an item belongs or it doesn't); written with capital letters such as A, B, C.
set operations
Union of two sets, A and B: set of elements that belong to at least one of the sets, written $A \cup B$.

Intersection of two sets, A and B: set of elements that belong to both sets, written $A \cap B$.

subset A is a subset of B if every member of first set is also member of second set; written $A \subseteq B$.

universal set Set of all possible elements, written U.

Venn diagram A diagram in which the universal set is represented by a rectangle and the individual sets are represented by circles within the rectangle.

REVIEW PROBLEMS

Solve Problems 1–8 with your calculator (to three places of accuracy).

1. $23.79 + 0.54 - 6$

2. $-3.7 - (-7.09) + 0.14$

3. $(-4)7.3 \div 22$

4. $(1.84 + 0.39)\frac{2}{5} \div 14$

5. $\frac{1}{6} + \frac{4}{15} - \frac{3}{10}$

6. $\left(\frac{4}{9} \div \frac{3}{14}\right)\frac{2^5}{3}$

7. $\left(-\frac{13}{28} + \frac{3}{8}\right) \div \frac{15}{12}$

8. $\left(\frac{11}{18} + \frac{7}{12} - \frac{4}{15}\right) \div \left(\frac{9}{10} \cdot \frac{17}{75}\right)$

Let $A = \{1, 3, 6, 9\}$, $B = \{1, 2, 4, 8\}$, $C = \{1, 9\}$, and $D = \{2, 7\}$. Also, $U = \{1, 2, 3, 4, 5, 6, 7, 8, 9\}$. For Problems 9–26, indicate which are true or false.

9. $A \subset U$

10. $A \subseteq B$

11. $C \not\subset U$

12. $C \subset D$

13. $\varnothing \subseteq D$

14. $\varnothing \subseteq U$

15. $\{1, 6\} \subseteq A$

16. $A \cup D = \{1, 3, 6, 7, 9\}$

17. B and C are disjoint

18. C has exactly three proper subsets

19. C has exactly three subsets

20. A has exactly 15 subsets

21. $\{2, 7\}$ is a subset of D

22. $A' = \{2, 4, 7, 8\}$

23. $A - C = \{3, 6, 9\}$

24. $A \cap (B \cap D) = \{1\}$

25. $(C \cup D)'$ is the empty set

26. $C \cup (A \cap B) = C$

27. A survey of 100 students at the local high school showed that

66 take English
58 take chemistry
45 take French
32 take English and chemistry
30 take English and French
19 take chemistry and French
6 take all three subjects

How many students

a) Take only English?
b) Take only chemistry?

c) Take only French?
d) Take only English and chemistry?
e) Take chemistry and French but not English?
f) Do not take chemistry?
g) Do not take English, chemistry or French?

Draw a Venn diagram and show the set that is appropriate for each of these questions.

trefoils
old fashioned
shortbread
Do Not Drop
10254
trefoils
old fashioned shortbread
"great cookies
for a great cause"
chocolate
chunks
juliettes
12 pkgs. 602
12 pkgs. 601
trefoils
old fashioned shortbread
GIRL SCOUTS

2

ALGEBRA

a) A rectangular box has a square base of x inches on a side and a height of h inches. There is no top. Express the total surface area as a polynomial in x and h.

b) A box is to be made from a piece of cardboard 12 inches square. Squares are to be cut out of the corners and the sides folded up. Express the volume of the box as a polynomial. (To be solved as Example 23, Section 2.2.)

PROBLEM FROM LIFE

INTRODUCTION

In this chapter we'll review some of the important principles of algebra that you will need throughout the book. We'll begin our review with a discussion of the basics of algebra (variables and exponents). We'll then review polynomials and operations involving polynomials, including addition, subtraction, and multiplication. Next we'll discuss factoring. We'll close the chapter with a review of algebraic fractions. We don't intend to teach you a full year of algebra in this chapter. In fact, if this material is unfamiliar to you, we advise you to take a background course in algebra.

2.1 BASICS OF ALGEBRA

Variables and Exponents

In Chapter 1 we stated that mathematics has become increasingly important for students of business, and we emphasized the role of mathematics in problem solving with models. Your ability to construct and manipulate mathematical models will be based to a great degree on your knowledge of algebra. To put it another way, algebra is fundamental to the study of mathematics.

The reason algebra is so important is that it enables us to construct and analyze *general* models rather than models that work only for a given set of numbers. Let's see how this can be done. Consider the following statements:

$$3 + 5 = 8 \qquad \text{is a true statement,}$$
$$3 + 5 = 9 \qquad \text{is a false statement;}$$

but $3 + x = 10$ might be true or false, depending upon what x is. In this statement, the symbol x represents some number. A symbol, such as x, that represents a number is called a **variable.** (Of course, we know that x had better stand for 7 or the statement $3 + x = 10$ will be false!) Unless you are told otherwise, you can assume the number is real. Although x is perhaps the most commonly used symbol, others will be used as well.

An important point to remember; Since a variable represents a real number, it will have all the properties and obey all the laws of real numbers. Thus we can add, subtract, multiply, and divide variables just as we do real numbers. For example, $x + 5$ represents the sum of (the real number) x and 5, while $2x - 4$ represents the difference between twice x and 4. We can even multiply variables by other variables (and divide variables by variables). For example, a statement such as "a variable times itself" would require us to write something like $x \cdot x$.

When x is multiplied by itself a number of times, it is much easier to write in the following form:

$$x^n = x \cdot x \cdot x \cdots x.$$

Thus $x^1 = x$, $x^2 = x \cdot x$, and so on. We call x the **base** and the number n an **exponent.** An exponent is nothing more than shorthand for repeated multiplication. If you remember this, the properties of exponents become much easier to remember. For example, what should we do with something like $x^2 \cdot x^3$? We can start by writing what the shorthand stands for:

$$x^2 = x \cdot x \quad \text{and} \quad x^3 = x \cdot x \cdot x,$$

so

$$x^2 \cdot x^3 \quad \text{must mean} \quad x \cdot x \cdot x \cdot x \cdot x \quad \text{or} \quad x^5.$$

Thus to multiply x^2 by x^3 we *added* the exponents. In fact, it is true in general that $x^i \cdot x^j$, where i and j are natural numbers, is given by x^{i+j}.

Let's see what else we can determine about working with exponents. What would an expression like $(x^3)^2$ mean? Again, let's just write out everything and see where that leads us. First of all, consider the part within parentheses. We know that x^3 means $x \cdot x \cdot x$. The exponent of 2 *outside* the parentheses must mean that whatever is *inside* the parentheses is multiplied by itself. Then, $(x^3)^2$ must mean $(x \cdot x \cdot x)(x \cdot x \cdot x)$ or $x \cdot x \cdot x \cdot x \cdot x \cdot x$, which is x^6. In general, it is true that $(x^i)^j$ is $x^{i \cdot j}$.

Now let's think about division. Consider $x^4 \div x^2$. Writing everything out gives us the following:

$$\begin{aligned} \frac{x^4}{x^2} &= \frac{x \cdot x \cdot x \cdot x}{x \cdot x} \\ &= x \cdot x = x^2, \end{aligned}$$

which suggests that we should subtract the exponents (especially since we added exponents when multiplying). In fact, in general it is true that $x^i \div x^j$ is x^{i-j}.

What should we do with something like $x^3 \div x^5$? If we follow the rule above, we would get $x^3 \div x^5 = x^{3-5} = x^{-2}$. Great, but what does it mean? Let's see:

$$\begin{aligned} \frac{x^3}{x^5} &= \frac{x \cdot x \cdot x}{x \cdot x \cdot x \cdot x \cdot x} \\ &= \frac{1}{x \cdot x} = \frac{1}{x^2} = x^{-2}. \end{aligned}$$

Thus, x^{-i} must mean $1/x^i$ for our rules to be consistent.

Finally, we know that $x^i \div x^i = 1$. But following our rule for dividing, we have $x^i \div x^i = x^{i-i} = x^0 = 1$, which must mean that $x^0 = 1$. In fact, it does, with the exception that 0^0 is *not* defined. These results and some examples are summarized as follows.

Properties of Exponents

We assume x and y are real numbers and i and j are integers.

Property	Example
1. $x^i = \underbrace{x \cdot x \cdots x}_{i \text{ times}}, \quad i > 0$	$2^3 = 2 \cdot 2 \cdot 2 = 8$
2. $x^0 = 1$ (0^0 is not defined)	$4^0 = 1$
3. $x^{-i} = 1/x^i, \quad x \neq 0$	$3^{-2} = 1/(3 \cdot 3) = 1/9$
4. $x^i \cdot x^j = x^{i+j}$	$2^3 \cdot 2^2 = 2^{3+2} = 2^5 = 32$
5. $(x^i)^j = x^{i \cdot j}$	$(2^2)^4 = 2^{2 \cdot 4} = 2^8 = 256$
6. $(xy)^i = x^i y^i$	$(2 \cdot 3)^3 = 2^3 3^3 = 8 \cdot 27 = 216$
7. $(x/y)^i = x^i/y^i = x^i y^{-i}$, $y \neq 0$ (by Property 3)	$(2/3)^3 = 2^3 \div 3^3 = 8/27$
8. $x^i/x^j = x^{i-j} = 1/x^{j-i}, \quad x \neq 0$	$3^5/3^8 = 3^{5-8} = 3^{-3} = 1/27$

Let's take a look at some more complicated calculations.

EXAMPLE 1

Simplify the following expressions and express the results using positive exponents only.

a) $\dfrac{y^{-3}}{x^{-7}}$ **b)** $(x^{-2}y^{-3})^4$

c) $8a^{-4}b^{-4} \div 6a^{-2}b^{-8}$ **d)** $(2m^4m^{-3} \div 3n^{-2})^{-2}$

SOLUTION

a) $\dfrac{y^{-3}}{x^{-7}} = \dfrac{x^7}{y^3}$

b) $(x^{-2}y^{-3})^4 = (x^{-2})^4(y^{-3})^4 = x^{-8}y^{-12} = \dfrac{1}{(x^8y^{12})}$

c) $8a^{-4}b^{-4} \div 6a^{-2}b^{-8} = 4a^{-4-(-2)}b^{-4-(-8)} \div 3 = \dfrac{4b^4}{3a^2}$

d) $(2m^4m^{-3} \div 3n^{-2})^{-2} = (2m^{4-3} \div 3n^{-2})^{-2} = \left(\dfrac{2m^1n^2}{3}\right)^{-2}$

$$= \frac{9m^{-2}n^{-4}}{4} = \frac{9}{4m^2n^4} \quad \square$$

Practice Problem

Reduce the following expressions and show the results using positive exponents.

a) $(5x^3)(3x^3)$ **b)** $(5x^3)^3$ **c)** $(a^{-3}a^7)^{-5}$

Answer: **a)** $15x^6$; **b)** $125x^9$; **c)** $1/a^{20}$

Practice Problem

Reduce the following expressions and show the results using positive exponents.

a) $6x^3x^{-5} \div 3x^{-8}x^2$ **b)** $(3u^7v^{-5} \div u^{-3}v^{-4})^{-3}$

Answer: **a)** $2x^4$; **b)** $\dfrac{v^3}{27u^{30}}$

Thus far we have discussed only exponents that are integers. Let's see what happens when the exponents are allowed to be rational numbers. (As you know, there is one remaining group of real numbers: the irrationals. However, the study of irrational exponents requires concepts that will be developed in the calculus portion of this text. Thus we must defer a discussion of irrational exponents.)

How can we assign a value to something like $x^{1/n}$? We would like to define $x^{1/n}$ so that it satisfies all the properties of exponents we discussed before. Since $(x^i)^j = x^{i \cdot j}$ when i and j are integers and we want this law to hold even when i and j are not integers, we can let $i = 1/n$ and $j = n$ to give us

$$(x^{1/n})^n = x^{(1/n)n} = x^{n/n} = x^1 = x.$$

In other words, the nth power of $x^{1/n}$ is x. Thus $x^{1/n}$ is the ***n*th root of *x*.** For example, $x^{1/2}$ is the square root of x and $x^{1/3}$ is the cube root of x. Recall that there are two square roots of a positive number: a positive value and a negative value. For example, both 2 and -2 are square roots of 4. To avoid confusion, the symbol $x^{1/n}$ is reserved for the **principal root.** When the root is even, the principal root of a positive number is positive and the principal root of a negative number is undefined. When the root is odd, the principal root of a number is positive when the number is positive and is negative when the number is negative.

EXAMPLE 2

Find the principal roots for the following.

a) $4^{1/2}$ **b)** $125^{1/3}$ **c)** $(-8)^{1/3}$

SOLUTION

a) $4^{1/2} = 2$ b) $125^{1/3} = 5$ c) $(-8)^{1/3} = -2$ □

WARNING

BE CAREFUL WITH PARENTHESES

In part (c) of Example 2 we had $(-8)^{1/3}$, which means to find the third root of -8. In other words, the parentheses tell us that the number is -8. On the other hand, $-4^{1/2}$ means to find the square root of 4 and then multiply it by -1, which gives -2.

A word of caution. There is no real number x such that $x^2 = -4$; in other words, $(-4)^{1/2}$ is not a real number. Therefore we must restrict $x^{1/n}$ to $x \geq 0$ whenever n is even.

We can now extend the definition of exponents to cover all rational values as follows.

Definition of Rational Exponents

For all real numbers x such that the appropriate root exists and for any rational number i/j,

$$x^{i/j} = (x^{1/j})^i = (x^i)^{1/j}.$$

EXAMPLE 3

Evaluate the following.

a) $169^{1/2}$ b) $64^{1/2}$ c) $625^{1/4}$

SOLUTION

a) $169^{1/2} = 13$ b) $64^{1/2} = 8$ c) $625^{1/4} = 5$ □

EXAMPLE 4

Evaluate the following.

a) $81^{1/4}$ b) $(-32)^{1/5}$ c) $(-27)^{1/3}$

SOLUTION

a) $81^{1/4} = 3$ b) $(-32)^{1/5} = -2$
c) $(-27)^{1/3} = -3$ □

Practice Problem

Evaluate the following.

a) $243^{1/5}$ b) $128^{1/7}$ c) $16^{1/2}$ d) $16^{1/4}$

Answer: **a)** 3; **b)** 2; **c)** 4; **d)** 2

EXAMPLE 5

Evaluate the following.

a) $(-64)^{1/2}$ b) $4^{3/2}$ c) $27^{2/3}$

SOLUTION

a) $(-64)^{1/2}$ is not a real number

b) $4^{3/2} = (4^{1/2})^3 = 2^3 = 8$

c) $27^{2/3} = (27^{1/3})^2 = 3^2 = 9$ □

EXAMPLE 6

Evaluate the following.

a) $125^{4/3}$ b) $(-32)^{3/5}$

SOLUTION

a) $125^{4/3} = (125^{1/3})^4 = 5^4 = 625$

b) $(-32)^{3/5} = (-32^{1/5})^3 = (-2)^3 = -8$ □

Practice Problem

Evaluate the following.

a) $(-256)^{1/4}$ b) $-256^{1/4}$ c) $(-243)^{3/5}$ d) $(-216)^{2/3}$

Answer: **a)** not a real number; **b)** -4; **c)** -27; **d)** 36

You probably learned to write the square root of a number with a radical sign, for example, $\sqrt{4}$. In general, we can write the nth root of x with a radical sign as $\sqrt[n]{x}$. Using radical signs, we have

$$x^{i/j} = (\sqrt[j]{x})^i = \sqrt[j]{x^i} \qquad \text{assuming the root exists.}$$

However, the use of the radical sign is awkward. We recommend strongly that you use rational exponents instead, such as $x^{i/j}$.

EXAMPLE 7

Write the following in radical form.

a) $25^{3/2}$ b) $16^{-3/4}$ c) $32x^{-4/5}$

SOLUTION

a) $25^{3/2} = (\sqrt{25})^3$

b) $16^{-3/4} = \dfrac{1}{16^{3/4}} = \left(\dfrac{1}{\sqrt[4]{16}}\right)^3$

c) $32x^{-4/5} = \dfrac{32}{x^{4/5}} = \dfrac{32}{(\sqrt[5]{x})^4}$ □

EXAMPLE 8

Write the following in exponential form.

a) $(\sqrt[3]{81})^2$ b) $\dfrac{1}{(\sqrt[3]{125})^4}$ c) $\dfrac{2}{(\sqrt{x})^5}$

SOLUTION

a) $(\sqrt[3]{81})^2 = 81^{2/3}$

b) $\dfrac{1}{(\sqrt[3]{125})^4} = (\sqrt[3]{125})^{-4} = 125^{-4/3}$

c) $\dfrac{2}{(\sqrt{x})^5} = 2(\sqrt{x})^{-5} = 2x^{-5/2}$ □

Practice Problem

Convert from exponential to radical form:

a) $5^{5/3}$ b) $16^{-6/5}$ c) $125x^{-3/8}$

Convert from radical to exponential form:

d) $(\sqrt[3]{64})^4$ e) $3/(\sqrt[5]{8x})^3$ f) $(\sqrt[5]{3y^3/2x^2})^4$

Answer: **a)** $(\sqrt[3]{5})^5$; **b)** $1/(\sqrt[5]{16})^6$; **c)** $125/(\sqrt[8]{x})^3$; **d)** $64^{4/3}$; **e)** $3(8x)^{-3/5}$; **f)** $(3y^3/2x^2)^{4/5}$

Using the x^2, $\sqrt{x}$, and y^x Keys on Your Calculator

Many calculators, including business and scientific models, have a variety of keys in addition to the basic keys we described in Chapter 1. The x^2 key squares the current display, and the $\sqrt{x}$ key takes the square root of the current display. (Note: on some calculators, you enter INV x^2. Other calculators may require other operations. Again, check *your* owner's manual.) Of course, the $\sqrt{x}$ key will work only for nonnegative numbers. If you attempt to take the square root of a negative number, your calculator will give you an error indication by either blinking the display or by spelling out "error" or by some other unusual display.

The y^x key enables you to raise a positive number, y, to any exponent, x. For example, to compute 5^4, simply enter the following: 5 y^x 4 =, and the result (625) will appear in the display. The y^x key is particularly useful for finding roots. For example, to calculate the cube root of 2, enter 2 y^x 3 $1/x$ =, and the result (1.25992, to five places) will appear. (Notice that we used the $1/x$ key to calculate the decimal equivalent of 1/3, so that we didn't have to worry about the accuracy of our entry.)

When using the y^x key, it is necessary that y be positive. This is because of the way that your calculator computes the result. Thus, although a negative value of y is certainly meaningful in many cases, your calculator will give you an error indication if you attempt to enter a negative y.

EXAMPLE 9

Calculate 3^2.

SOLUTION

Enter 3 x^2. The display shows 9. □

EXAMPLE 10

Calculate the (principal) square root of 110.

SOLUTION

Enter 110 $\sqrt{x}$. The display shows 10.48809 (to five places). Again, you don't need the = key. □

EXAMPLE 11

Calculate the square root of 1/6 + 1/9.

SOLUTION

Enter 6 $1/x$ + 9 $1/x$ = $\sqrt{x}$. The display shows .52705 (to five places). □

EXAMPLE 12

Calculate $4^4 - 5^3$.

SOLUTION

Enter 4 y^x 4 = − 5 y^x 3 =, with a result of 131. □

EXAMPLE 13

Find $(-8.5)^{1/3}$.

SOLUTION

Here's a case in which a little thought before we begin pushing keys will prove to be useful. We know the calculator will *not* accept -8.5 as a y value for the [y^x] key. However, we also know that $(-8.5)^{1/3} = -8.5^{1/3}$. Thus, enter the following: 8.5 [y^x] 3 [$1/x$] [=] [+/−], with a result of -2.04083. □

EXAMPLE 14

Calculate $(7 + \frac{2}{3} - \frac{5}{6})^3 - 4.5^2$.

SOLUTION

If you know that your calculator has a built-in hierarchy (order) of calculations, this calculation is straightforward. However, if it doesn't, you need to use memory or parentheses. We will use memory to solve it. Enter

2 [÷] 3 [=] [STO] 5 [+/−] [÷] 6 [=] [+] [RCL] [+] 7 [=] [y^x] 3 [=] [−] 4.5 [x^2] [=].

The result is 298.8287. □

EXAMPLE 15

Calculate $(27.5^{1/3}) \cdot (25.2^{1/2})$.

SOLUTION

Enter

27.5 [y^x] 3 [$1/x$] [=] [×] 25.2 [$\sqrt{x}$] [=].

The display shows 15.15227. □

EXAMPLE 16

Calculate $(27.5^{1/3}) \cdot (25.2^{1/3})$.

SOLUTION

This looks very similar to the previous problem, doesn't it? However, by noting that $(27.5^{1/3}) \cdot (25.2^{1/3}) = (27.5 \cdot 25.2)^{1/3}$, we can calculate the answer quite easily. Enter 27.5 [×] 25.2 [=] [y^x] 3 [$1/x$] [=]. The display shows 8.84934 (to five places). In this case, a little thought *before* we started pushing keys on our calculator resulted in an easy solution. □

Practice Problem

Solve the following on your calculator to five places of accuracy.

a) 2.1^5 **b)** $0.25^{1/2}$ **c)** 0.725^{-5}
d) $3.2^2 - 2.54^3$ **e)** $296^{2/3} + (-421)^{3/5}$

Answer: **a)** 40.84101; **b)** 0.5; **c)** 4.99241; **d)** −6.14706; **e)** 6.86833

EXERCISE SET 2.1

Solve the following problems using your calculator.

1. $(1.3)^5$ **2.** $(6.3)^4$ **3.** $(9.2)^3$ **4.** $(-2.5)^3$
5. $(17.1)^{10}$ **6.** 2^{32} **7.** $(2.8)^{1/5}$ **8.** $(625)^{1/3}$
9. $(14)^{1/6}$ **10.** $(0.25)^{1/3}$ **11.** $(0.98)^{25}$ **12.** $(0.5)^{1/10}$
13. $(-9.2)^{-3}$ **14.** $(2.5 + 3.31)^{2/3} - 7.42^{1/3}$
15. $(27.2 - 4.2^3)^{1/3}$ **16.** $[24.2^{1/2} - 2(-45)^{1/3}]^{3/4}$

Simplify each expression so that it contains only positive exponents. You may find it necessary to use your calculator.

17. $(x^{2/3})(x^{1/3})$ **18.** $(2.5x^5)(3x^4)$
19. $x^3 \div x^7$ **20.** $y^7 \div y^4$
21. $27x^{10} \div 3x^4$ **22.** $40x^3 \div 10x^8$
23. $721x^{11} \div 2.8x^3$ **24.** $16x^4 \div 32x^8$
25. $176x^{2/3} \div 16x^{1/3}$ **26.** $24x^{-5} \div x^{-2}$
27. $8x^{2/3} \div 2x^{1/4}$ **28.** $15x^{-7/8} \div 3x^{4/3}$
29. $(x^2)^9$ **30.** $(2x^{1/3})^3$
31. $(x^{2/3})^6$ **32.** $(15x^{1/3})^{1/4}$
33. $(7x^{-1/4})^{-1/3}$ **34.** $(0.98^{-1/4})^5$
35. $25x^{-4/3}$ **36.** $(x^2y)(2x^5y^4)$
37. $(x^{4/3}y)(x^{2/3}y^3)$ **38.** $(3x^{-5}y^5)(25x^{-4}y)$
39. $(x^{-2/3}y^{1/4})(x^{1/6}y^{1/8})$ **40.** $(2x^4y^3)^4$
41. $(3a^{1/3}b^{1/5})^{12}$ **42.** $(x^{-1/3}y^{-2/5})^{15}$
43. $(u^{1/4}v^{1/5})^{-20}$ **44.** $(p^3 \div q^2)^3$
45. $(x^{1/3} \div y^{2/3})^9$ **46.** $(m^5 \div n^{-6})^{-30}$
47. $(x^{-3/2} \div y^{-2})^4$ **48.** $(x^2y^3 \div 4xy)^3$
49. $(r^6s^3 \div r^{12}s^3)^{1/3}$ **50.** $(36x^{-2}y \div 6x^5y^4)^3$
51. $(9x^{4/3}y^{1/6} \div 6x^{2/3}y^{1/8})^{1/2}$ **52.** $x^{n+1} \div x^{n-2}$
53. $y^{m+7} \div y^{2m-3}$ **54.** $(v^{j+1} \div v^{j+3})^{-2}$
55. $(x^{3k+2} \div 4x^{k+5})^{1/2}$

2.2

POLYNOMIALS AND ALGEBRAIC EXPRESSIONS

Addition and Subtraction

A **monomial** in x is an expression of the form ax^n, where a is a real number and n is a natural number or 0. A **polynomial** is a sum (or difference) of monomials. Similarly, a **monomial** in x and y is an expression of the form ax^my^n, where m and n are natural numbers or zero. A **polynomial** in x and y is the sum (or difference) of monomials in x and y. (We can define monomials and polynomials in three or more variables in a similar fashion.)

EXAMPLE 17

Classify the following expressions as monomials or polynomials, and name the variables.

a) $3x^3$ **b)** $6u^3v^2$ **c)** $2c^2 + 5c + 17$
d) $8y^3 - 4y^2z - 9yz^2 + 3z^3$ **e)** 4

SOLUTION

a) $3x^3$ is a monomial with variable x.
b) $6u^3v^2$ is a monomial with variables u and v.
c) $2c^2 + 5c + 17$ is a polynomial with variable c.
d) $8y^3 - 4y^2z - 9yz^2 + 3z^3$ is a polynomial with variables y and z.
e) 4 is a monomial in which the variable has an exponent of 0. □

A **term of a polynomial** is any one of the monomials that are added (or subtracted) together to obtain the polynomial. The polynomial in part c) of Example 17 has three terms, and the polynomial in part d) has four terms. A polynomial with two terms is commonly called a **binomial** and a polynomial with three terms is called a **trinomial.** For example, $4x^3 + 3x$ is a binomial but $2x^2 - 6x + 4$ is a trinomial.

The **degree** of the term (monomial) ax^n is n (assuming $a \neq 0$). A nonzero constant (for example, 5) has a degree of zero when it occurs as a term. The constant 0 has no degree assigned to it. The **degree of a monomial in two or more variables** is the sum of the exponents of the variables. For example, the monomial $3x^2y^4$ has degree six. The **degree of a polynomial** is the highest degree of the monomials in it. A polynomial is often written so that the terms are in decreasing order of degree. In other words, the term with the highest degree is first, the term with the second highest degree is second, and so on. A general form for a polynomial in x of degree n is

$$a_nx^n + a_{n-1}x^{n-1} + \cdots + a_1x + a_0,$$

where n is a positive integer and the **coefficients** $a_0, a_1, \ldots, a_n$ are real numbers. A polynomial is **linear** if it has degree 1, **quadratic** if it has degree 2 and **cubic** if it has degree 3.

EXAMPLE 18

The following are polynomials:

a) $3x^4 - 5x^3 + 2x^2 - 5x + 1$ has degree 4.
b) $6x^2 - 7x + 4$ has degree 2 (quadratic).
c) $7x$ has degree 1 (linear).
d) 14 has degree 0. □

EXAMPLE 19

The following are *not* polynomials:

a) $x + 1/x$ is not a polynomial because $1/x = x^{-1}$, which is written with a negative integer (all exponents of variables must be natural numbers).
b) $5x^{3/2} + 4x + 2$ is not a polynomial because the variable x has an exponent of 3/2, which is not a natural number.
c) $(4x - 2)^{1/2}$ is not a polynomial because of the exponent of 1/2 and because we do not allow the sum or difference of terms to be raised to a power.
d) $(3x^2 + 2x + 1)/(5x - 3)$ is not a polynomial because it is the ratio of two polynomials. □

Practice Problem

Classify the following as polynomials or nonpolynomials. Give the degree if the expression is a polynomial.

a) $15x^3 - 6x^2 + 5$
b) $13x^{5/2} - 5x^3 + 4x$
c) $7x^2 + 20$
d) $(12x^3 - 15)/(4x - 30)$
e) $5x^2 + 2x - 3 + 4x^{-1}$

Answer: **a)** polynomial of degree 3; **b)** nonpolynomial; **c)** polynomial of degree 2; **d)** nonpolynomial; **e)** nonpolynomial

Consider the following addition:

$$(x^2 + 5x + 4) + (2x^3 + 4x^2 - 6x - 1).$$

How can we deal with this? The answer is actually quite easy. First, remove the parentheses, which gives us

$$x^2 + 5x + 4 + 2x^3 + 4x^2 - 6x - 1.$$

Next, gather all terms of the same degree together (placing the terms in descending order of degree), which results in

$$2x^3 + x^2 + 4x^2 + 5x - 6x + 4 - 1.$$

Now we can combine terms of the same degree by using the distributive property. This property tells us, for instance, that $x^2 + 4x^2$ can be written as $(1 + 4)x^2$ or $5x^2$. Applying the distributive property gives us

$$2x^3 + 5x^2 - x + 3.$$

We can add and subtract algebraic expressions by utilizing the same basic procedure we just used. An **algebraic expression** is the result of combining numbers and symbols with the operations of addition, subtraction, multiplication, division, raising to a power, and taking roots.

Simplifying Algebraic Expressions

1. Remove all parentheses by performing the indicated operations. Work from the innermost grouping to the outermost groupings. If necessary, convert subtraction to addition: $x - y$ is equal to $x + (-y)$.
2. Use the distributive property to combine terms of like degree.

EXAMPLE 20

Simplify the following.

a) $3(5x - 2) + 2(x^2 + 2x + 1)$

b) $2x^2 - (3x - 2)$

c) $5y^2 + 2y - y(2y - 3)$

SOLUTION

a) $$\begin{aligned} 3(5x - 2) + 2(x^2 + 2x + 1) &= 15x - 6 + 2x^2 + 4x + 2 \\ &= 2x^2 + 19x - 4 \end{aligned}$$

b) $$\begin{aligned} 2x^2 - (3x - 2) &= 2x^2 + (-1)(3x - 2) \\ &= 2x^2 + (-3x + 2) \\ &= 2x^2 - 3x + 2 \end{aligned}$$

c) $5y^2 + 2y - y(2y - 3) = 5y^2 + 2y - (2y^2 - 3y)$
$= 5y^2 + 2y + (-1)(2y^2 - 3y)$
$= 5y^2 - 2y^2 + 2y + 3y$
$= 3y^2 + 5y$ □

EXAMPLE 21

Simplify the following.

a) $2x - [7x - 4(3x - 2)]$ **b)** $4x^2 - [3x - x(5x + 2)]$

SOLUTION

a) $2x - [7x - 4(3x - 2)] = 2x - [7x - (12x - 8)]$
$= 2x - (7x - 12x + 8)$
$= 2x - 7x + 12x - 8$
$= 7x - 8$

b) $4x^2 - [3x - x(5x + 2)] = 4x^2 - [3x - (5x^2 + 2x)]$
$= 4x^2 - [3x - 5x^2 - 2x]$
$= 4x^2 - [-5x^2 + x]$
$= 4x^2 + [5x^2 - x]$
$= 4x^2 + 5x^2 - x$
$= 9x^2 - x$ □

Practice Problem

Remove any parentheses and simplify.

a) $4(5x^3 - 2x^2 + 7x - 3) - 3(2x^3 + x^2 - 3x - 4)$

b) $2x(4x^2 + 8x + 1) - 3x(2x^2 - x - 2)$

c) $2x[4x^2 - x(3x - 1)] + x[3x^2 - 2x(-3x + 2)]$

Answer: **a)** $14x^3 - 11x^2 + 37x$; **b)** $2x^3 + 19x^2 + 8x$; **c)** $11x^3 - 2x^2$

The same procedures work with polynomials and expressions involving more than one variable.

EXAMPLE 22

Simplify the following expressions.

a) $(a^2 - 3b^2 + 4c^2) - (5a^2 + 2b^2 - c^2)$

b) $(9y^2 - 14yz - 8z^2) - 3z(4y - 5z) + 2y(3y + 2z)$

SOLUTION

a) $(a^2 - 3b^2 + 4c^2) - (5a^2 + 2b^2 - c^2)$
$= a^2 - 3b^2 + 4c^2 - 5a^2 - 2b^2 + c^2$
$= -4a^2 - 5b^2 + 5c^2$

b) $(9y^2 - 14yz - 8z^2) - 3z(4y - 5z) + 2y(3y + 2z)$
$= 9y^2 - 14yz - 8z^2 - 12yz + 15z^2 + 6y^2 + 4yz$
$= 15y^2 - 22yz + 7z^2$ □

Practice Problem

Simplify the following expressions.

a) $10uv(u + 2v + 3u^2v^2) - 3uv[2u - 5(u - v)]$

b) $3xyz[2x(-7yz) - 3y(2xz) + 4z(5xy - 6z)]$

Answer: **a)** $19u^2v + 5uv^2 + 30u^3v^3$; **b)** $-72xyz^3$

EXAMPLE 23

Problem from Life **a)** A rectangular box has a square base of x inches on a side and a height of h inches. There is no top. Express the total surface areas as a polynomial in x and h.

b) A box is to be made from a piece of cardboard 12 inches square. Squares are to be cut out of the corners and the sides folded up. Express the volume of the box as a polynomial.

SOLUTION

a) We can begin solving part (a) by first reading it carefully. Clearly, a box is involved and our task is to develop a polynomial that describes its surface area. Recalling our four-step approach to problem solving (see Chapter 1), our next step is to generate a mathematical model. Normally we begin by defining a set of variables, but in this case that has already been done. In many problems, and especially in this one, drawing a picture can be a big help. (See Fig. 2.1.)

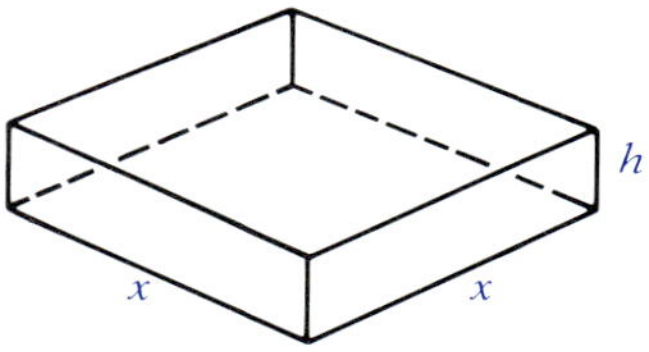

Figure 2.1 **A box with a square base and no top.**

The term *surface area* refers to the total area of the sides and bottom of the box. Since we know the area of a square with dimensions x by x is x^2, and the area of a rectangle with dimensions x by h is xh, our problem is quite straightforward. Looking at Fig. 2.1, we can see that the box includes

- one bottom of dimensions x by x, and
- four sides, each of dimensions x by h.

Thus the total surface area must be given by the polynomial

$$x^2 + 4xh.$$

b) Once again, we should begin by reading the problem carefully. We must construct a box by cutting the corners out of a square piece of cardboard and then folding up the sides.

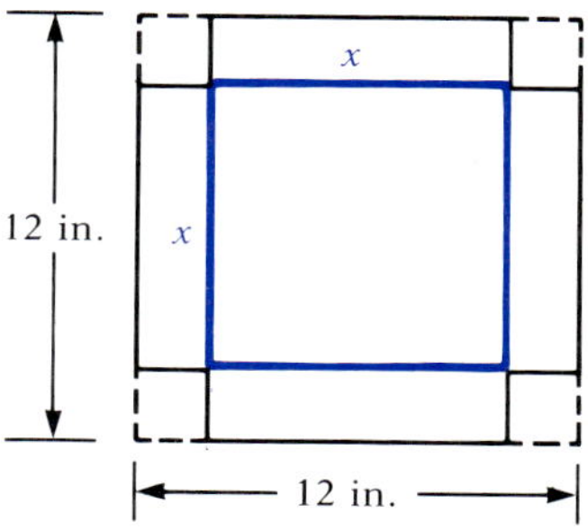

Figure 2.2 **Making a box by cutting out the corners and folding.**

The next step is to define a mathematical model. We must first define a variable. Let x be the length of cardboard remaining after the corners are cut out. (See Fig. 2.2.) Looking at Fig. 2.2, we can see that the resulting box will have a square base of dimensions x by x. Let's see if we can determine how high the sides will be. Now, we started with a piece of cardboard 12 in. square. After cutting the corners out, we still have a piece 12 in. across. Part of this (x) will form the base. The remainder must go into making up *two* sides. In other words, we make up two sides from a length of $12 - x$. Hence, the height of each side must be $\frac{1}{2}(12 - x) = 6 - x/2$. But now we know the dimensions of our box! It will have a square base of size x by x and sides with a height of $6 - x/2$. The volume must be given by the expression

$$x \cdot x \cdot \left(6 - \frac{x}{2}\right) = 6x^2 - \frac{x^3}{2}. \quad \square$$

Practice Problem

Suppose, in part (b) of Example 23, that we started with a piece of cardboard 20 in. square. What would be the volume of the resulting box?

Answer: $10x^2 - x^3/2$

Multiplication

We can multiply two polynomials by making use of the distributive property. In other words, we multiply each term of one by each term of the other and then simplify.

EXAMPLE 24

a) Multiply $(x^2 + 2)$ by $(2x^2 - 3)$

b) Multiply $(2y - 3)$ by $(5y^2 + 3y - 1)$

c) Multiply $(2x^2 - 2xy + y^2)$ by $(x^2 - xy - 2y^2)$

SOLUTION

a)
$$\begin{aligned}(x^2 + 2)(2x^2 - 3) &= x^2(2x^2 - 3) + 2(2x^2 - 3)\\ &= 2x^4 - 3x^2 + 4x^2 - 6\\ &= 2x^4 + x^2 - 6\end{aligned}$$

b)
$$\begin{aligned}(2y - 3)(5y^2 + 3y - 1) &= 2y(5y^2 + 3y - 1) - 3(5y^2 + 3y - 1)\\ &= 10y^3 + 6y^2 - 2y - 15y^2 - 9y + 3\\ &= 10y^3 + 6y^2 - 15y^2 - 2y - 9y + 3\\ &= 10y^3 - 9y^2 - 11y + 3\end{aligned}$$

c) $(2x^2 - 2xy + y^2)(x^2 - xy - 2y^2)$
$= 2x^2(x^2 - xy - 2y^2) - 2xy(x^2 - xy - 2y^2) + y^2(x^2 - xy - 2y^2)$
$= 2x^4 - 2x^3y - 4x^2y^2 - 2x^3y + 2x^2y^2 + 4xy^3 + x^2y^2 - xy^3 - 2y^4$
$= 2x^4 - 4x^3y - x^2y^2 + 3xy^3 - 2y^4$ □

Practice Problem

Perform the following multiplications.

a) $(3x - 2)(4x + 1)$
b) $(6u + 3)(2u - 1)$
c) $(4r - 3s)(5s + 2r)$
d) $(4y^2 - 3yz + 2z)(2y^2 + 3yz - 5z)$

Answer: **a)** $12x^2 - 5x - 2$; **b)** $12u^2 - 3$;
c) $8r^2 + 14rs - 15s^2$;
d) $8y^4 + 6y^3z - 16y^2z - 9y^2z^2 + 21yz^2 - 10z^2$

The following list of special products can be derived from the distributive property.

Special Products

1. $x(y + z) = xy + xz$
2. $(x + a)(x + b) = x^2 + (a + b)x + ab$
3. $(ax + b)(cx + d) = acx^2 + (ad + bc)x + bd$
4. $(x + a)^2 = x^2 + 2ax + a^2$ **Square of a binomial**
5. $(x - a)^2 = x^2 - 2ax + a^2$ **Square of a binomial**
6. $(x + a)(x - a) = x^2 - a^2$ **Product of a sum and difference**

EXAMPLE 25

Perform the following multiplications.

a) $(x + 3)(x - 2)$ **b)** $(2y + 5)(3y + 7)$ **c)** $(x + 5)^2$
d) $(u - 3)^2$ **e)** $(t - 4)(t + 4)$

SOLUTION

a) $(x + 3)(x - 2) = (x + 3)[x + (-2)]$
$= x^2 + (3 - 2)x + 3(-2)$ **Property 2**
$= x^2 + x - 6$

b) $(2y + 5)(3y + 7) = 2 \cdot 3y^2 + (2 \cdot 7 + 5 \cdot 3)y + 5 \cdot 7$ **Property 3**
$= 6y^2 + 29y + 35$

c) $(x + 5)^2 = x^2 + 2 \cdot 5x + 5^2$ **Property 4**
$= x^2 + 10x + 25$

d) $(u - 3)^2 = u^2 - 2 \cdot 3u + 3^2$ **Property 5**
$= u^2 - 6u + 9$

e) $(t - 4)(t + 4) = t^2 - 4^2$ **Property 6**
$= t^2 - 16$ □

Practice Problem

Perform the following multiplications.

a) $(z - 5)(z + 3)$ **b)** $(4a - 3)(5a + 2)$
c) $(x + 7)^2$ **d)** $(t - 6)^2$
e) $(y + 8)(y - 8)$

Answer: **a)** $z^2 - 2z - 15$; **b)** $20a^2 - 7a - 6$; **c)** $x^2 + 14x + 49$; **d)** $t^2 - 12t + 36$; **e)** $y^2 - 64$

EXAMPLE 26

Perform the following multiplications.

a) $(3r + 2s)(4r - s)$ **b)** $(x + 4y)^2$

SOLUTION

a) $(3r + 2s)(4r - 3)$
$= 3 \cdot 4r^2 + [(3)(-1s) + 4(2s)r] + (2s)(-s)$ **Property 3**
$= 12r^2 + 5rs - 2s^2$

b) $(x + 4y)^2 = x^2 + 2(4y)x + (4y)^2$ **Property 4**
$= x^2 + 8xy + 16y^2$ □

Practice Problem

Perform the following multiplications.

a) $(5c - 2d)(3c + 4d)$ **b)** $(x - 12y)^2$ **c)** $(x + 3yz)(x - 3yz)$

Answer: **a)** $15c^2 + 14cd - 8d^2$; **b)** $x^2 - 24xy + 144y^2$; **c)** $x^2 - 9y^2z^2$

Factoring

A very important process closely related to multiplication is factoring. When two or more expressions are multiplied together, the expressions are called *factors* of the product. For example, since $3 \cdot 2 = 6$, then 3 and 2 are both factors of 6. To **factor** an expression means to write an equivalent expression that is a product. In fact, factoring is the reverse of multiplication.

You should always check the validity of any factoring you do by simply carrying out the multiplication and making sure the result is equal to the expression with which you started.

The key to factoring is the distributive property. For example, we can write the expression $9x^2 + 6x$ as a product by factoring out the common factor $3x$:

$$9x^2 + 6x = 3x(3x + 2).$$

Going from left to right we used the distributive property to factor, while going from right to left we used the property to multiply.

The first step in factoring is to look for common factors. A common factor may be a single number or term as just shown, or it may be any common expression. When you look for a common factor, always factor out the *largest* common factor.

EXAMPLE 27

Factor the following expressions by finding factors common to all terms.

a) $5x^3 + 10x^2y + 15xy^2 + 5xy^3$ **b)** $10u^2v + 20uv^2 + 30u^3v^3$

SOLUTION

a) $5x^3 + 10x^2y + 15xy^2 + 5xy^3 = 5x\,(x^2 + 2xy + 3y^2 + y^3)$

Factor out 5*x*.

b) $10u^2v^2 + 20uv^2 + 30u^3v^3 = 10uv\,(u + 2v + 3u^2v^2)$

Factor out 10*uv*. □

EXAMPLE 28

Factor the following by finding common factors.

a) $3x^2(5x - 2) + 6x(5x - 2) - 3(5x - 2)$

b) $(x - 2)(x + 5) + (x - 2)(x + 8)$

SOLUTION

a) $3x^2(5x - 2) + 6x(5x - 2) - 3(5x - 2)$

$= (5x - 2)(3x^2 + 6x - 3)$ **Factor out (5*x* − 2).**

$= 3(5x - 2)(x^2 + 2x - 1)$ **Factor out 3.**

Note that the common factor here is a binomial.

b) $(x - 2)(x + 5) + (x - 2)(x + 8)$

$= (x - 2)[(x + 5) + (x + 8)]$ **Factor out (*x* − 2).**

$= (x - 2)(x + 5 + x + 8)$

$= (x - 2)(2x + 13)$ □

Practice Problem

Factor out any common factors.

a) $4u^2 + 2u$ **b)** $5x^4 + 20x^3$

c) $8xy + 10xz - 14xw$ **d)** $(m-4)(m+3)+(m-4)(m-3)$

Answer: **a)** $2u(2u + 1)$; **b)** $5x^3(x + 4)$;
c) $2x(4y + 5z - 7w)$; **d)** $2m(m - 4)$

What can we do after we've looked for common factors? Since factoring is the reverse of multiplication, we should look at the remaining expression to see if we recognize it. There are several factorizations that occur often enough that you should be thoroughly familiar with them:

Commonly Occurring Factorizations

1. $xy + xz = x(y + z)$ **Common factors**
2. $x^2 + 2ax + a^2 = (x + a)^2$ **Square of a sum**
3. $x^2 - 2ax + a^2 = (x - a)^2$ **Square of a difference**
4. $x^2 - a^2 = (x + a)(x - a)$ **Difference of two squares**
5. $x^2 + (a + b)x + ab = (x + a)(x + b)$
6. $acx^2 + (ad + bc)x + bd = (ax + b)(cx + d)$

Notes:

1. When factoring polynomials with integer coefficients, we almost always choose factors that are themselves polynomials with integer coefficients. For example, we do not write
$$x - 4 = (x^{1/2} + 2)(x^{1/2} - 2).$$
2. Always factor completely. For example,
$$3x^2 - 27 = 3(x^2 - 9) = 3(x + 3)(x - 3).$$

EXAMPLE 29

Factor the following.

a) $x^2 + 2x + 1$ **b)** $4x^2 + 4x + 1$ **c)** $4x^2 + 20x + 25$

SOLUTION

a) $x^2 + 2x + 1 = (x + 1)^2$ **Property 2**

b) $4x^2 + 4x + 1 = (2x + 1)^2$ **Property 2**

c) $4x^2 + 20x + 25 = (2x + 5)^2$ **Property 2** □

EXAMPLE 30

Factor the following.

a) $x^2 - 2x + 1$ **b)** $9x^2 - 6x + 1$ **c)** $9x^2 - 30x + 25$

SOLUTION

a) $x^2 - 2x + 1 = (x - 1)^2$ **Property 3**
b) $9x^2 - 6x + 1 = (3x - 1)^2$ **Property 3**
c) $9x^2 - 30x + 25 = (3x - 5)^2$ **Property 3** □

EXAMPLE 31

Factor the following.

a) $x^2 - 4$ **b)** $4x^2 - 1$ **c)** $25x^2 - 16$

SOLUTION

a) $x^2 - 4 = (x + 2)(x - 2)$ **Property 4**
b) $4x^2 - 1 = (2x + 1)(2x - 1)$ **Property 4**
c) $25x^2 - 16 = (5x + 4)(5x - 4)$ **Property 4** □

Practice Problem

Factor the following.

a) $x^2 + 6x + 9$ **b)** $9x^2 + 12xy + 4y^2$
c) $u^2 - 12u + 36$ **d)** $16a^2 - 24ab + 9b^2$
e) $m^2 - 9$ **f)** $16r^2 - 9s^2$

Answer: **a)** $(x + 3)^2$; **b)** $(3x + 2y)^2$; **c)** $(u - 6)^2$; **d)** $(4a - 3b)^2$; **e)** $(m + 3)(m - 3)$; **f)** $(4r + 3s)(4r - 3s)$

Of course, it is not unusual for common factors to be present that must be removed before one of the commonly occurring factorizations becomes apparent.

EXAMPLE 32

Factor the following.

a) $3x^2 + 6x + 3$ **b)** $5x^3 - 10x^2 + 5x$

SOLUTION

a) $3x^2 + 6x + 3 = 3(x^2 + 2x + 1)$ **Factor out 3.**
$= 3(x + 1)^2$ **Property 2**

b) $5x^3 - 10x^2 + 5x = 5x(x^2 - 2x + 1)$ **Factor out 5x.**
$= 5x(x - 1)^2$ **Property 3** □

EXAMPLE 33

Factor the following.

a) $2x^4 - 8x^2y^4$ **b)** $2ab^4 - 4ab^2c^2 + 2ac^4$

SOLUTION

a) $2x^4 - 8x^2y^4 = 2x^2(x^2 - 4y^4)$ **Factor out $2x^2$.**
$= 2x^2(x - 2y^2)(x + 2y^2)$ **Property 4**

b) $2ab^4 - 4ab^2c^2 + 2ac^4 = 2a(b^4 - 2b^2c^2 + c^4)$ **Factor out 2a.**
$= 2a(b^2 - c^2)^2$ **Property 3**
$= 2a[(b + c)(b - c)]^2$ **Property 5**
$= 2a(b + c)^2(b - c)^2$ □

Study the last example carefully. As you can see, we're not necessarily limited to dealing only with quadratic polynomials.

Practice Problems

Factor the following polynomials.

a) $16x^3 + 16x^2 + 4x$ **b)** $25a^2b^2 - 100a^2c^2$

c) $12mn^2 - 84mn + 147m$

Answer: **a)** $4x(2x + 1)^2$; **b)** $25a^2(b + 2c)(b - 2c)$
c) $3m(2n - 7)^2$

Using factorizations 5 and 6 is harder because they involve a trial-and-error process, especially number 6. The key to using them is being able to multiply the possibilities quickly and discard the incorrect ones.

EXAMPLE 34

Factor the following.

a) $x^2 - x - 6$ **b)** $x^2 + 4x - 12$

SOLUTION

a) By Property 5, we want a pair of factors of -6 that add up to -1. The only possibilities for a product of -6 are

$$(6)(-1) \qquad (-6)(1) \qquad (3)(-2) \qquad (-3)(2).$$

Since only -3 and 2 add up to -1, this must be the correct pair. Therefore,

$$\begin{aligned} x^2 - x - 6 &= [x + (-3)](x + 2) \\ &= (x - 3)(x + 2). \end{aligned}$$

b) Here we want a pair of factors of -12 that add up to 4. The possibilities for a product of 12 are

$$(12)(-1) \qquad (-12)(1) \qquad (6)(-2) \qquad (-6)(2) \qquad (4)(-3) \qquad (-4)(3)$$

The correct pair is $(6)(-2)$, so

$$x^2 + 4x - 12 = (x + 6)(x - 2). \quad \square$$

Practice Problem

Factor the following polynomials.

a) $x^2 - 2x - 15$ **b)** $x^2 - 7x + 6$

Answer: **a)** $(x - 5)(x + 3)$; **b)** $(x - 6)(x - 1)$

EXAMPLE 35

Factor the following.

a) $4x^2 + 7x + 3$ **b)** $6x^2 + 6x - 12$

SOLUTION

a) We can factor 4 as $4 \cdot 1$ and $2 \cdot 2$, but the only way to factor 3 is $3 \cdot 1$. (The middle term has a positive sign, so we need not consider the negatives of these pairs.) The possible factorings are $(4x + 1)(x + 3)$, $(4x + 3)(x + 1)$, and $(2x + 1)(2x + 3)$. The correct answer is

$$4x^2 + 7x + 3 = (4x + 3)(x + 1).$$ **Check by multiplication**

b) To begin, notice there is a common factor of 6. Hence

$$6x^2 + 6x - 12 = 6(x^2 + x - 2).$$

Next, the possible factorings of -2 are $2(-1)$ and $(-2)1$. A little trial and error should demonstrate that the correct answer is

$$6x^2 + 6x - 12 = 6(x - 1)(x + 2). \quad \square$$

Practice Problem

Factor the following.

a) $x^2 + 13x + 12$ **b)** $x^2 + 8x + 12$

c) $y^2 - 6y - 16$ **d)** $3a^2 + 8a - 3$

e) $6q^2 + 23q - 4$ **f)** $8c^2 - 14c + 3$

Answer: **a)** $(x + 12)(x + 1)$; **b)** $(x + 6)(x + 2)$; **c)** $(y - 8)(y + 2)$; **d)** $(3a - 1)(a + 3)$; **e)** $(6q - 1)(q + 4)$; **f)** $(4c - 1)(2c - 3)$

Let's see what can happen when we put everything together.

EXAMPLE 36

Factor the following.

a) $3x^3 + 24x^2 + 36x$ **b)** $4x^3y + 12x^3y + 8xy^3$

SOLUTION

a) $3x^3 + 24x^2 + 36x = 3x(x^2 + 8x + 12)$ **Property 1**

$= 3x(x + 6)(x + 2)$ **Property 5**

b) $4x^3y + 12x^2y^2 + 8xy^3 = 4xy(x^2 + 3xy + 2y^2)$ **Property 1**

$= 4xy(x + 2y)(x + y)$ **Property 5** □

EXAMPLE 37

Factor the following:

a) $18a^3b^2 + 42a^2b^3 - 36ab^4$

b) $(4a^2 - b^2)(2a^2 + 3b) - (4a^2 - b^2)(6ab - b^2)$

SOLUTION

a) $18a^3b^2 + 42a^2b^3 - 36ab^4 = 6ab^2(3a^2 + 7ab - 6b^2)$ **Property 1**

$= 6ab^2(3a - 2b)(a + 3b)$ **Property 6**

b) $(4a^2 - b^2)(2a^2 + 3ab) - (4a^2 - b^2)(6ab - b^2)$

$= (4a^2 - b^2)(2a + 3ab - 6ab + b^2)$ **Property 1**

$= 4a^2 - b^2)(2a^2 - 3ab + b^2)$

$= (2a + b)(2a - b)(2a^2 - 3ab + b^2)$ **Property 4**

$= (2a - b)(2a + b)(2a - b)(a - b)$ **Property 6**

$= (2a - b)^2(2a + b)(a - b)$ □

EXERCISE SET 2.2

Perform the indicated operations and simplify by combining like terms

1. $2(a + b)$

2. $a(x + y)$

3. $2x^2(a^2 + b^2)$

4. $5x^2(3x^3 + 2x)$

5. $(x + y)(x + y)$

6. $(2x + y)(x - y)$

7. $(x^2 + y)(x + y)$

8. $(2x + 3y)(4x - 5y)$

9. $x(x + y) + y(x + y)$

10. $2x^3(x + y^2) - x^2(x^2 - y)$

11. $(c + 4d)(c - 3d)$

12. $(7xy - 4)(6xy - 3)$

13. $(7m^2 - 3n^2)(4m^2 + 5n^2)$

14. $(x^2 + 3)(x^2 + 4y + z)$

15. $(x + y)(xy + xy^2 - y^3)$

Factor the following.

16. $6x^2 - 5x - 4$

17. $8x^2 - 2xy - 3y^2$

18. $8x^2 - 2xy - 15y^2$

19. $2 + x^4 - 36x^8$

20. $c^2 - 2c - 15$

21. $16x^2 - 9y^2$

22. $4x^4 - y^4$

23. $16x^2 + 40xy + 25y^2$

24. $4x^2 - 12xy + 9y^2$

25. $9x^3y - 4xy^3$

26. $3d^3 - 6d^2 - 30d$

27. $m^2 + 4m - 45$

28. $9 - 6u - 8u^2$

29. $6c^2u^2 + 5cuz^3 - 6z^6$

30. $2x^2 - 14x - 36$

31. $3c^2 + 9c - 120$

32. $2x^3 + x^2 - x$

33. $3c^2u + 7u + 22cu$

34. $4ax^2 - 12axy + 9ay^2$

35. $49x^3 - 36x^3z^2$

36. $9x^2 - y^2 + 2xz - z^2$

37. $4x^2 - y^2 - 2yz - z^2$

38. $81mnp^2 - 25mnq^4$

39. $x^6 - y^6$

40. $x^4 + x^2y^2 - 20y^4$

41. $6u^2 - 2uv - 28v^2$

42. $4x^4 - 12x^2y^2 + 9y^4$

43. $100x^6 - 64x^2y^4$

2.3 RATIONAL EXPRESSIONS

A **rational expression** is simply a ratio (fraction) in which both numerator and denominator are polynomials. We will be concerned with simplifying rational expressions or a series of rational expressions that are to be combined using the arithmetic operations (addition, subtraction, multiplication, and division). Before you begin, make sure you review the material in Section 1.2 dealing with the properties of fractions.

Simplifying Rational Expressions

To begin with, let's simply consider a single rational expression. We simplify such expressions by using the simplification rule for fractions.

Simplification Rule

$$\frac{ac}{bc} = \frac{a}{b}$$

We apply the rule by factoring both the numerator and the denominator and then dividing out (canceling out) any terms common to both numerator and denominator.

EXAMPLE 38

Simplify the following rational expressions.

a) $\dfrac{x^2 + 2xy + y^2}{x + y}$ **b)** $\dfrac{x^2 - y^2}{x^2 - 2xy + y^2}$

SOLUTION

a) $$\frac{x^2 + 2xy + y^2}{x + y} = \frac{\cancel{(x + y)}(x + y)}{\cancel{(x + y)}} = x + y$$

b) $$\frac{x^2 - y^2}{x^2 - 2xy + y^2} = \frac{(x + y)\cancel{(x - y)}}{(x - y)\cancel{(x - y)}} = \frac{x + y}{x - y}$$ □

EXAMPLE 39

Simplify the following rational expressions.

a) $\dfrac{x^3 + 5x^2 + 6x}{x^3 + x^2 - 6x}$ **b)** $\dfrac{4a^2x + 8abx - 60b^2x}{5a^2y - 30aby + 45b^2y}$

SOLUTION

a) $$\frac{x^3 + 5x^2 + 6x}{x^3 + x^2 - 6x} = \frac{x(x^2 + 5x + 6)}{x(x^2 + x - 6)} = \frac{\cancel{x}\cancel{(x + 3)}(x + 2)}{\cancel{x}\cancel{(x + 3)}(x - 2)} = \frac{x + 2}{x - 2}$$

b) $$\frac{4a^2x + 8abx - 60b^2x}{5a^2y - 30aby + 45b^2y} = \frac{4x(a^2 + 2ab - 15b^2)}{5y(a^2 - 6ab + 9b^2)} = \frac{4x(a + 5b)\cancel{(a - 3b)}}{5y(a - 3b)\cancel{(a - 3b)}} = \frac{4x(a + 5b)}{5y(a - 3b)}$$ □

Practice Problem

Simplify the following rational expressions.

a) $\dfrac{2x^2 + 4x}{8x^2 + 2x}$ **b)** $\dfrac{x^2 - 1}{2x^2 - x - 1}$ **c)** $\dfrac{12x^2 - 12y^2}{9x^2 + 6xy - 3y^2}$

Answer: **a)** $\dfrac{x+2}{4x+1}$; **b)** $\dfrac{x+1}{2x+1}$; **c)** $\dfrac{4(x-y)}{3x-y}$

WARNING

YOU MUST FACTOR BEFORE SIMPLIFYING

If you attempt to simplify without factoring, you are likely to make mistakes such as

$$\frac{\cancel{x}+5}{\cancel{x}} = 5 \quad \text{or} \quad \frac{\overset{2}{\cancel{4x}}+3}{\cancel{2x}} = 2+3 = 5.$$

Remember: You can simplify only by canceling out *factors,* not individual terms.

Multiplying and Dividing Rational Expressions

To multiply or divide two rational expressions, we simply use the rules for dealing with fractions involving real numbers (from Section 1.2). After multiplying or dividing, remember to simplify if necessary.

Multiplying and Dividing Rules

Multiplying: $\dfrac{a}{b}\cdot\dfrac{c}{d} = \dfrac{ac}{bd}$

Dividing: $\dfrac{a}{b}\div\dfrac{c}{d} = \dfrac{ad}{bc}$

EXAMPLE 40

Perform the indicated operations.

a) $\dfrac{4xy}{7x^2}\cdot\dfrac{7x}{2y}$

b) $\dfrac{x^2+2xy+y^2}{x-y}\cdot\dfrac{x^2-2xy+y^2}{6x+6y}$

c) $\dfrac{x^2+5x+4}{4x^2+12x-16}\cdot\dfrac{x^2-1}{3x+12}$

SOLUTION

a) $\dfrac{4xy}{7x^2} \cdot \dfrac{7x}{2y} = \dfrac{28\cancel{x^2y}}{14\cancel{x^2y}} = 2$

b) $$\begin{aligned}\frac{x^2 + 2xy + y^2}{x - y} \cdot \frac{x^2 - 2xy + y^2}{6x + 6y} &= \frac{(x^2 + 2xy + y^2)(x^2 - 2xy + y^2)}{(x - y)(6x + 6y)} \\ &= \frac{(x + y)\cancel{(x + y)}\cancel{(x - y)}(x - y)}{6\cancel{(x - y)}\cancel{(x + y)}} \\ &= \frac{(x + y)(x - y)}{6}\end{aligned}$$

c) $$\begin{aligned}\frac{x^2 + 5x + 4}{4x^2 + 12x - 16} \cdot \frac{x^2 - 1}{3x + 12} &= \frac{(x^2 + 5x + 4)(x^2 - 1)}{(4x^2 + 12x - 16)(3x + 12)} \\ &= \frac{\cancel{(x + 4)}(x + 1)(x + 1)\cancel{(x - 1)}}{4\cancel{(x + 4)}\cancel{(x - 1)}[3(x + 4)]} \\ &= \frac{(x + 1)(x + 1)}{12(x + 4)} = \frac{(x + 1)^2}{12(x + 4)}\end{aligned}$$ □

EXAMPLE 41 Perform the indicated operation.

a) $\dfrac{x^2 - 4x - 5}{3x - 6} \div \dfrac{3x - 15}{x^2 + 2x + 1}$

b) $\dfrac{x^3 - 3x^2 - 10x}{x^4 - 1} \div \dfrac{x^2 - 10x + 25}{x^2 - 1}$

SOLUTION

a) $$\begin{aligned}\frac{x^2 - 4x - 5}{3x - 6} \div \frac{3x - 15}{x^2 + 2x + 1} &= \frac{(x^2 - 4x - 5)(x^2 + 2x + 1)}{(3x - 6)(3x - 15)} \\ &= \frac{\cancel{(x - 5)}(x + 1)(x + 1)(x + 1)}{3(x - 2)3\cancel{(x - 5)}} \\ &= \frac{(x + 1)^3}{9(x - 2)}\end{aligned}$$

b) $$\begin{aligned}&\frac{x^3 - 3x^2 - 10x}{x^4 - 1} \div \frac{x^2 - 10x + 25}{x^2 - 1} \\ &= \frac{(x^3 - 3x^2 - 10x)(x^2 - 1)}{(x^4 - 1)(x^2 - 10x + 25)} \\ &= \frac{x\cancel{(x - 5)}(x + 2)\cancel{(x + 1)}\cancel{(x - 1)}}{\cancel{(x + 1)}\cancel{(x - 1)}(x^2 + 1)\cancel{(x - 5)}(x - 5)} \\ &= \frac{x(x + 2)}{(x^2 + 1)(x - 5)}\end{aligned}$$ □

Practice Problem

Perform the indicated calculations.

a) $\dfrac{x+5}{(x^2-16)} \cdot \dfrac{2x-8}{(x^2-25)}$

b) $\dfrac{m^2+4mn+4n^2}{bm^2-9bn^2} \cdot \dfrac{bm^2+6bmn+9bn^2}{m+2n}$

c) $\dfrac{a^2-b^2}{6a+12b} \div \dfrac{a^2-2ab+b^2}{a^2+4ab+4b^2}$

d) $\dfrac{-x^2+16x+36}{x^2-9} \div \dfrac{x^2-4}{3x^2+9x}$

Answer: **a)** $\dfrac{2}{(x+4)(x-5)}$; **b)** $\dfrac{(m+2n)(m+3n)}{(m-3n)}$;
c) $\dfrac{(a+b)(a+2b)}{6(a-b)}$; **d)** $\dfrac{-3x(x-18)}{(x-3)(x-2)}$

Adding and Subtracting Rational Expressions

To add or subtract rational expressions, we must first establish a common denominator. For example, to add $\frac{1}{3} + \frac{1}{4}$, we determine that 12 is a common denominator (since $4 \cdot 3 = 12$). Next, we convert each fraction to the equivalent fraction with a denominator of 12:

$$\frac{1}{3} \cdot \frac{4}{4} = \frac{4}{12} \quad \text{and} \quad \frac{1}{4} \cdot \frac{3}{3} = \frac{3}{12}.$$

Finally, we can add the equivalent fractions because they have the same denominator:

$$\frac{4}{12} + \frac{3}{12} = \frac{4+3}{12} = \frac{7}{12}.$$

Notice that we used the least common denominator of 12. Actually, any integer multiple of 12 would also have worked (try it with a denominator of 24). However, if we had used a denominator larger than 12 we would have had to simplify the result. Thus, in this case, using the least common denominator saved us from having to do subsequent simplification. (Note: you may have to simplify an answer, even though you used the LCD.)

To add or subtract rational expressions, we proceed exactly as we did earlier.

Adding and Subtracting Rational Expressions

1. Begin by factoring all the denominators. Form the LCD by using each factor the greatest number of times it occurs in any one denominator.
2. Convert each rational expression to the equivalent expression using the least common denominator.
3. Perform the appropriate additions or subtractions.
4. Simplify the result.

EXAMPLE 42

Perform the addition.

$$\frac{5x}{24y^2z} + \frac{5}{6xyz}$$

SOLUTION

Our first step is to compute the least common denominator. Factoring the denominators gives us

$$24y^2z = 3 \cdot 2 \cdot 2 \cdot 2 \cdot y \cdot y \cdot z \quad \text{and} \quad 6xyz = 3 \cdot 2 \cdot x \cdot y \cdot z.$$

The factors are 3, 2, x, y, and z. Does this mean that $3 \cdot 2 \cdot x \cdot y \cdot z = 6xyz$ is the least common denominator? No! Remember that we must use each factor the *most* number of times it occurs in any one denominator. Hence, the LCD is

$$3 \cdot 2 \cdot 2 \cdot 2 \cdot x \cdot y \cdot y \cdot z = 24xy^2z.$$

The next step is to convert each expression to the equivalent expression with the LCD as its denominator. This gives

$$\frac{5x}{24y^2z} \cdot \frac{x}{x} = \frac{5x^2}{24xy^2z} \quad \text{and} \quad \frac{5}{6xyz} \cdot \frac{4y}{4y} = \frac{20y}{24xy^2z}.$$

Our next step is to add the equivalent expressions above. We have

$$\frac{5x^2}{24xy^2z} + \frac{20y}{24xy^2z} = \frac{5x^2 + 20y}{24xy^2z}.$$

No further simplification is possible (though we could factor the numerator. □

EXAMPLE 43

Solve the following problems.

a) $\dfrac{x-4}{x-1} + \dfrac{x-1}{x+1}$ **b)** $\dfrac{x}{x-1} - \dfrac{x-1}{x+1}$

SOLUTION

$$\textbf{a)}\quad \frac{x-4}{x-1} + \frac{x-1}{x+1} = \frac{x-4}{x-1}\cdot\frac{x+1}{x+1} + \frac{x-1}{x+1}\cdot\frac{x-1}{x-1}$$
$$= \frac{x^2-3x-4}{(x-1)(x+1)} + \frac{x^2-2x+1}{(x-1)(x+1)}$$
$$= \frac{2x^2-5x-3}{(x-1)(x+1)}$$
$$= \frac{(2x+1)(x-3)}{(x-1)(x+1)}$$

$$\textbf{b)}\quad \frac{x}{x-1} - \frac{x-1}{x+1} = \frac{x}{x-1}\cdot\frac{x+1}{x+1} - \frac{x-1}{x+1}\cdot\frac{x-1}{x-1}$$
$$= \frac{x^2+x}{(x-1)(x+1)} - \frac{x^2-2x+1}{(x-1)(x+1)}$$
$$= \frac{x^2+x-(x^2-2x+1)}{(x-1)(x+1)}$$
$$= \frac{3x-1}{(x-1)(x+1)} \quad \square$$

EXAMPLE 44

Solve the problem.

$$\frac{2x}{x-3} + \frac{x}{x^2-x-6} - \frac{1}{x+2}$$

SOLUTION

$$\frac{2x}{x-3} + \frac{x}{x^2-x-6} - \frac{1}{x+2}$$
$$= \frac{2x}{x-3} + \frac{x}{(x-3)(x+2)} - \frac{1}{x+2}$$
$$= \frac{2x}{x-3}\cdot\frac{x+2}{x+2} + \frac{x}{(x-3)(x+2)} - \frac{1}{x+2}\cdot\frac{x-3}{x-3}$$
$$= \frac{2x^2+4x+x-(x-3)}{(x-3)(x+2)}$$
$$= \frac{2x^2+4x+3}{(x-3)(x+2)} \quad \square$$

Practice Problem

Solve the following problems.

a) $\dfrac{7}{x} + \dfrac{4}{xy}$

b) $\dfrac{x}{x+2} + \dfrac{2x}{x+1}$

c) $\dfrac{4x}{x-3} - \dfrac{2x^2}{x^2-5x+6}$

d) $\dfrac{1}{x} + \dfrac{5x}{x+1} + \dfrac{3x}{x-1} - 8$

Answer: **a)** $\dfrac{7y+4}{xy}$; **b)** $\dfrac{x(3x+5)}{(x+2)(x+1)}$;
c) $\dfrac{2x(x-4)}{(x-3)(x-2)}$; **d)** $\dfrac{-x^2+8x-1}{x(x+1)(x-1)}$

EXERCISE SET 2.3

Which of the following statements are true?

1. $\dfrac{x}{x+1} = \dfrac{1}{2}$

2. $\dfrac{1}{x} + \dfrac{3}{x} = \dfrac{4}{x}$

3. $\dfrac{1}{x} + \dfrac{3}{y} = \dfrac{4}{x+y}$

4. $\dfrac{x+3}{x} = 1 + \dfrac{3}{x}$

5. $\dfrac{x^2-9}{x+3} = x - 3$

6. $\dfrac{x^2+x}{x+1} = x$

Perform the indicated operations and simplify.

7. $(a \div b)b^3$

8. $(4 \div a)b$

9. $\dfrac{3}{x} \div \dfrac{4}{x}$

10. $\dfrac{4xy}{7x^2} + \dfrac{2y}{7x}$

11. $\dfrac{a^2-9}{a+5} \cdot \dfrac{a^2-25}{2a-6}$

12. $\dfrac{x+y}{y} \div \dfrac{x-y}{y}$

13. $\dfrac{x^2-y^2}{x^2-4y^2} \cdot \dfrac{x+2y}{x-y}$

14. $\dfrac{x^2-100}{x^2-9} \cdot \dfrac{x-3}{x-10}$

15. $\dfrac{25a^2-y^2}{16a^2-9y^2} \div \dfrac{5a-y}{4a-3y}$

16. $\dfrac{x^2-49}{(a+b)^2-c^2} \div \dfrac{x+7}{(a+b)-c}$

17. $\dfrac{x^2+3x+2}{x^2+5x+6} \cdot \dfrac{x^2+7x+12}{x^2+9x+20}$

18. $\dfrac{x}{ax-ay} - \dfrac{y}{ax-ay}$

19. $\dfrac{x+1}{2} + \dfrac{x-3}{5}$

20. $\dfrac{5x}{9} - \dfrac{3x}{2} + \dfrac{x}{6}$

21. $\dfrac{5x-11}{3} - \dfrac{2x-1}{10} - \dfrac{11x-5}{15}$

22. $\dfrac{2x+1}{4x} - \dfrac{x+3}{10x}$

23. $\dfrac{x}{x-b} - \dfrac{x^2}{x^2-b^2}$

24. $\dfrac{4x^2+y^2}{4x^2-y^2} - \dfrac{2x+y}{2x-y}$

25. $\dfrac{y+1}{y+2}+\dfrac{y-2}{y-3}+\dfrac{2y+7}{y^2-y-6}$

26. $\dfrac{x^2}{x^2-2x+1}+\dfrac{1}{3x-3}-\dfrac{1}{3}$

27. $\dfrac{x}{x^2-x-2}-\dfrac{1}{x^2-5x+6}+\dfrac{2}{x^2-2x-3}$

28. $\dfrac{x+y}{x^2-6xy-16y^2}-\dfrac{x-y}{x^2-16xy+64y^2}$

29. $\dfrac{3a}{a^2-7a+12}+\dfrac{2}{a-4}-\dfrac{1}{a-3}$

30. $\dfrac{x^2-y^2}{x^2-3xy+2y^2}\cdot\dfrac{xy-2y^2}{x^2+xy}\cdot\dfrac{(x-y)^2}{x(x-y)}$

31. $\left(\dfrac{(x+y)^2-a^2}{x^2+xy-xa}\div\dfrac{x^2y^2a^2}{x^2+xy-xa}\right)\div\left(\dfrac{y^2a^2}{xya}\right)$

32. $\left(\dfrac{a^2+7ab+10b^2}{a^2+6ab+5b^2}\div\dfrac{a+1}{a^2+4a+4}\right)\cdot\left(\dfrac{1}{a+2}\right)$

33. $\left(\dfrac{1}{x}-\dfrac{2}{x^2}+\dfrac{1}{x^3}\right)\div\left(\dfrac{(1-x)^2}{x^2}\right)$

34. $x^2-x-6\div\left(1-\dfrac{4}{x^2}\right)$

CHAPTER REVIEW

IMPORTANT TERMS AND SYMBOLS

2.1 Basics of Algebra

base and exponent In the form x^n, x is the base and n is the exponent.
exponent properties

1. $x^i = \underbrace{x\cdot x\cdots x}_{i\text{ times}}$
2. $x^0 = 1 \quad (x \neq 0)$
3. $x^{-i} = 1/x^i \quad (x \neq 0)$
4. $x^i \cdot x^j = x^{i+j}$
5. $(x^i)^j = x^{i\cdot j}$
6. $(xy)^i = x^i y^i$
7. $(x/y)^i = x^i/y^i = x^i y^{-i} \quad (y \neq 0)$
8. $x^i/x^j = x^{i-j} = 1/x^{j-i} \quad (x \neq 0)$

nth root of x The real number a, such that $a^n = x$.
principal root of x The nth root with the same sign as x.
variable A symbol that represents a real number.

2.2 Polynomials and Algebraic Expressions

algebraic expression Result of combining numbers and symbols with the operations of addition, subtraction, multiplication, division, and raising to a power.
Simplifying algebraic expressions:

1. Remove all parentheses by performing the indicated operations. Work from innermost grouping to the outermost groupings. If necessary, convert subtraction to addition.
2. Use the distributive property to combine terms of like degree.

degree of a polynomial Highest degree of monomials in it.
Linear polynomial: has degree 1.
Quadratic polynomial: has degree 2
Cubic polynomial: has degree 3.
degree of a term Exponent of the variable.
factoring Write an equivalent expression as a product.
Commonly occurring factorizations:

1. $xy + xz = x(y + z)$
2. $x^2 + 2ax + a^2 = (x + a)^2$
3. $x^2 - 2ax + a^2 = (x - a)^2$
4. $x^2 - a^2 = (x + a)(x - a)$
5. $x^2 + (a + b)x + ab = (x + a)(x + b)$
6. $acx^2 + (ad + bc)x + bd = (ax + b)(cx + d)$

monomial An expression of the form ax^n.
multiplication of two polynomials Multiply each term of one by each term of the other and then simplify.
Special products:

1. $x(y + z) = xy + xz$
2. $(x + a)(x + b) = x^2 + (a + b)x + ab$
3. $(ax + b)(cx + d) = acx^2 + (ad + bc)x + bd$
4. $(x + a)^2 = x^2 + 2ax + a^2$
5. $(x - a)^2 = x^2 - 2ax + a^2$
6. $(x + a)(x - a) = x^2 - a^2$

polynomial Sum of two or more monomials.
term of a polynomial One of the monomials of the polynomial.
Binomial: polynomial with two terms.
Trinomial: polynomial with three terms.

2.3 Rational Expressions

rational expression Ratio of two polynomials.
Simplifying rational expressions: use simplification rule

$$\frac{ac}{bc} = \frac{a}{b}$$

Multiplying and dividing rational expressions: use rules

Multiplying: $\frac{a}{b} \cdot \frac{c}{d} = \frac{ac}{bd}$

Dividing: $\frac{a}{b} \div \frac{c}{d} = \frac{ad}{bc}$

REVIEW PROBLEMS

Evaluate the following.

1. $(4.2)^3$

2. $(3.1)^{-4}$

3. $(-5.0)^{1/3}$

4. $(1.06)^{30}$

Simplify the following expressions so they contain only positive exponents.

5. $b^2 \div b^5$

6. $y^4 \div y^{-6}$

7. $(x^{-3} \div x^{-7})^3$

8. $(14x^3y^{-4} \div 21x^{-2}y^8)^{-4}$

9. $(x^{4m+3} \div x^{-4+2m})^{1/3}$

Perform the indicated operations with the following polynomials.

10. $(x^2 - 4x + 5) + (17 - 4x - 3x^2)$

11. $(4x^3 - 3x^2 + 6x - 1) - (x^3 - 6x^2 + 3x - 8)$

12. $(2x - 3)(6 - 3x)$

13. $(x + 2)(x^2 - 2x + 3)$

14. $(x^2 - 1)(x^2 + 2x + 1)$

15. $(x^3 - 2x^2 + 2x + 1)(x^3 - 3x + 2)$

Factor the following polynomials.

16. $4xy + 16xz$

17. $9abc - 27a^2b^2$

18. $24x^2 + 16xy + 12xy^2$

19. $x^2 - 81$

20. $36x^2 - 25$

21. $x^2 + 6x + 9$

22. $9x^2 - 12xy + 4y^2$

23. $2x^2 - 9xy - 5y^2$

24. $6x^2 + xy - 12y^2$

25. $6x^3 + 15x^2 - 9x$

Perform the indicated operations and simplify.

26. $\frac{2x}{3} \cdot \frac{3}{8x^2}$

27. $\frac{18xy^3}{4x^2} \cdot \frac{9x}{2y}$

28. $\frac{3u^{-2}v^4}{7u^2v^2} \div \frac{14u^6v^{-7}}{4uv^2}$

29. $\frac{36x^2 - 16}{x^2 - 4} \cdot \frac{x + 2}{6x - 4}$

30. $\frac{x^2 - 3x + 2}{x - 5} \cdot \frac{x^2 - 3x - 10}{x - 2}$

31. $\frac{x^2}{y} + \frac{2x^2 - 4x + 1}{3y^2}$

32. $\dfrac{x}{x-2} - \dfrac{x^2}{x^2-4}$

33. $\dfrac{2m}{m^2-m-6} + \dfrac{m}{m-3} - \dfrac{4}{3m+6}$

34. $\dfrac{x^2+x-6}{4-x} + \dfrac{x^2-16}{2x+6}$

35. $\dfrac{8x^3+16x^2+8x}{16y^2-x^2} + \dfrac{4x+4}{x^2+8xy+16y^2}$

3 SOLVING EQUATIONS

a) You are planning a trip home during semester break. You estimate that you can average 50 miles per hour; your home is 420 miles away. How long do you expect the trip home to take? (To be solved in Example 2, Section 3.1.)
b) At sea level, water freezes at 0° Celsius and 32° Fahrenheit and boils at 100° Celsius and 212° Fahrenheit. Derive the linear equation that relates temperatures expressed in degrees Celsius to those expressed in degrees Fahrenheit. (To be solved in Example 46, of Section 3.4.)
c) The manual that came with the refrigerator in your new apartment states that the food compartment temperature should be set at 5°C. What is the corresponding temperature in degrees Fahrenheit? (To be solved in Example 11, Section 3.2.)

PROBLEM FROM LIFE

INTRODUCTION

We'll begin this chapter with a discussion of how to use equations to model problems: for example, part (a) of the Problem from Life. We'll continue with a discussion of linear and quadratic equations and inequalities in one variable. We'll conclude the chapter with a discussion of the graphs of linear equations and inequalities in two variables.

3.1

MODELING WITH EQUATIONS

Equations

First, let's define what we mean by the term *equation*. Then, after we look at a few examples, we'll discuss how you can translate words into equations. An **equation** is a statement that says that two expressions are equal. The two expressions are generally referred to as the **sides** (or members) of the equation. They are separated by the **equality sign** $=$.

EXAMPLE 1

Which of the following equations are true?

a) $4 + 5 = 9$

b) $3 + 8 = 7$

c) $x + 3 = 7$

d) $w + z = 10$

e) $\dfrac{g}{2 - g} = 14$

SOLUTION

Statement (a) is an equation that is true. Statement (b) is an equation that is false. Statements (c), (d), and (e) are equations that contain one or more variables. They may be true or false depending upon the values of the variables. In statement (e), we require that g never be equal to 2, because this would make the left side of the equation undefined. □

In Example 1, statement (c) is an equation using the variable x (sometimes we say an equation "in x"), statement (d) is an equation using the variables w and z, and statement (e) is an equation using the variable g. We often refer to variables that occur in equations as **unknowns.**

Once we've derived (or have been given) an equation, we want to know the values of the variables that make the equation true. We *solve* an equation that contains one or more variables by determining all values of the variables for which the equation is true. These values are called **solutions** of the equation; we often say the solutions *satisfy* the equation. The set of all solutions to an equation is called the **solution set** of the equation.

Translating Words Into Equations

In Chapter 1 we discussed a four-step approach to solving problems quantitatively. The four steps are (briefly)

1. Define the problem.
2. Build a mathematical model of the problem.
3. Analyze the model.
4. Implement the solution.

We often accomplish the second step by using one or more equations to model the problem. How is this done? Let's take a look at a few examples, and then we'll show you a general approach.

EXAMPLE 2

Problem from Life, part (a) You are planning a trip home during semester break. You estimate that you can average 50 miles per hour; your home is 420 miles away. How long do you expect the trip home to take?

SOLUTION

The first step in solving a problem is to define it, which includes a thorough reading of the problem statement. This one is simple enough: We want to determine how long your trip home would take, given a distance of 420 miles and an average speed of 50 miles per hour.

The next step is to build a mathematical model. To accomplish this, we must begin by defining all the necessary variables. Let

$$t = \text{time to make trip},$$
$$d = \text{distance of trip},$$
$$r = \text{average rate at which you travel}.$$

If you went 50 miles at 50 miles per hour, it would take you 1 hour to make that trip. If you went 100 miles at 50 miles per hour, it would take you 2 hours. It seems reasonable, then, that the time to make the trip is given by the equation

$$t = \frac{d}{r},$$

and in this case,

$$t = \frac{420}{50} = 8.4 \text{ hours}.$$

Step 3 of the problem-solving approach involves analyzing the model, which we just did by inserting the data. Step 4 is implementing the solution, which means driving at an average of 50 miles per hour. □

EXAMPLE 3

Yvonne and George Nickerson want to rent a rug shampooer to clean the rugs in their apartment. The shampooer rents for \$3.50 per hour and the shampoo costs \$2.75 per bottle. Yvonne and George have estimated that it will take them 6 hours to do the cleaning, and they will need two bottles of shampoo. How much will it cost them?

SOLUTION

This problem requires us to compute the total cost of renting a shampooer and buying the shampoo. We can do so by determining the cost of the rug shampooer and the cost of the shampoo and then adding the two. The cost of the shampooer should be \$3.50 × number of hours, and the cost of the shampoo should be \$2.75 × number of bottles. Let

$$\begin{aligned} h &= \text{number of hours rug shampooer is used,} \\ b &= \text{number of bottles of shampoo,} \\ c &= \text{total cost.} \end{aligned}$$

Then we have

$$c = 3.5h + 2.75b.$$

In this specific case, with $h = 6$ and $b = 2$, we have

$$\begin{aligned} c &= 3.5(6) + 2.75(2), \\ c &= \$26.50. \quad \square \end{aligned}$$

EXAMPLE 4

Sam Neilson has just fenced in his yard, using a total of 650 feet of material. His yard is rectangular, and he knows that the yard is 75 feet longer than it is wide. What is the total area of his yard?

SOLUTION

The problem here is to determine the area of Sam Neilson's yard. Since we know that the yard is rectangular, this means that we must find the dimensions of the yard (its length and width), because the area of a rectangle is given by the product of its dimensions. Thus the real problem is to find the length and width of the yard. Moreover, since the yard is 75 feet longer than it is wide, once we know the width we'll also know the length. (See Fig. 3.1.)

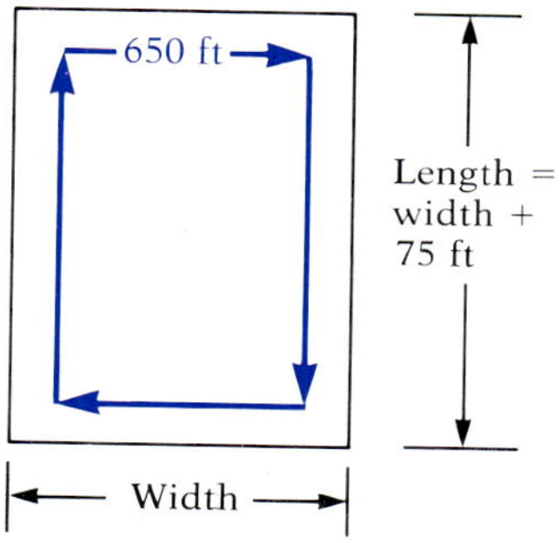

Figure 3.1 **Sam Neilson's yard.**

Now we can assign variables. Let

$$w = \text{width of the yard},$$
$$w + 75 = \text{length of the yard}.$$

Then the area of the yard is given by

$$w(w + 75) = w^2 + 75w.$$

That's great, assuming we know w. But how can we determine it? Don't forget, we also said that Sam Neilson used 650 feet of material to fence in the yard. That means the total distance around the yard, its perimeter, is 650. But the perimeter of a rectangle is given by the sum of twice each dimension. Thus, for Sam's yard, we have

$$650 = 2w + 2(w + 75).$$

Now the procedure is clear. First, we solve the equation just given for the value of w. Then we can use that value to determine the area by evaluating the expression $w^2 + 75w$. □

What conclusions can we draw from the preceding examples? First, notice that it is important to understand the problem before attempting to assign symbols to variables. If drawing a picture will help, do so. Next, the best way to attack almost any problem is to think about how you would do it if you had all the data you needed. Write out the relationships in words. Now you can assign variables and translate the relationships into symbols. At this point, think about your model. Does it make sense? If not, try to determine where it falls apart and reconstruct your model. Once your model seems to be reasonable, then you can analyze it. Another test of your model is to think about the real-world implications of the results of your analysis. If the answers don't seem to make sense, your model may be incorrect. Try to determine why the answers don't work and reconstruct your model.

We can summarize this discussion as follows.

Modeling With Equations

1. Make sure you understand the problem.
2. Draw a picture when appropriate.
3. Think about how you would proceed if you had all the data.
4. Write out the relationships in words.
5. Assign variables.
6. Translate the relationships into symbols.

EXERCISE SET 3.1

Translate the following problems into models.

1. A babysitting service charges \$2.00 per day plus \$1.50 per hour. What is the total cost of babysitting? How much would 6 hours of babysitting cost (assuming that they take place on the same day)?
2. Ace Electronics manufactures a variety of electronic components and related parts, including power supply filters for microcomputers. The production facility for making the filters costs Ace \$5000 per day to operate, and each filter costs \$40. What is the total cost of producing power supply filters? how much will it cost Ace to make 200 filters (assuming they can be manufactured in one day)?
3. Fine's Clothiers is having a sale. Suits advertised at 40% off are selling for \$120. How much did the suits cost originally?
4. Fine's Clothiers has certain items with its own label. What should the *original* price be on a suit so that Fine's can sell it for 25% off and still make a profit equal to 20% of the cost of the suit, given that the suit costs $\$C$ to make?
5. Rich is 32 years older than Janet. The sum of their ages is 42. How old is each?
6. A gardener wants to use 70 feet of fence material to make three plots, as shown in Fig. 3.2. What are the dimensions of each plot?

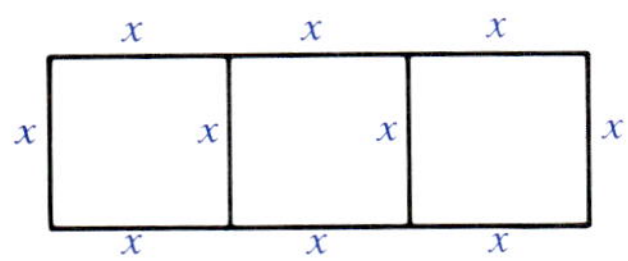

Figure 3.2 **Three garden plots.**

3.2

LINEAR EQUATIONS AND INEQUALITIES IN ONE VARIABLE

Linear Equations in One Variable

A **linear equation** in one variable is an equation that can be written in the form

$$ax + b = c,$$

where $a \neq 0$ and a, b, and c are constants.

EXAMPLE 5

a) $2x + 3 = 5$ is a linear equation in x.
b) $-7y - 2 = 4$ is a linear equation in y.
c) $-3a + 2 = -1$ is a linear equation in a. □

Let's see how we can solve a linear equation in one variable. Our approach will be to perform operations on the original equation that will give us an equivalent equation with an obvious solution. Two equations are **equivalent** if they have the same solution set. For example, both of the equations

$$2x + 3 = 5 \quad \text{and} \quad x = 1$$

have the same solution (which we'll show shortly). Thus we want to convert the original equation into an equivalent equation with the variable on one side of the equal sign and a number on the other side. We can use the following operations to produce equivalent equations.

Valid Operations for Linear Equations

1. Addition property: If $a = b$, then $a + c = b + c$.
2. Subtraction property: If $a = b$, then $a - c = b - c$.
3. Multiplication property: If $a = b$, then $ac = bc$.
4. Division property: If $a = b$, then $a/c = b/c$ for $c \neq 0$.

Note that we cannot divide both sides of an equation by zero. (In addition, we don't want to multiply by zero because this would reduce our original equation to $0 = 0$, which is true but not helpful.)

After solving an equation, check your work by replacing the variable with the solution. The resulting equation should be true. For example, $x = 2$ cannot be the solution to the equation

$$2x + 10 = 20$$

because if we replace x by 2 we obtain

$$2(2) + 10 = 14, \text{ which is a false statement.}$$

EXAMPLE 6

Solve the equation $2x + 3 = 5$ and check the solution.

SOLUTION

$$2x + 3 = 5$$

$$2x + 3 - 3 = 5 - 3 \qquad \textbf{Subtraction property}$$

$$\frac{2x}{2} = \frac{2}{2} \qquad \textbf{Division property}$$

$$x = 1$$

Check the solution by replacing x by the number 1 in the *original* equation:

$$2(1) + 3 \stackrel{?}{=} 5$$

$$5 \stackrel{\checkmark}{=} 5 \quad \square$$

EXAMPLE 7

Solve the equation $4(x - 2) + 5 = 2x - 2$ and check the solution.

SOLUTION

$$\begin{aligned} 4(x - 2) + 5 &= 2x - 2 \\ 4x - 8 + 5 &= 2x - 2 && \text{Distributive property} \\ 4x - 3 &= 2x - 2 && \text{Associative property} \\ 4x - 3 - 2x &= 2x - 2 - 2x && \text{Subtraction property} \\ 2x - 3 + 3 &= -2 + 3 && \text{Addition property} \\ \frac{2x}{2} &= \frac{1}{2} && \text{Division property} \\ x &= \frac{1}{2} \end{aligned}$$

Check:

$$4\left(\frac{1}{2} - 2\right) + 5 \stackrel{?}{=} 2\left(\frac{1}{2}\right) - 2$$

$$-1 \stackrel{\checkmark}{=} -1 \quad \square$$

EXAMPLE 8

Solve the equation $6x - 4(x - 1) = 2(x + 1) - 3x$ and check.

SOLUTION

$$\begin{aligned} 6x - 4(x - 1) &= 2(x + 1) - 3x \\ 6x - 4x + 4 &= 2x + 2 - 3x && \text{Distributive property} \\ 2x + 4 &= -1x + 2 && \text{Associative property} \\ 3x &= -2 && \text{Add } 1x \text{ to both sides.} \\ x &= -\frac{2}{3} && \text{Divide both sides by 3.} \end{aligned}$$

Check:

$$6\left(-\frac{2}{3}\right) - 4\left(-\frac{2}{3} - 1\right) \stackrel{?}{=} 2\left(-\frac{2}{3} + 1\right) - 3\left(-\frac{2}{3}\right)$$

$$\frac{8}{3} \stackrel{\checkmark}{=} \frac{8}{3} \quad \square$$

Practice Problem

Solve the following equations and check your solutions.

a) $4x - 3 = 17$ **b)** $6y + 8 = 9y + 5$

c) $3w + 5 = 5(2w + 8)$

Answers: **a)** $x = 5$; **b)** $y = 1$; **c)** $w = -5$

We can also apply the same rules to solving equations that include fractions. The first step in these problems is to remove the fractions by multiplying both sides of the equation by a common denominator.

EXAMPLE 9

Solve the equation $2(x + 2) - 5 = \frac{3x}{4} - 2$ and check.

SOLUTION

$$2(x + 2) - 5 = \frac{3x}{4} - 2$$

$$8(x + 2) - 20 = 3x - 8 \qquad \text{Multiply both sides by 4.}$$

$$8x + 16 - 20 = 3x - 8 \qquad \text{Distributive property}$$

$$8x - 4 = 3x - 8 \qquad \text{Associative property}$$

$$5x - 4 = -8 \qquad \text{Subtract } 3x \text{ from both sides.}$$

$$5x = -4 \qquad \text{Add 4 to both sides.}$$

$$x = -\frac{4}{5} \qquad \text{Divide both sides by 5.}$$

Check:

$$2\left(-\frac{4}{5} + 2\right) - 5 \stackrel{?}{=} \frac{3(-4/5)}{4} - 2$$

$$-\frac{13}{5} \stackrel{\checkmark}{=} -\frac{13}{5} \qquad \square$$

EXAMPLE 10

Solve the equation $\frac{2x - 2}{3} - \frac{x}{2} = 1$ and check.

SOLUTION

$$\frac{2x - 2}{3} - \frac{x}{2} = 1$$

$$2(2x - 2) - 3x = 6 \qquad \text{Multiply both sides by LCD (6).}$$

$$4x - 4 - 3x = 6 \qquad \text{Distributive property}$$

$$x - 4 = 6 \qquad \text{Associative property}$$

$$x = 10 \qquad \text{Add 4 to both sides.}$$

Check:

$$\frac{2(10) - 2}{3} - \frac{10}{2} \stackrel{?}{=} 1$$

$$6 - 5 \stackrel{\checkmark}{=} 1 \qquad \square$$

Practice Problem

Solve the following equations and check your solutions.

a) $\frac{3m - 2}{2} - \frac{m}{3} = \frac{2m}{3} + 1$ **b)** $y - \frac{y}{3} + \frac{y}{2} - \frac{2}{5} = \frac{y}{2}$

Answers: **a)** $m = 4$; **b)** $y = \frac{3}{5}$

A very common use of linear equations in algebra is solving formulas (equations) for one variable in terms of several others.

EXAMPLE 11

Problem from Life, part (c) Given the equation

$$C = \frac{5(F - 32)}{9},$$

which expresses temperature in degrees Celsius (C) given the temperature in degrees Fahrenheit (F), solve for F in terms of C.

SOLUTION

$$C = \frac{5(F - 32)}{9}$$

$$\frac{5(F - 32)}{9} = C \quad \textbf{Reverse the equation.}$$

$$5(F - 32) = 9C \quad \textbf{Multiply both sides by 9.}$$

$$5F - 160 = 9C \quad \textbf{Distributive property}$$

$$5F = 9C + 160 \quad \textbf{Add 160 to both sides.}$$

$$F = \frac{9C + 160}{5} \quad \textbf{Divide both sides by 5.}$$

which is often written

$$F = \frac{9}{5}C + 32.$$

Now we can answer part (c) of the Problem from Life. According to the equation given, the Fahrenheit that corresponds to 5°C is given by

$$F = \frac{9}{5}(5) + 32,$$

so

$$F = 41.$$

Set the thermostat to 41°F. □

EXAMPLE 12

Given the equation $A = P(1 + rt)$ (which tells us how to compute simple interest and will be discussed in Chapter 10), solve for P and then t in terms of the other variables.

SOLUTION

First, P:

$$A = P(1 + rt)$$

$$P(1 + rt) = A \qquad \textbf{Reverse the equation.}$$

$$P = \frac{A}{1 + rt} \qquad \textbf{Divide both sides by } (1 + rt).$$

Next, let's solve for t:

$$A = P(1 + rt)$$

$$P(1 + rt) = A \qquad \textbf{Reverse the equation.}$$

$$P + Prt = A \qquad \textbf{Distributive property}$$

$$Prt = A - P \qquad \textbf{Subtract } P \textbf{ from both sides.}$$

$$t = \frac{A - P}{Pr} \qquad \textbf{Divide both sides by } Pr. \quad \square$$

EXAMPLE 13

Given the equation $P = I^2R$, solve for R.

SOLUTION

$$P = I^2R$$

$$I^2R = P \qquad \textbf{Reverse the equation.}$$

$$R = \frac{P}{I^2} \qquad \textbf{Divide both sides by } I^2. \quad \square$$

Practice Problem

Solve for the indicated variables.

a) $PV = NRT$ for T

b) $A = 2LH + 2HW + 2LW$ for L

Answers: **a)** $T = \dfrac{PV}{NR}$; **b)** $L = \dfrac{A - 2HW}{2(H + W)}$

You may be puzzled by Example 13 because the equation doesn't appear to be linear. Actually you are partially correct. The equation is indeed not linear in I. However, it is linear in terms of R because we can make believe that all variables other than R are constants. Thus the factor I^2 is a constant with respect to R. (After all, if the equation were $P = QR$, then we would have $R = P/Q$. This is exactly equivalent to making believe that $I^2 = Q$.)

A major use of equations is in an area of application called **break-even analysis.** A typical problem would involve a product or service being sold. Given that the sale price is more than the cost of the product or service, how many units must be sold in order to meet fixed costs (break even) associated with the product or service? Let's study some examples.

EXAMPLE 14

A sandwich shop sells a variety of sandwiches. It sells each sandwich for \$0.75 more than it costs to make it. Given that it cost \$24,000 to open the business, how many sandwiches must be sold in order to break even?

SOLUTION

Let x = number of sandwiches to be sold. Then the shop breaks even if

$$\begin{aligned} \$0.75x &= \$24{,}000, \\ x &= 32{,}000. \quad \square \end{aligned}$$

EXAMPLE 15

Ann Lopez is running for city council. To help raise money, she is considering staging a concert by a popular band. Although the band will play for free, she must rent the auditorium for \$1500. Announcements and other publicity will cost another \$750. How many tickets at \$5.00 each must be sold for the concert to break even, given that each ticket costs approximately \$0.05 to print?

SOLUTION

Let

$$\begin{aligned} x &= \text{number of tickets to be sold,} \\ c &= \text{cost of staging the concert,} \\ r &= \text{revenue from sale of tickets.} \end{aligned}$$

The concert will break even if $r = c$. But

$$c = \text{fixed costs} + \text{variable costs},$$

so

$$c = \$1500 + \$750 + \$0.05x$$

and

$$r = \$5.00x.$$

Equating r and c gives us

$$\begin{aligned} \$5.00x &= \$2250 + \$0.05x, \\ 4.95x &= 2250, \\ x &= 454.55 \qquad \text{(approximately).} \end{aligned}$$

Thus 455 tickets must be sold for the concert to break even. (Check the answer!) □

Practice Problem

In Example 15, how many tickets must be sold at \$6.50 each for the concert to break even?

Answer: approximately 349 tickets

Linear Inequalities in One Variable

Recall from our discussion in Chapter 1 that if a and b are real numbers, then either

$$a < b \qquad (a \text{ is less than } b),$$

or

$$a = b,$$

or

$$a > b \qquad (a \text{ is greater than } b).$$

The symbols $<$ and $>$ are inequality symbols. There are two other commonly used inequality symbols:

$\leq$ is read "less than or equal to"

$\geq$ is read "greater than or equal to"

We say that $a \leq b$ only if $a < b$ or $a = b$. Similarly, we say $a \geq b$ only if $a = b$ or $a > b$.

A **linear inequality** in one variable is an inequality that can be written in the form

$$ax + b < c \quad (\text{or} \quad \leq, >, \text{or} \geq),$$

where $a \neq 0$ and a, b, and c are constants. For example,

$$3x + 4 < 22$$

is a linear inequality in the variable x.

As with linear equations in one variable, we will be interested in finding the solution set. However, with an inequality, the solution set generally contains more than one number.

We will use the same approach with linear inequalities that we used with linear equations: Generate an equivalent inequality with the variable isolated on one side of the inequality by using the appropriate operations. Here we must be careful, because the properties of inequalities are not quite the same as those of equations.

Valid Operations for Inequalities

1. Addition property: If $a < b$, then $a + c < b + c$.
2. Subtraction property: If $a < b$, then $a - c < b - c$.
3. Multiplication properties:
   i. If $a < b$ and c is positive, then $ac < bc$.
   ii. If $a < b$ and c is negative, then $ac > bc$.
4. Division properties:
   i. If $a < b$ and c is positive, then $\frac{a}{c} < \frac{b}{c}$.
   ii. If $a < b$ and c is negative, then $\frac{a}{c} > \frac{b}{c}$.

(These properties are also true for the inequality signs $>$, $\leq$ and $\geq$.)

WARNING

BE CAREFUL WHEN MULTIPLYING OR DIVIDING INEQUALITIES

Notice the multiplication and division properties. When we multiply or divide an inequality by a positive number, the sense or direction of the inequality is unchanged. *When we multiply or divide by a negative number, the direction of the inequality is changed!* For example, suppose we have the inequality

$$5 < 7.$$

Multiplying both sides by 2 gives

$$10 < 14,$$

which has the same direction as the original inequality. However, if we multiply the original inequality by -1 we have

$$-5 > -7.$$

Here we must change the direction of the inequality for it to be true.

Sometimes we use the **double inequality** $a < x < b$. This is actually equivalent to the two inequalities $a < x$ and $x < b$. Thus you could break up a double inequality and solve each of the resulting inequalities. The solution set to the original double inequality would then be the *intersection* of the solution sets of the two resulting inequalities (because the variable would have to satisfy *both* of the inequalities). However, it is easier to analyze the entire double inequality, as Example 17 shows.

EXAMPLE 16

Solve and graph the solution set on the number line for the following inequalities.

a) $3x + 4 < 22$ **b)** $-3(x + 1) \geq 2(x + 4)$

SOLUTION

a) $3x + 4 < 22$

$3x < 18$ **Subtraction property**

$x < 6$ **Division property (positive)**

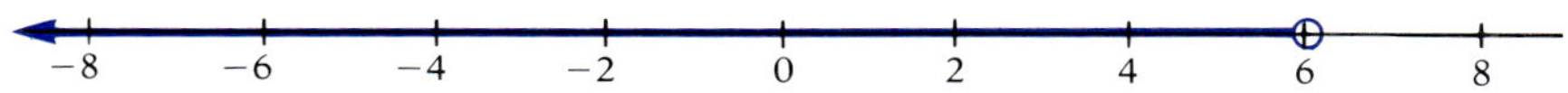

(Note that in our graphs an open circle implies that the endpoint is *not* in the solution set, while a full circle implies that the endpoint *is* in the solution set.)

b) $-3(x + 1) \geq 2(x + 4)$

$-3x - 3 \geq 2x + 8$

$-5x - 3 \geq 8$ **Subtraction property**

$-5x \geq 11$ **Subtraction property**

$x \leq \dfrac{-11}{5}$ **Multiplication property (negative)** □

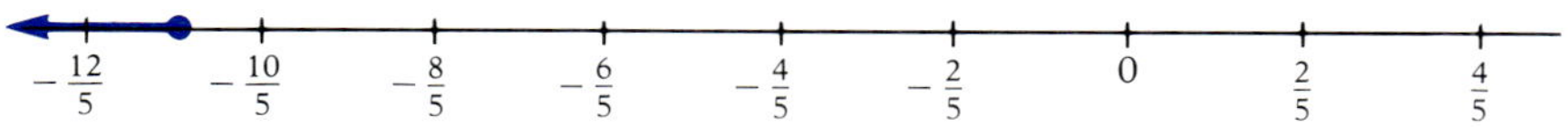

EXAMPLE 17

Solve and graph the solution set for the inequality

$$2 < 4(x + 2) < 12.$$

SOLUTION

This is a double inequality. Rather than attempt to isolate x on the left, let's isolate it in the middle.

$2 < 4(x + 2) < 12$	
$2 < 4x + 8 < 12$	**Distributive property**
$-6 < 4x < 4$	**Subtraction property:** **subtract 8 from all three parts of the inequality.**
$-\frac{3}{2} < x < 1$	**Division property (positive):** **divide all three parts of the inequality by 4.** □

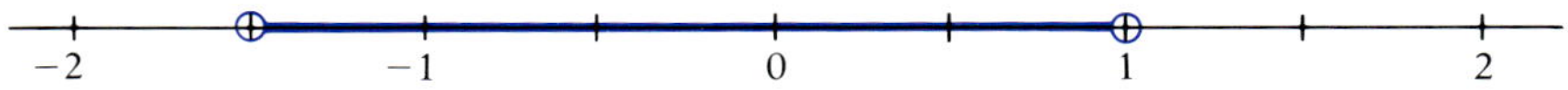

EXAMPLE 18

Solve and graph the solution set for the inequality $4(4 - x) \leq 2(3 - x)$.

SOLUTION

$4(4 - x) \leq 2(3 - x)$	
$16 - 4x \leq 6 - 2x$	**Distributive property**
$16 - 2x \leq 6$	**Addition property**
$-2x \leq -10$	**Subtraction property**
$x \geq 5$	**Division property (negative)** □

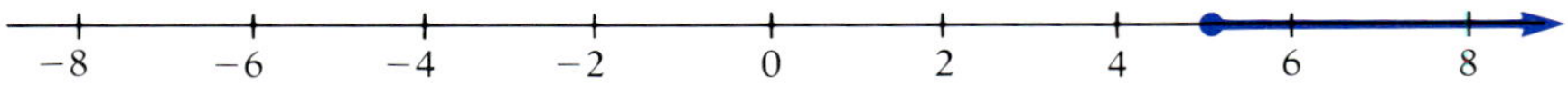

Practice Problem

Solve the following inequalities and graph their solution sets.

a) $6x - 2 > 3$ **b)** $4(6 - x) \leq 3(x + 1)$

c) $12 < 3(7 - 2x) < 24$ **d)** $3[2 - 5(x + 1)] \leq 2[2 + 3(x + 1)]$

Answers: **a)** $x > \frac{5}{6}$

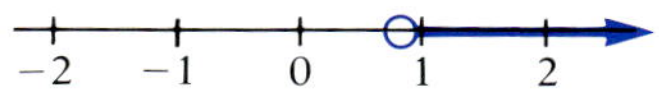

b) $x \geq 3$

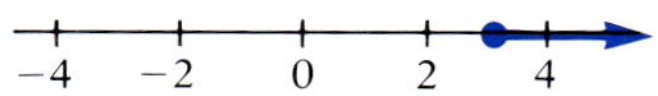

c) $-\frac{1}{2} < x < \frac{3}{2}$

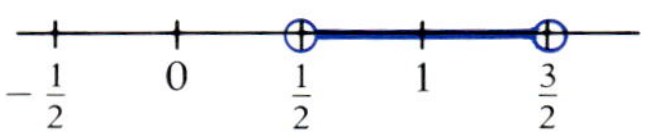

d) $x \geq -\frac{19}{21}$

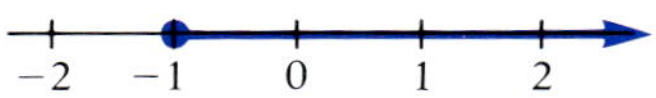

EXERCISE SET 3.2

Solve the following equations and inequalities. Graph the solution sets for the inequalities.

1. $3x - 1 = 2x + 2$

2. $3(x - 1) = 2x + 2$

3. $2(1 - x) = 6(x - 2)$

4. $x - 4 < -3$

5. $4 - y \geq 3y$

6. $3(m - 2) \leq -5(1 - m)$

7. $2[3 - 2(x - 1)] = 5(2 - x)$

8. $\dfrac{3a}{4} - \dfrac{a}{3} = 6$

9. $0.5(t - 3) = 0.75(2t + 4)$

10. $-2(1 - c) = 4[3 - 2(1 - c)]$

11. $\dfrac{2(3 - n)}{5} = \dfrac{4n}{5} - \dfrac{3(1 - n)}{10}$

12. $y - \dfrac{1}{5} > \dfrac{3}{5} - \dfrac{y}{5}$

13. $\dfrac{4w}{9} + \dfrac{3(2 - 2w)}{9} - 1 \geq \dfrac{w}{9} - \dfrac{7w}{9}$

14. $\dfrac{0.1t}{3} + \dfrac{0.4(1 - t)}{4} < 5 + \dfrac{0.3(2 + 3t)}{3}$

15. $-3 < 7x - 2 < 14$

16. $12 \leq 3(1 - 2x) < 15$

17. $0.3 < 0.5(1 - 3y) \leq 1$

18. $\dfrac{1}{7} < \dfrac{3(3 - x)}{7} < 11$

Solve the following equations for the indicated variable.

19. $3x + 2y = 15$, for y

20. $4s - 6t = -3$, for t

21. $C = \dfrac{5}{9}(F - 32)$, for F

22. $3(1 - y^2)x + x = 1$, for x

23. $b = 2(a - 2c + d)$, for c

24. $-3(1 - x) = 9(y + 3z)$, for z

25. Temperatures in Richmond, Virginia, range from a record low of -24.44°C to a record high of 41.67°C. What is the equivalent range in degrees Fahrenheit? (See Example 11.)

26. At what temperature are the Celsius and Fahrenheit values equal? (See Example 11.)

27. A company manufactures home computer systems at a cost of \$435 each and sells them for \$500 each. The company spent \$150,000 in order to begin manufacturing the computers. How many systems must it sell in order to break even?

28. The company described in Problem 27 has the opportunity to introduce a more sophisticated computer system that would sell for \$1500 each. It would cost an additional \$100,000 to modify the production line. Given that the new systems would cost \$1375 each to manufacture, how many of the new systems would have to be sold to break even?

29. Janet Smith and Cheryl Jones must drive to Washington, D.C., to attend a training session and then return to Richmond, Virginia. They have decided to rent a car for the trip. Hurtz has advertized a flat rate of \$97.30 per week (no mileage fee), while Avarice wants \$60 per week and \$0.08 per mile. The round-trip mileage is 225 miles and the trip will last two weeks. Determine the additional number of miles the women must drive for the two companies to be equally costly.

30. A retail merchant pays a flat weekly rental fee of \$175. She is considering adding an additional product line. She can sell each unit of the new line for \$2.50 profit. However, if she adds the new line, she must rent additional space at an added cost of \$75 weekly. In addition, she estimates her utility bill will go up by \$15 weekly (because of the new space). How many units of the new product must she sell each week to break even?

3.3 QUADRATIC EQUATIONS AND INEQUALITIES

Quadratic Equations

An equation of the form

$$ax^2 + bx + c = 0, \qquad a \neq 0,$$

where a, b, and c are constants, is called a **quadratic equation.** There are three approaches to solving quadratic equations: taking the square root of both sides, factoring, and the quadratic formula. Let's consider taking the square root of both sides first.

EXAMPLE 19

Solve the equation $x^2 = 4$.

SOLUTION

Since the left side is a perfect square and x does *not* appear on the right side, we can take the square root of both sides:

$$x = \pm 2,$$

which is equivalent to the *two* solutions

$$x = 2 \quad \text{or} \quad x = -2. \quad \square$$

EXAMPLE 20

Solve the equation $(y - 4)^2 = 6$.

SOLUTION

Again, the left side is a perfect square and the variable does not appear on the right, so we can take the square root of both sides;

$$y - 4 = \pm\sqrt{6},$$

so

$$y = 4 \pm \sqrt{6},$$

which is equivalent to the two solutions

$$y = 4 + \sqrt{6} \quad \text{or} \quad y = 4 - \sqrt{6}. \quad \square$$

EXAMPLE 21

Solve the equation $x^2 = -3$.

SOLUTION

This equation has no real solution because there is no real number that is negative when squared. □

Practice Problem

Solve the following equations.

a) $y^2 = 16$ **b)** $(x + 2)^2 = 8$
c) $(a - 3)^2 = -2$

Answers: **a)** $y = \pm 4$; **b)** $x = -2 \pm \sqrt{8}$; **c)** no real solution

WARNING

POSITIVE NUMBERS HAVE TWO SQUARE ROOTS

Don't forget to include both the positive and negative square root when solving a quadratic equation by taking the square root of both sides.

Now let's take a look at how we can use factoring to solve quadratic equations.

EXAMPLE 22

Solve the equation $x^2 + 4x + 3 = 0$.

SOLUTION

Notice that we can factor the left side of the equation:

$$x^2 + 4x + 3 = (x + 3)(x + 1).$$

That means we can write the equation

$$(x + 3)(x + 1) = 0.$$

Now, when the product of two different numbers is zero, one of them *must* be zero. Thus either $(x + 3) = 0$, in which case we have $x = -3$, or it must be true that $x + 1 = 0$, in which case the solution is $x = -1$. Therefore the solution set for the equation contains the solutions $x = -1$ or $x = -3$. (We leave it to you to check these solutions.) □

EXAMPLE 23

Solve the equation $x^2 - 4x + 4 = 0$.

SOLUTION

Once again we can factor the left side of the equation. This gives us

$$(x - 2)(x - 2) = 0.$$

This means that $x = 2$ is the only member of the solution set. □

Practice Problem

Solve the following equations by factoring.

a) $x^2 + 2x + 1 = 0$ **b)** $9y^2 - 36 = 0$

c) $2a^2 - a - 1 = 0$

Answers: **a)** $x = -1$; **b)** $y = 2$ or $y = -2$;

c) $a = 1$ or $a = -\frac{1}{2}$

Factoring is a good approach to use to solve a quadratic equation when indeed you can factor it. However, we'll soon derive the quadratic formula, which can be used regardless of whether factoring is possible. Hence we offer you the following:

WARNING

BEWARE OF SPENDING EXCESSIVE TIME TRYING TO FACTOR

Not all quadratic equations can be solved by factoring. Moreover, since another approach is available, we recommend that you not spend more than one or two minutes trying to solve a quadratic equation by factoring.

You should have noticed in Examples 22 and 23 that the first equation had two different solutions and the second equation had only one. (Theoretically, the second equation had two solutions. They just happened to be identical.) It is also possible for a quadratic equation to have no real solution (the solution set is empty). Consider the equation

$$x^2 + 1 = 0.$$

By subtracting 1 from both sides we can derive the equivalent equation

$$x^2 = -1.$$

Since the square of a real number is always nonnegative, this equation has no real solution.

Now let's see how we can use a technique called **completing the square** to solve quadratic equations. Consider the equation

$$x^2 + 6x + 6 = 0.$$

A little thought should convince us that we can't factor the left side of this equation using only integers. However, if we had $x^2 + 6x + 9$ on the left side, we would have a perfect square, because $x^2 + 6x + 9$ factors into $(x + 3)^2$. How can we get something like this? The trick is to subtract the given constant 6 from both sides and then add the desired constant, 9, to both sides:

$$x^2 + 6x + 6 - 6 = 0 - 6$$

or

$$x^2 + 6x = -6.$$

Now we add 9 to both sides:

$$x^2 + 6x + 9 = -6 + 9$$

or

$$x^2 + 6x + 9 = 3.$$

By factoring, we can write this as

$$(x + 3)^2 = 3.$$

Since we have a perfect square on the left, let's take the square root of both sides. Be careful here! A positive number has two square roots. We indicate this by writing

$$x + 3 = \pm\sqrt{3}.$$

Our last step is to subtract 3 from both sides, giving us

$$x = -3 \pm \sqrt{3}.$$

Thus the equation has two solutions:

$$x = -3 + \sqrt{3} \quad \text{or} \quad x = -3 - \sqrt{3}.$$

Let's think about what we just did. We converted the left side of the equation into a perfect square by first subtracting the original constant from both sides of the equation and then adding the desired constant to both sides. How do we know what to add? The answer is to add exactly the square of one half the coefficient of the x term (as long as the coefficient of the squared term is 1), because we know

$$(x + a)^2 = x^2 + 2ax + a^2.$$

Now let's formalize this a bit, which will enable us to derive an extremely useful result.

We want to find the values of x that will satisfy the equation $ax^2 + bx + c = 0$.

1. Subtract the constant c from both sides of the equation and then divide both sides by a, which results in

$$x^2 + \frac{bx}{a} = \frac{-c}{a}.$$

2. Next, let's make the left side a perfect square. We can do this by adding the square of one-half b/a to both sides, which gives us

$$x^2 + \frac{bx}{a} + \left(\frac{b}{2a}\right)^2 = \frac{-c}{a} + \left(\frac{b}{2a}\right)^2.$$

3. Now we can factor the left side (which is the reason for step 2) and rearrange the right side (which takes a bit of doing):

$$\left(x + \frac{b}{2a}\right)^2 = \frac{b^2 - 4ac}{4a^2}.$$

4. Take the square root of both sides to give us

$$x + \frac{b}{2a} = \pm\frac{\sqrt{b^2 - 4ac}}{2a}.$$

5. Finally, subtract $b/2a$ from both sides, which gives the following formula.

Quadratic Formula

$$x = \frac{-b \pm \sqrt{b^2 - 4ac}}{2a}$$

We can use the quadratic formula with any quadratic equation by simply replacing a, b, and c with the appropriate values. Note that the symbol $\pm$ actually implies the existence of two solutions:

$$x = \frac{-b - \sqrt{b^2 - 4ac}}{2a} \quad \text{or} \quad x = \frac{-b + \sqrt{b^2 - 4ac}}{2a}.$$

EXAMPLE 24 Solve the equation $2x^2 + 5x - 1 = 0$.

SOLUTION

Now the left side cannot be factored (try it!), so we use the quadratic formula with $a = 2$, $b = 5$, and $c = -1$:

$$x = \frac{-5 \pm \sqrt{5^2 - 4(2)(-1)}}{2(2)} \quad \text{or} \quad x = \frac{-5 \pm \sqrt{33}}{4}.$$

In other words, the solutions are

$$x = \frac{-5 + \sqrt{33}}{4} \quad \text{or} \quad x = \frac{-5 - \sqrt{33}}{4}. \quad \square$$

EXAMPLE 25

Solve the equation $3x^2 + 54x + 237 = 0$.

SOLUTION

We can't factor the left side so let's apply the quadratic formula. Then $a = 3$, $b = 54$, and $c = 237$, which gives us

$$x = \frac{-54 \pm \sqrt{54^2 - 4(3)(237)}}{2(3)}$$

or

$$x = \frac{-54 \pm \sqrt{72}}{6} = \frac{-54 \pm \sqrt{36 \cdot 2}}{6}$$

$$= \frac{-54 \pm 6\sqrt{2}}{6} = -9 \pm \sqrt{2}.$$

You may have noticed that we can divide both sides of the equation

$$3x^2 + 54x + 237 = 0$$

by three. This would result in

$$x^2 + 18x + 79 = 0.$$

Applying the quadratic formula with $a = 1$, $b = 18$, and $c = 79$, we have

$$x = \frac{-18 \pm \sqrt{18^2 - 4(1)(79)}}{2(1)}$$

or

$$x = \frac{-18 \pm \sqrt{8}}{2} = \frac{-18 \pm \sqrt{4 \cdot 2}}{2}$$

$$= \frac{-18 \pm 2\sqrt{2}}{2} = -9 \pm \sqrt{2},$$

which is the same result we obtained earlier. Reducing an equation has no effect on the answer, but it lessens the work involved in deriving the answer. $\square$

EXAMPLE 26

Solve the equation $-x^2 + 3x - 7 = 0$.

SOLUTION

Again, we cannot factor the left side. Applying the quadratic formula with $a = -1$, $b = 3$, and $c = -7$, we have

$$x = \frac{-3 \pm \sqrt{3^2 - 4(-1)(-7)}}{2(-1)} \quad \text{or} \quad x = \frac{-3 \pm \sqrt{-19}}{-2}.$$

Because -19 has no real square roots, there are *no* real numbers that solve this equation! □

EXAMPLE 27

Solve the equation $4x^2 - 12x + 9 = 0$.

SOLUTION

Here we could factor the left side. However, let's solve the equation using the quadratic formula to see what happens. We have $a = 4$, $b = -12$, and $c = 9$, so

$$x = \frac{-(-12) \pm \sqrt{(-12)^2 - 4(4)(9)}}{2(4)} \quad \text{or} \quad x = \frac{12 \pm \sqrt{144 - 144}}{8},$$

which means that $x = \frac{12}{8} = 1.5$. Thus this equation has only one solution. □

Practice Problem

Solve the following quadratic equations by using the quadratic formula.

a) $x^2 + 6x + 4 = 0$ **b)** $x^2 - x - 1 = 0$

c) $-2x^2 - 4x + 13 = 0$ **d)** $x^2 + x + 1 = 0$

Answers: **a)** $x = \dfrac{-6 \pm \sqrt{20}}{2} = -3 \pm \sqrt{5}$; **b)** $x = \dfrac{1 \pm \sqrt{5}}{2}$;

c) $x = \dfrac{-(4 \pm \sqrt{120})}{4}$; **d)** no real solution

Take another look at Examples 24–27. Notice that the equations in Examples 24 and 25 each had two real solutions, the equation in Example 26 had no real solution, and the equation in Example 27 had one real solution. The part of the quadratic formula under the radical sign, $b^2 - 4ac$, is called the **discriminant** and it will tell us how many real solutions there will be.

Number of Real Solutions to a Quadratic Equation

$$\text{When } b^2 - 4ac \begin{cases} > 0, \text{ there are two real solutions,} \\ = 0, \text{ there is one real solution,} \\ < 0, \text{ there is no real solution.} \end{cases}$$

Now let's take a look at some applications that involve solving quadratic equations.

EXAMPLE 28

The Crest View Apartments contain 100 units. At \$300 per month, every apartment is rented. For every \$10 increase in rent, there will be one less apartment rented. The owner wants to receive \$32,000 in rents each month. What should the rent for each apartment be? (If there are two answers, the owner prefers the one that results in the larger occupancy.)

SOLUTION

We can actually model this problem two different ways. The first method is more obvious but leads to a more complicated model. The second method requires more thought initially but leads to a simpler model.

Method 1 We want to know the new rent. Let

$$r = \text{monthly rent for each apartment.}$$

Now,

$$\text{total rent} = (\text{monthly rent})(\text{number of apartments rented})$$

and

$$\text{number of apartments rented} = 100 - (\text{number of \$10 increases}).$$

Since

$$\text{increase in rent} = \text{new rent} - \text{old},$$

the number of \$10 increases must be given by (increase in rent) ÷ 10. We can now convert everything to symbols:

$$\text{increase in rent} = r - 300,$$

so

$$\text{the number of \$10 increases} = \frac{r - 300}{10}$$

and therefore

$$\text{total rent} = r\left[100 - \left(\frac{r - 300}{10}\right)\right],$$

so

$$\text{total rent} = r\left(\frac{1000 - r + 300}{10}\right),$$

or

$$32{,}000(10) = r(1300 - r),$$

$$r^2 - 1300r + 320{,}000 = 0.$$

Using the quadratic formula with $a = 1$, $b = -1300$, and $c = 320{,}000$, we have

$$r = \frac{-(-1300) \pm \sqrt{(-1300)^2 - 4(1)(320{,}000)}}{2(1)} = \frac{1300 \pm 640.31}{2}.$$

The rent for each apartment can be either \$329.85 or \$970.16. Here the second solution would cause more vacant apartments, so we conclude that the rent should be \$329.85, which we could round off to an even \$330.

Method 2 Now let's try another approach. Let

$$n = \text{the number of \$10 increases in rent.}$$

Then the new rent would be \$300 + \$10n and the number of vacancies would be n. Since we still have

$$\text{total rent} = (\text{monthly rent})(\text{number of apartments rented}),$$

$$\text{total rent} = (300 + 10n)(100 - n),$$

so we have

$$32{,}000 = (300 + 10n)(100 - n),$$

$$32{,}000 = 30{,}000 + 700n - 10n.$$

Rearranging gives us

$$10n^2 - 700n + 2000 = 0.$$

We can divide both sides by 10, so we now have

$$n^2 - 70n + 200 = 0.$$

Using the quadratic formula with $a = 1$, $b = -70$, and $c = 200$ gives

$$n = \frac{-(-70) \pm \sqrt{(-70)^2 - 4(1)(200)}}{2(1)} = \frac{70 \pm 64.03}{2},$$

which gives us the solutions $n = 2.985$ and $n = 67.015$. Converting these to monthly rents gives us the values

$$\text{rent} = 300 + 10(2.985) = 329.85,$$
$$\text{rent} = 300 + 10(67.015) = 970.15.$$

We conclude that the first solution is more desirable, so the new rent should be \$329.85 (which we would round off to \$330). □

EXAMPLE 29

Modern Mechanical Products has decided to convert some open land into a garden around which will be benches and chairs for its employees to use. The original plot of land is 120 feet by 100 feet. The company would like to remove an equal amount of land around each of the borders so that two thirds of the original area will be left. How wide should the border area be?

SOLUTION

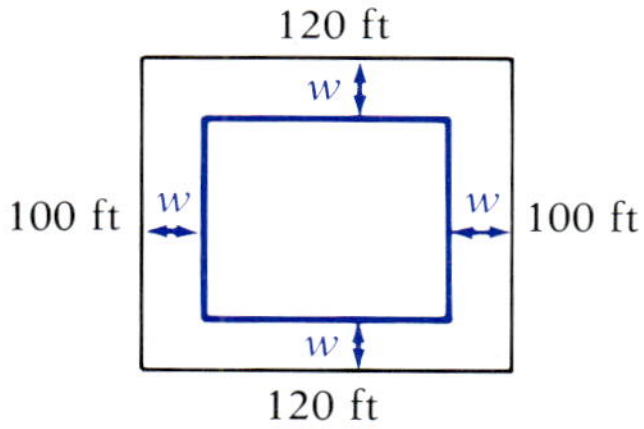

Figure 3.3 Putting a border around a garden.

Since this problem involves geometry, we can benefit by drawing a picture of it. As Fig. 3.3 shows, our task requires us to determine the width of the border area. Let

$$w = \text{width of the border area.}$$

Then the new garden will have dimensions of $120 - 2w$ by $100 - 2w$. So, we have the following areas:

$$\text{area of original land} = 120 \cdot 100 = 12{,}000$$
$$\text{area of garden} = (120 - 2w)(100 - 2w).$$

Finally, the area of the garden is to be two thirds of the area of the original area, which gives us the equation

$$(120 - 2w)(100 - 2w) = \frac{2}{3}(12{,}000),$$

which is

$$4w^2 - 440w + 12{,}000 = 8000$$

or

$$4w^2 - 440w + 4000 = 0$$

or

$$w^2 - 110w + 1000 = 0.$$

This equation is easily factored:

$$(w - 100)(w - 10) = 0,$$

which means that the solutions are $w = 100$ or $w = 10$. However, the first solution makes no sense practically. We cannot reduce the original dimensions of 120 feet by 100 feet by removing 100 feet along each border. (The results would be negative.) We conclude that we should use a width of 10 feet. □

Quadratic Inequalities

An inequality of the form

$$ax^2 + bx + c < 0 \qquad (\text{or } \le, >, \text{or } \ge), \quad a \ne 0,$$

where a, b, and c are constants is called a **quadratic inequality.** For example,

$$x^2 + 2x + 1 > 0, \qquad 3y^2 - 12y - 36 \le 0, \qquad -5a^2 + 10a \ge 0,$$

are all quadratic inequalities. As the next two examples show, we can solve quadratic inequalities by making use of the number line.

EXAMPLE 30

Solve the quadratic inequality $x^2 - x - 2 > 0$.

SOLUTION

We can factor the left side:

$$x^2 - x - 2 = (x + 1)(x - 2),$$

so the given inequality is equivalent to

$$(x + 1)(x - 2) > 0.$$

Thus we want the product of the factors $(x + 1)$ and $(x - 2)$ to be positive. This will happen only if both factors are positive or if both factors are negative. Looking at the factors we have

$$\begin{aligned} x + 1 &> 0, & x - 2 &> 0, \\ x &> -1, & x &> 2. \end{aligned}$$

The factors are positive when both of these inequalities are true, which occurs when $x > 2$. Similarly

$$\begin{aligned} x + 1 &< 0, & x - 2 &< 0, \\ x &< -1, & x &< 2. \end{aligned}$$

The factors are negative when both of these inequalities are true, which occurs when $x < -1$. See the **sign graph** in Fig. 3.4.

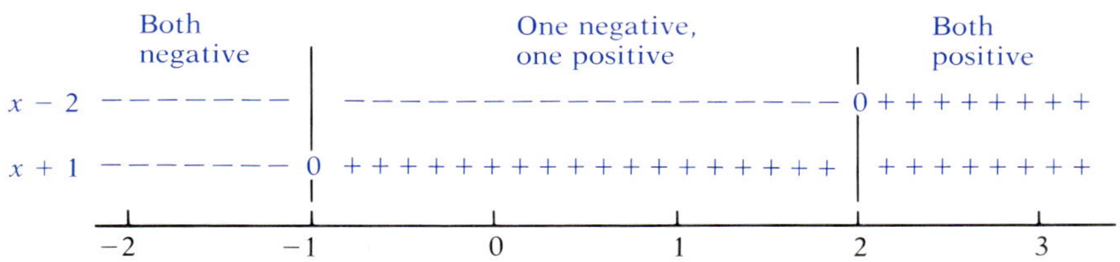

Figure 3.4 Sign graph for the factors $(x + 1)$ and $(x - 2)$.

When the factors have opposite signs, their product will be negtative. As the sign graph in Fig. 3.4 shows, this occurs when $-1 < x < 2$. The solution to the given inequality is therefore

$$x < -1 \quad \text{or} \quad x > 2. \quad \square$$

EXAMPLE 31 Solve the quadratic inequality $2y^2 + 5y \leq 3$.

SOLUTION

We must rewrite the inequality so the right side is zero:

$$2y^2 + 5y - 3 \leq 0.$$

Next, factor the left side:

$$(2y - 1)(y + 3) \leq 0.$$

The product of two factors is negative only when the factors have opposite signs. We also know the product is zero only when one of the factors is zero. As the sign graph in Fig. 3.5 shows, the solution is therefore

$$-3 \leq y \leq \frac{1}{2}.$$

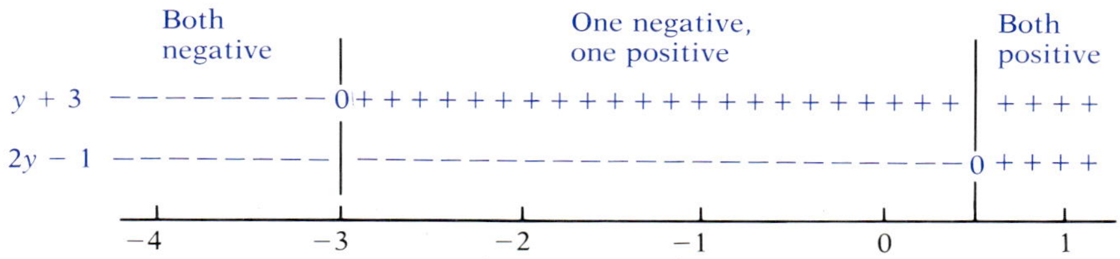

Figure 3.5 **Sign graph for the factors $(2y - 1)$ and $(y + 3)$.** $\square$

Practice Problem

Solve by developing a sign graph for each of the following inequalities.

a) $x^2 - 3x + 2 > 0$ **b)** $n^2 - 7n \leq 8$

Answers: **a)** $x < 1$ or $x > 2$

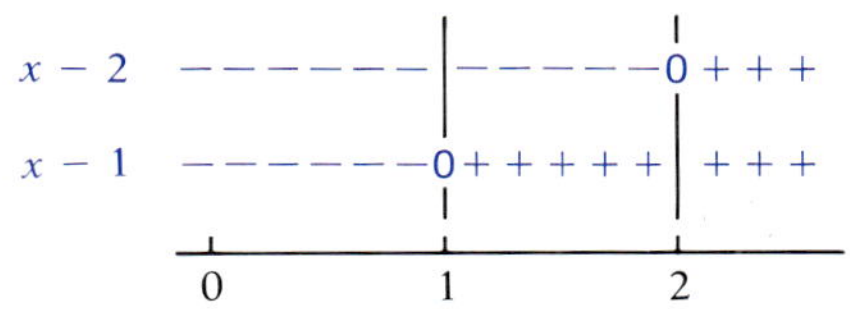

b) $-1 \leq n \leq 8$

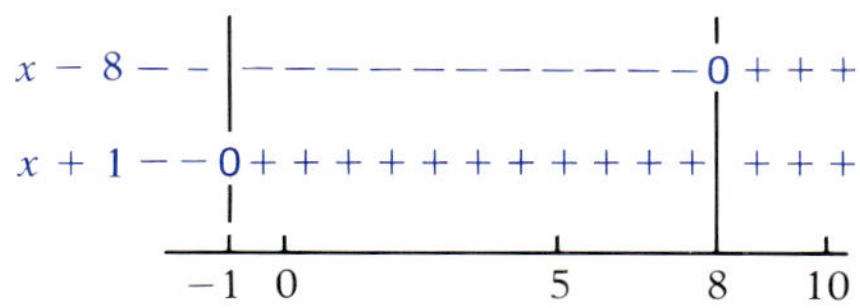

EXERCISE SET 3.3

Solve the following equations.

1. $x^2 - 4 = 0$

2. $2x^2 + 3 = 0$

3. $4y^2 - 7 = 0$

4. $(w + 3)^2 = 0$

5. $x^2 + 4x - 4 = 0$

6. $x^2 + 5x + 6 = 0$

7. $-y^2 - y + 12 = 0$

8. $15u^2 + 13u - 6 = 0$

9. $-3v^2 - 3v + 16 = 0$

10. $k^2 - 10k + 20 = 0$

11. $5x^2 + 4x - 1 = 0$

12. $y^2 - 16 = 0$

13. $9x^2 - 49 = 0$

14. $25 - 16c^2 = 0$

15. $6m^2 + 2m - 12 = 0$

16. $4b^2 + 24b + 5 = 0$

Solve the following quadratic inequalities and graph each solution.

17. $x^2 - 7x + 12 < 0$

18. $2u^2 - 8u - 10 \geq 0$

19. $a^2 - 5a + 6 > 0$

20. $2b^2 - 2 \leq 0$

21. $2c^2 + 7c + 3 < 0$

22. $x^2 + 6x + 9 > 0$

23. $9y^2 + 6y + 1 < 0$

24. $m^2 - 10m + 24 \geq 0$

25. The monthly revenue of a company is given by the equation

$$R = 10p + 8p^2,$$

where p is the price (in dollars) of the product being manufactured. What must the price be for the revenue to be \$12,000?

26. Super Candy has been losing money. One of the suggested measures is to cut the volume of their popular Super-Choc bar by 10%. The candy bar is currently 4 inches long, 3 inches wide, and 0.2 inch thick. What would the dimensions of the new candy bar be if

a) The length and width are reduced by equal amounts and the thickness is unchanged?

b) the width and the thickness are reduced by equal amounts and the length is unchanged?

3.4

LINEAR EQUATIONS AND INEQUALITIES IN TWO VARIABLES

Earlier we studied linear equations and inequalities in one variable. In the next section, we'll extend our discussion to include linear equations and inequalities in two variables. Equations and inequalities of this type have solution sets that consist of infinitely many solutions. We'll show you how to represent these solution sets graphically. In addition, we'll also show you how to derive the equation of a line when given certain information concerning its graph. Before we begin, we'll present a review of how we graph with two variables.

Cartesian Coordinate System

We can extend the concept of the number line to a plane by using two mutually perpendicular number lines, one of which is horizontal and the other vertical. We call the number lines **axes** or **coordinate axes.** We usually refer to the horizontal axis as the **x-axis** and to the vertical axis as the ***y*-axis.** The point where the axes cross is called the **origin,** which corresponds to the point 0 on both number lines (axes). (See Fig. 3.6.) Each point in the plane is uniquely identified by an ordered pair of numbers. The first number is always the horizontal coordinate (the location on the horizontal number line) and the second number is always the vertical coordinate (the location on the vertical number line). (See Fig. 3.7.) Note that the point A (3, 1) and the point B (1, 3) are different. In fact, that's why we call them *ordered pairs:* the order tells us which number goes with which axis. We often refer to the horizontal coordinate as the **abscissa** and to the vertical coordinate as the **ordinate.**

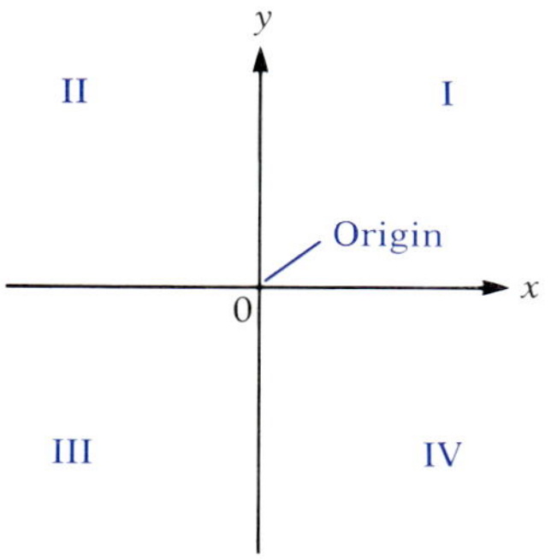

Figure 3.6 Cartesian coordinate system.

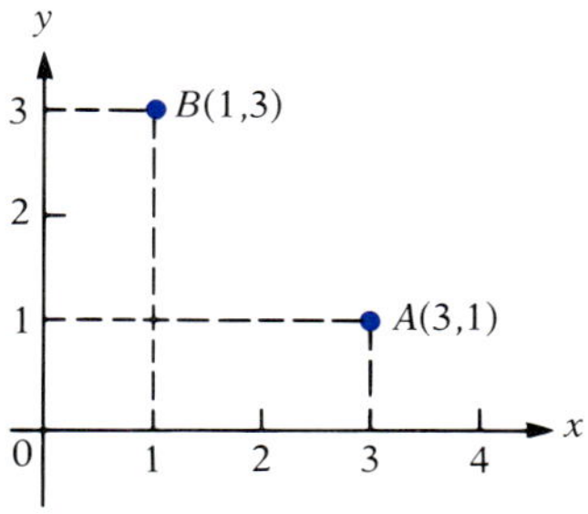

Figure 3.7 Locating points using Cartesian coordinates.

The coordinate axes divide the plane into four regions called **quadrants.** We number the quadrants I, II, III, and IV as shown in Fig. 3.6.

We refer to this system of axes as the **Cartesian** (or **rectangular**) coordinate system, named after the French mathematician René Descartes (1596–1650).

Linear Equations in Two Variables

The standard form of a linear equation in two variables is

$$Ax + By = C,$$

where A, B, and C are constants and we do not allow A and B to both be zero. For example, the following are linear equations in two variables:

$$3x + 2y = 6, \qquad x = -2, \qquad y = 3.$$

A solution to a linear equation involving two variables means specifying a value for each variable. Consider an equation such as $3x + 2y = 6$. Suppose we pick a value for y: for example, 0. If we substitute that value into the equation, we have $3x + 2(0) = 6$ or $3x = 6$, which is a linear equation in one variable. Thus every time we choose a value for y, we could compute the accompanying value for x. However, there is absolutely nothing that limits our choice of values for y. In other words, we can pick any value we wish for y, and compute the appropriate value for x. Therefore linear equations in two variables have an infinite (unlimited) number of solutions (as opposed to linear equations in one variable, which have one solution). A convenient way to represent all the solutions to an equation in two variables is to draw a picture of them, a process we call graphing the equation.

A linear equation in standard form (or that can be transformed to this form) has a straight line graph in Cartesian coordinates, which is why we call it *linear*. Moreover, any straight line in Cartesian coordinates can be written as a linear equation in standard form.

Since linear equations graph as straight lines, they are easy to graph. We need only determine two points that lie on the line.

EXAMPLE 32

Graph the equation $3x + 2y = 6$.

SOLUTION

We must find two points that lie on the graph of the equation (two solutions of the equation). From our preceding discussion we know $y = 0$ means $x = 2$, so $(2, 0)$ must be a solution. Suppose we let $x = 0$. Then we have $3(0) + 2y = 6$, which means $y = 3$. Thus $(0, 3)$ must also lie on the line. Now we can draw the graph as shown in Fig. 3.8.

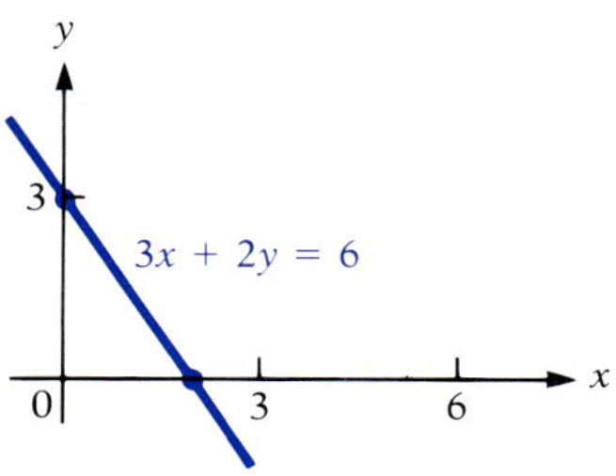

Figure 3.8 The graph of $3x + 2y = 6$. □

EXAMPLE 33

Graph the equation $2x - 4y = 8$.

SOLUTION

Let's try the same approach we just used in Example 32 to get two points. First, set $x = 0$. This gives us $2(0) - 4y = 8$, which means we have $y = -2$, so the point $(0, -2)$ is on the line. Now set $y = 0$, which gives us $2x - 4(0) = 8$, so $x = 4$. The other point is $(4, 0)$. (See Fig. 3.9.)

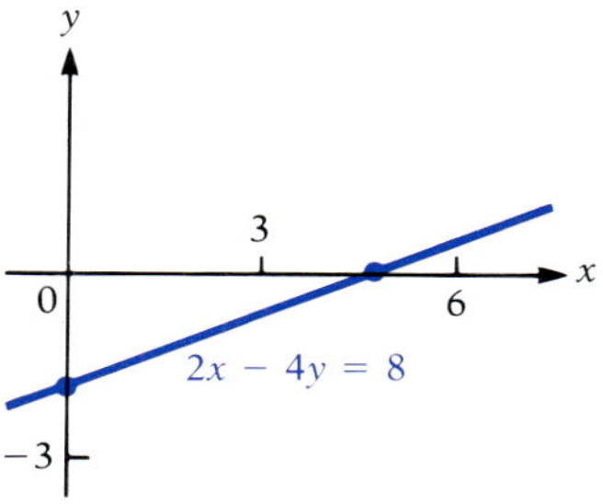

Figure 3.9 **The graph of $2x - 4y = 8$.** □

EXAMPLE 34

Graph the equation $x = -2$.

SOLUTION

This equation is different from the first two we've just graphed because only one variable, x, appears. The implication of this equation is that no matter what y is, x is always -2. That means, for example, that $(-2, 0)$ and $(-2, 1)$ both lie on the line. (See Fig. 3.10.)

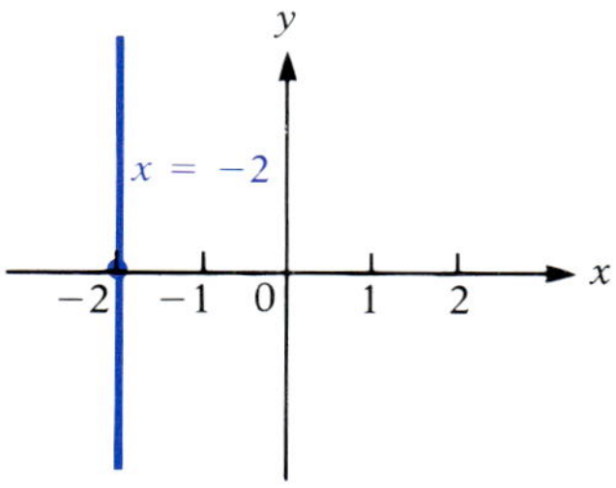

Figure 3.10 **The graph of $x = -2$.** □

EXAMPLE 35

Graph the equation $y = 3$.

SOLUTION

Again only one variable appears in this equation, this time, y. The implication is that no matter what x is, y is always 3. Thus two points that lie on the line are $(0, 3)$ and $(1, 3)$. (See Fig. 3.11.)

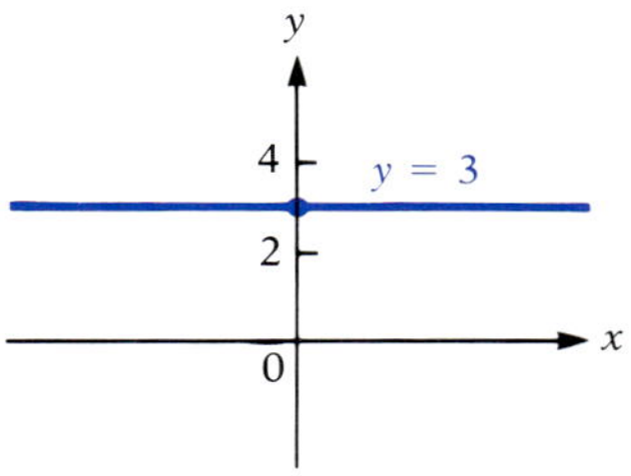

Figure 3.11 **The graph of $y = 3$.** □

Practice Problem

Graph the following equations.

a) $2x + 3y = 6$ b) $x = 4$ c) $y = -1$

Answers:

a) b) c)

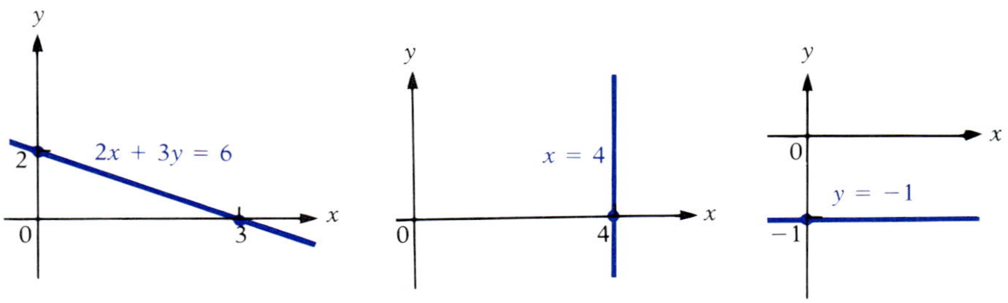

Let's see if we can generalize a bit. To graph an equation in which both variables appear, we must find two points on the line by choosing values for one variable and then solving for the other. We found it convenient to let each variable be zero in turn. You should notice (see Figs. 3.8 and 3.9) that these points always appeared on one of the axes. We call the points where the equation crosses the axes **intercepts.** The **x-intercept** is the point on the x-axis (horizontal axis) through which the line goes. It is found by setting the value of y to zero and solving for x. The **y-intercept** is the point on the y-axis through which the line goes; it is found by setting x to zero and solving for y. Let's see how this works.

EXAMPLE 36

Graph the equation $2x - 3y = 6$.

SOLUTION

See Fig. 3.12.

x	y	
3	0	(x-intercept)
0	−2	(y-intercept)

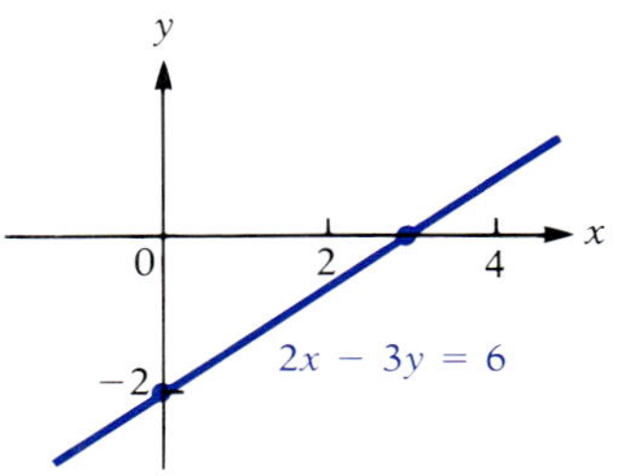

Figure 3.12
The graph of $2x - 3y = 6$.

□

EXAMPLE 37

Graph the equation $y = \frac{3}{2}x + 6$.

SOLUTION

See Fig. 3.13. Although we could manipulate this equation to put it into standard form, we need not.

x	y	
-4	0	(x-intercept)
0	6	(y-intercept)

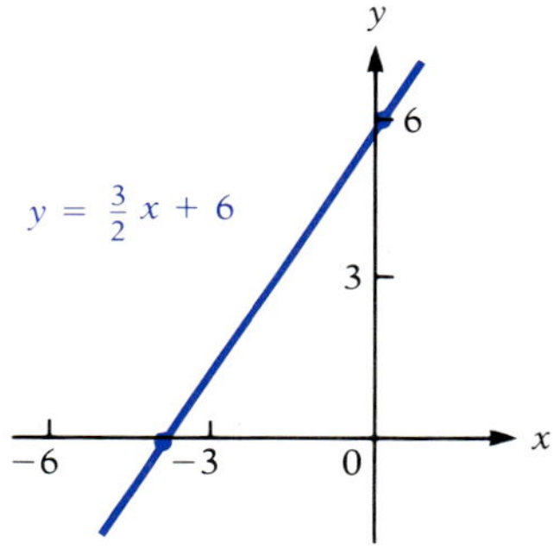

Figure 3.13
The graph of $y = \frac{3}{2}x + 6$.

□

EXAMPLE 38

Graph the equation $y = 2x$.

SOLUTION

See Fig. 3.14. Here we have a bit of a problem. Notice that both the x-intercept and the y-intercept are zero. In other words, the graph goes through the origin. In this case, we must determine another point on the graph. We can do this easily by arbitrarily choosing a value for x, say 1. Then since $y = 2x$, we must have $y = 2(1) = 2$.

x	y
0	0
1	2

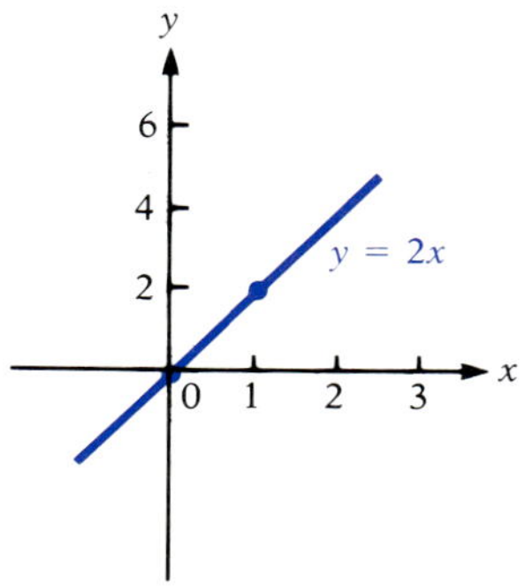

Figure 3.14
The graph of $y = 2x$.

□

Practice Problem

Graph the following equations.

a) $y = x$ **b)** $3x + y = 0$

Answers:

a)

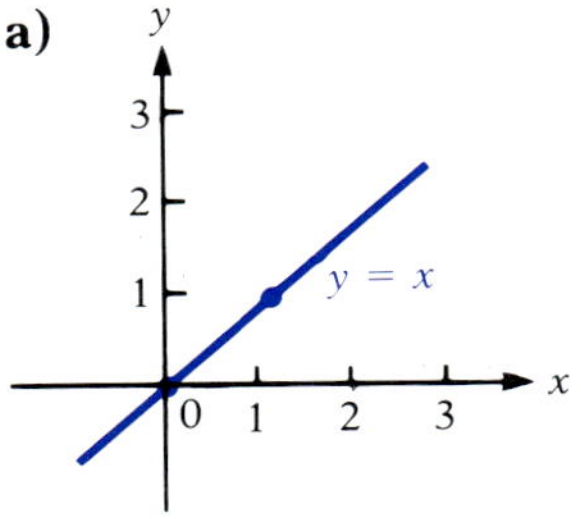

b)

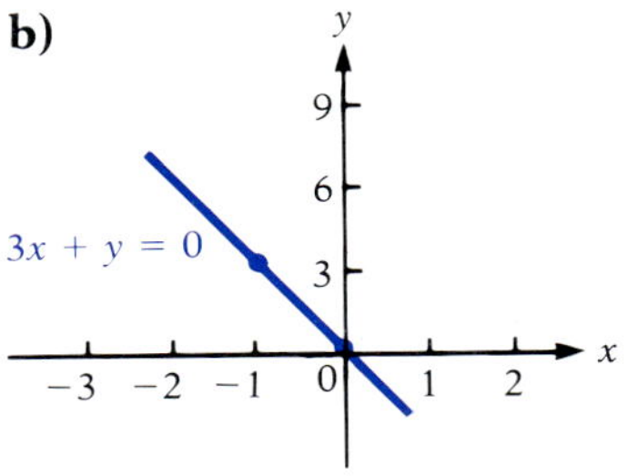

We also saw some equations in which only one variable appeared. If only x appears, the graph is a vertical line ($x = c$) and all points have the same value of x. If only y appears, the graph is a horizontal line ($y = c$) and all points have the same value of y. To summarize:

Graphing Linear Equations in Two Variables

1. If both variables appear in the equation, it can be graphed by finding the x-intercept (set $y = 0$ and solve for x) and the y-intercept (set $x = 0$ and solve for y). Draw the line through the two intercepts. (This approach will not work when the constant is zero. See Example 38.)
2. If only x appears, the line is vertical. Pick any two values for y (with the given value for x). Plot the points and draw a line through them.
3. If only y appears, the line is horizontal. Pick any two values for x (with the given value for y). Plot the points and draw a line through them.

Practice Problem

Graph the following equations.

a) $5x - 3y = 15$

b) $7x + 2y = 14$

c) $x = 5$

d) $y = 1$

Answers:

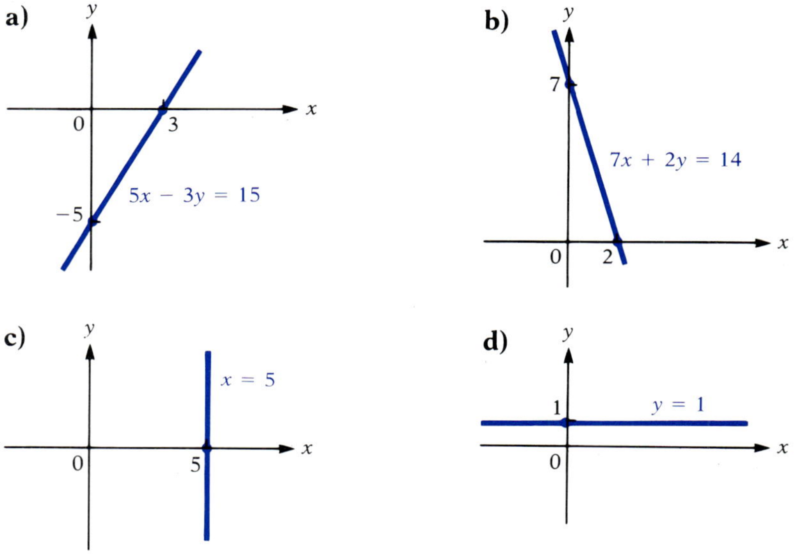

Straight Lines and Slopes

An important concept in dealing with a straight line is its steepness or slope. The **slope** of a straight line is defined to be

$$m = \text{slope} = \frac{y_2 - y_1}{x_2 - x_1} = \frac{\text{rise}}{\text{run}},$$

where (x_2, y_2) and (x_1, y_1) are two points on the line. The slope of a line is the same no matter which two points are used to compute the slope. (See Fig. 3.15.)

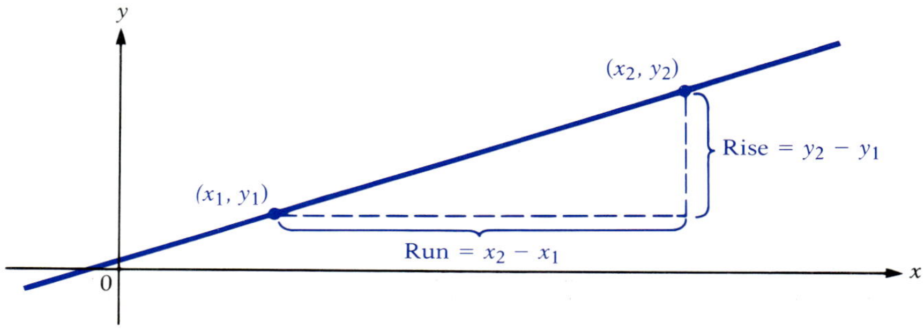

Figure 3.15 **The definition of slope.**

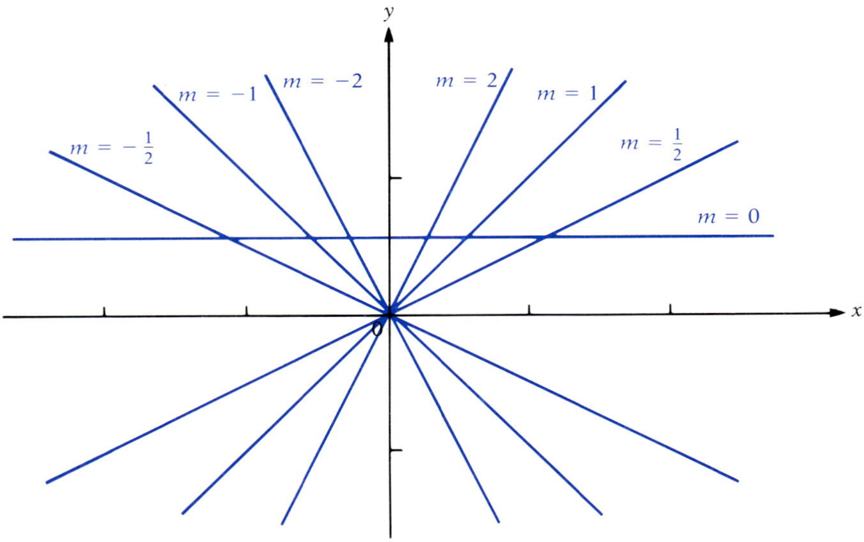

Figure 3.16 **Different slopes.**

The slope of a line can be positive, negative, or zero. When we examine a line from left to right, positive slope means the line is increasing (going up). A line with negative slope is decreasing (going down) from left to right. A line with a slope of zero is horizontal. (See Fig. 3.16.) The slope of a vertical line is not defined. Why?

EXAMPLE 39

Find the slope of the line through the points (2, 3) and (4, 6).

SOLUTION

Let $(x_2, y_2) = (4, 6)$ and $(x_1, y_1) = (2, 3)$. Then

$$m = \frac{y_2 - y_1}{x_2 - x_1} = \frac{6 - 3}{4 - 2} = \frac{3}{2} = 1.5.$$

Notice that we can reverse the definitions of the points: in other words, we could let $(x_2, y_2) = (2, 3)$ and $(x_1, y_1) = (4, 6)$ and we would get exactly the same slope. Try it! □

Practice Problem

Find the slopes of the lines through

a) (3, 4) and (6, 2) **b)** (2, 6) and (2, 0)
c) (1, 3) and (3, 3)

Answers: **a)** $m = -\frac{2}{3}$; **b)** m = undefined; **c)** $m = 0$

By making use of slope, there are alternative ways we can write the equation of a straight line. The first is called the **slope-intercept** form:

$$y = mx + b \qquad \textbf{Slope-intercept form}$$

because m is the slope of the line and b is the value of the y-intercept. Notice that when x is zero, then y is b, which means the line goes through the point $(0, b)$.

Suppose we know the line goes through the point (x_1, y_1) with a slope of m. By definition, the slope is rise ÷ run, so if we let (x, y) be any other point on the line, then we must have

$$m = \frac{y - y_1}{x - x_1} \quad \text{or} \quad y - y_1 = m(x - x_1)$$

which is called the **point-slope** form.

$$y - y_1 = m(x - x_1) \qquad \textbf{Point-slope form}$$

The slope-intercept form is especially useful for graphing a line, while the point-slope form is useful both for graphing a line and finding the equation of a line.

One last fact relating to slopes: Two lines are **parallel** if and only if they have the same slope. For example, the equations $y = 2x + 5$ and $y = 2x - 5$ have the same slope so they are parallel, while the lines $y = 2x + 5$ and $y = x + 5$ are not parallel. (See Fig. 3.17.)

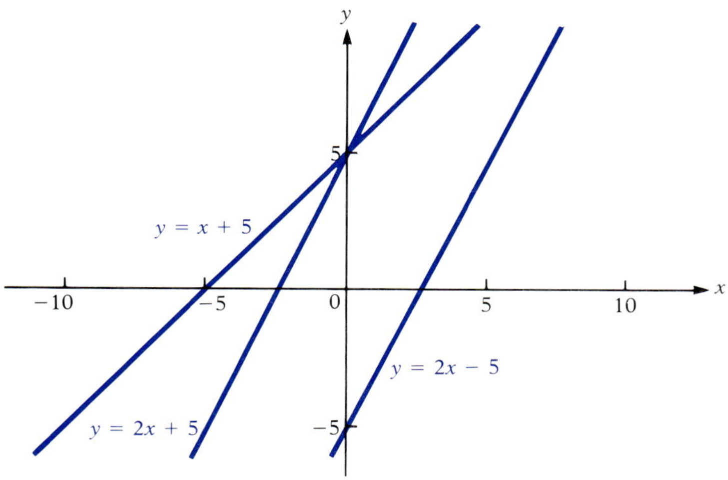

Figure 3.17 Parallel lines have the same slope.

Let's summarize the various forms we have for a linear equation in two variables.

Linear Equations in Two Variables

1. Standard form: $Ax + By = C$
2. Slope-intercept form: $y = mx + b$
3. Point-slope form: $y - y_1 = m(x - x_1)$
4. Vertical line: $x = c$
5. Horizontal line: $y = b$

Now let's take a look at how we can use the slope-intercept and point-slope forms.

EXAMPLE 40

Graph the equation $y = 2x + 1$.

SOLUTION

This equation has a slope of 2 and y-intercept (0, 1). A slope of 2 means that for a unit increase in x (a move of 1 unit to the right) y must increase by two units (a move of 2 units up). Thus the line must go through the y-intercept (0, 1) and the point (1, 3). (See Fig. 3.18.)

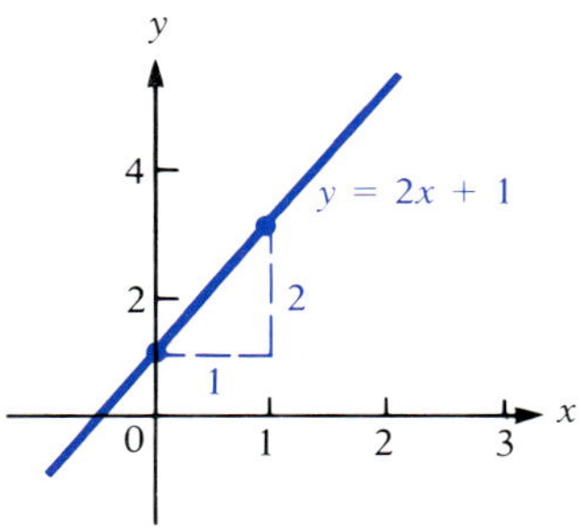

Figure 3.18 **The graph of $y = 2x + 1$.** □

EXAMPLE 41

Graph the equation $4x + 5y = 20$.

SOLUTION

First, let's put the equation in slope-intercept form by solving it for y. We obtain

$$y = -\frac{4}{5}x + 4,$$

which means the line has a slope of $-\frac{4}{5}$ and y-intercept (0, 4). A slope of $-\frac{4}{5}$ means that y must decrease 4 units for every 5 units that x increases. Since the line goes through the point (0, 4) (why?), it must also go through the point $(0 + 5, 4 - 4) = (5, 0)$. (Since the slope is $-\frac{4}{5}$, simply add 5 units to the x-coordinate and -4 units to the y-coordinate of the y-intercept.) (See Fig. 3.19.)

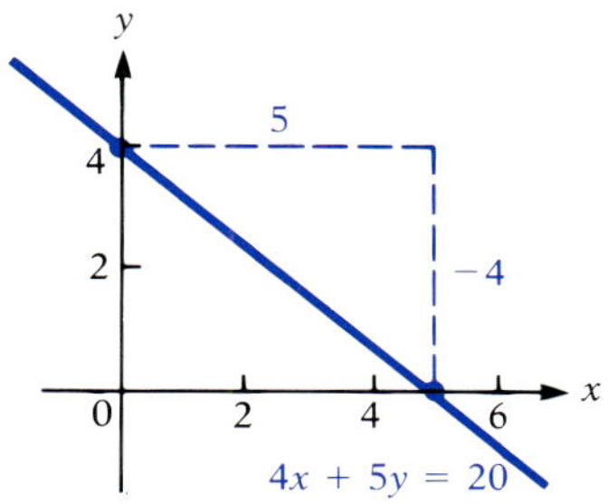

Figure 3.19 **The graph of $4x + 5y = 20$.** □

EXAMPLE 42

Find the equation of the line through the point (3, 3) with a slope of 2.

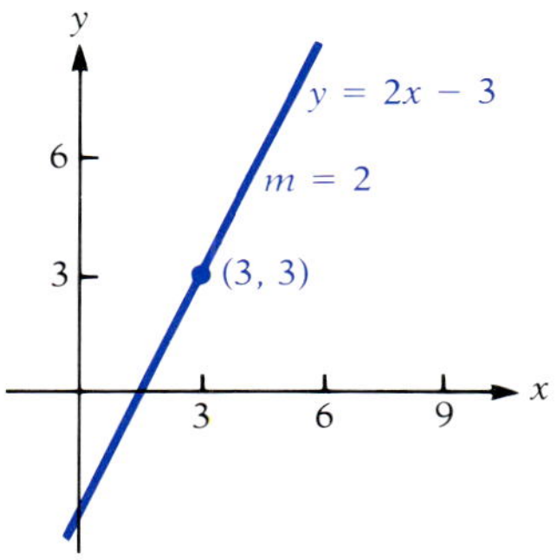

Figure 3.20 **The line through (3, 3) with $m = 2$.**

SOLUTION

We know the slope so we can use the point-slope form. We have $m = 2$ and $(x_1, y_1) = (3, 3)$ so

$$y - y_1 = m(x - x_1)$$

or

$$y - 3 = 2(x - 3).$$

We can also write this equation as

$$y = 2x - 3$$

or

$$2x - y = 3$$

by doing some algebra. (See Fig. 3.20.) □

EXAMPLE 43

Find the equation of the line through the points (3, 7) and (5, 1).

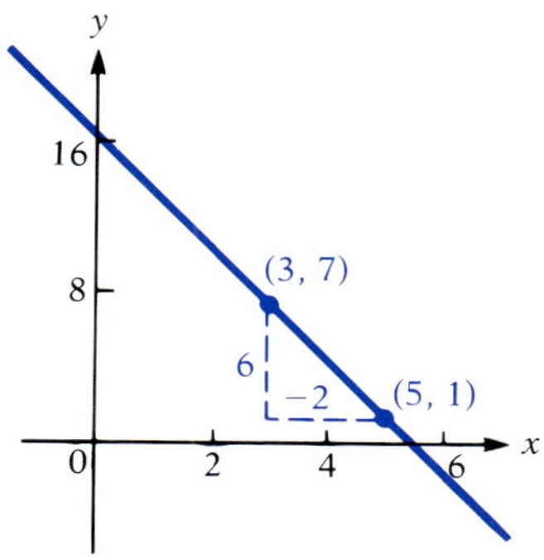

Figure 3.21 **The line through (3, 7) and (5, 1).**

SOLUTION

We would like to use the point-slope form of the line, but because we don't know the slope, we must first compute it. Then we can use the slope and pick one of the points to use in the point-slope form. Let

$$(x_2, y_2) = (5, 1) \quad \text{and} \quad (x_1, y_1) = (3, 7),$$

which tells us $m = -3$. Then we have

$$y - 7 = -3(x - 3).$$

This is equivalent to $y = -3x + 16$ and also to $3x + y = 16$. (See Fig. 3.21.) □

EXAMPLE 44

Find the equation of the line through the points (2, 6) and (4, 6).

SOLUTION

Again, we must compute the slope of the line before we use the point-slope form. Letting $(x_2, y_2) = (4, 6)$ and $(x_1, y_1) = (2, 6)$, we find that $m = 0$. Then we have

$$y - 6 = 0(x - 2),$$

which is, of course, equivalent to $y = 6$. This line is horizontal, which is obvious from Fig. 3.22. In fact, we could simply note the line is horizontal with a y-intercept of 6, which would give us the equation $y = 6$ directly, without computing the slope. It is always worthwhile to draw the line, especially when you are trying to determine its equation.

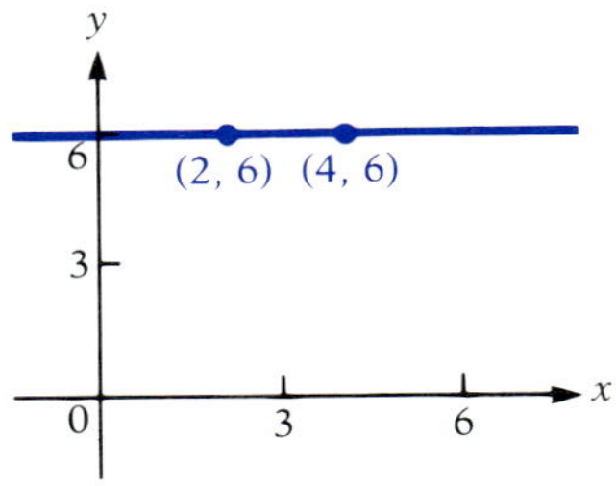

Figure 3.22 **The line through (2, 6) and (4, 6).** □

EXAMPLE 45

Find the equation of the line through the points (3, 2) and (3, 5).

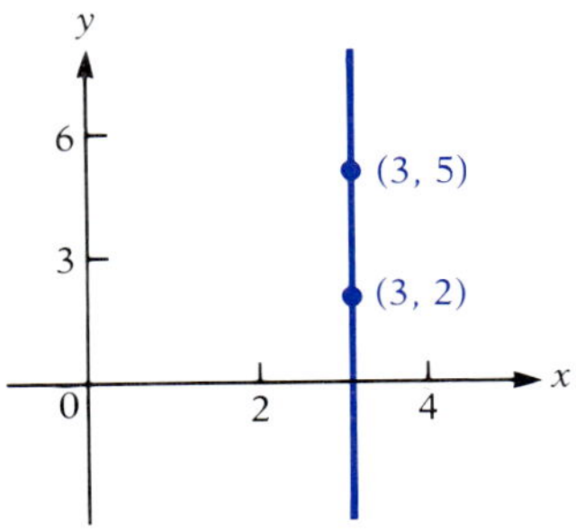

SOLUTION

If we proceed blindly to compute the slope of the line, we will shortly be in trouble, because we will find $m = 3 \div 0$, which is undefined. However, if we first draw the line, we can see very easily that it is vertical. (See Fig. 3.23.) Recalling the equation of a vertical line to be $x = c$, the equation for this line must be $x = 3$.

Figure 3.23 **The line through (3, 5) and (3, 2).** □

Practice Problem

Solve the following.

a) Graph the line $2x - 4y = 8$ by using the slope-intercept form.
b) Find the equation of the line through the points (2, 2) and (4, 1).
c) Find the equation of the line through the points (1, 3) and (−2, 3).
d) Find the equation of the line through the points (−4, −2) and (−4, 3).

Answers:

a) The slope-intercept form is $y = \frac{1}{2}x - 2$.

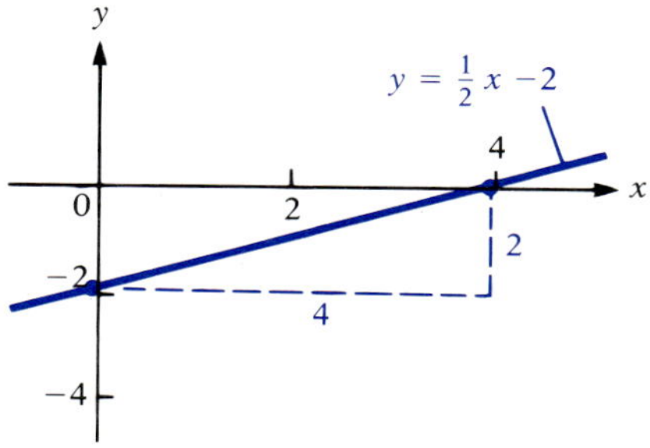

b) The equation is $y = -\frac{1}{2}x + 3$.

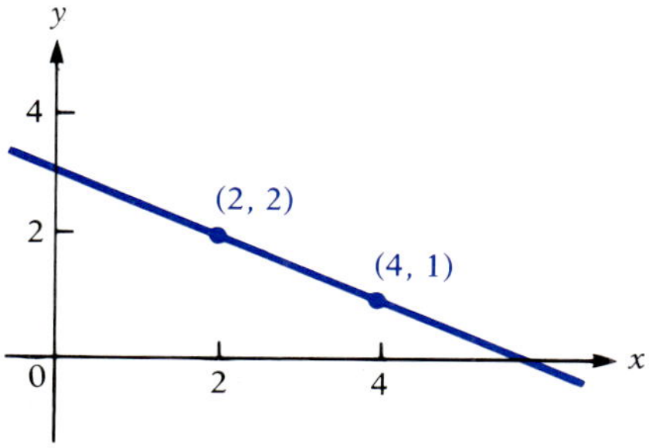

c) The equation is $y = 3$.

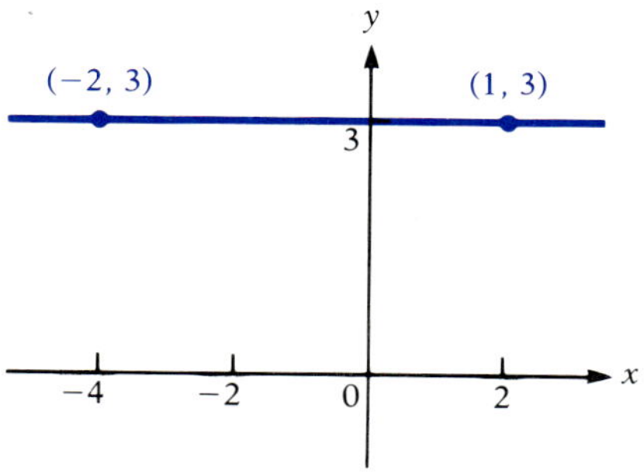

d) The equation is $x = -4$.

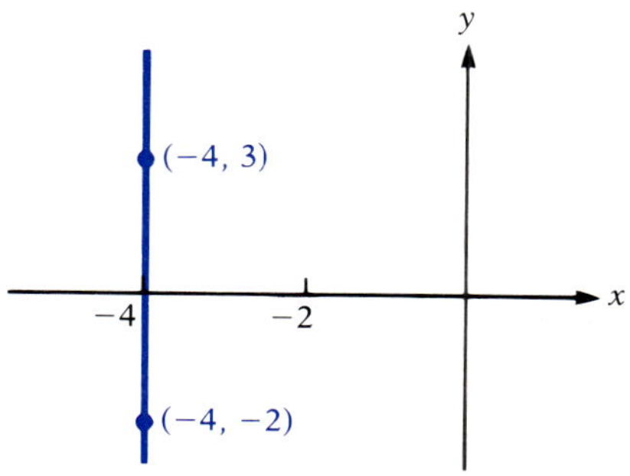

Now we can solve the last part of the Problem from Life.

EXAMPLE 46

Problem from Life, part (b) At sea level, water freezes at 0° Celsius and 32° Fahrenheit and boils at 100° Celsius and 212° Fahrenheit. Derive the linear equation that relates temperatures expressed in degrees Celsius to those expressed in degrees Fahrenheit.

SOLUTION

We want to derive an equation that relates temperatures on one scale to temperatures on another. The best way to begin is to draw a graph to help us locate the points. Let's put degrees F (Fahrenheit) on the horizontal axis and degrees C (Celsius) on the vertical scale. (See Fig. 3.24.)

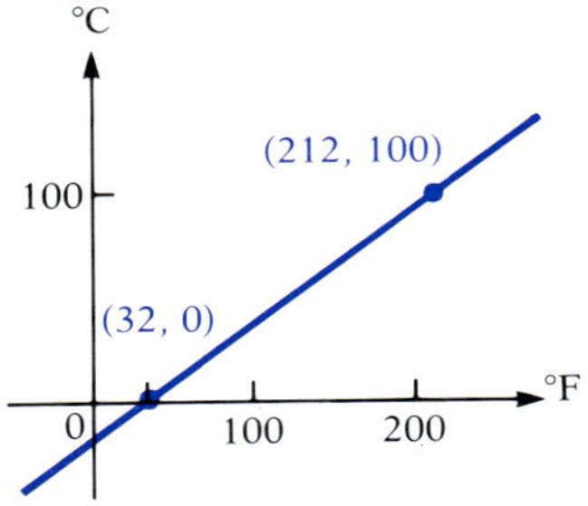

Figure 3.24 **Relating temperatures in Fahrenheit degrees to Celsius degrees.**

Now we can recognize our problem as one requiring us to determine the equation of a line through two points. One point is (32, 0) (freezing) and the other is (212, 100) (boiling). As we did earlier, we must first determine the slope. Let $(F_1, C_1) = (32, 0)$ and $(F_2, C_2) = (212, 100)$, which gives us $m = \frac{5}{9}$. Using the point-slope form with (F_1, C_1) and m gives us

$$C - 0 = \frac{5}{9}(F - 32)$$

or

$$C = \frac{5}{9}(F - 32). \quad \square$$

Linear Inequalities in Two Variables

We studied earlier how to deal with linear inequalities in one variable. Let's turn our attention to inequalities in two variables, for example, $3x - 2y > 15$ and $y \geq 6x - 12$. As with equations in two variables, the solution set of a linear inequality in two variables contains many points. Our desire is to teach you how to graph these solution sets. This turns out to be easy, once you've learned how to deal with linear equations in two variables.

It can be shown (although we won't) that any line divides a plane into two halves. For example, the graph of a linear equation in two variables divides the plane into the half above the line, called the **upper half-plane,** and the half below the line, called the **lower half-plane.** (Vertical lines divide the plane into left and right half-planes. (See Fig. 3.25.)

The solution set of a linear inequality in two variables is one of the half-planes. Thus graphing the solution set of such an inequality involves determining which half-plane (upper or lower) is appropri-

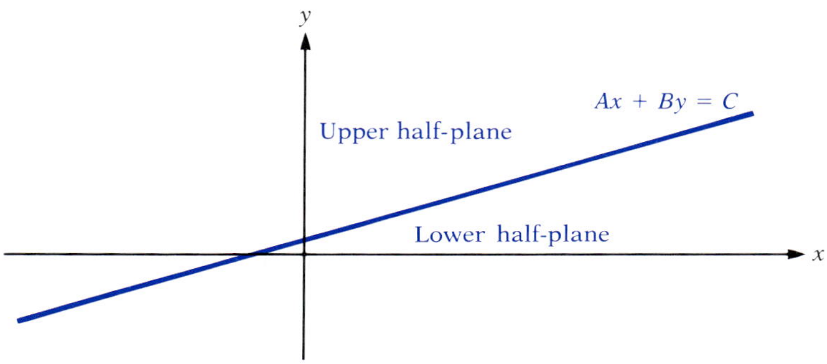

Figure 3.25 **Lines divide planes into half-planes.**

ate. We can do so by simply choosing an arbitrary point in the plane that is *not* on the dividing line and checking to see whether it satisfies the given inequality. If the test point does satisfy the inequality, it must lie in the correct half-plane. If the test point does not satisfy the inequality, the other half-plane is the correct one. Although any point will work as a test point, the origin (0, 0) is usually the easiest point to use.

With respect to the solution set, what is the difference between "strict" inequalities (those written with $<$ or $>$) and "less than or equal to" or "greater than or equal to" inequalities (those written with $\leq$ or $\geq$)? The difference is that the solution set for a strict inequality does *not* include the line dividing the plane, while for a $\leq$ or $\geq$ inequality the solution set does include the dividing line.

EXAMPLE 47

Graph the inequality $3x - 2y > 15$.

SOLUTION

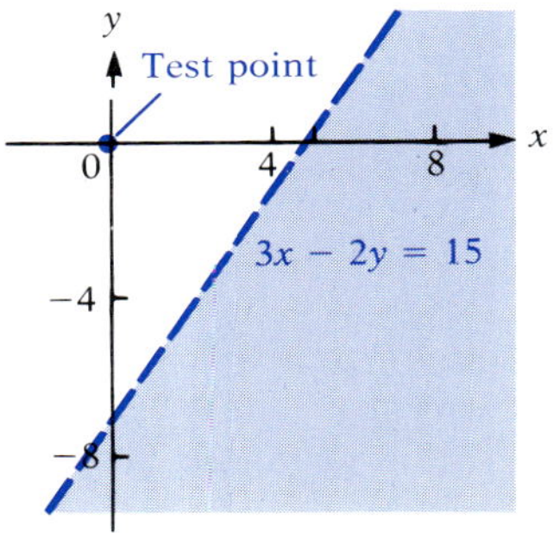

First, graph the equation $3x - 2y = 15$ to determine the half-planes. (See Fig. 3.26.) Since the given inequality is strict (in this case, $>$) we use a broken line to indicate that the solution set does not include points on the boundary (points that satisfy the equation). Next, let's use the point (0, 0) as our test point, noting that (0, 0) is in the upper half-plane. Substituting (0, 0) into the inequality gives us $3(0) - 2(0) = 0 \not> 15$. In other words, (0, 0) does not satisfy the inequality. Therefore, the solution set is the lower half-plane.

Figure 3.26 **The inequality $3x - 2y > 15$.** □

EXAMPLE 48

Graph the inequality $y \geq 6x - 12$.

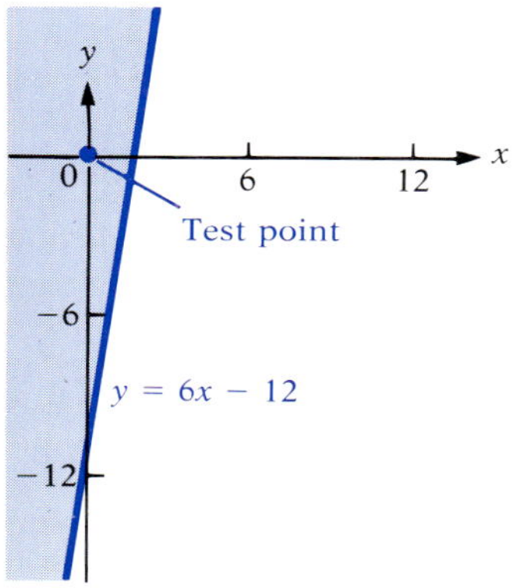

SOLUTION

Again, we begin by graphing an equation, in this case, $y = 6x - 12$. (See Fig. 3.27.) Using the test point (0, 0) we have $0 \geq 6(0) - 12$ so the point (0, 0) satisfies the inequality. Since (0, 0) lies in the upper half-plane, the solution set is the upper half-plane.

Figure 3.27 **The inequality $y \geq 6x - 12$.** □

We can summarize the procedure as follows.

Graphing Linear Inequalities in Two Variables

1. Make believe the given inequality is an equation. Draw the graph of the equation to determine the two half-planes. If the original inequality is "strict," the equation should be drawn with a broken line. Otherwise, use a solid line.
2. Choose a test point anywhere in the plane *not* on the line of the equation [we recommend the point (0, 0)] and substitute it into the inequality. [If (0, 0) is on the line of the equation, choose a different point.]
3. If the test point satisfies the inequality, the solution set is the half-plane containing the test point. Otherwise, the solution set is the half-plane not containing the test point.

Practice Problem

Graph the following.

a) $2x - 3y < 6$ **b)** $y \leq -3x + 5$

Answers:

a)

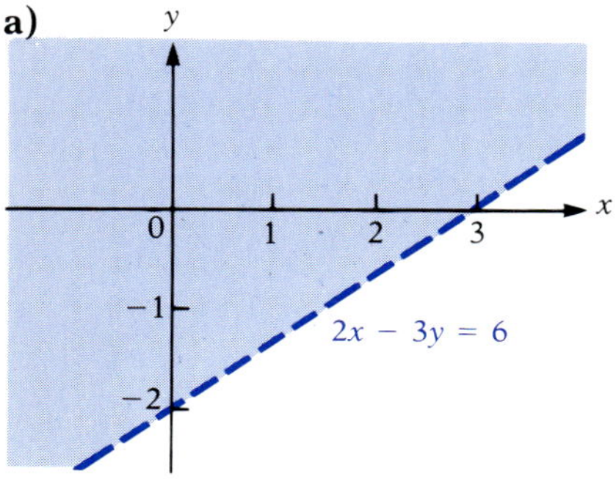

b)

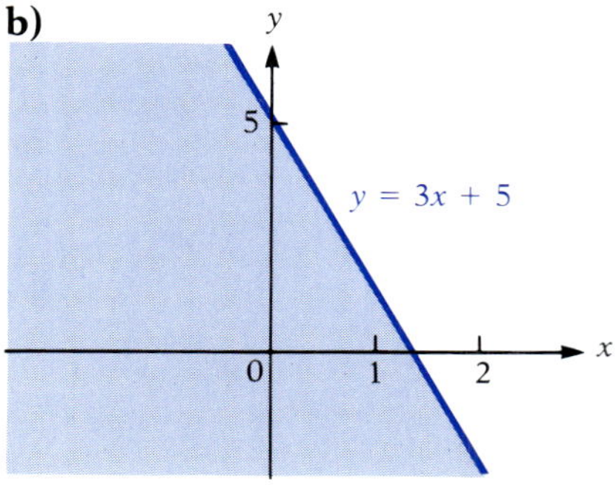

EXERCISE SET 3.4

For each of the following, give the slope and the y-intercept. In addition, interpret the slope (for example, the line goes up from left to right) and graph the equation.

1. $y = -2x + 3$

2. $y = -5x$

3. $y = 4x + 1$

4. $y = -6x + 30$

5. $y = 7x + 3$

6. $x + y = 8$

7. $x - y = 15$
8. $2x - 3y = 7$
9. $\frac{2}{3}x - \frac{4}{3}y = -\frac{7}{3}$
10. $5x + 3y = -6$
11. $x = -2$
12. $y = -4$
13. $2 - y = 0$
14. $-x - 4 = 3$
15. $3x + 4y - 25 = 0$
16. $10 - 20x - 5y = 0$
17. $8 - y = 4x$
18. $-20 + 15y - 10x = 0$

For each of the following, find the equation of the line (in both slope-intercept and standard form) with the given characteristics.

19. $m = 2$ through the point $(3, 5)$
20. $m = \frac{1}{3}$ through the point $(2, 5)$
21. $m = -1$ through the point $(4, 3)$
22. $m = -3$ through the point $(-2, 1)$
23. $m = -\frac{1}{5}$ through the point $(7, -2)$
24. $m = 3$ through the point $(-2, -3)$
25. $m = 1$ through the point $(5, 8)$
26. $m = \frac{1}{4}$ through the point $(-3, 5)$
27. $m = 0$ through the point $(3, 3)$
28. $m = 0$ through the point $(-2, 3)$

For each of the following, find the slope and the equation of the line through the points (in both slope-intercept and standard forms).

29. $(1, 2)$ and $(3, 4)$
30. $(-1, -2)$ and $(3, 4)$
31. $(-2, 3)$ and $(-5, 4)$
32. $(-4, -3)$ and $(3, -2)$
33. $(-5, -3)$ and $(2, -4)$
34. $(9, 5)$ and $(6, -2)$
35. $(7, 5)$ and $(0, 0)$
36. $(0, 5)$ and $(7, 1)$
37. $(3, 1)$ and $(5, 1)$
38. $(-5, 2)$ and $(-5, -1)$
39. $(3, -3)$ and $(-2, -3)$
40. $(2, 3)$ and $(5, 8)$
41. $(4, -2)$ and $(4, -4)$
42. $(6, -2)$ and $(7, -3)$

Graph the following inequalities.

43. $3x + y \leq 7$
44. $x + 2y \leq 16$
45. $x - y \geq 0$
46. $-x + 2y \leq 2$
47. $2x + 2y \geq 4$
48. $x - y \geq -2$
49. $2x - y < 6$
50. $-3x + 2y > 6$

51. Rich bought a computer system for his home business for \$5500. The system will have a value of \$0 after 5 years. For tax purposes, Rich can claim the yearly decline in value as a business expense called **depreciation.**

a) Write the linear equation that relates the value of the computer system in dollars to its age in years. [*Hint:* Find the equation of the line that goes through the points (0, 5500) and (5, 0)].

b) What is the value of the system after 3 years?

c) Graph the equation for values of age from 0 through 5 years.

52. Sue and Joan bought an apartment building for \$150,000. They expect it to be worth \$30,000 after 30 years.

a) Write the linear equation that states the value of the building in dollars given its age in years.

b) What is the value of the building after 5 years? After 20 years?

c) Graph the equation for values of age from 0 through 30 years.

53. Economists in a certain city have determined that they can predict the amount of coffee people will buy given its price, as follows:

Price	*Demand*
\$2.00	1000 lb
\$2.25	800 lb
\$2.50	600 lb
\$2.75	400 lb

a) Determine the equation of the line that passes through these points. This is called a **demand equation.**

b) How much coffee will be purchased at a price of \$2.65? At a price of \$2.15?

c) What must the price be for the amount demanded to be zero? (It is unlikely that this would really occur at this price. In many real cases, the demand equation will not be linear. However, if we restrict ourselves to reasonable price levels, our linear equation may be a good approximation of the true situation.)

54. In the same town, economists have determined the amount of coffee that stores will be willing to supply at a given price. This is called a **supply equation.** The data are

Price	*Amount Supplied*
\$2.00	400 lb
\$2.20	500 lb
\$2.40	600 lb
\$2.60	700 lb

a) Determine the supply equation.

b) How much coffee will be supplied at \$2.45? At \$2.65?

c) Draw the supply equation and the demand equation from Problem 53 on the same graph. Do they cross? Where? What do you think will happen at the point where they cross?

CHAPTER REVIEW

IMPORTANT TERMS AND SYMBOLS

3.1 Modeling Equations

equation Statement that two expressions are equal; often uses variables (unknowns).

modeling with equations

1. Understand the problem.
2. Draw a picture when appropriate.
3. Think about how you would proceed if you had all the data.
4. Write out the relationships in words.
5. Assign variables.
6. Translate the relationships into symbols.

solution set of an equation Set of all values of the variable(s) that satisfy it (make it true).

solve an equation Find its solution set.

3.2 Linear Equations and Inequalities in One Variable

linear equation in one variable Has the standard form $ax + b = c$. Valid operations for linear equations:

1. Addition property: If $a = b$, then $a + c = b + c$.
2. Subtraction property: If $a = b$, then $a - c = b - c$.
3. Multiplication property: If $a = b$, then $ac = bc$.
4. Division property: If $a = b$, then $a/c = b/c \quad (c \neq 0)$.

linear inequality in one variable Has standard form $ax + b < c$ or $ax + b \leq c$ or $ax + b > c$ or $ax + b \geq c$. Valid operations for linear inequalities:

1. Addition property: If $a < b$, then $a + c < b + c$.
2. Subtraction property: If $a < b$, then $a - c < b - c$.
3. Multiplication properties:
   i. If $a < b$ and c is positive, then $ac < bc$.
   ii. If $a < b$ and c is negative, then $ac > bc$.
4. Division properties:
   i. If $a < b$ and c is positive, then $a/c < b/c$.
   ii. If $a < b$ and c is negative, then $a/c > b/c$.

(These properties also hold for $\leq$, $>$, and $\geq$ relationships.)

Double inequality: involves two inequality signs; e.g., $3 < x < 5$.

3.3 Quadratic Equations and Inequalities

quadratic equation Has standard form $ax^2 + bx + c = 0$.
Solution approaches:

1. Take square root of both sides.
2. Factor the left side.
3. Quadratic formula: $x = \dfrac{-b \pm \sqrt{b^2 - 4ac}}{2a}$

Discriminant ($b^2 - 4ac$) determines number of solutions to quadratic equation:

$$\text{When} \begin{cases} b^2 - 4ac > 0, \text{ there are two real solutions,} \\ b^2 - 4ac = 0, \text{ there is one real solution,} \\ b^2 - 4ac < 0, \text{ there is no real solution.} \end{cases}$$

quadratic inequality Has standard form $ax^2 + bx + c <$ (or $\le$) 0.
sign graph Used to solve quadratic inequalities that can be factored; plot on a number where each factor is positive and negative.

3.4 Linear Equations and Inequalities in Two Variables

Cartesian (rectangular) coordinate system A system of two mutually perpendicular number lines, called the coordinate axes, which is commonly used when graphing.
coordinate axes The two mutually perpendicular lines in a Cartesian coordinate system:
x-axis: the horizontal axis,
y-axis: the vertical axis.
horizontal line Has the equation $y = b$, where b is a constant.
linear equation in two variables Is usually written in one of three forms:

Standard form: $Ax + By = C$, where $A > 0$,
Slope-intercept form: $y = mx + b$,
Point-slope form: $y - y_1 = m(x - x_1)$.

linear inequality in two variables Has standard form $Ax + By < C$ (or $\le$).
Graphing linear inequalities in two variables:

1. Make believe the inequality is an equation and graph the equation. If the original inequality is "strict," draw the equation with a broken line. Otherwise, use a solid line.
2. Choose a test point anywhere in the plane *not* on the line of the equation [we recommend (0, 0)] and substitute it into the inequality.
3. If the test point satisfies the inequality, the solution set is the half-plane containing the test point. Otherwise, the solution set is the half-plane *not* containing the test point.

lower half-plane A linear inequality in two variables divides the plane into two regions, commonly called the lower-half plane and the upper half-plane.
origin The point in a Cartesian coordinate system where the axes cross; it has coordinates (0, 0).

quadrant One of the four regions into which the plane is divided by Cartesian coordinates; numbered I, II, III, and IV, proceeding counterclockwise from the upper-right quadrant.

slope of a nonvertical straight line (represented by m)

$$m = \frac{\text{rise}}{\text{run}} = \frac{y_2 - y_1}{x_2 - x_1}$$

where (x_1, y_1) and (x_2, y_2) are two points on the line.

upper half-plane See lower half-plane.

vertical line Has the equation $x = c$, where c is a constant.

REVIEW PROBLEMS

Solve the following.

1. $5x + 4 = 3x - 1$

2. $7(2 - x) = 3(x + 1)$

3. $\frac{4c}{3} + \frac{3c}{2} = 15c + 1$

4. $-4(2 - 3y) = 3[-2 - 5(1 - y)]$

5. $y + \frac{1}{10} > \frac{3}{10} - \frac{y}{5}$

6. $1 \le 3x - 4 \le 6$

7. $\frac{0.6w}{5} + 2(3 - w) \ge \frac{w}{5} - 7\left(1 - \frac{w}{10}\right)$

8. $-2 \le -3(2 - 5s) \le 4$

9. $x^2 - 5x - 14 = 0$

10. $(x - 4)^2 = 1$

11. $12x^2 - 36x + 12 = 0$

12. $-(x - 3)^2 + 2x - 1 = 0$

13. $5x^2 - 7x - 3 = 0$

14. $x^2 - 9 \ge 0$

15. $20x^2 + 2x - 2 \le 0$

16. $5x^2 + 8x + 3 < 0$

Graph the following equations. Determine the slope and both the x- and y-intercepts.

17. $y = 4x + 2$

18. $y = (-2x/3) + 1$

19. $2x + y = 5$

20. $3(x - y) - 15 = 0$

Find the equation of the lines (in both slope-intercept and standard form) with the following characteristics.

21. $m = 3$ through the point (1, 1)

22. $m = -4$ through the point (4, −2)

23. $m = 0$ through the point (5, 5)

24. through the points (1, 1) and (4, 7)

25. through the points (−10, 5) and (5, −10)

26. through the points (−3, −2) and (−3, 17)

Graph the following inequalities.

27. $3x + 2y \le 12$

28. $x + y > -2$

29. $2x - 3y - 4 \ge 0$

30. $-2(x - 3y) + 15 < 0$

4 FUNCTIONS AND GRAPHS

a) The cost of mailing a first-class letter that weighs x ounces is given by the function

$$\text{cost} = f(x) = \begin{cases} 22, & 0 < x \leq 1, \\ 39, & 1 < x \leq 2, \\ 56, & 2 < x \leq 3, \\ & \text{etc.} \end{cases}$$

In other words, the postage is 22¢ for the first ounce (or fraction thereof) plus 17¢ for each additional ounce (or fraction thereof). Graph this function and determine the cost of mailing an envelope weighing 3.5 ounces. (To be solved in Example 29, Section 4.2.)

b) A crew of painters can paint a house in 45 working hours. How many hours will it take if one of the painters quits so that the crew is reduced to four painters? (To be solved in Example 32, Section 4.3.)

PROBLEM FROM LIFE

INTRODUCTION

In the previous chapter we talked about how we could use equations in modeling, and we also discussed how we could graph linear equations and inequalities in two variables. In this chapter we'll discuss functions, how functions relate to equations and modeling, and how to graph functions. Next we'll discuss variation, which will help us to model and solve a variety of problems (for example, part (b) of the problem from life). We'll close the chapter with a discussion of some very special functions called sequences.

4.1 FUNCTIONS

Relations and Functions

The concept of a correspondence between the members of two sets is simple yet quite powerful. For example, associated with each car is a license plate, and associated with every driver is a driver's license. Sometimes there may be a general rule that will enable us to state in a simple fashion the correspondence between two sets. For example, a line manager in a factory would like to know the relationship between the number of units produced and the amount of labor necessary to produce the next unit. A home buyer would like to know the relationship between interest rate and monthly mortgage payment. A chemist would like to know the relationship between the temperature and the rate at which a chemical reaction proceeds. A basketball coach would like to know if there is a relationship between biorhythms and performance in a game situation. The common thread running through all these examples is *the matching of members from one set with those of another.* We refer to the first set as the **domain** of the correspondence and to the second set as the **range.** A **relation** is a correspondence between a first set (the domain) and a second set (the range) such that to each member of the domain there corresponds one or more members in the range.

EXAMPLE 1

Consider the correspondence shown in Table 4.1.

SOLUTION

For the relation in Table 4.1, to each member of the domain (the different people) there corresponds exactly one member of the range (a weight). Here the domain contains five members while the range contains four members. Notice that two (or more) members of the domain can correspond to the same member of the range. For instance, both Jane Doe and Sam Eller correspond to the weight 132.

TABLE 4.1 Students and their weights

Name	Weight
Alan Adams	150
Sally Barnes	120
Robert Carpenter	185
Jane Doe	132
Sam Eller	132

Now consider the correspondence shown in Table 4.2. In this relation each member of the domain (the people) corresponds to one or more members of the range (current courses). Alan Adams, Sally Barnes, and Jane Doe are all taking more than one course so that each of these people corresponds to two or more members of the range. (Once again, notice there can be more than one member of

TABLE 4.2 Students and their semester courses

Name	Courses this term
Alan Adams	Anthropology Sociology College mathematics
Sally Barnes	Chemistry Calculus Physics
Robert Carpenter	Accounting
Jane Doe	Geology Physical education
Sam Eller	Physical education

the domain that corresponds to a given member of the range. For example, both Jane Doe and Sam Eller correspond to the course Physical education). □

As Example 1 demonstrates, a relation links members from one set (the domain) to members of another set (the range). Sometimes each member of the domain is linked (corresponds) to only one member of the range (as in Table 4.1), while in other cases a relation can link a member of the domain to several members of the range (as in Table 4.2).

Next, consider a home buyer who wants to know the relationship between interest rates and the resulting monthly mortgage payments. Ideally, each member of the domain (a specific interest rate) should correspond to only one member of the range (the monthly mortgage payment). Otherwise, the home buyer would be very confused. Similarly, a chemist trying to relate temperature to the rate at which a reaction proceeds would like the relationship to link each member of the domain (the temperature) to only one member of the range (a reaction rate). If this didn't happen, the chemist would not know the rate of reaction, which could lead to disastrous results (such as an explosion).

In general, models require a relation between the domain and the range in which each member of the domain is related to only one member of the range. In other words, we want our models to give us specific predictions. A **function** is a relation between its domain and range such that to each member of the domain there corresponds exactly one member of the range. Thus all functions are relations, but some relations are not functions. (See Fig. 4.1.)

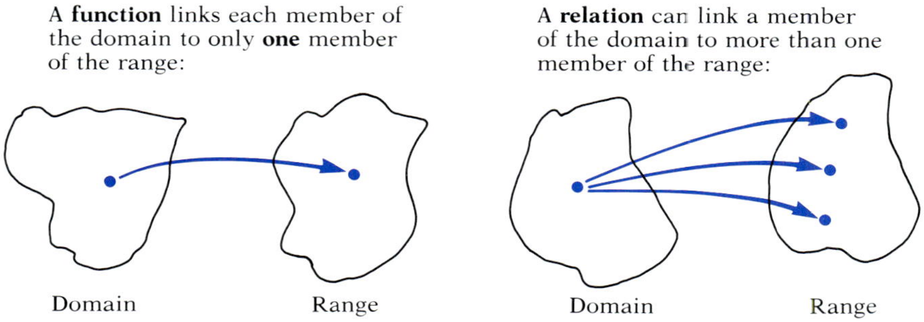

Figure 4.1 Functions and relations.

Functional Notation

Since a function (or relation) is a correspondence between the members of one set and the members of another, the sets need not consist of real numbers. For example, consider the relation between students and their courses in Example 4.1. However, we are primarily interested in functions that relate real numbers in the domain to real numbers in the range. When a function links numbers (in the domain) to numbers (in the range), it is generally stated in the form of an equation

$$y = f(x),$$

which is read "y equals f of x." This states that y is the member of the range that corresponds to the member x of the domain. We call x the **independent variable** and y the **dependent variable.**

When there is no explicit mention made of the domain, we assume that the domain is the largest subset of real numbers that will result in values of the range that are real numbers. (No, we are not being picky. We can't let every domain be the set of all real numbers because in some cases this won't work. For example, let $y = f(x) = \sqrt{x}$. In this case, we must restrict the domain to be only nonnegative real numbers.)

A simple way to think of a function is to picture it as a machine. We put the independent variable into the function machine, push the button to turn it on, and out comes the value of the dependent variable. To evaluate $f(x)$ for a particular value, say 4, we replace x in the definition of the function by 4. Thus, as Fig. 4.2 shows, when we evaluate the function

$$f(x) = 3x - 2$$

for $x = 4$, we replace x by 4 to obtain

$$f(4) = 3 \cdot 4 - 2 = 10.$$

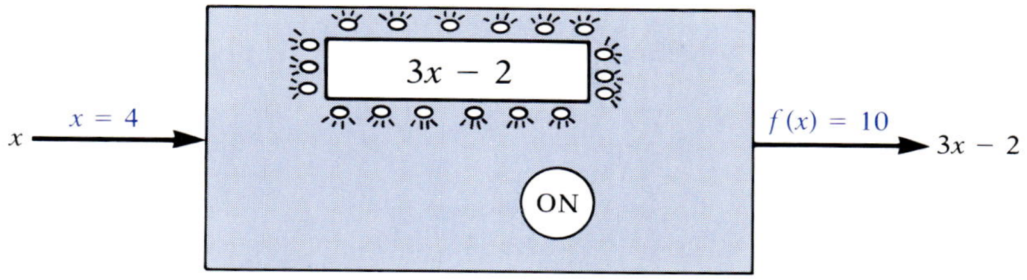

Figure 4.2 **The function $3x - 2$ pictured as a machine.**

Now let's look at some different functions.

EXAMPLE 2

Consider the function $y = f(x) = 3$. What are its domain and range?

SOLUTION

This function is called a **constant function** because the value of y is always 3, no matter what the value of x is. Thus, for instance, $f(0) = 3$ and so would $f(4)$ and $f(-10)$. The domain of this function is the set of all real numbers and its range is the number 3. □

EXAMPLE 3

Now consider the function $y = f(x) = 2x + 3$:

When $x = 0$, $y = f(0) = 2(0) + 3 = 3$.
When $x = 1$, $y = f(1) = 2(1) + 3 = 5$.
When $x = 5$, $y = f(5) = 2(5) + 3 = 13$.
When $x = -2.5$, $y = f(-2.5) = 2(-2.5) + 3 = -2$.

What is its domain?

SOLUTION

The domain of this function is the set of all real numbers. It is not obvious in this example, or in general, what the range of a function is. □

EXAMPLE 4

In the United States, it is quite common to see temperatures reported in both degrees Fahrenheit and Celsius. As we showed in Chapter 3 (without using functional notation), there is a very simple functional relationship between the two scales of measurement:

$$C = T(F) = \frac{5}{9}(F - 32).$$

For example, under normal conditions water freezes at 32°F. The corresponding Celsius temperature is

$$\begin{aligned} C &= T(32) \\ &= \frac{5}{9}(32 - 32) = 0. \end{aligned}$$

We might consider a temperature of 72°F to be very pleasant. The corresponding Celsius temperature is

$$\begin{aligned} C = T(72) &= \frac{5}{9}(72 - 32) \\ &= 22.22 \quad \text{(approximately)}. \end{aligned}$$

Since the lowest possible temperature (absolute zero) is approximately −456°F, we will consider the domain of this function is the set of all real numbers greater than or equal to −456.

What is this function's range?

SOLUTION

The range is the set of all real numbers greater than or equal to −273 (the Celsius temperature that corresponds to absolute zero).

Notice we referred to the independent variable (the temperature in degrees Fahrenheit) as F and to the dependent variable (the corresponding temperature in degrees Celsius) as C. It is not necessary that we restrict ourselves to the letters x and y for the independent and dependent variables. Moreover, it is not necessary that we call the function f. □

EXAMPLE 5

Of course, a functional relationship need not be so simple. Consider the function

$$y = f(x) = 3x^3 - 2x^2 + 2x - 4.$$

Some of its functional values are

$$\begin{aligned} f(1) &= 3(1)^3 - 2(1)^2 + 2(1) - 4 = -1, \\ f(4) &= 3(4)^3 - 2(4)^2 + 2(4) - 4 = 164, \\ f(-2) &= 3(-2)^3 - 2(-2)^2 + 2(-2) - 4 = -40. \end{aligned}$$

What is the domain of this function?.

SOLUTION

The domain of this function is the set of all real numbers. □

Practice Problem

Evaluate the following functions at the points $x = -5, -3, -1, 1, 3$, and 5.

a) $y = f(x) = 2x + 1$ **b)** $y = f(x) = 7$

c) $y = f(x) = 3x^2 - 1$ **d)** $y = f(x) = x^2 - 5x + 6$

Answers: **a)** $-9, -5, -1, 3, 7, 11$; **b)** 7, 7, 7, 7, 7, 7; **c)** 74, 26, 2, 2, 26, 74; **d)** 56, 30, 12, 2, 0, 6

In Example 4, we noted that we need not restrict ourselves to just using x and y and f in writing the definition of a function. The reason for this is simple: The letters (or other symbols) we use do *not* matter. We must use some symbols in order to define the function, but our choice is immaterial. For example, the functions $y = f(x) = 7x + 2$ and $y = f(w) = 7w + 2$ are identical. In each case, the independent variable is multiplied by 7 and then 2 is added to the result. Thus as long as we use the same value for x and w (for the independent variable), the computed value of y (the dependent variable) will be the same. The fact that we use the letters x and y to represent the variables is simply by convention. In fact, we could use any letters or symbols. For example, we could just as easily write the function $y = f(\&) = 7\& + 2$. The important point is that we know how to evaluate the function for different values of the independent variable: We simply replace its general symbol by a specific value. When the variables have some real meaning (for example, the temperature in degrees Fahrenheit and in degrees Celsius), we generally use symbols that help us relate to them (for example, F and C rather than x and y).

When we evaluate a function, we don't always restrict ourselves to replacing the independent variable by a real number.

EXAMPLE 6

What are some of the functional values of $y = f(x) = 3x^2 - 2x - 4$?

SOLUTION

Some examples of the different functional values are

$$f\left(\frac{1}{y}\right) = 3\left(\frac{1}{y}\right)^2 - 2\left(\frac{1}{y}\right) - 4 = \frac{3}{y^2} - \frac{2}{y} - 4,$$

$$f(t^2) = 3(t^2)^2 - 2(t^2) - 4 = 3t^4 - 2t^2 - 4,$$

$$\begin{aligned} f(x + h) &= 3(x + h)^2 - 2(x + h) - 4 \\ &= 3(x^2 + 2xh + h^2) - 2(x + h) - 4 \\ &= 3x^2 + 6xh + 3h^2 - 2x - 2h - 4. \quad \square \end{aligned}$$

Practice Problem

For the function $g(x) = -2x^2 + 3x + 6$, evaluate

a) $g(3y)$ **b)** $g(w^3)$ **c)** $g(x + h)$

Answers: **a)** $-18y^2 + 9y + 6$; **b)** $-2w^6 + 3w^3 + 6$; **c)** $-2x^2 - 4xh - 2h^2 + 3x + 3h + 6$

WARNING

BE CAREFUL WITH FUNCTIONAL NOTATION

1. $f(x)$ does *not* mean f times x.
2. $f(x + h)$ means replace x by $x + h$. It does *not* mean $f(x) + h$. It does *not* mean $f(x) + f(h)$.

In the last chapter we discussed linear equations in two variables, for example, $2x + y = 6$. We also talked about how to write such equations in slope-intercept form, for example, $y = -2x + 6$. How does this differ and how is it similar to functions? Actually, there is little difference; it is mostly a change of viewpoint. When we use functional notation to write

$$y = f(x) = -2x + 6,$$

we are emphasizing the fact that x is the independent variable and, moreover, the value of y depends solely upon the value of x. In the standard form

$$2x + y = 6$$

it is not clear which variable is the independent variable and which one is the dependent variable. When we are modeling problems, we generally wish to emphasize the functional relationships. In other words, we want to make it clear how the dependent variable depends upon the independent variable.

Operations with Functions

We can combine functions to create new functions.

EXAMPLE 7

Suppose $f(x) = 3x + 6$ and $g(x) = x^2 + 1$. How can these functions be combined?

SOLUTION

We can add $f(x)$ and $g(x)$ to obtain

$$f(x) + g(x) = 3x + 6 + (x^2 + 1) = x^2 + 3x + 7$$

(assuming that x is in the domain of both functions). Similarly, we can also subtract, multiply, and divide the functions:

$$f(x) - g(x) = 3x + 6 - (x^2 + 1)$$
$$= -x^2 + 3x + 5$$
$$f(x) \cdot g(x) = (3x + 6)(x^2 + 1)$$
$$= 3x^3 + 6x^2 + 3x + 6$$
$$\frac{f(x)}{g(x)} = \frac{3x + 6}{x^2 + 1}, \quad \text{when } g(x) \neq 0 \quad \square$$

Practice Problem

Given the functions $f(x) = 2x^2 - 3x + 1$ and $g(x) = 2x - 3$, find the new functions $f(x) + g(x)$, $f(x) - g(x)$, $f(x) \cdot g(x)$ and $f(x) \div g(x)$.

Answers: $f(x) + g(x) = 2x^2 - x - 2$; $f(x) - g(x) = 2x^2 - 5x + 4$; $f(x) \cdot g(x) = 4x^3 - 12x^2 + 11x - 3$; $\dfrac{f(x)}{g(x)} = \dfrac{2x^2 - 3x + 1}{2x - 3}$

There is also another way in which we can combine two functions. Consider the function

$$h(x) = (2x - 7)^2.$$

Certainly $h(x)$ is a function in its own right. However, we can also think of it as a special combination of the simple functions f and g, where

$$y = f(u) = u^2 \quad \text{and} \quad u = g(x) = 2x - 7.$$

By using $g(x)$ first, we can transform x into $u = 2x - 7$. Then, by computing $f(u) = f(2x - 7)$, we can obtain $h(x)$. In other words,

$$h(x) = f(u) = f[g(x)].$$

We call the function h a **composite function.** The domain of $h(x)$ is the set of all real numbers. However, that won't be the case for all composite functions.

EXAMPLE 8

Let $y = f(x) = 3x - 6$ and $g(x) = x^{1/2}$. Find

a) $f[g(x)]$ **b)** $g[f(x)]$

SOLUTION

a) Here we will use $g(x)$ first and then use the result as the **argument** of (input to) $f(x)$. Therefore

$$f[g(x)] = f[x^{1/2}] = 3x^{1/2} - 6.$$

For example,

$$\begin{aligned} f[g(4)] &= 3(4)^{1/2} - 6 \\ &= 3 \cdot 2 - 6 \\ &= 0. \end{aligned}$$

The domain of $f[g(x)]$ is the set of all nonnegative real numbers, which is the set $x \geq 0$.

b) To find $g[f(x)]$ we use $f(x)$ first and then use the result as the argument for $g(x)$. Then

$$\begin{aligned} g[f(x)] &= g[3x - 6] \\ &= (3x - 6)^{1/2}. \end{aligned}$$

Here,
$$\begin{aligned} g[f(4)] &= (3 \cdot 4 - 6)^{1/2} \\ &= 6^{1/2}. \end{aligned}$$

The domain of $g[f(x)]$ is the set of real numbers for which $3x - 6$ is nonnegative, that is, $x \geq 2$. □

Practice Problem

Given the functions $f(x) = x^{1/3}$ and $g(x) = x - 1$, find

a) $f[g(9)]$ **b)** $g[f(8)]$
c) $f[g(x)]$ **d)** $g[f(x)]$

Answers: **a)** 2; **b)** 1; **c)** $(x - 1)^{1/3}$; **d)** $x^{1/3} - 1$

As Example 8 indicates, the domain of the composite of two functions depends not only on the functions but also on the order in which the composition is taken. In general, the domain of the composite function $f[g(x)]$ is the set of all real numbers in the domain of $g(x)$ such that $g(x)$ is in the domain of $f(x)$. (See Figure 4.3.)

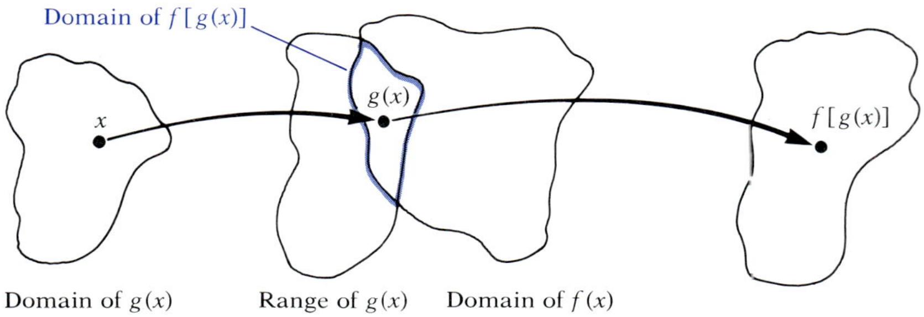

Figure 4.3 The domain of composite functions.

EXERCISE SET 4.1

For each of the following, indicate the name of the function, the independent variable, and the dependent variable.

1. $y = f(x) = 5x + 2$
2. $i = h(x) = -3x + 4$
3. $z = g(y) = 18y + 4$
4. $b = z(a) = 3a - 5$
5. $l = g(h) = -3h^2 + h$
6. $y = h(x) = x^3 + x^2 - 4$

Given that $f(x) = x + 5$, $g(x) = x^2 - 1$, and $h(x) = x^2 + 2x + 1$, evaluate the following.

7. $h(3)$
8. $g(5)$
9. $h(5)$
10. $f(3)$
11. $g(2)$
12. $f(1)$
13. $g(-1)$
14. $f[g(3)]$
15. $g[f(2)]$
16. $h[f(2)]$
17. $g(a)$
18. $h(b)$
19. $f(a + 2)$
20. $g(2) - f(2)$
21. $h(3) - g(3)$
22. $h(6) + f(3)$
23. $h(4)/g(3)$
24. $f(1)/g(2)$
25. $f(2) \cdot h(3)$
26. $h(2) \cdot g(1)$
27. $f(a + 2) - f(a)$
28. $g(2 + 2) - g(2)$
29. $h(3 + 2) - h(3)$
30. $g(a + 2) - g(a)$
31. $h(x + 2) - h(x)$
32. $f(x + 3) - f(x)$
33. $[f(a + 2) - f(a)]/2$
34. $[g(a + 2) - g(a)]/2$
35. $[h(a + 2) - h(a)]/2$
36. $f[g(x)]$
37. $g[f(x)]$
38. $h[f(x)]$
39. $f[f(x)]$
40. $g[g(x)]$

For Problems 41–44, find functions $f(x)$ and $g(x)$ such that $h(x) = f[g(x)]$.

41. $h(x) = (x^2 + 2x + 1)^3$
42. $h(x) = (x^2 - 9)^{1/3}$
43. $h(x) = (2x^4 - x^2)^3 + 3(2x^4 - x^2)$
44. $h(x) = \dfrac{(x^2 + 5x + 6) - 9}{3(x^2 + 5x + 6)^2}$

4.2

GRAPHING FUNCTIONS AND RELATIONS

Basic Approach

Thus far, we've used tables and equations to represent functions and relations. When a relation or function is a correspondence between two sets of real numbers, we can graph it using Cartesian coordinates. This is an extremely valuable tool because it is often easier to determine the nature of a function (or relation) by looking at its graph than by looking at its defining equation. In fact, by looking at a graph, we can determine easily whether a correspondence is a relation or a function.

The following procedure will enable us to graph a function.

Graphing Functions

1. Determine the domain D of the function.
2. Determine several pairs of values consisting of members of the domain and their corresponding values in the range. For each pair, the member from the domain will be the x-coordinate and the member from the range will be the y-coordinate.
3. Plot the x and y values on a Cartesian coordinate system with an appropriate scale for each axis.
4. Draw a smooth curve connecting the points you plotted in step 3, unless the domain and range are restricted to be integers or otherwise consist of isolated points. (See Example 11.)

EXAMPLE 9

Graph the function $y = f(x) = 3x + 2$.

SOLUTION

In this case, there are no restrictions on the value of the independent variable, so the domain is the set of all real numbers. As we saw in Chapter 3, the graph of this function is a straight line. For that reason, we say this function is **linear.** The graph of the function is shown in Fig. 4.4.

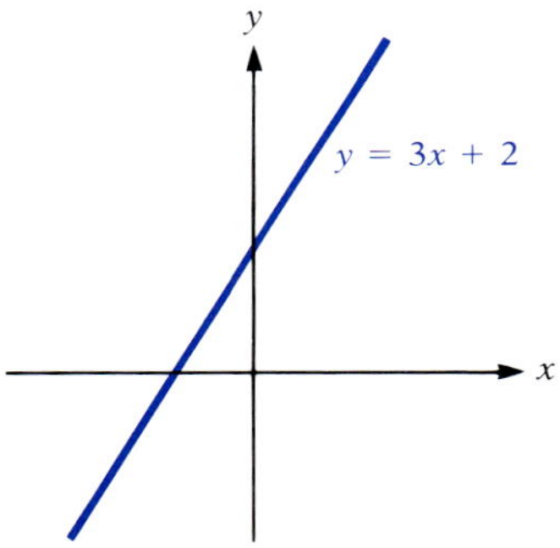

Figure 4.4 The graph of the linear function $y = f(x) = 3x + 2$. □

EXAMPLE 10

Graph the function $y = f(x) = x^2$.

SOLUTION

Again, there are no restrictions on the value of x, so the domain is the set of real numbers. Pairs of values from the domain and range

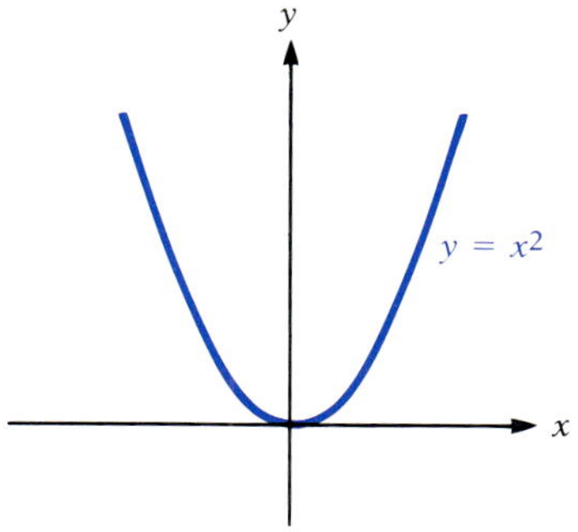

are shown in the following table.

x	-3	-2	-1	0	1	2	3
y	9	4	1	0	1	4	9

The function is shown in Fig. 4.5. This function is said to be **nonlinear** because its graph is not a straight line.

Figure 4.5 **The graph of the nonlinear function $y = f(x) = x^2$.** □

EXAMPLE 11

Consider a hobbyist who goes to a computer store to determine the price of a particular home computer system. The manager gives her the following data:

Number of systems purchased	1–4	5–9	10 or more
Cost per system	\$500	\$450	\$425

Graph this function.

SOLUTION

The data in the table define a function with the independent variable being the number of systems purchased and the dependent variable being the cost per system. The function is graphed in Fig. 4.6, with n representing the independent variable and c the dependent variable. Note that the domain consists of the natural numbers because we cannot purchase a fractional system or a negative number of systems. Thus we do *not* connect the points plotted in this graph.

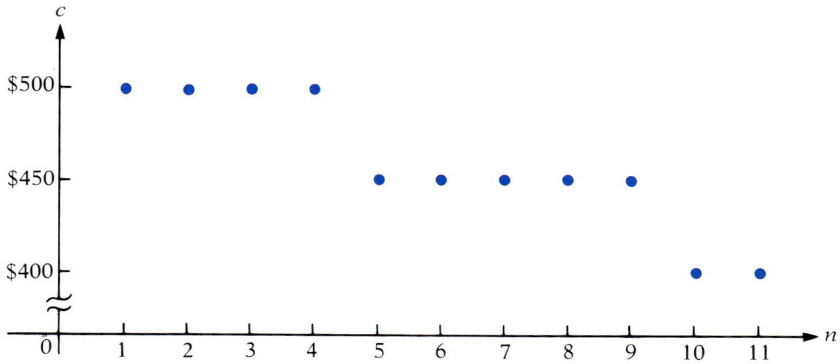

Figure 4.6 **Cost of a computer system versus the number of systems purchased.** □

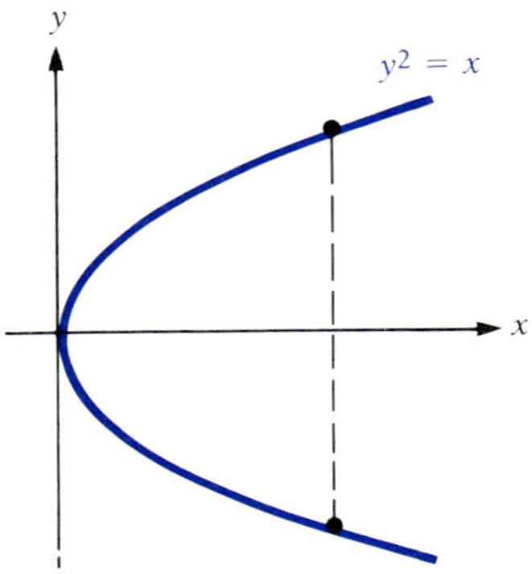

Figure 4.7 The correspondence $y^2 = x$ is not a function.

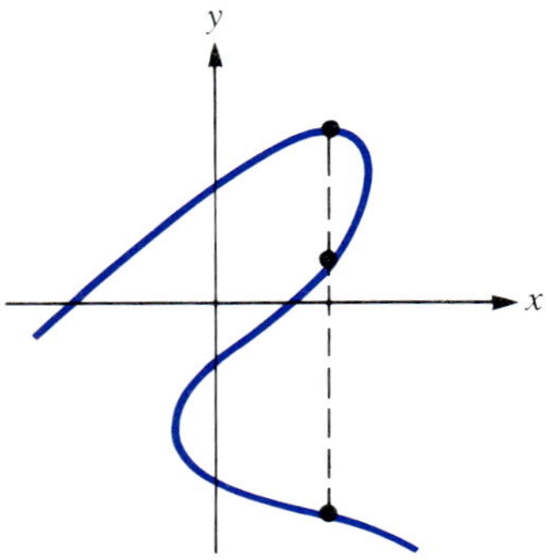

Figure 4.8 Another correspondence that is not a function.

Consider the correspondence from x to y given by $y^2 = x$. In this case, the domain must consist of only nonnegative real numbers because by taking the square root of both sides we see that $y = \pm\sqrt{x}$ and we know the square root of a negative number is not defined. Pairs of values from the domain and range are as follows.

x	0	1	4	9	16
y	0	-1 and $+1$	-2 and $+2$	-3 and $+3$	-4 and $+4$

Note that for each value of x (except at $x = 0$), there are *two* values of y because $y = \pm\sqrt{x}$. Thus this correspondence must be a relation but not a function because the definition of a function states that to each member of the domain there must correspond a *unique* member of the range. This is certainly not true here. The graph is shown in Fig. 4.7.

Similarly, the correspondence shown in Fig. 4.8 is also a relation because there are several values of x at which a vertical line encounters more than one value of y. Again, there are members of the domain that do not correspond to unique members of the range.

We can generalize the above discussion in the form of a rule.

Graphical Test for Functions

A correspondence (from x to y) between two sets of real numbers is a function only if the vertical line drawn at each value of x in the domain intersects the graph at at most one point.

By this test, the correspondence shown in Fig. 4.7 cannot be a function because a vertical line intersects the graph at more than one point. The same conclusion is true for the correspondence shown in Fig. 4.8.

Practice Problem

Which of the following graphs show functions from x to y?

a

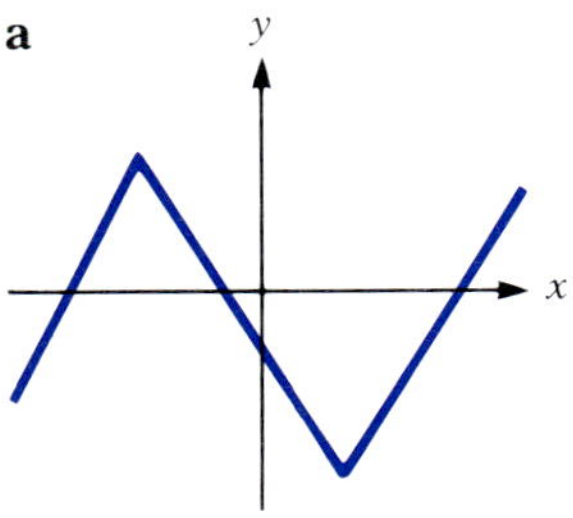

b

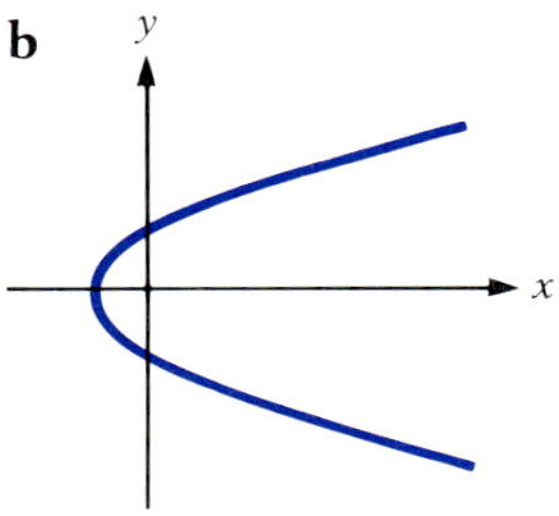

c

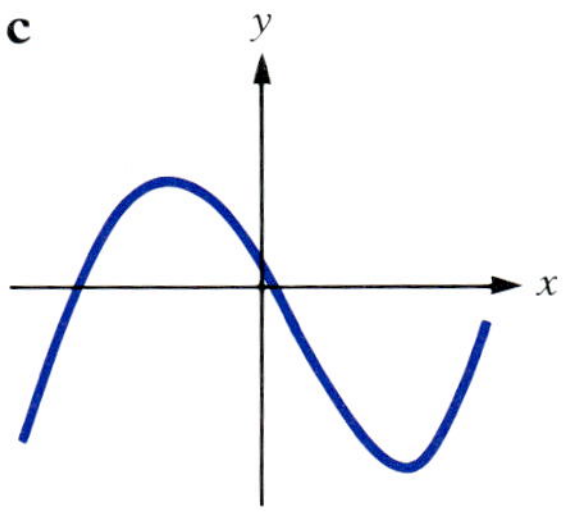

d

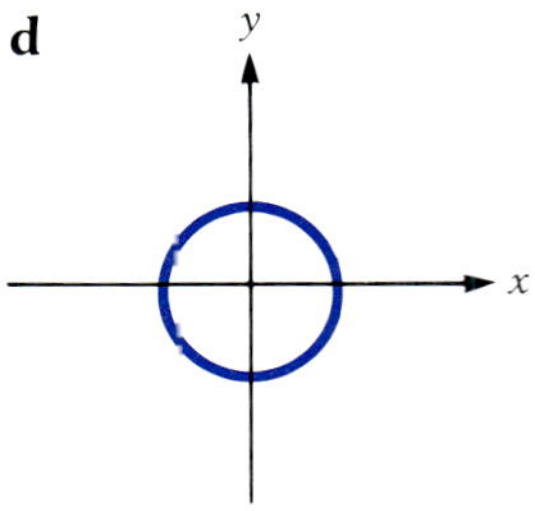

Answers: **a)** function; **b)** not a function; **c)** function; **d)** not a function

Linear Functions

In the previous section we discussed a general procedure for graphing functions and relations. Essentially, to graph a function or relation, choose several representative pairs of points and plot them on a Cartesian coordinate system. The trick is choosing "representative" pairs of points. In the next four sections we'll discuss some of the functions you are most likely to encounter. We'll begin with linear functions.

A **linear function** is a function of the form

$$y = f(x) = mx + b, \qquad \textbf{Linear function}$$

where m is the slope and b is the value of the y-intercept [the line goes through the point $(0, b)$].

You learned in Chapter 3 how to graph linear equations in two variables, which means you also learned how to graph linear functions. As we said earlier, when we write $y = f(x)$, we are emphasizing that the value of y depends upon the value of x. Note that when m is zero, we obtain a horizontal line. In other words, as we mentioned in Chapter 3, a horizontal line has the equation $y = b$. (We often refer to horizontal lines as **constant** functions.) We cannot have functions that are vertical lines (by the graphical test for functions), so equations of the form $x = c$ do not describe functions.

Since the same procedure works for graphing linear functions as for graphing linear equations in two variables, we'll review it quickly.

Graphing Linear Functions

1. The graph of the function $y = f(x) = mx + b$ is a straight line. (When $m = 0$, it is a horizontal line.)
2. When $m \neq 0$, the x-intercept is found by solving the equation $y = 0$.
3. The y-intercept occurs at the point $(0, b)$. If $b = 0$, then the function's x-intercept and y-intercept **both** occur at the origin; $(0, 0)$. Pick a value of x other than zero and determine the corresponding value of y.
4. Plot the points determined in steps 2 and 3 and draw a straight line through them.

EXAMPLE 12

Graph the function $y = 2x - 2$.

SOLUTION

Following the procedure outlined, the x-intercept is found by solving the equation $y = 0$, which gives us $2x - 2 = 0$, which has the solution $x = 1$. Hence the x-intercept is the point $(1, 0)$. The y-intercept is the point $(0, -2)$. Plot the points and connect them with a straight line. (See Fig. 4.9.)

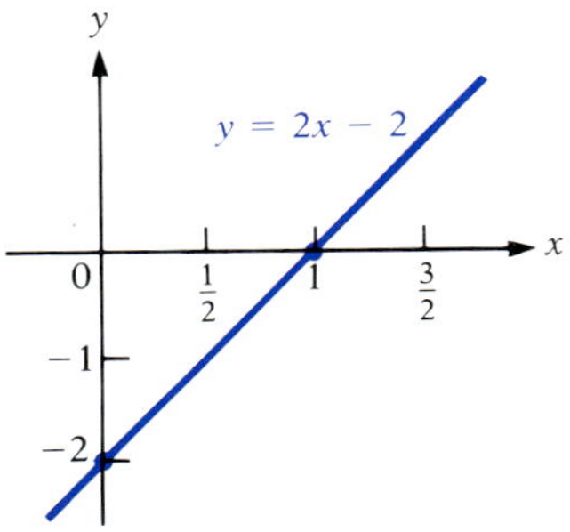

Figure 4.9 **The linear function $y = 2x - 2$.** □

EXAMPLE 13

Graph the function $y = -\frac{5}{3}x + 5$.

SOLUTION

The x-intercept is the point $(3, 0)$ and the y-intercept is the point $(0, 5)$. (See Fig. 4.10.)

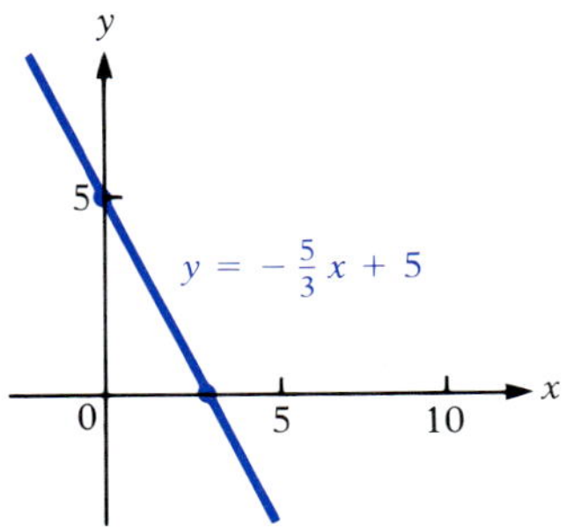

Figure 4.10 **The linear function $y = -\dfrac{5}{3}x + 5$.** □

Now let's take a look at some applications of linear functions.

EXAMPLE 14

In many cases, the cost of producing a product is the sum of a **fixed** cost (which is independent of the number of units produced) and a **variable** cost (which depends only upon the number of units produced). For example, suppose calculators can be made for \$5 each and there is a weekly fixed cost of \$1500. The weekly cost function is then

$$\text{cost} = f(x) = 5x + 1500,$$

where x is the number of calculators produced per week. (See Fig. 4.11.)

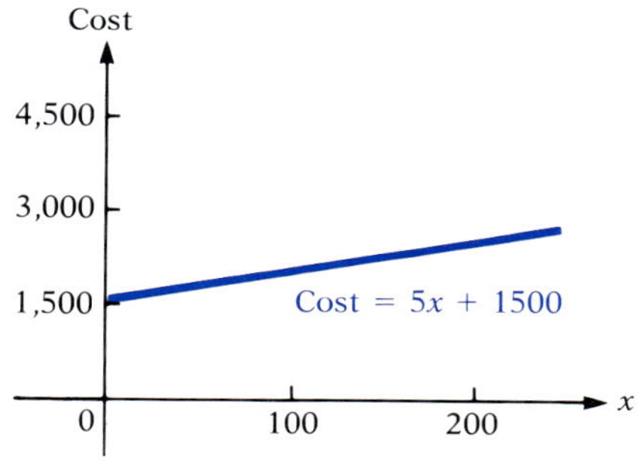

Figure 4.11 **A linear cost function.** □

EXAMPLE 15

Suppose a company expects the sales of a new product to be 5000 units in the first three months after its introduction. Sales are then expected to grow by 2500 units per quarter for the next two years. We can represent this projection by the **trend line**

$$\text{sales} = f(t) = 2500t + 5000,$$

where t is the number of quarters (three-month periods) after the introduction of the product. (Note that t is restricted: $t \leq 8$.) (See Fig. 4.12.)

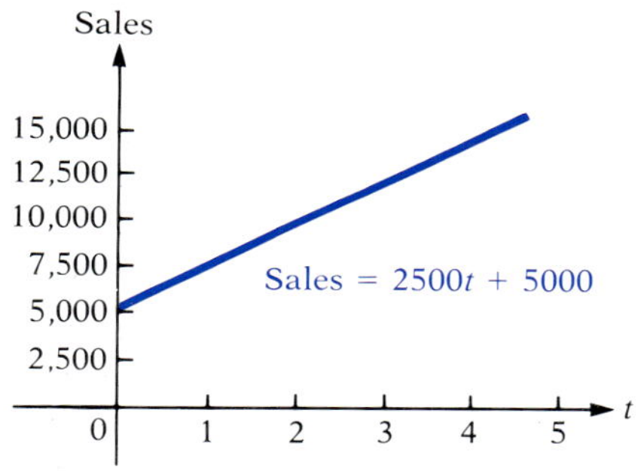

Figure 4.12 **A linear trend line.** □

The Absolute Value Function

We denote the **absolute value** of x by the symbol $|x|$, which we define as follows:

$$y = f(x) = |x| = \begin{cases} x, & \text{if } x \geq 0, \\ -x, & \text{if } x < 0. \end{cases}$$

Notice that the definition of $|x|$ depends upon whether x is negative or not. This may seem strange to you but it is perfectly valid, because to each member x of the domain, there corresponds only one value y of the function. (The absolute value of zero is defined by the upper line in our definition, not both lines.) Moreover, note that $|x|$ is never negative. Its graph is shown in Fig. 4.13.

x	$\|x\|$
−3	3
−2	2
−1	1
0	0
1	1
2	2
3	3

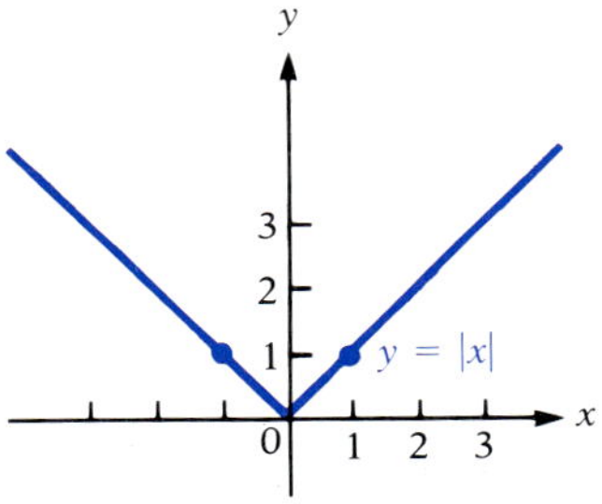

Figure 4.13 **The absolute value function $y = |x|$.**

Practice Problem

Graph the function $y = |x - 2|$. (Hint: determine a table of values first.)

Answer:

x	$x - 2$	$\|x - 2\|$
−3	−5	5
−2	−4	4
−1	−3	3
0	−2	2
1	−1	1
2	0	0
3	1	1
4	2	2

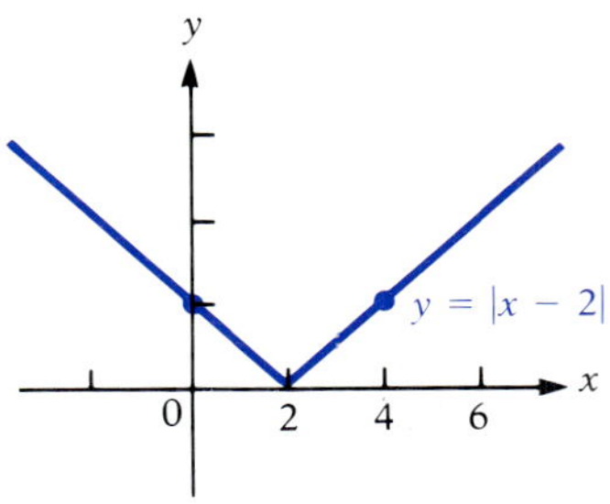

Let's see what happens when we encounter the absolute value function in an equation.

EXAMPLE 16 Solve the equation $|x| - 3 = 0$.

SOLUTION

Remember the definition of the absolute value function depends upon whether x is negative or not. This means that there are two cases to consider.

Case 1 $x \geq 0$. Then $|x| = x$, so we have $x - 3 = 0$ or $x = 3$.

Case 2 $x < 0$. Then $|x| = -x$, so we have $-x - 3 = 0$ or $x = -3$. Since both cases make sense, there are two answers: $x = 3$ and $x = -3$.

We can also think of the problem of solving the given equation as trying to determine when the function $y = |x| - 3$ takes on the value $y = 0$. As Fig. 4.14 shows, this occurs when $x = \pm 3$.

x	$\lvert x\rvert - 3$
−6	3
−5	2
−4	1
−3	0
−2	−1
−1	−2
0	−3
1	−2
2	−1
3	0
4	1
5	2
6	3

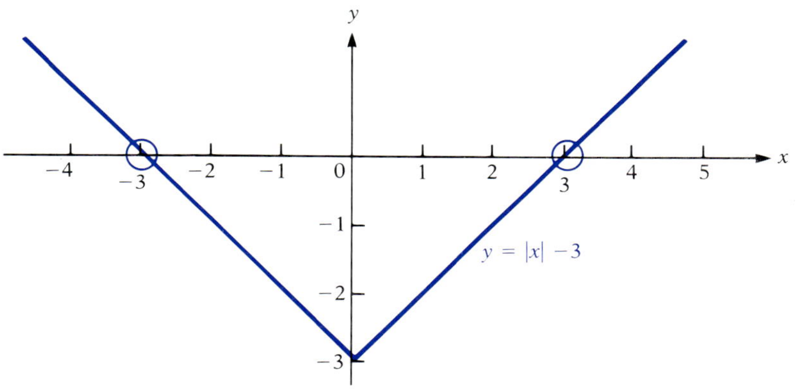

Figure 4.14 **Finding values of x so that $y = |x| - 3 = 0$.** □

EXAMPLE 17

Solve the equation $|x| + 2 = 0$.

SOLUTION

We can rearrange this equation to read $|x| = -2$. By now you should be familiar enough with the absolute value function to realize it can *never* be negative. Thus this equation has no solution. □

EXAMPLE 18

Solve the equation $|x - 1| + x = 2$.

SOLUTION

Case 1 $x - 1 \geq 0$. Then we have $x - 1 + x = 2$ or $x = \frac{3}{2}$. Since $x - 1 = \frac{3}{2} - 1$ is indeed nonnegative, $x = \frac{3}{2}$ is a solution.

Case 2 $x - 1 < 0$. Then we have $-(x - 1) + x = 2$ or $1 = 2$, which is nonsense. The only solution must be $x = \frac{3}{2}$.

As before, we can think of this problem as finding the values of x for which $y = |x - 1| + x$ is equal to 2. As Fig. 4.15 shows, this occurs when $x = \frac{3}{2}$.

x	$\|x - 1\| + x$
-4	$5 + (-4) = 1$
-3	$4 + (-3) = 1$
-2	$3 + (-2) = 1$
-1	$2 + (-1) = 1$
0	$1 + 0 = 1$
1	$0 + 1 = 1$
$\frac{3}{2}$	$\frac{1}{2} + \frac{3}{2} = 2$
2	$1 + 2 = 3$
3	$2 + 3 = 5$
4	$3 + 4 = 7$

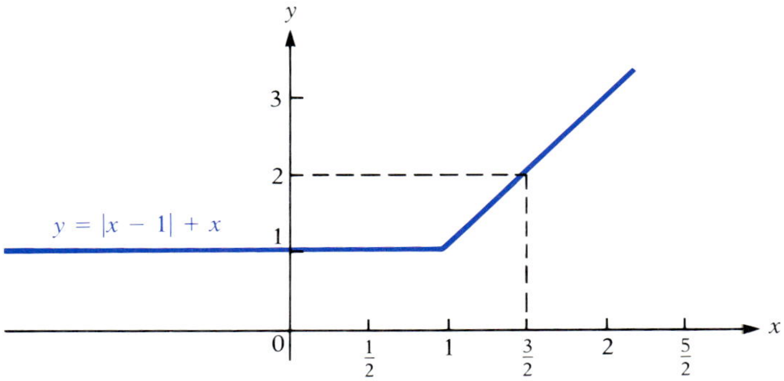

Figure 4.15 **Finding the value of x for which $y = |x - 1| + x = 2$.** □

As these examples show, when you try to solve equations involving the absolute value function, there will always be two cases: one in which the portion between the absolute value symbols is nonnegative and one in which it is negative. And, as we showed, you could also solve the equation graphically (assuming you have the time and the ability to graph accurately).

Practice Problem

Solve the following equations.

a) $|x - 1| - 4 = 0$ **b)** $2x + |4 - x| = 3$

Answers: **a)** $x = 5$ or $x = -3$; **b)** $x = -1$ is the only solution

Quadratic Function

Let's see how we can use our knowledge of solving quadratic equations (see Section 3.3) to help us graph quadratic functions. A **quadratic function** is a function of the form

$$y = ax^2 + bx + c, \quad a \neq 0. \qquad \textbf{Quadratic function}$$

We want to avoid having to make a large table of x and y values to be able to graph a quadratic function. The following approach is much more efficient.

The graph of the quadratic function $y = ax^2 + bx + c$ is called a **parabola.** Figure 4.16 shows two different parabolas. The parabola in Fig. 4.16(a) is said to **open up,** while the one in Fig. 4.16(b) is said to **open down.** A parabola will open up when $a > 0$ and open down when $a < 0$. Notice that each parabola can be cut in half and folded over, with the two halves matching exactly. We say that a parabola is **symmetric** about a vertical line, called the **axis of symmetry,** drawn through its middle. It can be shown (rather easily, in fact) that the high point or low point (depending upon whether the parabola opens down or up) of a parabola occurs on the vertical line

$$x = \frac{-b}{2a},$$

which is the axis of symmetry. We can compute the value of the function on the axis of symmetry which will tell us the coordinates of the **vertex** (the maximum or minimum value of the function):

$$y = a\left(\frac{-b}{2a}\right)^2 + b\left(\frac{-b}{2a}\right) + c = \frac{-b^2}{4a} + c.$$

In other words, we have

$$\left(\frac{-b}{2a}, \frac{-b^2}{4a} + c\right).$$ **Vertex of a Parabola**

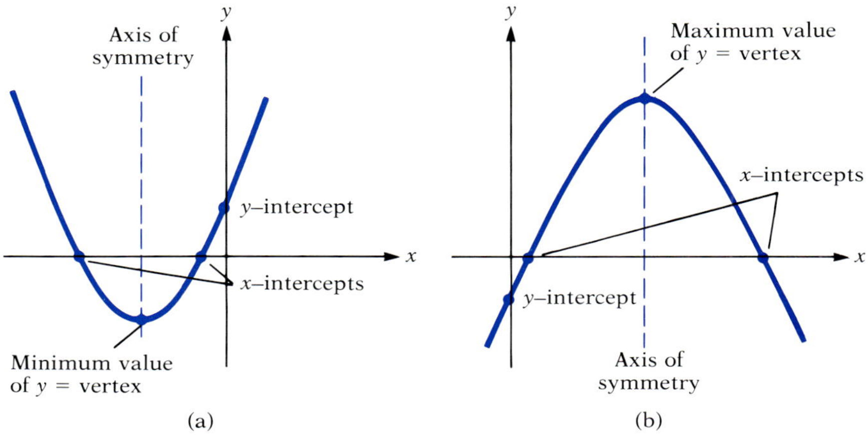

Figure 4.16 **(a) A parabola that opens up. (b) A parabola that opens down.**

The other key facts we must have to draw a parabola are its x-intercept(s) and its y-intercept. We can find the x-intercept(s) by solving the quadratic equation

$$y = f(x) = ax^2 + bx + c = 0.$$

As you know, this equation can have two solutions (in which case there are two x-intercepts), one solution (one x-intercept), or no solution (no x-intercept). The y-intercept is found by setting $x = 0$, which gives us

$$y = f(0) = a(0)^2 + b(0) + c$$

or

$$y = c.$$

Now let's graph some quadratic functions.

EXAMPLE 19

Graph the function $y = f(x) = x^2 + 4x + 3$.

SOLUTION

First, let's find the x-intercepts. The equation is

$$y = f(x) = x^2 + 4x + 3 = 0.$$

The left side can be factored, so we have

$$(x + 3)(x + 1) = 0,$$

which has the solutions $x = -3$ or $x = -1$. This means the function has x-intercepts when $x = -3$ and $x = -1$. In addition, since $a = 1$, it must open up. The minimum value of y occurs on the axis of symmetry

$$x = \frac{-b}{2a}$$

or

$$x = \frac{-4}{2 \cdot 1} = -2.$$

The y-coordinate at the vertex is

$$y = \frac{-b^2}{4a} + c$$

or

$$y = \frac{-4^2}{4 \cdot 1} + 3 = -1.$$

Finally, the y-intercept occurs at the point (0, 3). (See Fig. 4.17.) □

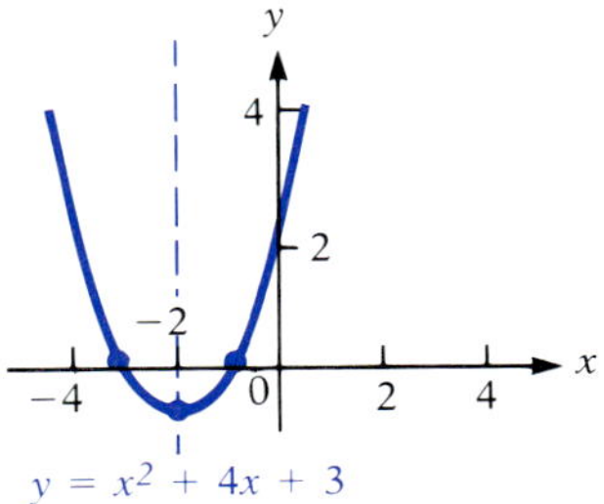

Figure 4.17 The graph of $y = x^2 + 4x + 3$.

EXAMPLE 20

Graph the function $y = f(x) = x^2 - 4x + 4$.

SOLUTION

To find the x-intercept(s) we must solve the equation

$$y = f(x) = x^2 - 4x + 4 = 0,$$

which has the single solution $x = 2$; because the left side factors we have

$$(x - 2)^2 = 0.$$

Therefore, the only x-intercept occurs when $x = 2$. Again, $a = 1$, so this parabola opens up. The axis of symmetry is the vertical line

$$x = \frac{-b}{2a} = \frac{-(-4)}{2 \cdot 1} = 2.$$

But that means the x-intercept occurs on the axis of symmetry. The reason there is only one x-intercept is because the parabola just touches the x-axis at its low point. This means the minimum value of y, which occurs on the axis of symmetry, should be zero. As predicted,

$$y = \frac{-b^2}{4a} + c = \frac{-(-4)^2}{4 \cdot 1} + 4 = 0.$$

The y-intercept is the point (0, 4). (See Fig. 4.18.) □

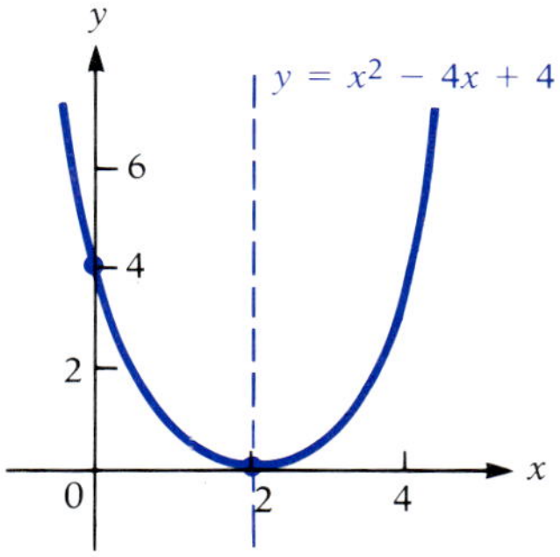

Figure 4.18 **The graph of $y = x^2 - 4x + 4$.**

EXAMPLE 21

Graph the function $y = f(x) = 2x^2 + 5x - 1$.

SOLUTION

By solving the equation $y = f(x) = 0$, we can determine that the x-intercepts occur when $x = (-5 + \sqrt{33})/4$ (approximately 0.186) and $x = (-5 - \sqrt{33})/4$ (approximately -2.686). Since $a = 2$, the parabola opens up. The axis of symmetry is the line

$$x = \frac{-b}{2a} = \frac{-5}{2 \cdot 2} = -1.25.$$

The minimum value of y is given by

$$y = \frac{-b^2}{4a} + c = \frac{-5^2}{4 \cdot 2} + (-1) = -4.125.$$

The y-intercept occurs at the point $(0, -1)$. (See Fig. 4.19.)

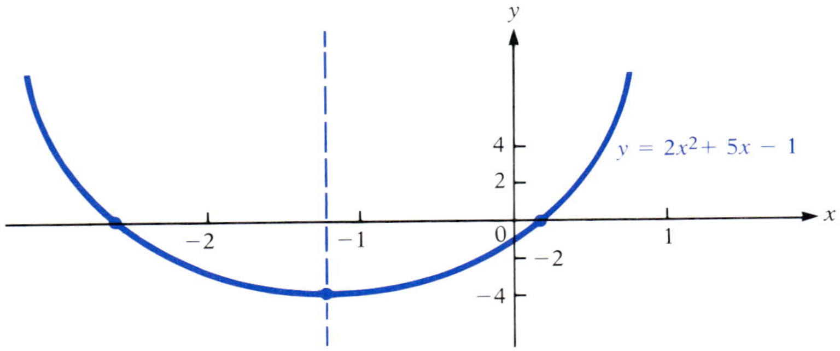

Figure 4.19 **The graph of $y = 2x^2 + 5x - 1$.**

□

EXAMPLE 22

Graph the function $y = f(x) = -x^2 + 3x - 7$.

SOLUTION

This function has no x-intercepts because the equation $y = f(x) = 0$ has no solution. The quadratic formula gives us

$$x = \frac{-b \pm \sqrt{b^2 - 4ac}}{2a} = \frac{-3 \pm \sqrt{3^2 - 4(-1)(-7)}}{2(-1)},$$

which has no real solution. By graphing this function, we can see that it has no x-intercept. Since $a = -1$, this parabola must open down. The maximum value of y occurs on the axis of symmetry

$$x = \frac{-b}{2a} = \frac{-3}{2(-1)} = 1.5.$$

The maximum value of y is given by

$$y = \frac{-b^2}{4a} + c = \frac{-3^2}{4(-1)} + (-7) = -4.75.$$

Since this is the maximum value of the function, it must not cross the x-axis (because on the x-axis, $y = 0$) and so the function cannot have any x-intercepts. The y-intercept is the point $(0, -7)$. (See Fig. 4.20.)

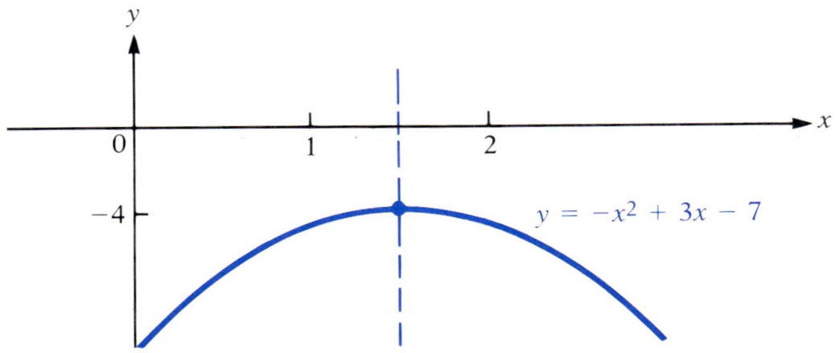

Figure 4.20 **The graph of $y = -x^2 + 3x - 7$.**

□

We can summarize the key facts for graphing quadratic functions as follows.

Graphing Quadratic Functions

1. The graph of the function $y = ax^2 + bx + c$ $(a \neq 0)$ is a parabola.
2. The parabola opens up if $a > 0$ and opens down if $a < 0$.
3. The parabola has an axis of symmetry about the vertical line

$$x = \frac{-b}{2a}.$$

4. The maximum (if $a < 0$) or minimum (if $a > 0$) value of the function occurs on the axis of symmetry at the point

$$x = \frac{-b}{2a}, \qquad y = \frac{-b^2}{4a} + c.$$

5. The x-intercepts are determined by solving the equation $y = 0$.
6. The y-intercept occurs at the point $(0, c)$.

Practice Problem

For each of the following quadratic functions, find the maximum or minimum values and graph the function.

a) $y = f(x) = x^2 - 6x + 9$
b) $y = f(x) = -2x^2 + 8x - 4$
c) $y = f(x) = 3x^2 - 5x - 7$

Answers:

a) minimum at $(3, 0)$

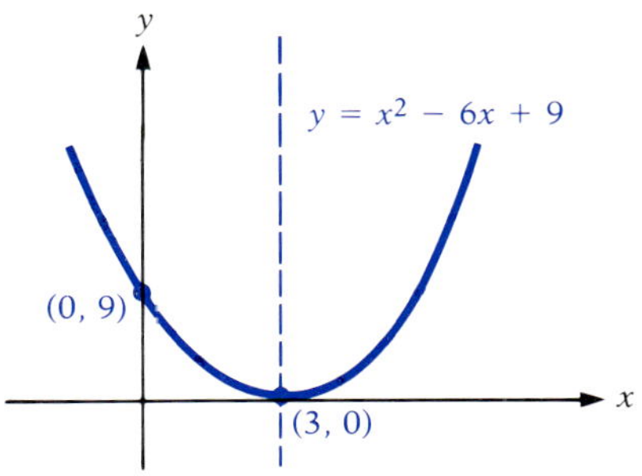

b) maximum at $(2, 4)$

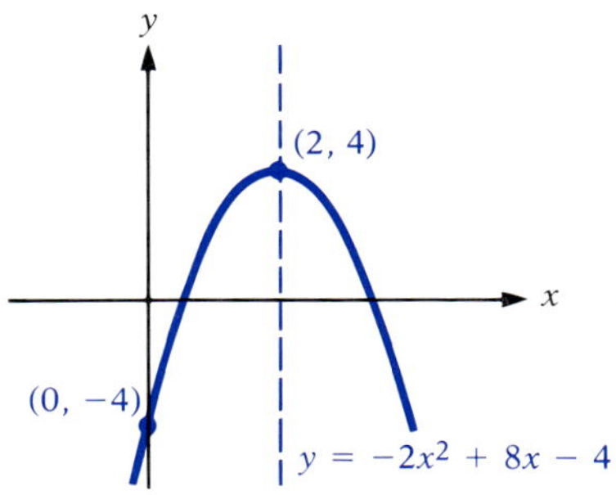

c) minimum at $(\frac{5}{6}, -\frac{109}{12})$

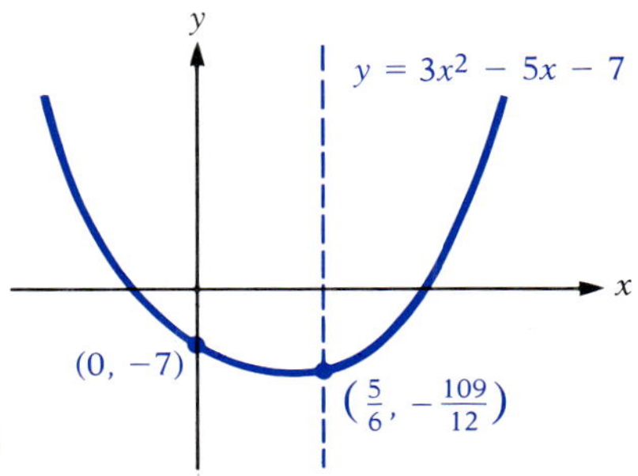

Polynomial and Rational Functions

A function of the form

$$y = a_n x^n + a_{n-1}x^{n-1} + \cdots + a_1 x + a_0, \qquad a_n \neq 0$$

is called a **polynomial function of degree *n*.** We've already discussed functions of degree one (linear) and of degree two (quadratic). Here we will discuss some simple functions of degrees three and four. An analysis of higher degree functions is best done by calculus. If you have access to a computer, you may be able to use a "plot routine" to graph functions for you.

EXAMPLE 23

Graph the function $y = f(x) = x^3$.

SOLUTION

The graph is shown in Fig. 4.21. Notice that we have simply chosen seven pairs of x,y-values and graphed them.

x	y
−3	−27
−2	−8
−1	−1
0	0
1	1
2	8
3	27

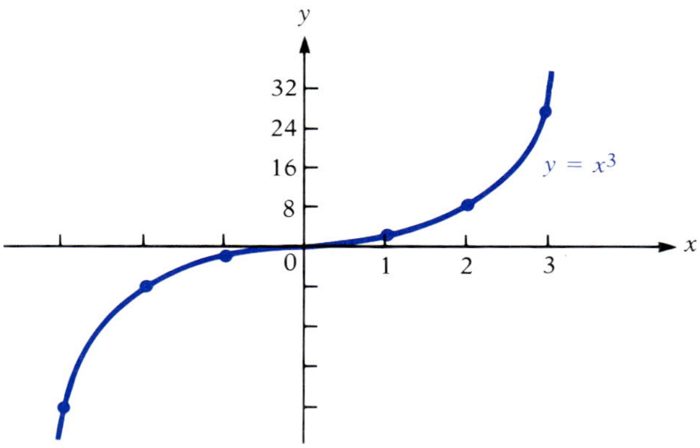

Figure 4.21 **The graph of $y = x^3$.** □

EXAMPLE 24

Graph the function $y = f(x) = 2x^3 - 3x^2 - 12x - 1$.

SOLUTION

The function is graphed in Fig. 4.22. Again, we have chosen several pairs of x,y-values and plotted them.

x	y
-3	-46
-2	-5
-1	6
0	-1
1	-14
2	-21
3	-10
4	31

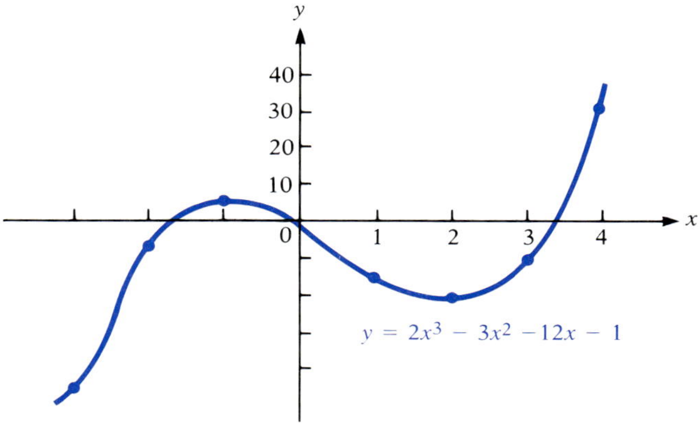

Figure 4.22 **The graph of $y = 2x^3 - 3x^2 - 12x - 1$.**

□

The graph of the function in Example 24 has not one point where it reaches as maximum or minimum (as a quadratic function does) but two points, called **turning points.** In general, a polynomial of degree n can have at most $n - 1$ turning points. Thus a (cubic polynomial of degree three) can have at most two turning points. Note the domain and range of this function are the set of real numbers.

EXAMPLE 25

Graph the function $y = f(x) = 3x^4 - 8x^3 - 6x^2 + 24x - 8$.

SOLUTION

The function is graphed in Fig. 4.23. This is a fourth-degree polynomial, so it can have at most $4 - 1 = 3$ turning points, which it does. The domain is the set of all real numbers; the range is the set $\{y \mid y \geq -27\}$ (not proven here but reasonable given the graph).

x	y
-2	32
-1.5	-15.3125
-1	-27
0	-8
1	5
1.5	2.6875
2	0
3	37

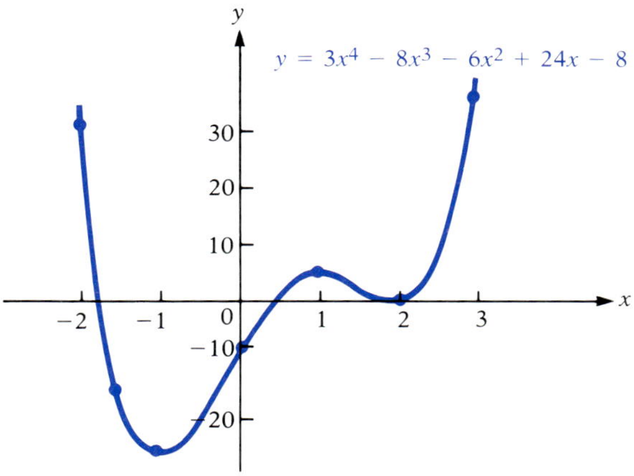

Figure 4.23 **The graph of $y = 3x^4 - 8x^3 - 6x^2 + 24x - 8$.**

□

Practice Problem

Graph the following functions.

a) $y = x^3 + 2$ **b)** $y = 2x^3 - 4x^2 - 10x + 12$

c) $y = 3x^4 + 8x^3 - 6x^2 - 24x + 2$

Answers:

a

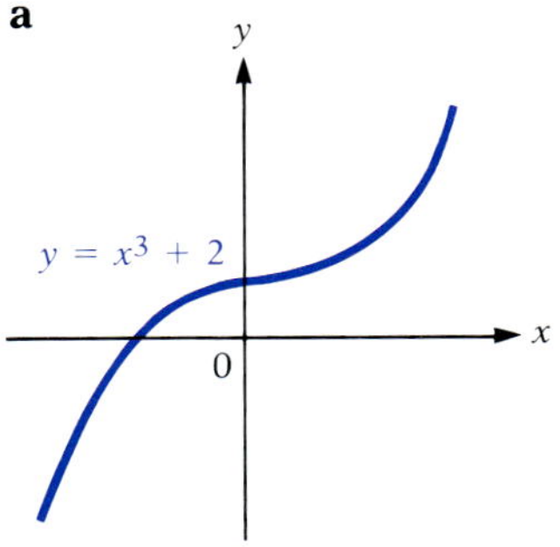

b

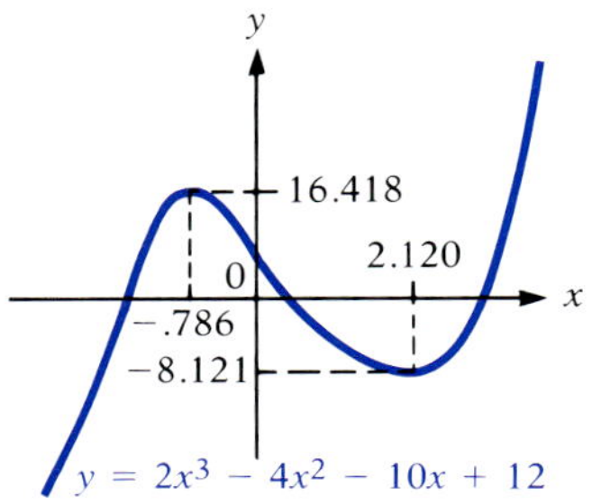

c

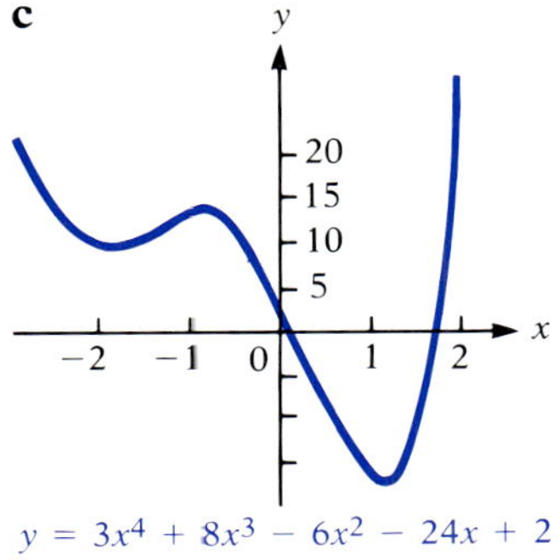

Functions of the form

$$y = \frac{f(x)}{g(x)}, \qquad g(x) \neq 0,$$

where $f(x)$ and $g(x)$ are polynomial functions of x, are called **rational functions.** You must be careful with rational functions because there may be values of x for which the function is undefined [because $g(x)$ is zero].

Rational functions often display asymptotic behavior. An **asymptote** is, crudely speaking, a straight line that a graph approaches but never reaches. Asymptotes may be horizontal or vertical or even oblique (slanted). (See Fig. 4.24.) We will restrict ourselves to functions with vertical or horizontal asymptotes.

Figure 4.24 **Asymptotes can be (a) horizontal, (b) vertical, or (c) oblique.**

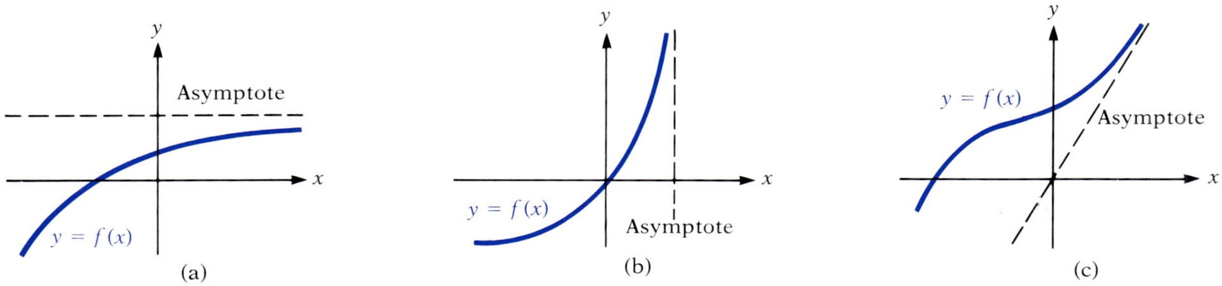

EXAMPLE 26

Graph the function $y = \dfrac{1}{2 + x}$.

SOLUTION

This function is undefined for $x = -2$ because the denominator is zero, so the graph cannot cross a vertical line drawn at $x = -2$. Note the behavior of y as x gets very close to -2. When $x < -2$, y is negative and large in absolute value, whereas when $x > -2$, then y is positive and large. This is shown in Table 4.3.

TABLE 4.3 **Values of x and y for the function $y = \dfrac{1}{2 + x}$**

x	$2 + x$	$y = 1/(2 + x)$
−1	1	1
−1.5	0.5	2
−1.75	0.25	4
−1.9	0.1	10
−1.99	0.01	100
−3	−1	−1
−2.5	−0.5	−2
−2.25	−0.25	−4
−2.1	−0.1	−10
−2.01	−0.01	−100

Note that when x is negative and very large in absolute value, y is negative but very close to zero. Also, when x is very large and positive, then y is also positive and very close to zero. Thus, as the graph in Fig. 4.25 shows, this function has a vertical asymptote at -2 and a horizontal asymptote on the x-axis.

x	y
−5	−0.33
−4	−0.5
−3	−1
−2.5	−2
−1.5	2
−1	1
0	0.5
1	0.33

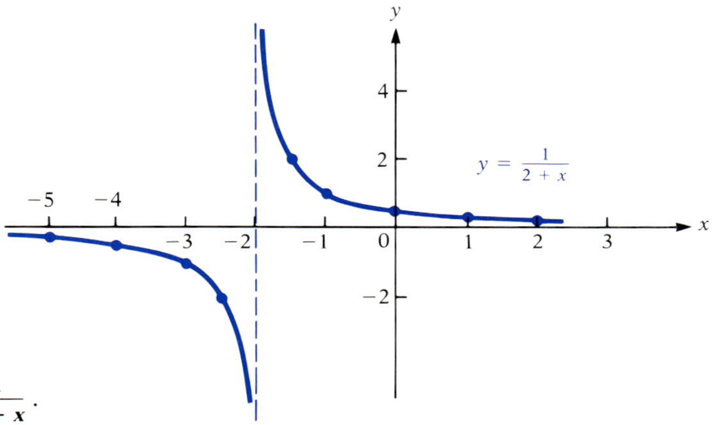

Figure 4.25 **The graph of $y = \dfrac{1}{2 + x}$.** □

EXAMPLE 27

Graph the function $y = \dfrac{2x+1}{2x-1}$.

SOLUTION

Because the denominator is zero when $x = \frac{1}{2}$, this function must have a vertical asymptote at $\frac{1}{2}$. (We should also check to see whether $x = \frac{1}{2}$ makes the numerator zero. If it does, then there may not be an asymptote at $\frac{1}{2}$. That isn't the case here.) We can check for a horizontal asymptote by letting x get very large and positive and then very large in absolute value and negative. A little effort with a calculator should convince you that the line $y = 1$ is a horizontal asymptote for the graph. (See Fig. 4.26.)

x	$y = \dfrac{2x+1}{2x-1}$
−3	0.714
−2	0.6
−1	0.333
0	−1
0.25	−3
0.75	5
1	3
2	1.667
3	1.4

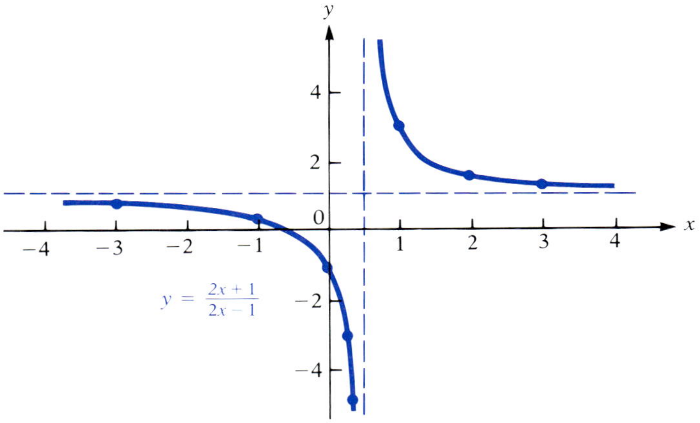

Figure 4.26 **The graph of $y = \dfrac{2x+1}{2x-1}$.** □

EXAMPLE 28

Graph the function $y = \dfrac{x^2 - x - 2}{x^2 - 1}$

SOLUTION

The denominator is zero when $x = 1$ or when $x = -1$ because these are the solutions of the equation $x^2 - 1 = 0$. The function has a vertical asymptote at $x = 1$. However, there is **no** vertical asymptote at $x = -1$. See Table 4.4. The line $y = 1$ is a horizontal asymptote. The graph of the function is shown in Fig. 4.27.

TABLE 4.4 **Values of x and y for the function $y = \dfrac{x^2 - x - 2}{x^2 - 1}$**

x	$y = \dfrac{x^2 - x - 2}{x^2 - 1}$	x	$y = \dfrac{x^2 - x - 2}{x^2 - 1}$
−5	1.167	0	2
−4	1.2	0.5	3
−3	1.25	0.9	11
−2	1.333	0.99	105.477
−1.5	1.4	1.01	−99
−1.1	1.476	1.10	−9
−1.01	1.498	1.5	−1
−0.99	1.502	2	0
−0.9	1.526	3	0.5
−0.5	1.667	4	0.667
		5	0.75

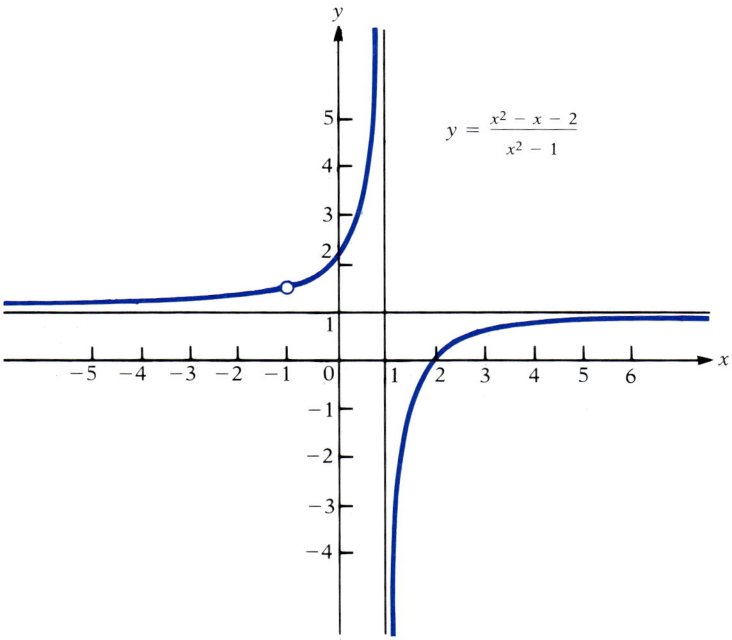

Figure 4.27 **The graph of $y = \dfrac{x^2 - x - 1}{x^2 - 1}$.** □

Practice Problem

Graph the following functions.

a) $y = \dfrac{5}{2 - x}$ **b)** $y = \dfrac{3x - 3}{2x + 2}$ **c)** $y = \dfrac{3 - x}{x + 2}$

Answers:

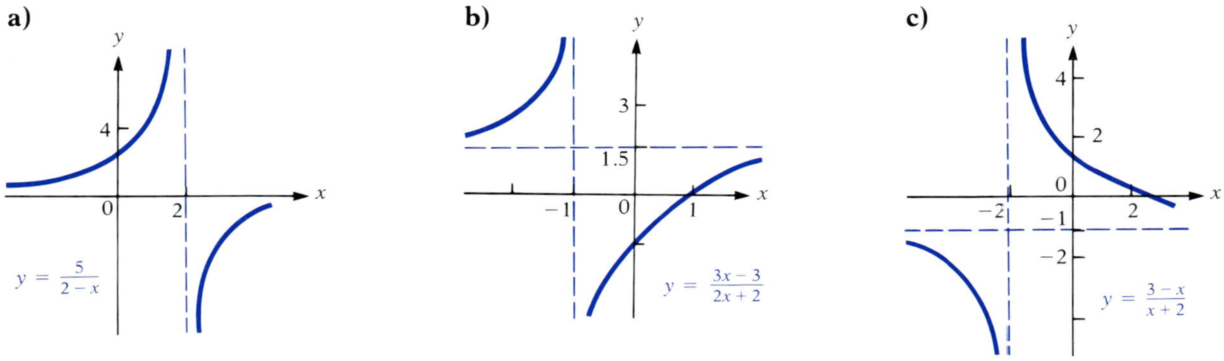

Other Functions

In addition to the functions we've discussed, here are some examples of others you may encounter. We'll start with part (a) of the Problem from Life.

EXAMPLE 29

Problem from Life, part (a) The cost of mailing a first-class letter that weighs x ounces is given by the function

$$\text{cost} = f(x) = \begin{cases} 22 & 0 < x \leq 1, \\ 39 & 1 < x \leq 2, \\ 56 & 2 < x \leq 3, \\ & \text{etc.} \end{cases}$$

In other words, the postage is 22¢ for the first ounce (or fraction thereof) plus 17¢ for each additional ounce (or fraction thereof.) Graph this function and determine the cost of mailing an envelope weighing 3.5 ounces.

SOLUTION

This function is an example of a **step function.** (See Fig. 4.28.) Note that the function looks like a series of steps (hence the name.)

An envelope weighing 3.5 ounces would cost \$0.73 to mail because we would have to go to the step defined by $3 < x \leq 4$.

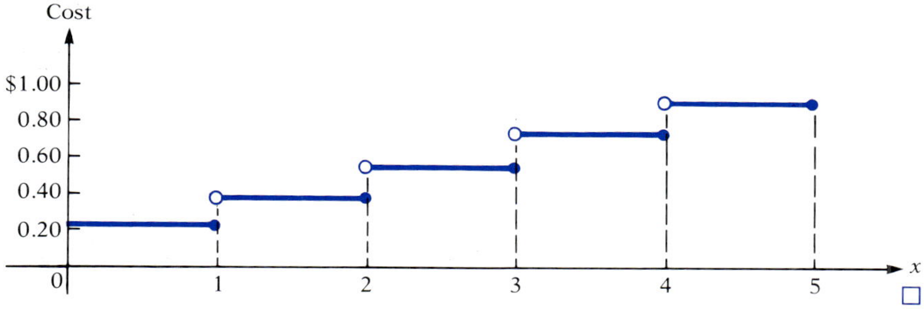

Figure 4.28 First-class mail rates form a step function.

EXAMPLE 30

Consider a company that produces tables. Normally, each table produced and sold will have a profit of $50. However, if the demand is larger than 100 per day, the company must add a second shift. Tables produced on the second shift will have a profit of only $40 because a shift premium must be paid. Graph the total daily profit function.

SOLUTION

Figure 4.29 shows the total daily profit (up to the limits of daily production). We have drawn the function with straight lines to make it easier for you to study. Strictly speaking, the function makes sense for only integer values. Note that the function is made up of two different straight-line segments. This function is called **piecewise linear.** We can define it using functional notation as follows:

$$\text{total profit} = f(t) = \begin{cases} 50t & 0 \leq t \leq 100, \\ 40t + 5000 & 100 < t \leq 200, \end{cases}$$

where t is the total number of tables produced per day. (We have assumed that second-shift production capacity is an additional 100 tables.)

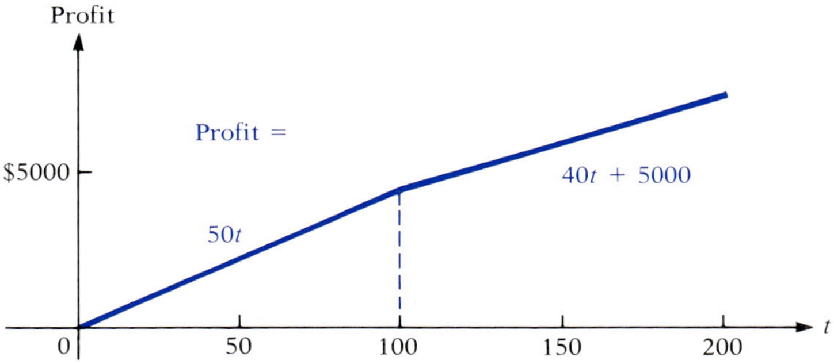

Figure 4.29 Total profits represented by a piecewise linear function.

EXERCISE SET 4.2

Solve the following equations.

1. $|x - 3| = 4$

2. $|3 - x| - 6 = 0$

3. $|-2 - x| = 12$

4. $x + |x| = 3$

5. $3x - |x - 3| = -4$

6. $2x - 3|x - 1| - 6 = 0$

Graph the following functions.

7. $y = |2x|$

8. $y = |3 - x|$

9. $y = |-2 - 3x|$

10. $y = x - |x|$

11. $y = 2x - |x - 1|$

12. $y = 3x - |4 - x|$

13. $y = x^2 - 4$

14. $y = 2x^2 + 3$

15. $y = 4x^2 - 7$

16. $y = (x + 3)^2$

17. $y = (x - 3)^2$

18. $y = -(x + 1)^2$

19. $y = -2(x + 3)^2$

20. $y = x^2 + 4x - 4$

21. $y = x^2 + 5x + 6$

22. $y = -x^2 - x + 12$

23. $y = 15x^2 + 13x - 6$

24. $y = -3x^2 - 3x + 16$

25. $y = \dfrac{2}{x}$

26. $y = \dfrac{2}{x - 4}$

27. $y = \dfrac{4}{1 - x}$

28. $y = \dfrac{3}{3 - x}$

29. $y = \dfrac{x}{x - 2}$

30. $y = \dfrac{x}{2 - x}$

31. $y = \dfrac{x + 1}{x - 1}$

32. $y = \dfrac{3x + 1}{2x}$

33. $y = \dfrac{4x - 2}{2x - 2}$

34. $y = \dfrac{4 - 2x}{4 - x}$

35. $y = f(x) = \begin{cases} 3x + 2 & 0 \le x < 2 \\ 2x^2 & 2 \le x \le 4 \end{cases}$

36. $y = f(x) = \begin{cases} 3x & 0 < x \le 150 \\ 2x + 450 & 150 < x \le 300 \end{cases}$

4.3 VARIATION

In this section we'll discuss a topic called variation, which will help us model and solve a variety of problems.

Linear Variation

Consider a part-time worker who makes \$5 per hour. In 1 hour, he will earn \$5, in 2 hours, he will earn \$10, etc. Letting t be the number of hours worked and m be the money earned, we have

$$m = f(t) = 5t.$$

This is an example of **direct variation.** In general, we have

$$y = f(x) = kx, \qquad k > 0, \qquad \text{Direct variation}$$

where k is called the **constant of variation.**

By utilizing knowledge that y varies directly with x, we can solve certain problems.

EXAMPLE 31

The Smith family has four members. During a normal meal they consume 24 ounces of hamburger meat. The Jones family, with three members, is visiting. Assuming that the consumption of hamburger meat varies directly with the number of people eating, how much hamburger meat is needed?

SOLUTION

We can solve this problem using two different approaches.

Approach 1 The amount of hamburger consumed should vary directly with the number of people. Let n = the number of people and h = amount of hamburger. We have

$$h = kn,$$

and moreover,

$$24 = k \cdot 4$$

so

$$k = 6.$$

With the Jones family visiting, there will be seven people, so

$$h = 6 \cdot 7,$$
$$h = 42 \text{ ounces}.$$

Approach 2 We can construct a proportion as follows.

$$\frac{h_1}{h_2} = \frac{n_1}{n_2},$$

where h_1 and h_2 are the amounts of hamburger consumed by the Smith family and the combined families, respectively, and n_1 and n_2 are the number of people in the Smith family and the combined families, respectively. Since $h_1 = 24$, $n_1 = 4$, and $n_2 = 7$, we can solve for h_2:

$$h_2 = h_1\left(\frac{n_2}{n_1}\right),$$

or

$$h_2 = 24\left(\frac{7}{4}\right)$$

$$h_2 = 42 \quad \square$$

In general, when y varies directly with x we have $y = kx$. Suppose (x_1, y_1) and (x_2, y_2) are two different solutions of the equation. Then we can write the **proportion**

$$\frac{y_1}{y_2} = \frac{kx_1}{kx_2}$$

so

$\dfrac{y_1}{y_2} = \dfrac{x_1}{x_2}.$	**Proportion, direct variation**

Practice Problem

The power consumed by an electrical circuit (in watts) varies directly with the current (in amperes). If the circuit uses 100 watts when the current is 4 amperes, how much power will be consumed when the current is 6 amperes?

Answer: 150 watts

In some cases we may have **inverse variation:** the larger x becomes, the smaller y becomes. For example, consider a car that makes a trip of 400 miles. If 20 gallons of gas were consumed, the car averaged 20 miles per gallon, whereas if 25 gallons of gas were used, the car averaged 16 miles per gallon. In general, when y varies inversely with x, we have

$y = \dfrac{k}{x}, \quad k > 0.$	**Inverse variation**

In this case, the proportion is

$$\frac{y_1}{y_2} = \frac{k/x_1}{k/x_2}$$

or

$$\frac{y_1}{y_2} = \frac{x_2}{x_1}. \qquad \textbf{Proportion, inverse variation}$$

We can use this to solve part (b) of the Problem from Life.

EXAMPLE 32

Problem from Life, part (b) A crew of painters can paint a house in 45 working hours. How many hours will it take if one of the painters quits, reducing the crew to four painters?

SOLUTION

Again, we could solve this problem by determining the constant of variation or by using a proportion. The second approach is more efficient, so we'll show it. Let p = the number of painters and h = the number of hours required to paint the house. We have $p_1 = 5$, $h_1 = 45$, and $p_2 = 4$. Then

$$\frac{h_2}{h_1} = \frac{p_1}{p_2},$$

so

$$h_2 = h_1\left(\frac{p_1}{p_2}\right)$$

or

$$h_2 = 45\left(\frac{5}{4}\right),$$

$$h_2 = 56.25 \text{ hours.} \quad \square$$

Practice Problem

The resistance (in ohms) in an electrical circuit varies inversely with the current (in amperes). A rear-window defogger measures 1 ohm when the current is 5 amperes. What is the resistance when the current is 7.5 amperes?

Answer: $\frac{2}{3}$ ohm

Quadratic Variation

Sometimes variation may be based upon the square of a variable. For instance, the area of a square varies directly as the square of the length of a side. In general, we have

$y = kx^2, \qquad k > 0.$	**Quadratic variation**

when y varies directly as the square of x. Using this information to write a proportion for the points (x_1, y_1) and (x_2, y_2) we have

$$\frac{y_1}{y_2} = \frac{kx_1^2}{kx_2^2}$$

or

$\dfrac{y_1}{y_2} = \dfrac{x_1^2}{x_2^2}.$	**Proportion, quadratic variation**

EXAMPLE 33

The surface area of a cube varies directly with the square of the length of a side. A cube with a side of 2 inches has a surface area of 24 square inches. Find the surface area of a cube with a side of 3 inches.

SOLUTION

Let L = length of a side and s = the surface area. Then $L_1 = 2$, $s_1 = 24$, and $L_2 = 3$. Since

$$\frac{s_1}{s_2} = \frac{L_1^2}{L_2^2},$$

$$s_2 = s_1\left(\frac{L_2^2}{L_1^2}\right),$$

$$s_2 = 24\left(\frac{3^2}{2^2}\right),$$

$$s_2 = 24\left(\frac{9}{4}\right),$$

$$s_2 = 54 \text{ square inches.} \quad \square$$

Practice Problem

The area of a circle varies directly with the square of its diameter. A circle with a diameter of 4 inches has an area of 4π square inches. What is the area of a circle with a diameter of 5 inches?

Answer: 6.25π square inches.

There can be situations in which a variable varies inversely with the square of a variable, in which case we have

$$y = \frac{k}{x^2}. \quad \textbf{Inverse quadratic variation}$$

Using the points (x_1, y_1) and (x_2, y_2), we can write the proportion in this case as

$$\frac{y_1}{y_2} = \frac{k/x_1^2}{k/x_2^2}$$

or

$$\frac{y_1}{y_2} = \frac{x_2^2}{x_1^2}. \quad \textbf{Proportion, inverse quadratic variation}$$

EXAMPLE 34

The weight of an object varies inversely as the square of its distance from the center of the earth (when the object is above the surface of the earth). A person weights 200 pounds on the surface (approximately 3977 miles from the center of the earth). Compute his weight at the top of a mountain 2 miles high, using a proportion.

SOLUTION

Let d = the distance from the center of the earth and w = the person's weight. Then $d_1 = 3977$, $d_2 = 3979$, and $w_1 = 200$. Since

$$\frac{w_1}{w_2} = \frac{d_2^2}{d_1^2},$$

$$w_2 = w_1\left(\frac{d_1^2}{d_2^2}\right),$$

$$w_2 = 200\left(\frac{3977^2}{3979^2}\right)$$

$$w_2 = 199.799 \text{ pounds}$$

(which is not a significant change; traveling to the tops of mountains will not replace dieting). □

Practice Problem

In Example 34, suppose a 200-pound person traveled to a position 200 miles above the surface of the earth in the space shuttle. Find his weight assuming the shuttle is motionless with respect to the surface.

Answer: 181.306 pounds

EXERCISE SET 4.3

1. The voltage in an electrical circuit varies directly with the current. At 3 amperes of current a circuit has a voltage across it of 120. What is the voltage for a current of 4 amperes?
2. The number of hits by a baseball player varies directly with the number of games played. A player has 110 hits after 90 games. How many hits will he have in 162 games (assuming he continues to produce hits at the same rate)?
3. A typing pool generated 120 pages of material that contained 144 errors. Assuming the number of errors varies directly with the number of pages, how many errors would there be in 300 pages of material?
4. A new tire has a tread depth of 1 centimeter. After 15,000 miles the tread depth is 0.6 centimeter. How long will the tire last? (Assume that tire wear varies directly with mileage.)
5. The time required to travel a fixed distance varies inversely with the speed. It takes 45 minutes for Fred to drive to work averaging 45 mph. How long will it take Sally to drive the same distance at 55 mph?
6. A crew of 5 people can cut 3 cords of wood in 6 hours. How long will it take 3 people to cut 3 cords?
7. A shot putter can put a 12-pound shot 72 feet. Assuming that the distance varies inversely with the weight of the shot, how far can he put a 16-pound shot?
8. Light travels 186,000 miles per second. The nearest star is 4.3 light-years away. How long would it take to get there at 100,000 mph? at 100 mph? (Use 1 year = 365.24 days.)
9. The distance traveled by an accelerating object varies directly with the square of the time. A rocket travels 5 miles in the first 2 seconds after launch. How far will it travel in the first 5 seconds?
10. The kinetic energy of a moving object varies directly with the square of its speed. As an example, the stopping distance of a car (after the brakes are applied) varies directly with the square of the speed. A car traveling 30 mph can stop in 50 feet. How many feet will it take to stop a car traveling 50 mph? 70 mph?
11. The surface area of a sphere varies directly with the square of its radius. A ball with a radius of 1.5 inches has a surface area of 2.25π square inches. What is the surface area of a ball with a radius of 2 inches?

12. The intensity of a radio signal varies inversely with the square of the distance from it. Suppose the intensity of an FM station is 3 microvolts per square meter at 25 miles from its antenna. What is the intensity at a distance of 40 miles?
13. The intensity of light from a light bulb varies inversely with the square of the distance from it. After rearranging her desk, a student has moved her lamp 1.5 feet closer. The lamp was originally 4.5 feet away from the student. Assuming that a 100-watt bulb provided enough illumination originally, can the student now use a 75-watt bulb?

4.4 SEQUENCES AND SERIES

Sequences

At this point we've looked at a variety of functions. In this section we will study some very special functions which will be useful to us in the study of the mathematics of finance (Chapter 10) and also later on in several different places.

A **sequence** is a function whose domain is the set of natural numbers. For example,

$$a(n) = 2n + 1, \qquad n = 1, 2, 3, \ldots$$

is a sequence. It is quite common to use the letter n rather than x to emphasize the fact that the domain consists of the natural numbers. For the same reason, the letter a rather than f is used to name the function.

The values of the sequence are called the **terms** of the sequence. For the sequence defined above, we have

$$a(1) = 3, \qquad a(2) = 5, \qquad a(3) = 7, \text{ etc.}$$

Most people use subscripts rather than functional notation to indicate particular terms of a sequence. Thus we would write

$$a_1 = 3, \qquad a_2 = 5, \qquad a_3 = 7, \text{ etc.},$$

for the sequence we defined earlier. In addition, we usually define a sequence by giving a formula for the ***n*th** term (in other words, for a general term). For the sequence we've been using, we have

$$a_n = 2n + 1.$$

(Notice the difference between this approach and standard functional notation.) Of course, once we have a formula for the nth term, we can find any specific term we want by using the appropriate value of n. For instance, for the given sequence, $a_{10} = 21$ and $a_{50} = 101$.

EXAMPLE 35

Find the first four terms and the tenth term for the following sequences:

a) $a_n = 3n - 1$ **b)** $a_n = \dfrac{n}{n+1}$

SOLUTION

a) Replacing n by the appropriate natural number gives us

$$a_1 = 3(1) - 1 = 2,$$

and

$$a_2 = 3(2) - 1 = 5.$$

Similarly, $a_3 = 8$, $a_4 = 11$, and $a_{10} = 29$.

b) For $n = 1$ we have

$$a_1 = \frac{1}{1+1} = \frac{1}{2}.$$

Moreover,

$$a_2 = \frac{2}{2+1} = \frac{2}{3}, \qquad a_3 = \frac{3}{3+1} = \frac{3}{4}, \qquad a_4 = \frac{4}{4+1} = \frac{4}{5},$$

and

$$a_{10} = \frac{10}{10+1} = \frac{10}{11}. \quad \square$$

Practice Problem

Find the first four terms for the following sequences.

a) $a_n = -3n - 2$ **b)** $a_n = \dfrac{n+2}{2n-1}$

Answers: **a)** $-5, -8, -11, -14$; **b)** $3, \frac{4}{3}, 1, \frac{6}{7}$

Just as there are certain special functions (for example, linear and quadratic) that are worthy of extra attention, there are also special sequences. An **arithmetic sequence** is a sequence in which each term after the first is found by adding the same number to the preceding term. The sequence $a_n = 2n + 1$ is an arithmetic sequence because we add $+2$ to any term to get the next term. Similarly, the sequence of Example 35(a) is an arithmetic sequence in which each term can be found by adding 3 to the preceding term. The number that is added to any term to obtain the next term is called the

common difference (because we obtain this difference when we subtract any term from its successor). We denote the common difference by the letter d.

Suppose we know that a sequence is arithmetic. Since we represent the first term by a_1 and the common difference by d, we must have $a_2 = a_1 + d$. Moreover, $a_3 = a_2 + d = (a_1 + d) + d = a_1 + 2d$. Similarly, $a_4 = a_1 + 3d$. In general, we have the

nth Term of an Arithmetic Sequence

$$a_n = a_1 + (n - 1)d.$$

EXAMPLE 36

A town with a population of 25,000 is growing by 750 people per year. What will its population be 10 years from now?

SOLUTION

Since the population grows by 750 per year, the population figures form an arithmetic sequence with a constant difference of 750. Given that $a_1 = 25{,}000$, the population after 10 years is

$$a_{10} = a_1 + (10 - 1)750,$$

or

$$a_{10} = 25{,}000 + 9 \cdot 750,$$

or

$$a_{10} = 31{,}750. \quad \square$$

Practice Problem

Suppose a company with sales of \$150,000 experiences a decline of \$5000 in sales per year. What will its sales be after 5 years?

Answer: \$125,000

Another common type of sequence is the **geometric sequence,** in which each term after the first is found by multiplying the preceding term by the same number. For example,

$$1, -2, 4, -8, 16, -32, \ldots$$

is a geometric sequence in which each term is found by multiplying the preceding term by -2. The constant multiplier is called the **common ratio.**

As with the arithmetic sequence, we can develop a general formula for a geometric sequence. Letting a_1 be the first term and r be the common ratio, then $a_2 = a_1r$ defines the second term. Similarly, $a_3 = a_2r = (a_1r)r = a_1r^2$ and also $a_4 = a_1r^3$. In general, we have

nth Term of a Geometric Sequence

$$a_n = a_1r^{n-1}.$$

EXAMPLE 37

For each of the following sequences, find the indicated term.

a) 12, 36, 108, 324, . . . ; find a_8. **b)** 256, −128, 64, −32, . . . ; find a_9.

SOLUTION

a) Obviously, we have $a_1 = 12$. We can find r, the common ratio, by dividing any term (other than the first) by the preceding term. Using the first and second terms we have $r = \frac{36}{12} = 3$. To find the eighth term we simply use the formula for a geometric sequence with $n = 8$, $a_1 = 12$, and $r = 3$, giving

$$a_8 = a_1r^{8-1}$$

or

$$a_8 = 12(3^7),$$

so, since 3^7 is 2187,

$$a_8 = 12(2187) = 26{,}244.$$

b) Now we have $a_1 = 256$ and $r = -\frac{1}{2}$. Setting $n = 9$ gives us

$$a_9 = 256\left[\left(-\frac{1}{2}\right)^{9-1}\right]$$

or

$$a_9 = 256\left(\frac{1}{256}\right) = 1. \quad \square$$

EXAMPLE 38

Suppose the number of bacteria in a culture doubles every two hours. If 500 bacteria are present initially, how many bacteria would be present after 24 hours assuming unrestrained growth?

SOLUTION

After 2 hours, each of the initial bacteria will have doubled, so we have $a_1 = 1000$. *Notice that we are counting by two-hour periods!* The common ratio is 2. The period of 24 hours is exactly 12 two-hour periods, so we are looking for the value of a_{12} (*not* a_{24}). Using the formula,

$$a_{12} = 1000(2^{12-1}) = 1000(2048) = 2{,}048{,}000 \text{ bacteria.} \quad \square$$

Practice Problem

Find the indicated terms for the following sequences.

a) Find a_6 and a_9 given $a_1 = 5$ and $r = -3$.

b) Find a_7 and a_{10} given $a_1 = 10{,}000$ and $r = 0.25$.

Answers: **a)** -1215 and 32,805; **b)** 2.44141 and 0.03815 (to five places)

Practice Problem

Suppose a bacteria culture starting with 1000 bacteria triples every 4 hours. How many bacteria are there after 8 hours? after 24 hours?

Answer: 9000 and 729,000

EXAMPLE 39

A company's yearly sales grow by 15% per year. Assuming that sales in 1985 were \$100,000, what will the yearly sales be in 1990? in 1995?

SOLUTION

Since sales each year grow by 15%, we have $r = 115\%$ or 1.15 (*not* 0.15). In addition, let's call 1985 the point at which time is zero. Therefore, sales in 1990 correspond to a_5 and sales in 1995 correspond to a_{10}. By the same logic, a_1 corresponds to sales in 1986, which must be \$115,000. This gives us

$$a_5 = \$115{,}000(1.15)^{5-1} = \$115{,}000(1.749006) = \$201{,}136,$$

$$a_{10} = \$115{,}000(1.15)^{10-1} = \$115{,}000(3.517876) = \$404{,}556. \quad \square$$

Practice Problem

For the company in Example 39, what will the yearly sales be in 1987? in 1992?

Answer: \$132,250; \$266,002

Series

In many cases we will find it useful to be able to add the terms of a sequence; for instance, when the terms of a sequence are yearly profits, their sum represents total profits (over some time frame). A **series** is a sequence whose terms are found by adding up the terms of a given sequence. Let's see how we can portray this symbolically. Let $a_1, a_2, a_3, \ldots$ be the terms of some sequence. The terms of the associated series are represented by $s_1, s_2, s_3, \ldots$ and are found as follows:

$$s_1 = a_1,$$
$$s_2 = a_1 + a_2 = s_1 + a_2,$$
$$s_3 = a_1 + a_2 + a_3 = s_2 + a_3,$$

and in general

$$s_n = a_1 + a_2 + \cdots + a_n = s_{n-1} + a_n.$$

EXAMPLE 40

Given the sequence $a_n = n^2 - 2$, find the first four terms of the associated series.

We have $s_1 = a_1 = 1^2 - 2 = -1$. Also,

$$s_2 = s_1 + a_2 = -1 + (2^2 - 2) = 1,$$
$$s_3 = s_2 + a_3 = 1 + (3^2 - 2) = 8,$$
$$s_4 = s_3 + a_4 = 8 + (4^2 - 2) = 22. \quad \square$$

Practice Problem

Find the first four terms of the series associated with the sequence $a_n = -5n + 3$.

Answer: $-2, -9, -21, -38$

As you might guess, it is often possible to derive a formula that tells us the values of the terms in a series, in other words, the value of the sum of the first n terms of a sequence. For an arithmetic sequence with first term a_1, common difference d, and general term a_n, the general term of the associated sequence can be found with either of the following formulas:

Arithmetic Series

$$s_n = \left(\frac{n}{2}\right)[2a_1 + (n - 1)d]$$

or

$$s_n = \left(\frac{n}{2}\right)(a_1 + a_n).$$

EXAMPLE 41

Find the sum of the first 30 terms of the arithmetic sequence $-4, 3, 10, 17, \ldots$.

SOLUTION

For this sequence, $a_1 = -4$ and $d = 7$. The first formula is easier to use (because we don't have a_{30}), so we have

$$s_{30} = \left(\frac{30}{2}\right)[2(-4) + (30 - 1)7],$$
$$s_{30} = 15[-8 + 203],$$
$$s_{30} = 2925. \quad \square$$

EXAMPLE 42

Find the sum of the integers 1 through 2500.

SOLUTION

We can think of the integers 1 through 2500 as the terms of a sequence with $a_1 = 1$, $d = 1$, and $a_{2500} = 2500$. Using the second formula results in

$$s_{2500} = \left(\frac{2500}{2}\right)(1 + 2500),$$
$$s_{2500} = 1250(2501),$$
$$s_{2500} = 3{,}126{,}250. \quad \square$$

Practice Problem

Find the sum of

a) the first 35 terms of the arithmetic sequence $29, 25, 21, \ldots,$

b) the first 750 natural numbers.

Answers: **a)** -1365; **b)** 281,625

There is also a formula for the sum of the terms of a geometric sequence. Letting the first term be a_1 and the common ratio be r, then the general term of the associated series is

Geometric Series

$$s_n = \frac{a_1(r^n - 1)}{r - 1}.$$

EXAMPLE 43

Find the sum of the first ten terms of the geometric sequence 2, 6, 18, ...

SOLUTION

Here we have $a_1 = 2$ and $r = 3$, so applying the formula gives us

$$s_{10} = \frac{2(3^{10} - 1)}{3 - 1},$$

$$s_{10} = \frac{2(59049 - 1)}{2},$$

$$s_{10} = 59{,}048. \quad \square$$

EXAMPLE 44

Find the sum of the first six terms of the geometric sequence with $a_1 = 250$ and $r = -\frac{1}{2}$.

SOLUTION

Applying the formula directly yields

$$s_6 = \frac{250[(-\frac{1}{2})^6 - 1]}{-\frac{1}{2} - 1},$$

$$s_6 = \frac{250(0.015625 - 1)}{-\frac{3}{2}},$$

$$s_6 = 164.0625. \quad \square$$

Practice Problem

Find the appropriate term of the series associated with the given geometric sequence.

a) s_7 for the sequence 5, −15, 45, . . .

b) s_6 for the sequence with $a_1 = 900$ and $r = \frac{1}{3}$

Answers: **a)** 2735; **b)** 1348.14815

EXAMPLE 45

Consider a company whose yearly sales grow by 8%. Assuming that its sales were $150,000 in 1985, what will the total sales be over the period 1985 through 1989?

SOLUTION

We can use the value of a geometric series to answer this question. We have $r = 1.08$ and $a_1 = 150{,}000$. Also, in 1989, the value of n will be 5 (be careful!). Therefore total sales are given by

$$s_5 = \$150{,}000\left[\frac{1.08^5 - 1}{1.08 - 1}\right],$$

$$s_5 = \frac{\$150{,}000(1.469328 - 1)}{0.08},$$

$$s_5 = \$879{,}990. \quad \square$$

Practice Problem

For the company in Example 45, what will total sales be for the period 1985 through 1992?

Answer: $1,595,494

Series are often written with **sigma notation** (which is used very widely in mathematics and statistics). For example, we can use sigma notation to write the sum of the first five terms of the sequence given by $a_n = 2n$:

$$s_5 = \sum_{i=1}^{5} 2i$$

where $\sum$ is the capital Greek letter sigma. The letter i is called the **summation index.** It may start at any integer (in general) and end at any integer equal to or greater than the starting integer. The starting value of the index is often referred to as the **lower limit of the sum** and the ending value of the index is the **upper limit of the sum.** The interpretation of this notation is straightforward: Add up all of the terms that are formed by setting the summation index at the lower limit and incrementing it by 1 until it reaches its upper limit. Thus the notation

$$\sum_{i=1}^{5} 2i$$

actually represents the sum

$$\underbrace{2(1)}_{i\,=\,1} + \underbrace{2(2)}_{i\,=\,2} + \underbrace{2(3)}_{i\,=\,3} + \underbrace{2(4)}_{i\,=\,4} + \underbrace{2(5)}_{i\,=\,5} = 30.$$

EXAMPLE 46

$$\sum_{i=1}^{4} f(i) = f(1) + f(2) + f(3) + f(4)$$

$$\sum_{i=1}^{5} i^2 = 1^2 + 2^2 + 3^2 + 4^2 + 5^2 = 55$$

$$\sum_{i=1}^{6} 4 = 4 + 4 + 4 + 4 + 4 + 4 = 24$$

$$\sum_{i=1}^{4} (3 - 2i) = (3 - 2 \cdot 1) + (3 - 2 \cdot 2) + (3 - 2 \cdot 3) + (3 - 2 \cdot 4)$$
$$= -8$$

$$\sum_{i=1}^{n} f(i) = f(1) + f(2) + \cdots + f(n) \quad \square$$

Practice Problem

Write each of the following sums in expanded form (as in Example 46). You need not simplify.

a) $\sum_{i=1}^{6} g(i)$ **b)** $\sum_{i=1}^{5} 2i$

c) $\sum_{i=1}^{4} \frac{2i - 1}{i}$ **d)** $\sum_{i=1}^{7} -7$

Answers: **a)** $g(1) + g(2) + g(3) + g(4) + g(5) + g(6)$

b) $2 \cdot 1 + 2 \cdot 2 + 2 \cdot 3 + 2 \cdot 4 + 2 \cdot 5$

c) $\frac{2 \cdot 1 - 1}{1} + \frac{2 \cdot 2 - 1}{2} + \frac{2 \cdot 3 - 1}{3} + \frac{2 \cdot 4 - 1}{4}$

d) $(-7) + (-7) + (-7) + (-7) + (-7) + (-7) + (-7)$

Although we won't derive them, the following formulas can be used to evaluate sums.

Properties of Sums

1. $\sum_{i=1}^{n} c = nc,$ where c is a constant

2. $\sum_{i=j}^{n} ca_i = c \sum_{i=j}^{n} a_i$

3. $\sum_{i=j}^{n} [a_i \pm b_i] = \sum_{i=j}^{n} a_i \pm \sum_{i=j}^{n} b_i$ where a_i and b_i are general terms of two sequences

Let's look at some examples of these properties.

EXAMPLE 47

$$\begin{aligned}\sum_{i=1}^{5}(2i+3) &= \sum_{i=1}^{5}2i + \sum_{i=1}^{5}3 && \text{Property 3}\\ &= 2\sum_{i=1}^{5}i + \sum_{i=1}^{5}3 && \text{Property 2}\\ &= 2\left[\left(\frac{5}{2}\right)(1+5)\right] + 5\cdot 3 && \text{Property 1}\\ &= 30 + 15\\ &= 45 \quad \square\end{aligned}$$

The first term is the sum of an arithmetic sequence with $n = 5$, $a_1 = 1$, and $a_5 = 5$.

EXAMPLE 48

$$\begin{aligned}\sum_{i=1}^{10}\{(25\cdot 3^i) + 2[10\cdot(-4)^i]\} &= \sum_{i=1}^{10}(25\cdot 3^i) + 2\sum_{i=1}^{10}[10\cdot(-4)^i] && \text{Properties 3, 2}\\ &= \frac{75(3^{10}-1)}{(3-1)} + 2\left\{-40\frac{[(-4)^{10}-1]}{(-4-1)}\right\}\\ &= 2{,}214{,}300 + 2(8{,}388{,}600)\\ &= 18{,}991{,}500 \quad \square\end{aligned}$$

Each term is the sum of a geometric sequence. The first has $a_1 = 75$, $r = 3$, and $n = 10$, and the second has $a_1 = -40$, $r = -4$, and $n = 10$. (Be careful in computing a_1 for each sequence!)

Practice Problem

Evaluate the following sums.

a) $\sum_{i=1}^{10}(3i - 10)$ **b)** $\sum_{i=1}^{9}(6\cdot 2^i) + 1000\left(\frac{3}{5}\right)^i$

Answers: **a)** 65; **b)** 7616.88346

EXERCISE SET 4.4

In the following exercises, a formula for the general term of a sequence is given. Identify each sequence as arithmetic, geometric, or neither, and give its first four terms.

1. $a_n = 4n + 2$

2. $a_n = -3n + 6$

3. $a_n = \dfrac{3n}{n+1}$

4. $a_n = 3 - 7n$

5. $a_n = 4^n$

6. $a_n = 2n^2 - 3$

7. $a_n = (-5)^n$

8. $a_n = \dfrac{3}{5n - 1}$

9. $a_n = 25 \cdot (-6)^n$

10. $a_n = \dfrac{5^n}{n^2}$

11. $a_n = (-3)^{n+1}$

12. $a_n = 4 \cdot 2^{-n}$

Identify each of the following sequences as being arithmetic, geometric, or neither. If the sequence is arithmetic, give the common difference. If the sequence is geometric, give the common ratio.

13. 3, 7, 11, 15, 18, . . .

14. 1, 1, 3, 5, 8, . . .

15. 1, 2, 4, 8, 16, . . .

16. $-80, -65, -50, -35, \ldots$

17. 9, 3, 3, 9, 3, . . .

18. $6, 1, -4, -7, -10, \ldots$

19. $\frac{1}{3}, -1, 3, -9, 27, \ldots$

20. 5, 20, 100, 600, 4200, . . .

21. $3a, 5a, 7a, 9a, 11a, \ldots$

22. $3x, -6x, 12x, -24x, 48x, \ldots$

23. $2, 4x, 8x^2, 16x^3, 32x^4, \ldots$

24. $x, x/2, x/3, x/4, x/5, \ldots$

Find the appropriate term for the following arithmetic sequences.

25. $a_1 = 4, d = 6$; find a_{10}

26. $a_1 = 17, d = -3$; find a_{20}

27. $a_1 = 40, d = -6$; find a_{15}

28. $a_1 = 99, d = 5$; find a_{11}

29. a_{10} for the sequence $1, \frac{3}{2}, 2, \frac{5}{2}, \ldots$

30. a_8 for the sequence 6, 15, 21, 27, . . .

31. Which term in the sequence $7, 4, 1, -2, \ldots$ is equal to -50?

32. Which term in the sequence $\frac{1}{2}, \frac{5}{4}, 2, \frac{11}{4}, \ldots$ is equal to $21\frac{1}{2}$?

Find the appropriate sum for the following arithmetic sequences.

33. s_6 for the sequence with $a_1 = 4$ and $d = 6$

34. s_{10} for the sequence with $a_1 = 40$ and $d = -6$.

35. s_8 for the sequence 5, 10, 15, 20, . . .

36. s_{12} for the sequence $6, 1, -4, -9, \ldots$

37. s_5 for the sequence $-80, -65, -50, -35, \ldots$

38. s_{16} for the sequence $\frac{1}{2}, \frac{5}{4}, 2, \frac{11}{4}, \ldots$

Find the appropriate term of the following geometric sequences.

39. $a_1 = 1, r = 2$; find a_7

40. $a_1 = 48, r = \frac{1}{2}$; find a_6

41. $a_1 = 3, r = -3$; find a_5

42. $a_1 = 1000, r = -\frac{1}{5}$; find a_5

Find the appropriate sum for the following geometric sequences.

43. $a_1 = 1, r = 3$; find s_5

44. $a_1 = 243, r = \frac{2}{3}$; find s_4

45. $a_1 = 1000, r = -\frac{4}{5}$; find s_4

46. $a_1 = 729, r = -\frac{1}{3}$; find s_5

47. Find s_5 for the sequence $\frac{1}{18}, \frac{1}{6}, \frac{1}{2}, \frac{3}{2}, \ldots$.

48. Find s_6 for the sequence $6, -12, 24, -48, \ldots$.

Write the following summations in expanded form. You need not simplify.

49. $\sum_{i=1}^{7} i^{1/2}$

50. $\sum_{i=2}^{6} \left(3 + \frac{2}{i}\right)$

51. $\sum_{i=5}^{10} f(i + 3)$

52. $\sum_{i=1}^{5} \frac{h(i)}{i}$

53. $\sum_{i=2}^{5} \left(4 - \frac{3}{i}\right) i^3$

54. $\sum_{i=1}^{4} g(3 + i^2)\left(\frac{2}{i}\right)$

Evaluate the following summations.

55. $\sum_{i=1}^{5} 4.5$

56. $\sum_{i=2}^{7} (-3)$

57. $\sum_{i=1}^{7} 4i$

58. $\sum_{i=3}^{9} (4 + 2i)$

59. $\sum_{i=1}^{5} (3i - 9)$

60. $\sum_{i=1}^{5} (2 - 5i)$

61. $\sum_{i=1}^{4} 3^i$

62. $\sum_{i=2}^{5} 2000 \cdot \left(-\frac{1}{2}\right)^i$

63. $\sum_{i=3}^{7} (4^i + 6 - 3i)$

64. $\sum_{i=1}^{4} \left[729 \cdot \left(\frac{1}{3}\right)^i + 25i\right]$

65. Anne starts work as a programmer for \$17,500 per year. She has been promised that she will receive a \$750 raise every six months for the next two years.

a) What will her salary be at the end of two years?

b) How much total income will she earn over the two-year period?

c) Suppose Anne is offered a choice between a \$1500 raise each year for two years and the \$750 every six months. Which one should she choose?

66. Rob begins work at the same time as Anne (see Exercise 65) at an annual salary of \$15,000. He has been promised that he will receive a 5% raise every six months for the next two years. What will his salary be at the end of two years? How much total income will he earn over the two years?

67. A company buys a piece of equipment for \$25,000. Each year, the equipment is expected to lose 20% of the value it had at the beginning of the year. What will its value be after five years? after ten years?

68. At the beginning of 1985, the population of Middletown was 50,000. Given an annual growth rate of 4%, what will the population be at the beginning of the year 2000?

69. At the beginning of 1985, the population of Concord was also 50,000, with an annual growth rate of 2000. What will the population be at the beginning of the year 2000?

70. A company buys a piece of equipment in 1986. Annual operating costs are expected to increase by \$2500 per year. The initial operating costs are estimated to be \$25,000. What will the annual operating costs be in 1991? How long will it take for the annual costs to double? What will the total costs of operation be over the period 1986–1996?

71. A company's yearly sales are increasing by 12% per year. Given that sales were \$500,000 in 1983, what will yearly sales be in 1988? 1992? What will total sales be for the period 1983–1988? for 1983–1992?

72. A grandfather clock strikes the same number of times as the hour (once for one o'clock, twice for two o'clock, etc.) and also strikes once for the first quarter hour, twice for the second quarter hour, three times for the third quarter hour, and four times on the hour each hour. How many times does the clock strike (starting at 1 minute after midnight) in 12 hours? in 24 hours? in one year (365 days)? (*Hint:* Write an expression relating the number of strikes to the time.)

CHAPTER REVIEW

IMPORTANT TERMS AND SYMBOLS

4.1 Functions

argument The input to a function.

composite function A function formed by using one function as the argument of another function. For example, let $u = g(x)$ and $y = f(u)$. Then $h(x) = f[g(x)]$ is a composite function.

constant function A function of the form $y = b$, where b is a constant.

dependent variable Variable whose values come from the range of a function; if $y = f(x)$, then y is the dependent variable.

domain The set from which the values of the input to a function or relation must come.

evaluating a function Replacing the independent variable (or argument) of a function by a number or expression.

function A relation that matches only one member of the range to each member of the domain; represented by $y = f(x)$.

independent variable Variable whose values come from the domain of a function; if $y = f(x)$, then x is the independent variable

relation A correspondence that matches members of one set (domain) with members of another set (range).

4.2 Graphing Functions and Relations

absolute value function defined by $y = f(x) = |x| = \begin{cases} x, & \text{if } x \geq 0, \\ -x, & \text{if } x < 0. \end{cases}$

asymptote A straight line that the graph of a function approaches but never reaches; an asymptote can be horizontal, vertical, or oblique.

graphical test for functions A relation is a function only if the vertical line drawn at each value in the domain intersects the graph of the relation at at most one point.

graphing functions

1. Determine the domain of the function.
2. Determine several pairs of values from the domain and range. The member from the domain will be the x-coordinate and the member from the range will be the y-coordinate.
3. Plot the values on a Cartesian coordinate system.
4. Draw a smooth curve through the points.

linear function Function whose graph is a straight line; has form

$$y = mx + b.$$

nonlinear function Function whose graph is not a straight line.

piecewise linear function Function whose graph is made of different straight-line segments joined together.

polynomial function of degree *n* Has form

$$y = f(x) = a_n x^n + a_{n-1}x^{n-1} + \cdots + a_1 x + a_0, \qquad a_n \neq 0.$$

Turning points: points where a function reaches a maximum or minimum; a polynomial function of degree n has at most $n - 1$ turning points.

quadratic function Has form $y = f(x) = ax^2 + bx + c$.

Parabola: graph of a quadratic function; opens up or down.

Axis of symmetry: the vertical line $x = \dfrac{-b}{2a}$.

Vertex: the point of maximum or minimum value; has coordinates

$$\left(\frac{-b}{2a}, \frac{-b^2}{4a} + c\right).$$

rational function Has form $y = \dfrac{f(x)}{g(x)}$, $g(x) \neq 0$, where $f(x)$ and $g(x)$ are polynomial functions.

step function Function whose graph is a series of horizontal line segments.

***x*-intercept(s)** Point(s) where graph of function crosses x-axis; found by solving $y = f(x) = 0$.

***y*-intercept** Point where graph of function crosses y-axis, given by $f(0)$.

4.3 Variation

direct variation Has form $y = kx$, $k > 0$.

Proportion: $\dfrac{y_1}{y_2} = \dfrac{x_1}{x_2}$.

direct variation (inverse) Has form $y = \dfrac{k}{x}$, $k > 0$.

Proportion: $\dfrac{y_1}{y_2} = \dfrac{x_2}{x_1}$.

quadratic variation Has form $y = kx^2$, $k > 0$.

Proportion: $\dfrac{y_1}{y_2} = \dfrac{x_1^2}{x_2^2}$.

quadratic variation (inverse) Has form $y = \dfrac{k}{x^2}$, $k > 0$.

Proportion: $\dfrac{y_1}{y_2} = \dfrac{x_2^2}{x_1^2}$.

4.4 Sequences and Series

arithmetic sequence One in which each term after the first is found by adding the same number to each preceding term.

Common difference: the amount added to each term in an arithmetic sequence, denoted by d.

nth term of an arithmetic sequence: $a_n = a_1 + (n - 1)d$.

Sum of an arithmetic sequence:

$$s_n = \left(\frac{n}{2}\right)[2a_1 + (n - 1)d]$$
$$= \left(\frac{n}{2}\right)(a_1 + a_n).$$

geometric sequence Each term after the first is found by multiplying the preceding term by the same number.

Common ratio: the constant multiplier in a geometric sequence, denoted by r.

nth term of a geometric sequence: $a_n = a_1 r^{n-1}$.

Sum of a geometric sequence: $s_n = a_1 \dfrac{(r^n - 1)}{(r - 1)}$.

sequence Function whose domain is the set of natural numbers.

Terms of a sequence: denoted by $a_1, a_2, \ldots$.

series A sequence whose terms are found by adding up the terms of a given sequence.

Terms of a series: denoted by $s_1, s_2, \ldots$.

sigma notation Used to indicate summation, denoted by $\sum$.

Summation index: denoted by i.

Lower limit of the sum: initial value of i.

Upper limit of the sum: largest value of i.

Properties of sums:

1. $\sum_{i=1}^{n} c = nc$, where c is a constant.
2. $\sum_{i=j}^{n} ca_i = c \sum_{i=j}^{n} a_i$
3. $\sum_{i=j}^{n} [a_i \pm b_i] = \sum_{i=j}^{n} a_i \pm \sum_{i=j}^{n} b_i$, where a_i and b_i are the general terms of two sequences.

REVIEW PROBLEMS

Evaluate the function $f(x) = 3x^2 - 3x + 2$ at the following points.

1. $x = 0$

2. $x = -1$

3. $x = 3$

4. $x = 2 - h$

Given that $f(x) = 2x - 1$ and $g(x) = 2x^2 - 1$, evaluate the following.

5. $f(x) + g(x)$

6. $f(x) - g(x)$

7. $g(x) - f(x)$

8. $f(x) \cdot g(x)$

9. $f(x) \div g(x)$

10. $f[g(1)]$

11. $g[f(1)]$

12. $f[g(x)]$

13. $g[f(x - a]$

Solve the following equations.

14. $|2x - 1| = 8$

15. $2x - |x| = 6$

Graph the following functions.

16. $y = 4x + 2$

17. $y = \dfrac{-2x}{3} + 1$

18. $2x + y = 5$

19. $3(x - y) - 15 = 0$

20. $y = |2x - 1|$

21. $y = 2x - |3 - x|$

22. $y = 3x^2 - 5$

23. $y = -(x - 2)^2$

24. $y = -2x^2 + 5x - 2$

25. $y = -(4 - 3x)^2$

26. $y = \dfrac{-2}{x}$

27. $y = \dfrac{3x}{x - 1}$

28. $y = \dfrac{-(6 - 2x)}{4 - 3x}$

29. $y = \dfrac{x}{|x|}$

30. $y = x^3 - 2x^2 + 3x - 4$

31. $y = x^4 - x^3 + x^2 - x + 2$

32. The dosage of a certain antibiotic varies directly with the weight of the patient. If a child weighing 50 pounds should take 1 teaspoon, how much should a child weighing 30 pounds take?

33. The voltage in a device of fixed power consumption varies inversely with the resistance of the device. If a device with a resistance of 100 ohms requires 150 volts, what voltage will be required if the device has a resistance of 125 ohms?

34. The kinetic energy of a moving object varies directly with the square of its velocity. A bullet has a muzzle velocity of 2500 feet per second (fps). By how much does the kinetic energy increase if the muzzle is increased to 3000 fps?

35. The weight (actually the mass, but for this problem weight can be used) of a moving object of fixed kinetic energy varies inversely with the square of the velocity. A car weighing 2500 pounds and traveling 35 mph has a certain kinetic energy. How much should a car traveling 55 mph weigh to have the same kinetic energy?

Identify each of the following sequences as arithmetic, geometric, or neither. Give the first four terms of each sequence.

36. $a_n = 2n + 4$

37. $a_n = 6 - 2n$

38. $a_n = \dfrac{2 - n}{4n}$

39. $a_n = 3n^2 - 2n - 2$

40. $a_n = 2^n - 1$

41. $\dfrac{a_n = 5^{2n-1}}{n^3}$

Identify each of the following sequences as being arithmetic, geometric, or neither. If the sequence is arithmetic, give the common difference. If the sequence is geometric, give the common ratio.

42. $-2, 3, 8, 13, \ldots$

43. $\dfrac{2}{x}, -\dfrac{4}{x^2}, \dfrac{8}{x^3}, \dfrac{16}{x^4}, \ldots$

44. $15, -5, 10, -10, \ldots$

45. $100, -50, -300, -450, \ldots$

Find the appropriate term for the following sequences.

46. a_6 for the sequence $-2, 3, 8, 13, \ldots$

47. a_6 for the sequence with $a_1 = 10$ and $r = 1.05$

48. a_{12} for the sequence with $a_1 = 93$ and $d = -4$

49. a_{15} for the sequence with $a_{10} = 100$ and $d = 3$

50. a_{20} for the sequence with $a_{15} = 1000$ and $d = 0.95$

Find the appropriate sum.

51. s_6 for the sequence with $a_1 = 10$ and $r = 1.05$

52. s_{15} for the sequence with $a_1 = 93$ and $d = -4$

53. s_8 for the sequence with $a_1 = 1000$ and $d = 0.9$

54. s_5 for the sequence with $a_1 = -45$ and $d = 20$

Write the following in expanded form.

55. $\sum_{i=1}^{5} (2i - 3)^2$

56. $\sum_{i=1}^{6} \dfrac{f(i^2)}{i}$

57. $\sum_{i=2}^{5} g[f(i - 1)]$

58. $\sum_{i=6}^{10} \dfrac{g(i - 1)}{f(3 - i)}$

Evaluate the following.

59. $\sum_{i=4}^{11} (3 - 2i)$

60. $\sum_{i=3}^{6} 1000(1.1)^i$

61. $\sum_{i=1}^{5} (4i - 1) + [7 - 25(-2)^i]$

5

VECTORS, MATRICES, AND RELATED OPERATIONS

ABC Breweries operates plants in Lizard Lick and Ypsilanti. Each plant produces 7-ounce bottles, 12-ounce bottles, and cans for two brands (Green Label and Purple Label). Last week's production figures were as follows.

Lizard Lick

	Number of		
	7-oz	12-oz	Cans
Green Label	1500	1200	2500
Purple Label	2200	3000	2000

Ypsilanti

	Number of		
	7-oz	12-oz	Cans
Green Label	2250	2500	4000
Purple Label	6500	5000	6500

Selling prices are as follows.

	7-oz	12-oz	Cans
Green Label	\$0.40	\$0.65	\$0.60
Purple Label	\$0.30	\$0.60	\$0.55

a) What was last week's total production? (To be solved in Example 13 of Section 5.2.)
b) How much more than Lizard Lick did Ypsilanti produce? (To be solved in Example 17 of Section 5.2.)
c) What would total production be if Lizard Lick's production were doubled and Ypsilanti's halved? (To be solved in Example 15 of Section 5.2.)
d) How much revenue did each plant generate last week? What was the total revenue? (To be solved in Example 24 of Section 5.2.)

PROBLEM FROM LIFE

INTRODUCTION

The study of vectors and matrices has been important for mathematicians, statisticians, scientists, and engineers for some time, especially since the advent of the digital computer. People in business and the social sciences have recently become aware that the study of vectors and matrices can aid them in dealing with situations that involve more than one variable. The Problem from Life is an example of a situation that can be made easier to analyze by using matrices, as we will demonstrate.

You will find that knowledge of vectors and matrices will help you to solve systems of linear equations (Chapter 6), enable you to understand the theory of linear programming and the mechanics of the simplex method (Chapters 7 and 8), and aid in the discussion of Markov chains (Chapter 11). Those of you who study computer programming will also find the material in this chapter quite useful.

5.1 VECTORS AND VECTOR OPERATIONS

Vectors

A **vector** is an ordered set of numbers or **elements.** For example, the ordered pair (3, 2) is a vector of **dimension** (or size) two, while (5, 2, 3) is a vector of dimension three. We will represent vectors by lowercase boldface letters, such as **a, b, c,** and **x.**

When the dimension of a vector is two or three, we can portray it as an arrow going from the origin to a specific point. In Fig. 5.1, we show the vector $\mathbf{a} = (3, 2)$ and in Fig. 5.2 the vector $\mathbf{b} = (5, 3, 2)$.

In general, vectors have both a *direction,* which points from the origin to the point in space corresponding to the elements of the vector, and a *magnitude* (length). For example, in Fig. 5.2, the vector $\mathbf{b} = (5, 3, 2)$ is defined by the direction shown by the arrow and by the length of the arrow. We need both a direction and a magnitude because we must be able to distinguish between two vectors such as (1, 1) and (2, 2), which point in the same direction but have different magnitudes. (See Fig. 5.3.)

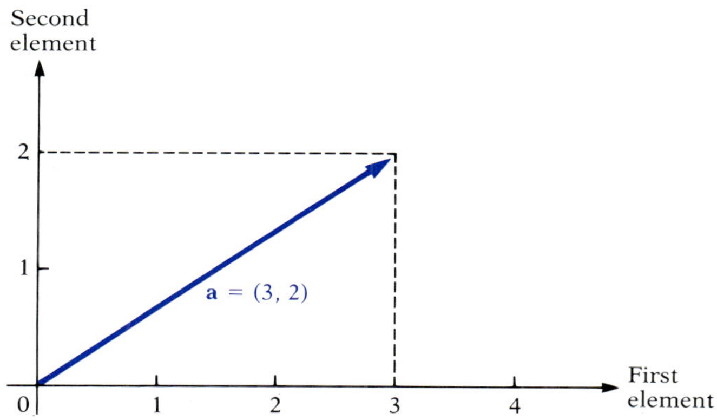

Figure 5.1 **Representing the vector a = (3, 2).**

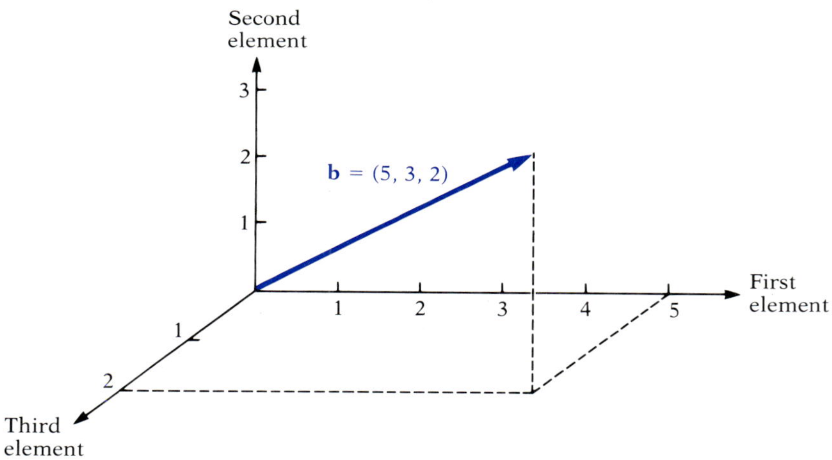

Figure 5.2 **Representing the vector b = (5, 3, 2).**

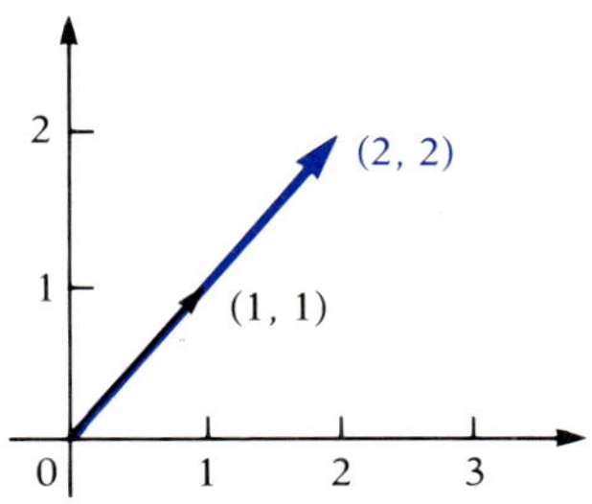

Figure 5.3 **Vectors have both direction and magnitude.**

We can use vectors to help us organize data as the following example shows.

EXAMPLE 1

Consider the ABC Tire Company. Suppose that January's production figures are shown in the following table. Represent these data by a vector.

Type of tire	January production
Radial	500
Truck	200

SOLUTION

We can represent these figures by the vector $\mathbf{j} = (500, 200)$ if we agree to let the first element represent the number of radial tires and the second represent the number of truck tires. □

The letter used to name a vector is generally related to the data being portrayed. In Example 1, we used $\mathbf{j}$ to represent the vector of tire production during January. The following examples show more instances in which vectors can be used to represent data that are ordered in some fashion.

EXAMPLE 2

After the last game at a baseball stadium, the following data on ticket sales were collected. Represent these data by a vector.

Type of ticket	Number sold
Box seat	1000
Reserve seat	2000
General admission	3500
Bleacher	5000

SOLUTION

We can represent these data by the vector

$$\mathbf{t} = (1000, 2000, 3500, 5000).$$

Here the elements represent the number of box seat, reserve seat, general admission, and bleacher tickets sold, respectively. □

EXAMPLE 3

At the end of the semester, a professor found that the grades in her mathematics class were as follows. Represent these data by a vector.

Grade	Number
A	5
B	11
C	25
D	18
F	8
W	15

SOLUTION

We can represent these data by the vector

$$\mathbf{g} = (5, 11, 25, 18, 8, 15),$$

in which the elements list the number of students who received grades of A, B, C, D, F, or W, respectively. □

Practice Problem

An ice cream store recorded the following data on the number of pints of its homemade ice cream sold during the last week.

Flavor	Number pints sold
Vanilla	50
Chocolate	75
Strawberry	15
Chocolate chip	100

Represent these data by a vector.

Answer: $\mathbf{i} = (50, 75, 15, 100)$

In many cases we will need to refer to a specific element of a vector. Mathematicians often use **subscripts** to enable them to refer to elements of a vector. For example, rather than "the fourth element of vector **a**," we have $\mathbf{a}_4$. The *i*th element of the vector **b** would be $\mathbf{b}_i$. We will make use of subscripts throughout the discussion of vectors and (later on) of matrices. You will find the use of subscripts is especially useful in computer programming. (Vectors are often referred to as *arrays* in many computer courses.)

Practice Problem

Given the vector $\mathbf{v} = (3, 4.7, -1, 3.25)$.

a) What is the value of $\mathbf{v}_3$?

b) What element of **v** has the value 3?

Answers: **a)** -1; **b)** $\mathbf{v}_1$

Vector Operations

As with real numbers, there are a variety of operations that can be performed with vectors.

Equality. Two vectors are equal if they have the same dimension (number of elements) and if their corresponding elements are equal.

> We say that vectors **a** and **b** of dimension n are *equal* if
>
> $$\mathbf{a}_i = \mathbf{b}_i \qquad \text{for } i = 1 \text{ to } n.$$ **Vector equality**

EXAMPLE 4

Suppose that ABC Tire produced 400 radials and 300 truck tires in February. We can represent these data by the vector $\mathbf{f} = (400, 300)$. The vectors $\mathbf{j} = (500, 200)$ and $\mathbf{f} = (400, 300)$ are not equal because $\mathbf{j}_1 \neq \mathbf{f}_1$ and $\mathbf{j}_2 \neq \mathbf{f}_2$ (neither the first nor second elements are equal). □

Practice Problem

a) Are the vectors $\mathbf{a} = (1, 1, 2)$ and $\mathbf{b} = (1, 1, 2, 0)$ equal?

b) Are the vectors $\mathbf{c} = (1, 0, 1)$ and $\mathbf{d} = (1, 1, 0)$ equal?

c) Are the vectors $\mathbf{e} = (2, 3)$ and $\mathbf{f} = (2, 3)$ equal?

Answers: **a)** No, because **a** has dimension three while **b** has dimension four; **b)** no, because $\mathbf{c}_2$ and $\mathbf{c}_3$ differ from $\mathbf{d}_2$ and $\mathbf{d}_3$, respectively; **c)** yes

Addition. We can add vectors of the same dimension by adding corresponding elements. We can represent vector addition by using subscripts as follows.

> Let **a** and **b** be two vectors of the same dimension, say n. Then $\mathbf{c} = \mathbf{a} + \mathbf{b}$ is given by
>
> $$\mathbf{c}_i = \mathbf{a}_i + \mathbf{b}_i, \qquad \text{for } i = 1 \text{ to } n.$$ **Vector addition**

EXAMPLE 5

The ABC Tire Company has produced a total of 900 radials and 500 truck tires in January and February. (See Examples 1 and 4.) Show this by adding the January and February production vectors.

SOLUTION

We have

$$\begin{aligned}\mathbf{j} + \mathbf{f} &= (500, 200) + (400, 300) \\ &= (500 + 400, 200 + 300) \\ &= (900, 500).\end{aligned}$$ □

Practice Problem

Add the vectors $\mathbf{v} = (3.5, 2.75, 4)$ and $\mathbf{w} = (3.1, -5, 1.3)$.

Answer: $\mathbf{v} + \mathbf{w} = (6.6, -2.25, 5.3)$

Multiplication by a Scalar. In the context of vectors (and matrices), we refer to a single real number as a **scalar.** To multiply a vector by a scalar, simply multiply each element of the vector by the scalar, which results in a new vector of the same dimension.

> Let $\mathbf{v}$ be a vector of dimension n and let s be a scalar. Then $s\mathbf{v}$ is given by
>
> $s\mathbf{v}_i$, for $i = 1$ to n. **Multiplication by a scalar**

EXAMPLE 6

Suppose we know that in March ABC produced exactly 1.5 times February's production. Compute the March production vector $\mathbf{m}$.

SOLUTION

$$\begin{aligned} \mathbf{m} &= 1.5\mathbf{f} \\ &= 1.5(400, 300) \\ &= (1.5 \cdot 400, 1.5 \cdot 300) \\ &= (600, 450) \quad \square \end{aligned}$$

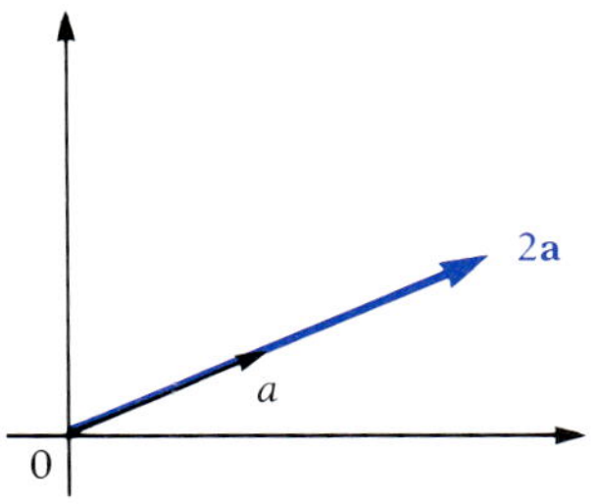

Figure 5.4 **Multiplying a vector by a positive scalar can change magnitude, not direction.**

Practice Problem

Multiply the vector $\mathbf{v} = (3, 6, 8)$ by the scalars

a) 2 **b)** -3 **c)** 0.5

Answers: **a)** (6, 12, 16); **b)** $(-9, -18, -24)$; **c)** (1.5, 3, 4)

When a vector is multiplied by a positive scalar, the *direction* of the resulting vector will be the same as the direction of the original vector; the *magnitude* of the resulting vector will be the magnitude of the original vector multiplied by the scalar. (See Fig. 5.4.) When a vector is multiplied by a negative scalar, the *direction* of the resulting vector will be the opposite of the direction of the original vector. The *magnitude* of the result will be the magnitude of the original vector multiplied by the absolute value of the scalar. (See Fig. 5.5.)

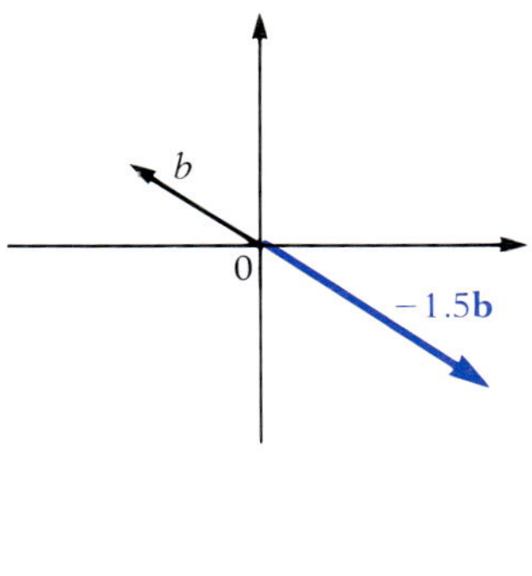

Figure 5.5 **Multiplying a vector by a negative scalar changes direction and magnitude.**

Subtraction of Vectors. Vectors of the same dimension can be subtracted by subtracting the elements of the second vector from the corresponding elements of the first vector.

> Let **a** and **b** be vectors of the same dimension, say n. Then $\mathbf{c} = \mathbf{a} - \mathbf{b}$ is given by
>
> $$\mathbf{c}_i = \mathbf{a}_i - \mathbf{b}_i, \qquad \text{for } i = 1 \text{ to } n.$$ **Subtraction of vectors**

EXAMPLE 7 Measure the change in ABC's tire production from January to February. (See Examples 1 and 4.)

SOLUTION

$$\begin{aligned}\mathbf{f} - \mathbf{j} &= (400, 300) - (500, 200)\\ &= (400 - 500, 300 - 200)\\ &= (-100, 100)\end{aligned}$$

We can interpret this to mean that ABC produced 100 fewer radial tires and 100 more truck tires in February as compared to January. □

Practice Problem

Perform the following vector subtractions.

a) $(4, 3) - (1, 2)$ **b)** $(4, 8, -5, 6) - (2, 3, -2, -1)$

Answers: **a)** $(3, 1)$; **b)** $(2, 5, -3, 7)$

Dot Product. Two vectors of the same dimension can be "multiplied" to yield a real number (scalar). This operation is called the **dot** (or **inner**) **product**. The resulting scalar is equal to the sum of the products of the corresponding elements.

> Let **a** and **b** be two vectors of the same dimension. Then
>
> $$\mathbf{a} \bullet \mathbf{b} = \mathbf{a}_1 \cdot \mathbf{b}_1 + \mathbf{a}_2 \cdot \mathbf{b}_2 + \cdots + \mathbf{a}_n \cdot \mathbf{b}_n.$$ **Dot product**

EXAMPLE 8 Suppose ABC Tire Company sells its radial tires for \$40 each and its truck tires for \$30 each. Then the selling prices can be represented by the vector $\mathbf{p} = (40, 30)$. Compute the revenue from sales in January and in February.

SOLUTION

The revenue from sales in January is given by the dot product

$$\begin{aligned}\mathbf{p} \bullet \mathbf{j} &= (40, 30) \bullet (500, 200)\\ &= 40 \cdot 500 + 30 \cdot 200\\ &= 20{,}000 + 6000\\ &= \$26{,}000.\end{aligned}$$

The sales income for February is given by the dot product

$$\begin{aligned}\mathbf{p} \bullet \mathbf{f} &= (40, 30) \bullet (400, 300)\\ &= 40 \cdot 400 + 30 \cdot 300\\ &= 16{,}000 + 9000\\ &= \$25{,}000. \quad \square\end{aligned}$$

Practice Problem

Compute the following dot products.

a) $(2, 1) \bullet (3, 2)$ **b)** $(3, 2) \bullet (2, 1)$ **c)** $(4.5, 2.5) \bullet (-1, 4)$

Answers: **a)** 8; **b)** 8; **c)** 5.5

The following summarizes the vector operations we've discussed.

Vector Operations

For vectors **a**, **b**, and **c** of order (dimension) n and s a scalar,

1. Equality: $\mathbf{a} = \mathbf{b}$ if $\mathbf{a}_i = \mathbf{b}_i$ for $i = 1$ to n
2. Addition: $\mathbf{c} = \mathbf{a} + \mathbf{b}$ is given by $\mathbf{c}_i = \mathbf{a}_i + \mathbf{b}_i$ for $i = 1$ to n
3. Multiplication by a scalar: $s\mathbf{a}$ is given by $s\mathbf{a}_i$ for $i = 1$ to n
4. Subtraction: $\mathbf{c} = \mathbf{a} - \mathbf{b}$ is given by $\mathbf{c}_i = \mathbf{a}_i - \mathbf{b}_i$ for $i = 1$ to n
5. Dot Product: $\mathbf{a} \bullet \mathbf{b} = \mathbf{a}_1 \cdot \mathbf{b}_1 + \mathbf{a}_2 \cdot \mathbf{b}_2 + \cdots + \mathbf{a}_n \cdot \mathbf{b}_n$

EXERCISE SET 5.1

What are the dimensions of the following vectors?

1. $(2, 6)$ **2.** $(5, 6, 4, 2)$ **3.** $(-2.3, 6.08)$
4. $(2, 65, -4)$ **5.** $(3, 4, 5)$ **6.** $(-5.57, 6, 5.34)$

Find the indicated sums.

7. $(1, 2, 3) + (2, 4, 6)$ **8.** $(2, -6) + (-5, 1.5)$
9. $(7, 4.3) + (-2, 5.1)$ **10.** $(1, 2, 4, 5) + (-3, 0.1, 0, 2)$

11. $(-0.9, 8) + (2.2, -3)$
12. $(1,0, -7,0.6) + (0.2, -4,9, -0.8)$

Which of the following are true?

13. $(2, 6) = (6, 2)$
14. $(-2, 6) = (-2, 6)$
15. $(3, -1) + (-2, 3) = (1, 2)$
16. $(3, 0.1, 0) + (2, 3) = (5, 3.1, 0)$

Find the results of the following scalar multiplications.

17. $2(1, 2)$
18. $-1(2, -4, 6)$
19. $2(1.5, 2.3, -3)$
20. $-3.1(7, 4)$
21. $4(2.2, -5.5)$
22. $5.5(-1, 0, 7, -6)$
23. $-2(2, 4, -3, 2)$
24. $-4(-1.3, 2.1, -3.6, 4)$

Find the indicated differences.

25. $(7, 4) - (2, 5)$
26. $(5, 1) - (-2, 6)$
27. $(2, 4, 6) - (-1, 0.2, 3)$
38. $(1.2, -3) - (2, 4.6)$
29. $(9, 8) - (-2, 3)$
30. $(3, 1, 0, 0.2) - (1, -2, 0.4, 0.5)$

Perform the following operations.

31. $3(7, 4) - 2(2, -5)$
32. $2(-7, 4) + 3(2, -5)$
33. $2(1, 2.5, 0.3) + 3(2, -4, 0.6)$
34. $3(1, 2.2, 0.33) - 2(2, -4.4, 0.66)$
35. $0.2(1, 2, 4, -5) + 3(-3, 0.1, 0. 12)$
36. $3(1, 2, 4, 5) - 2(0.33, 1.1, 0, 0.2)$
37. $-3(2, 5) + 2(7, 4)$
38. $-3(9, 8) + 2(-2, 0.6)$
39. $-0.3(1, 2, -3) + 0.2(2, -4, 3.6)$
40. $-3(1, 2, 4, 5) + 2.5(3, 1, 0, -2)$

Find the following dot products.

41. $(2, 6) \cdot (5, 1)$
42. $(9, 8) \cdot (-2, 3)$
43. $(1, -2, 3) \cdot (0.2, 4, 6)$
44. $(1, 2, -4, 5) \cdot (3, -1, 0, 2)$
45. $(0.5, 1) \cdot (2, 0.6)$
46. $(0.1, 0, 7, 6) \cdot (2, -4, 0.9, 8)$
47. $(-0.9, 0.8) \cdot (0.2, -0.3)$
48. $(0.1, -2, 0.3) \cdot (-2, -0.4, 6)$

Using the vectors $\mathbf{o} = (0, 0)$, $\mathbf{x} = (2, 3)$, $\mathbf{y} = (5, 2)$, and $\mathbf{z} = (6, 1)$, verify that the following are true.

49. $\mathbf{x} + \mathbf{o} = \mathbf{x}$ (additive identity for vectors)
50. $\mathbf{o} + \mathbf{y} = \mathbf{y}$ (additive identity for vectors)
51. $\mathbf{x} + \mathbf{y} = \mathbf{y} + \mathbf{x}$ (commutative property of vector addition)
52. $(\mathbf{x} + \mathbf{y}) + \mathbf{z} = \mathbf{x} + (\mathbf{y} + \mathbf{z})$ (associative property of vector addition)
53. $2(\mathbf{x} + \mathbf{z}) = 2\mathbf{x} + 2\mathbf{z}$ (distributive property of scalar multiplication)
54. $\mathbf{y} + (-\mathbf{y}) = \mathbf{o}$ (inverse property of vector addition)

55. $\mathbf{x} \cdot \mathbf{y} = \mathbf{y} \cdot \mathbf{x}$ (commutative property of dot product)

56. $\mathbf{x} \cdot (2\mathbf{y}) = 2(\mathbf{x} \cdot \mathbf{y})$

57. $\mathbf{x} \cdot (\mathbf{y} + \mathbf{z}) = (\mathbf{y} + \mathbf{z}) \cdot \mathbf{x} = \mathbf{x} \cdot \mathbf{y} + \mathbf{x} \cdot \mathbf{z}$

58. Which of the properties in Problems 49–57 do you think hold in general for all vectors?

59. John and Paul go to the local lumber yard. John buys ten 2 × 4s, three pieces of 4 ft × 8 ft plywood, and 1 pound of nails. Paul buys twenty 2 × 4s, five pieces of 4 ft × 8 ft plywood, and 2.5 pounds of nails.

a) Represent their purchases as vectors.

b) What is the vector representing their total purchases?

c) How much more than John did Paul buy?

d) Suppose 2 × 4s cost $2.00 each, 4 ft × 8 ft plywood costs $9.00 a sheet, and nails cost $0.60 per pound. How much did John spend? How much did Paul spend? How much did both spend?

5.2 MATRICES AND MATRIX OPERATIONS

Matrices

We can extend our discussion of vectors to include numbers arranged in a table. A **matrix** is a rectangular array (table) of numbers. The numbers are called the **elements** of the matrix. We will denote a matrix by enclosing the array in brackets and represent it by a capital letter. Thus

$$A = \begin{bmatrix} 1 & 2 & 3 & 4 \\ 5 & 6 & 7 & 8 \end{bmatrix}$$

is a matrix with two rows and four columns.

The **order** (or **dimension**) of a matrix is its number of rows and number of columns. The matrix A shown in the preceding paragraph is of order 2 × 4 (read "2 by 4") because it has two rows and four columns.

Practice Problem

What is the order of $\begin{bmatrix} 3 & 4 & 7 \\ 0 & 9 & 2 \end{bmatrix}$?

Answer: 2 × 3

A **square** matrix is a matrix with the same number of rows as columns. A matrix that is not necessarily square is called **rectangular**.

EXAMPLE 9

Let

$$A = \begin{bmatrix} 1 & 2 \\ 3 & 4 \end{bmatrix}, \qquad B = \begin{bmatrix} 2 & 4 & 8 \\ 3 & 9 & 27 \end{bmatrix},$$

$$C = \begin{bmatrix} 3 & 0 & 2 & 7 \\ 4 & 6 & 5 & 1 \\ 1 & 2 & 3 & 4 \end{bmatrix}, \qquad D = \begin{bmatrix} 1 & 0 & 0 & 0 \\ 0 & 1 & 0 & 0 \\ 0 & 0 & 1 & 0 \\ 0 & 0 & 0 & 1 \end{bmatrix}.$$

Determine the order of each matrix and whether it is square or rectangular.

SOLUTION

A is 2×2 and square. B is 2×3 and rectangular. C is 3×4 and rectangular, and D is 4×4 and square. □

EXAMPLE 10

We can use matrices to organize data into rows and columns. Based upon the examples in Section 5.1, we know that ABC Tire's first-quarter production figures are as shown in the table following. Represent these data by a matrix.

Tire production

	Month		
Type	**Jan.**	**Feb.**	**Mar.**
Radial	500	400	600
Truck	200	300	450

SOLUTION

We can represent these data with the matrix P, with the understanding that the rows refer to the type of tire (radial or truck) and the columns refer to the month (January, February, or March), as shown:

$$P = \begin{bmatrix} 500 & 400 & 600 \\ 200 & 300 & 450 \end{bmatrix}. \quad \square$$

One way to think of the matrix P is to notice that P is made up of the three monthly production vectors **j**, **f**, and **m** (see Examples 1, 4, and 6) listed side by side. We can also think of P as being made by "stacking" the two vectors $\mathbf{r} = (500, 400, 600)$ and $\mathbf{t} = (200, 300, 450)$, which give the production of radial and of truck tires, respectively. Although we could keep these vectors separate, it is often convenient to put the vectors into a matrix.

Practice Problem

An ice cream store has recorded the following data, which show the number of pints of ice cream sold during the last three months.

Ice cream sales

	Month		
Flavor	**Jun.**	**Jul.**	**Aug.**
Vanilla	50	60	55
Chocolate	75	90	105
Strawberry	15	40	50
Chocolate chip	100	120	110

Organize these data into a matrix. What do the rows and columns of your matrix represent?

Answer:

$$S = \begin{bmatrix} 50 & 60 & 55 \\ 75 & 90 & 105 \\ 15 & 40 & 50 \\ 100 & 120 & 110 \end{bmatrix}$$

where each row represents a flavor and each column represents a month.

As you can see, the best way to put data into a matrix is to first organize it and place it into a table. (In fact, in computer courses, matrices are often referred to as *tables*.) Any data that can be placed into a table can be represented by a matrix. We've seen two examples, which involved sales by type and by time period. Another example is the stock market report, where the rows represent the various stocks and the columns represent a variety of things, including prices and number of shares traded.

As with vectors, we can use subscripts to describe the elements in a matrix. Since the data in a matrix are classified by row and by column, we need two subscripts to refer to a particular element. For example, in the matrix

$$A = \begin{bmatrix} 3 & 4 & 5 \\ 8 & 2 & 6 \end{bmatrix}$$

we refer to the four in the first row and second column as $a_{1,2}$ and to the eight in the second row and first column as $a_{2,1}$. Note that we use a *small* letter when referring to the numbers (elements) of a matrix. Sometimes we write

$$A = [a_{i,j}] \qquad \text{where } i \text{ is the row and } j \text{ is the column}$$

to indicate the elements of A.

Practice Problem

Given the matrix

$$C = \begin{bmatrix} 3 & 0 & -2 & 0.7 \\ 4 & 0.6 & 5 & 1 \\ 9 & 2 & 0.3 & 0.4 \end{bmatrix}.$$

a) What is the value of $c_{3,3}$?

b) Which element has the value -2? 9?

Answers: **a)** $c_{3,3} = 0.3$; **b)** $c_{1,3} = -2$ and $c_{3,1} = 9$.

Simple Matrix Operations

Equality. Two matrices are **equal** if they are of the same order and their corresponding elements are equal.

> The $m \times n$ matrices A and B are equal if
>
> $a_{i,j} = b_{i,j}$ for $i = 1$ to m and $j = 1$ to n. **Equality of matrices**

EXAMPLE 11

Given the matrices

$$A = \begin{bmatrix} 2 & 4 \\ 3 & 1 \end{bmatrix}, \quad B = \begin{bmatrix} 2 & 4 \\ 3 & 1 \end{bmatrix}, \quad C = \begin{bmatrix} 4 & 2 \\ 1 & 3 \end{bmatrix},$$

$$D = \begin{bmatrix} 3 & 7 \\ 2 & 8 \end{bmatrix}, \quad E = \begin{bmatrix} 3 & x \\ 2 & y \end{bmatrix}, \quad F = \begin{bmatrix} 3 & 7 & 8 \\ 2 & 6 & 1 \end{bmatrix},$$

$$G = \begin{bmatrix} 3 & 7 \\ 2 & 6 \end{bmatrix} \text{ and } H = \begin{bmatrix} 3 & x \\ 2 & z \end{bmatrix}.$$

Answer the following.

a) Are A and B equal?

b) Are A and C equal?

c) Are D and E equal?

d) Are F and G equal?

e) Are F and H equal?

SOLUTION

a) Yes, $A = B$.

b) No, A and C are not equal because their corresponding elements aren't equal.

c) D and E are equal if $x = 7$ and $y = 8$.

d) No, F and G are not equal because they aren't of the same order.

e) No, F and H can never be equal because they aren't of the same order. □

Addition. We can add matrices of the same order by adding their corresponding elements.

> Let A and B be $m \times n$ matrices. Then $C = A + B$ is given by
>
> $c_{i,j} = a_{i,j} + b_{i,j}$ for $i = 1$ to m and $j = 1$ to n. **Matrix addition**

EXAMPLE 12

Given the matrices

$$A = \begin{bmatrix} 2 & 4 \\ 3 & 1 \end{bmatrix}, \quad B = \begin{bmatrix} 1 & 2 \\ 4 & 0 \end{bmatrix},$$

$$C = \begin{bmatrix} 1 & 2 & 3 \\ 4 & 5 & 6 \end{bmatrix}.$$

Find $A + B$ and $A + C$.

SOLUTION

$$A + B = \begin{bmatrix} 2 & 4 \\ 3 & 1 \end{bmatrix} + \begin{bmatrix} 1 & 2 \\ 4 & 0 \end{bmatrix} = \begin{bmatrix} 2+1 & 4+2 \\ 3+4 & 1+0 \end{bmatrix} = \begin{bmatrix} 3 & 6 \\ 7 & 1 \end{bmatrix}$$

We cannot add A and C. □

Practice Problem

Add the matrices

$$A = \begin{bmatrix} 4 & -2 & 3 \\ -1 & 2 & -0.1 \end{bmatrix}$$

and

$$B = \begin{bmatrix} 9 & 0.2 & -2 \\ 0.4 & -1 & 0.7 \end{bmatrix}.$$

Answer: $A + B = \begin{bmatrix} 13 & -1.8 & 1 \\ -0.6 & 1 & 0.6 \end{bmatrix}$

EXAMPLE 13

Problem from Life, part (a) ABC Breweries' production figures last week were as follows:

Lizard Lick

	Number of		
	7-oz	**12-oz**	**Cans**
Green Label	1500	1200	2500
Purple Label	2200	3000	2000

Ypsilanti

	Number of		
	7-oz	**12-oz**	**Cans**
Green Label	2250	2500	4000
Purple Label	6500	5000	6500

What was last week's total production?

SOLUTION

We can answer this question by organizing the data into matrices, using L to represent the production at Lizard Lick and Y the production at Ypsilanti. We have

$$L = \begin{bmatrix} 1500 & 1200 & 2500 \\ 2200 & 3000 & 2000 \end{bmatrix}$$

and

$$Y = \begin{bmatrix} 2250 & 2500 & 4000 \\ 6500 & 5000 & 6500 \end{bmatrix}.$$

Last week's total production is the sum of L and Y:

$$\begin{aligned} L + Y &= \begin{bmatrix} 1500 & 1200 & 2500 \\ 2200 & 3000 & 2000 \end{bmatrix} + \begin{bmatrix} 2250 & 2500 & 4000 \\ 6500 & 5000 & 6500 \end{bmatrix} \\ &= \begin{bmatrix} 1500 + 2250 & 1200 + 2500 & 2500 + 4000 \\ 2200 + 6500 & 3000 + 5000 & 2000 + 6500 \end{bmatrix} \\ &= \begin{bmatrix} 3750 & 3700 & 6500 \\ 8700 & 8000 & 8500 \end{bmatrix}. \quad \square \end{aligned}$$

Multiplication by a Scalar. We can multiply a matrix by a scalar by multiplying each element of the matrix by the scalar.

> Let A be an $m \times n$ matrix and s a scalar. Then sA is given by
>
> $sA = [sa_{i,j}]$ for $i = 1$ to m and $j = 1$ to n. **Multiplication by a scalar**

EXAMPLE 14

Let

$$A = \begin{bmatrix} 2 & 4 \\ 3 & 1 \end{bmatrix} \quad \text{and} \quad B = \begin{bmatrix} 1 & 2 & 3 \\ 4 & 5 & 6 \end{bmatrix}.$$

Find $2A$ and $-3B$.

SOLUTION

$$2A = 2\begin{bmatrix} 2 & 4 \\ 3 & 1 \end{bmatrix} = \begin{bmatrix} 2 \cdot 2 & 2 \cdot 4 \\ 2 \cdot 3 & 2 \cdot 1 \end{bmatrix} = \begin{bmatrix} 4 & 8 \\ 6 & 2 \end{bmatrix}$$

and

$$-3B = -3\begin{bmatrix} 1 & 2 & 3 \\ 4 & 5 & 6 \end{bmatrix} = \begin{bmatrix} -3 \cdot 1 & -3 \cdot 2 & -3 \cdot 3 \\ -3 \cdot 4 & -3 \cdot 5 & -3 \cdot 6 \end{bmatrix}$$

$$= \begin{bmatrix} -3 & -6 & -9 \\ -12 & -15 & -18 \end{bmatrix}. \quad \square$$

Practice Problem

Given

$$A = \begin{bmatrix} 3 & 6 \\ 1 & 2 \end{bmatrix} \quad \text{and} \quad B = \begin{bmatrix} -2 & 1.5 & 6 \\ 3 & 4 & 0.5 \\ 7 & 9 & 0.8 \end{bmatrix}.$$

Find **(a)** $2A$ and **(b)** $-1.5B$.

Answers: **a)** $\begin{bmatrix} 6 & 12 \\ 2 & 4 \end{bmatrix}$; **b)** $\begin{bmatrix} 3 & -2.25 & -9 \\ -4.5 & -6 & -0.75 \\ -10.5 & -13.5 & -1.2 \end{bmatrix}$

EXAMPLE 15

Problem from Life, part (c) What would total production be if Lizard Lick's production were doubled and Ypsilanti's halved? (See Example 13.)

SOLUTION

Here we must compute $2L + 0.5Y$, where

$$L = \begin{bmatrix} 1500 & 1200 & 2500 \\ 2200 & 3000 & 2000 \end{bmatrix} \quad \text{and} \quad Y = \begin{bmatrix} 2250 & 2500 & 4000 \\ 6500 & 5000 & 6500 \end{bmatrix}.$$

The new total production is

$$\begin{aligned} 2L + 0.5Y &= 2\begin{bmatrix} 1500 & 1200 & 2500 \\ 2200 & 3000 & 2000 \end{bmatrix} + 0.5\begin{bmatrix} 2250 & 2500 & 4000 \\ 6500 & 5000 & 6500 \end{bmatrix} \\ &= \begin{bmatrix} 3000 & 2400 & 5000 \\ 4400 & 6000 & 4000 \end{bmatrix} + \begin{bmatrix} 1125 & 1250 & 2000 \\ 3250 & 2500 & 3250 \end{bmatrix} \\ &= \begin{bmatrix} 4125 & 3650 & 7000 \\ 7650 & 8500 & 7250 \end{bmatrix}. \quad \square \end{aligned}$$

Subtraction of Matrices. We can subtract matrices of the same order by subtracting the elements of the second from the corresponding elements of the first.

> Let A and B be $m \times n$ matrices. Then $C = A - B$ is given by
>
> $c_{i,j} = a_{i,j} - b_{i,j}$ for $i = 1$ to m and $j = 1$ to n.
>
> **Subtraction of matrices**

EXAMPLE 16

Let

$$A = \begin{bmatrix} 2 & 4 \\ 3 & 1 \end{bmatrix}, \quad B = \begin{bmatrix} 1 & 2 \\ 4 & 0 \end{bmatrix}, \quad \text{and} \quad C = \begin{bmatrix} 1 & 2 & 3 \\ 4 & 5 & 6 \end{bmatrix}.$$

Find $A - B$, $B - A$, $A - C$, $C - A$, $B - C$, and $C - B$.

SOLUTION

$$A - B = \begin{bmatrix} 2 & 4 \\ 3 & 1 \end{bmatrix} - \begin{bmatrix} 1 & 2 \\ 4 & 0 \end{bmatrix} = \begin{bmatrix} 2-1 & 4-2 \\ 3-4 & 1-0 \end{bmatrix} = \begin{bmatrix} 1 & 2 \\ -1 & 1 \end{bmatrix},$$

$$B - A = \begin{bmatrix} 1 & 2 \\ 4 & 0 \end{bmatrix} - \begin{bmatrix} 2 & 4 \\ 3 & 1 \end{bmatrix} = \begin{bmatrix} 1-2 & 2-4 \\ 4-3 & 0-1 \end{bmatrix} = \begin{bmatrix} -1 & -2 \\ 1 & -1 \end{bmatrix},$$

but $A - C$, $C - A$, $B - C$, and $C - B$ are not defined because the matrices are of different dimensions. $\square$

EXAMPLE 17

Problem from Life, part (b) How much more than Lizard Lick did Ypsilanti produce?

SOLUTION

Here we must compute $Y - L$ (see Example 13). We have

$$Y - L = \begin{bmatrix} 2250 & 2500 & 4000 \\ 6500 & 5000 & 6500 \end{bmatrix} - \begin{bmatrix} 1500 & 1200 & 2500 \\ 2200 & 3000 & 2000 \end{bmatrix}$$
$$= \begin{bmatrix} 750 & 1300 & 1500 \\ 4300 & 2000 & 4500 \end{bmatrix}. \quad \square$$

Practice Problem

Given the matrices

$$A = \begin{bmatrix} 3.5 & 4 \\ -2 & 6 \end{bmatrix} \quad \text{and} \quad B = \begin{bmatrix} -1.5 & 4.1 \\ 3 & 1.5 \end{bmatrix}.$$

Find **(a)** $A - B$ and **(b)** $B - A$

Answers: **a)** $\begin{bmatrix} 5 & -0.1 \\ -5 & 4.5 \end{bmatrix}$; **b)** $\begin{bmatrix} -5 & 0.1 \\ 5 & -4.5 \end{bmatrix}$

Transpose of a Matrix. At times we will find it useful to interchange the rows and columns of a matrix, which results in a new matrix called the **transpose** of the original matrix. Let A be $m \times n$. Then the transpose of A, denoted A^T, is $n \times m$ and is

$$A^T = [a_{j,i}] \quad \text{for } j = 1 \text{ to } n \text{ and } i = 1 \text{ to } m. \qquad \textbf{Transpose of a matrix}$$

EXAMPLE 18

Let

$$A = \begin{bmatrix} 2 & 4 \\ 3 & 1 \end{bmatrix} \quad \text{and} \quad B = \begin{bmatrix} 1 & 2 & 3 \\ 4 & 5 & 6 \end{bmatrix}.$$

Find A^T and B^T. Give the order of each.

SOLUTION

$$A^T = \begin{bmatrix} 2 & 3 \\ 4 & 1 \end{bmatrix},$$

which is 2×2. (The first row of A is the first column of A^T and the second row of A is the second column of A^T.)

$$B^T = \begin{bmatrix} 1 & 4 \\ 2 & 5 \\ 3 & 6 \end{bmatrix},$$

which is 3×2. The rows of B have become the columns of B^{T} (and vice versa). □

Practice Problem

Given the matrices

$$A = \begin{bmatrix} 3.5 & 4 \\ -2 & 6 \end{bmatrix} \quad \text{and} \quad B = \begin{bmatrix} 4 & 9 & -1 & 2.5 \\ 0.4 & 2 & 8 & -4 \\ 1 & 3 & 5 & 7 \end{bmatrix}.$$

Find A^{T} and B^{T}.

Answers: $A^{\mathrm{T}} = \begin{bmatrix} 3.5 & -2 \\ 4 & 6 \end{bmatrix}$, which is 2×2, and

$$B^{\mathrm{T}} = \begin{bmatrix} 4 & 0.4 & 1 \\ 9 & 2 & 3 \\ -1 & 8 & 5 \\ 2.5 & -4 & 7 \end{bmatrix}, \text{ which is } 4 \times 3.$$

Some Special Matrices; Vectors and Matrices

A **zero** matrix is a matrix consisting of zeros; it is represented by 0. A zero matrix need not be square. Note that a zero matrix (of the correct order!) added to any matrix A results in A. Thus, the zero matrix is an *additive identity*.

EXAMPLE 19

Let

$$A = \begin{bmatrix} 3 & 2 & 6 \\ 4 & 5 & 1 \end{bmatrix} \quad \text{and} \quad 0 = \begin{bmatrix} 0 & 0 & 0 \\ 0 & 0 & 0 \end{bmatrix}.$$

Find $A + 0$ and $0 + A$.

SOLUTION

$$A + 0 = \begin{bmatrix} 3 & 2 & 6 \\ 4 & 5 & 1 \end{bmatrix} + \begin{bmatrix} 0 & 0 & 0 \\ 0 & 0 & 0 \end{bmatrix} = \begin{bmatrix} 3 & 2 & 6 \\ 4 & 5 & 1 \end{bmatrix} = A$$

$$0 + A = \begin{bmatrix} 0 & 0 & 0 \\ 0 & 0 & 0 \end{bmatrix} + \begin{bmatrix} 3 & 2 & 6 \\ 4 & 5 & 1 \end{bmatrix} = \begin{bmatrix} 3 & 2 & 6 \\ 4 & 5 & 1 \end{bmatrix} = A \quad □$$

The following is a summary of the matrix operations we've discussed so far.

Matrix Operations

Let A, B, and C be matrices of order (dimensions) $m \times n$ and let s be a scalar.

1. Equality: $A = B$ if $a_{i,j} = b_{i,j}$ for $i = 1$ to m and $j = 1$ to n.
2. Addition: $C = A + B$ is given by $c_{i,j} = a_{i,j} + b_{i,j}$ for $i = 1$ to m and $j = 1$ to n.
3. Multiplication by a scalar: $sA = [sa_{i,j}]$ for $i = 1$ to m and $j = 1$ to n.
4. Subtraction: $C = A - B$ is given by $c_{i,j} = a_{i,j} - b_{i,j}$ for $i = 1$ to m and $j = 1$ to n.
5. Transpose: $A^{\mathrm{T}} = [a_{j,i}]$ for $j = 1$ to n and $i = 1$ to m.

Remember: The zero matrix must be of the same order as the matrix to which it is added. What is the additive identity for the real numbers?

An **identity** matrix is a square matrix that contains 1's along the diagonal that runs from the upper left to the lower right (the *main diagonal*) and 0's elsewhere. The 2×2 identity matrix, denoted by I_2, is

$$I_2 = \begin{bmatrix} 1 & 0 \\ 0 & 1 \end{bmatrix},$$

and the 3×3 identity matrix is

$$I_3 = \begin{bmatrix} 1 & 0 & 0 \\ 0 & 1 & 0 \\ 0 & 0 & 1 \end{bmatrix}.$$

When the order of an identity matrix is understood from context or is not important, then we drop the subscript.

At this point, you may be a bit puzzled by the relationship between vectors and matrices. After all, is (500, 200) a vector? or is it a 1×2 matrix? The correct answer is both! For this reason, many books do not discuss vectors; they use *matrices* as a generic term to include both matrices and vectors. However, we believe you should know about vectors because you are quite likely to encounter them again as you progress in your studies. We will use the term *vector* as the preferred name for a $1 \times n$ (or $n \times 1$) matrix. At times it will be important to distinguish between vectors such as (3, 2) and $\begin{pmatrix} 3 \\ 2 \end{pmatrix}$. In order to do so, we refer to the first as a **row vector** and to the second as a **column vector.** (In other words, a row vector is $1 \times n$ and a column vector is $n \times 1$.)

Matrix Multiplication

Before we formally describe matrix multiplication, let's think a bit about how it should be performed. For instance, consider the ABC Tire Company. In Example 10, we determined that the matrix

$$P = \begin{bmatrix} 500 & 400 & 600 \\ 200 & 300 & 350 \end{bmatrix}$$

described total production of radial tires (first row) and truck tires (second row) during the three months January (first column), February (second column), and March (third column). Now, knowing the radial tires sell for \$40 each and truck tires for \$30 each, we should be able to compute the income from tire sales. In Example 8 we did this for the vectors **j** and **f**, which represented January and February sales, respectively, by forming dot products. We should be able to proceed in a similar fashion here by multiplying the selling price matrix (or row vector)

$$S = [40 \quad 30]$$

by P. In other words, we should be able to compute the matrix of monthly sales revenues, R, by the matrix multiplication

$$R = SP.$$

In fact, if we form the dot product of the row of S and each column of P, we will obtain our desired result:

$$R = \left[(40 \quad 30) \cdot \begin{pmatrix} 500 \\ 200 \end{pmatrix} \quad (40 \quad 30) \cdot \begin{pmatrix} 400 \\ 300 \end{pmatrix} \quad (40 \quad 30) \cdot \left(600350 \right) \right]$$

so

$$R = [40 \cdot 500 + 30 \cdot 200 \quad 40 \cdot 400 + 30 \cdot 300 \quad 40 \cdot 600 + 30 \cdot 350],$$
$$R = [26{,}000 \quad 25{,}000 \quad 34{,}500].$$

In general, we perform matrix multiplication as we did above: We form dot products with the rows of the first matrix and the columns of the second matrix. Let's think about what this means. Suppose the first matrix is $1 \times p$, which means its row has p elements. Then each column of the second matrix must also have p elements, so that we can form the dot products of the rows of the first and the columns of the second matrix. Therefore the second matrix must be $p \times n$ for some n. The total number of columns, n, is not restricted, because the resulting matrix will have one entry for each column of the second matrix (as in the earlier calculation). Thus the result of multiplying a $1 \times p$ matrix and a $p \times n$ matrix is a $1 \times n$ matrix.

In fact, we can multiply two matrices A and B in the order AB *only* when A has as many *columns* as B has *rows*. Suppose that A is

$A \cdot B = C$

must match

$(m \times p) \quad (p \times n) = m \times n$

size of result

Figure 5.6 **Two matrices can be multiplied when the number of columns of the first matches the number of rows of the second.**

$m \times p$ and B is $p \times n$. Then the matrix $C = AB$ is $m \times n$. In other words, when multiplying two matrices, the size of the product is given by the number of rows of the first and the number of columns of the second. (See Fig. 5.6.) As we indicated earlier, the entry in row i and column j of C (the result) is formed by the dot product of the ith row of A and the jth column of B. Using subscript notation, the entry in row i and column j of $C = AB$ is

$$c_{i,j} = a_{i,1} \cdot b_{1,j} + a_{i,2} \cdot b_{2,j} + \cdots + a_{i,p} \cdot b_{p,j}.$$

EXAMPLE 20

Let

$$A = \begin{bmatrix} 3 & 2 & 1 \\ 2 & 1 & 2 \end{bmatrix} \quad \text{and} \quad B = \begin{bmatrix} 1 & 3 \\ 2 & 1 \\ 1 & 2 \end{bmatrix}.$$

Find $C = AB$.

SOLUTION

A is 2×3 and B is 3×2. Then $C = AB$ should be 2×2. We have

Row 1 of *A* • Column 1 of *B*   **Row 1 of *A* • Column 2 of *B***

$$C = \begin{bmatrix} [3 \quad 2 \quad 1] \bullet \begin{bmatrix} 1 \\ 2 \\ 1 \end{bmatrix} & [3 \quad 2 \quad 1] \bullet \begin{bmatrix} 3 \\ 1 \\ 2 \end{bmatrix} \\ [2 \quad 1 \quad 2] \bullet \begin{bmatrix} 1 \\ 2 \\ 1 \end{bmatrix} & [2 \quad 1 \quad 2] \bullet \begin{bmatrix} 3 \\ 1 \\ 2 \end{bmatrix} \end{bmatrix}$$

Row 2 of *A* • Column 1 of *B*   **Row 2 of *A* • Column 2 of *B***

$$= \begin{bmatrix} 3 \cdot 1 + 2 \cdot 2 + 1 \cdot 1 & 3 \cdot 3 + 2 \cdot 1 + 1 \cdot 2 \\ 2 \cdot 1 + 1 \cdot 2 + 2 \cdot 1 & 2 \cdot 3 + 1 \cdot 1 + 2 \cdot 2 \end{bmatrix}$$

$$= \begin{bmatrix} 8 & 13 \\ 6 & 11 \end{bmatrix}$$

(which is indeed 2×2). □

By now you are probably thinking that matrix multiplication is rather strange. Actually, it's stranger than you may have suspected. Unlike multiplying real numbers, matrix multiplication is *not* commutative.

EXAMPLE 21

Let A and B be defined as in Example 20, so that

$$A = \begin{bmatrix} 3 & 2 & 1 \\ 2 & 1 & 2 \end{bmatrix} \quad \text{and} \quad B = \begin{bmatrix} 1 & 3 \\ 2 & 1 \\ 1 & 2 \end{bmatrix}.$$

A little thought should convince us that we can multiply in the order BA because B is 3×2 and A is 2×3, so let's find $D = BA$.

SOLUTION

$$D = \begin{bmatrix} 1 & 3 \\ 2 & 1 \\ 1 & 2 \end{bmatrix} \cdot \begin{bmatrix} 3 & 2 & 1 \\ 2 & 1 & 2 \end{bmatrix}$$

$$D = \begin{bmatrix} [1 \quad 3] \cdot \begin{bmatrix} 3 \\ 2 \end{bmatrix} & [1 \quad 3] \cdot \begin{bmatrix} 2 \\ 1 \end{bmatrix} & [1 \quad 3] \cdot \begin{bmatrix} 1 \\ 2 \end{bmatrix} \\ [2 \quad 1] \cdot \begin{bmatrix} 3 \\ 2 \end{bmatrix} & [2 \quad 1] \cdot \begin{bmatrix} 2 \\ 1 \end{bmatrix} & [2 \quad 1] \cdot \begin{bmatrix} 1 \\ 2 \end{bmatrix} \\ [1 \quad 2] \cdot \begin{bmatrix} 3 \\ 2 \end{bmatrix} & [1 \quad 2] \cdot \begin{bmatrix} 2 \\ 1 \end{bmatrix} & [1 \quad 2] \cdot \begin{bmatrix} 1 \\ 2 \end{bmatrix} \end{bmatrix}$$

$$= \begin{bmatrix} 1 \cdot 3 + 3 \cdot 2 & 1 \cdot 2 + 3 \cdot 1 & 1 \cdot 1 + 3 \cdot 2 \\ 2 \cdot 3 + 1 \cdot 2 & 2 \cdot 2 + 1 \cdot 1 & 2 \cdot 1 + 1 \cdot 2 \\ 1 \cdot 3 + 2 \cdot 2 & 1 \cdot 2 + 2 \cdot 1 & 1 \cdot 1 + 2 \cdot 2 \end{bmatrix}$$

$$= \begin{bmatrix} 9 & 5 & 7 \\ 8 & 5 & 4 \\ 7 & 4 & 5 \end{bmatrix}$$

Compare this result with the matrix C obtained in Example 20. Obviously C and D aren't equal; they aren't even the same size. □

Practice Problem

Perform the matrix multiplication AB, where

$$A = \begin{bmatrix} 3 & 4 & 5 \\ 1 & 3 & 2 \\ 2 & 1 & 1 \end{bmatrix} \quad \text{and} \quad B = \begin{bmatrix} 1 & 2 \\ 3 & 1 \\ 2 & 4 \end{bmatrix}.$$

Answer: $AB = \begin{bmatrix} 25 & 30 \\ 14 & 13 \\ 7 & 9 \end{bmatrix}$

Some pairs of matrices can only be multiplied in one order. Suppose F is 3×2 and G is 2×4. Then FG makes sense, but GF does not.

Finally, using square matrices still does not result in commutativity for matrix multiplication.

EXAMPLE 22

Let

$$A = \begin{bmatrix} 1 & 3 \\ 2 & 4 \end{bmatrix} \quad \text{and} \quad B = \begin{bmatrix} 2 & 3 \\ 4 & 6 \end{bmatrix}.$$

Find $C = AB$ and $D = BA$. Are C and D equal?

SOLUTION

$$C = AB = \begin{bmatrix} 14 & 21 \\ 20 & 30 \end{bmatrix} \qquad \text{(Why?)}$$

and

$$D = BA = \begin{bmatrix} 8 & 18 \\ 16 & 36 \end{bmatrix} \qquad \text{(Why?)}$$

and obviously $C \neq D$. □

You may be wondering why we defined the identity matrix: We did so because it is a multiplicative identity. In other words, $AI = IA = A$. *Caution:* Each matrix I must be the correct size! Unless A is square, the identity matrices will be of different sizes.

EXAMPLE 23

Let

$$A = \begin{bmatrix} 3 & 2 & 1 \\ 2 & 1 & 2 \end{bmatrix}$$

which is 2×3. Find AI and IA.

SOLUTION

The multiplication AI requires the identity to be 3×3:

$$\begin{aligned} AI = AI_3 &= \begin{bmatrix} 3 & 2 & 1 \\ 2 & 1 & 2 \end{bmatrix} \begin{bmatrix} 1 & 0 & 0 \\ 0 & 1 & 0 \\ 0 & 0 & 1 \end{bmatrix} \\ &= \begin{bmatrix} 3 \cdot 1 + 2 \cdot 0 + 1 \cdot 0 & 3 \cdot 0 + 2 \cdot 1 + 1 \cdot 0 & 3 \cdot 0 + 2 \cdot 0 + 1 \cdot 1 \\ 2 \cdot 1 + 1 \cdot 0 + 2 \cdot 0 & 2 \cdot 0 + 1 \cdot 1 + 2 \cdot 0 & 2 \cdot 0 + 1 \cdot 0 + 2 \cdot 1 \end{bmatrix} \\ &= \begin{bmatrix} 3 & 2 & 1 \\ 2 & 1 & 2 \end{bmatrix} \end{aligned}$$

which is just A and the multiplication IA requires the identity to be 2×2:

$$IA = I_2A = \begin{bmatrix} 1 & 0 \\ 0 & 1 \end{bmatrix} \begin{bmatrix} 3 & 2 & 1 \\ 2 & 1 & 2 \end{bmatrix}$$
$$= \begin{bmatrix} 1 \cdot 3 + 0 \cdot 2 & 1 \cdot 2 + 0 \cdot 1 & 1 \cdot 1 + 0 \cdot 2 \\ 0 \cdot 3 + 1 \cdot 2 & 0 \cdot 2 + 1 \cdot 1 & 0 \cdot 1 + 1 \cdot 2 \end{bmatrix}$$
$$= \begin{bmatrix} 3 & 2 & 1 \\ 2 & 1 & 2 \end{bmatrix} = A. \quad \square$$

Practice Problem

Let $A = \begin{bmatrix} 2 & 2 \\ 1 & 2 \\ 4 & 2 \end{bmatrix}$.

a) What is $\begin{bmatrix} 0.5 & 0 & 0 \\ 0 & 1 & 0 \\ 0 & 0 & 1 \end{bmatrix} A$? **b)** What is $\begin{bmatrix} 1 & 0 & 0 \\ -1 & 1 & 0 \\ 0 & 0 & 1 \end{bmatrix} A$?

Answers: **a)** $\begin{bmatrix} 1 & 1 \\ 1 & 2 \\ 4 & 2 \end{bmatrix}$; **b)** $\begin{bmatrix} 2 & 2 \\ -1 & 0 \\ 4 & 2 \end{bmatrix}$

EXAMPLE 24

Problem from Life, part (d) Given last week's production figures at Lizard Lick (L) and Ypsilanti (Y)

$$L = \begin{bmatrix} 1500 & 1200 & 2500 \\ 2200 & 3000 & 2000 \end{bmatrix}, \qquad Y = \begin{bmatrix} 2250 & 2500 & 4000 \\ 6500 & 5000 & 6500 \end{bmatrix}$$

(where the first row of each matrix represents production of Green Label beer and the second row production of Purple Label beer and the columns represent 7-ounce bottles, 12-ounce bottles, and cans, respectively) and the selling prices (in dollars)

$$P = \begin{bmatrix} 0.40 & 0.65 & 0.60 \\ 0.30 & 0.60 & 0.55 \end{bmatrix},$$

find the revenue generated at each plant and the total revenue.

SOLUTION

To find the revenue at each plant, we'll have to multiply P and L and also P and Y. However, we have to think about this a bit because we have to make sure the matrices have the correct dimensions. Both L and Y are 2×3, as is P. That means we *cannot* form the products PL and PY, nor the products LP and YP. What can we do?

Let's assume we want to multiply the matrices with the production levels in the first matrix and the prices in the second. This means that the rows of the production matrices must match the columns of the price matrix in length. Now, both L and Y have three entries in each row, one for each size of product. Therefore each column of the price matrix should have three entries. We can get the prices in this form by using the *transpose* of P. However, a little more thought will point out a serious mistake with this strategy! If we're not careful, we'll end up multiplying the production levels of Green Label beer and the prices of Purple Label beer, and vice versa.

What we really want to do is to construct a one-row matrix for production (at each location) and a one-column matrix for prices. This gives us

$$L = [1500 \quad 1200 \quad 2500 \quad 2200 \quad 3000 \quad 2000],$$
$$Y = [2250 \quad 2500 \quad 4000 \quad 6500 \quad 5000 \quad 6500],$$

and

$$P = \begin{bmatrix} 0.40 \\ 0.65 \\ 0.60 \\ 0.30 \\ 0.60 \\ 0.55 \end{bmatrix}$$

Now the revenues at Lizard Lick are given by LP and the revenues at Ypsilanti by YP. The calculations are

$$\begin{aligned} LP &= 1500 \cdot 0.40 + 1200 \cdot 0.65 + 2500 \cdot 0.60 + 2200 \cdot 0.30 \\ &\quad + 3000 \cdot 0.60 + 2000 \cdot 0.55 \\ &= \$6440 \qquad \text{(revenues at Lizard Lick)}, \\ YP &= 2250 \cdot 0.40 + 2500 \cdot 0.65 + 4000 \cdot 0.60 + 6500 \cdot 0.30 \\ &\quad + 5000 \cdot 0.60 + 6500 \cdot 0.55 \\ &= \$13{,}450 \qquad \text{(revenues at Ypsilanti)}. \end{aligned}$$

Therefore total revenues are \$6440 + \$13,450 = \$19,890. □

As Example 24 shows, you should always think *before* you calculate. Sometimes a fancy approach (using matrices with multiple rows and columns) doesn't work as well as a more simple approach (we essentially used the dot product of a row vector and a column vector).

Here's another odd aspect of matrix multiplication. Consider the multiplication of an $n \times 1$ matrix and a $1 \times n$ matrix (which is the opposite of the calculation we used in Example 24).

EXAMPLE 25 Let

$$A = \begin{bmatrix} 3 \\ 4 \\ 5 \end{bmatrix} \text{ and let } B = [2 \quad 5 \quad -1].$$

Find AB. What is the order of AB?

SOLUTION

$$AB = \begin{bmatrix} [3][2] & [3][5] & [3][-1] \\ [4][2] & [4][5] & [4][-1] \\ [5][2] & [5][5] & [5][-1] \end{bmatrix}$$
$$= \begin{bmatrix} 6 & 15 & -3 \\ 8 & 20 & -4 \\ 10 & 25 & -5 \end{bmatrix}$$

which is 3×3. □

Practice Problem

Given $A = \begin{bmatrix} -2 \\ 5 \end{bmatrix}$ and $B = [3 \quad 7]$, find **(a)** BA, and **(b)** AB.

Answers: **a)** $BA = 29$ (a real number); **b)** $AB = \begin{bmatrix} -6 & -14 \\ 15 & 35 \end{bmatrix}$

The following summarizes matrix multiplication.

> **Matrix Multiplication**
>
> Let A be $m \times p$ and B be $p \times n$ matrices. Then $C = AB$, which will be $m \times p$, is given by the dot product of the *rows* of A with the *columns* of B:
>
> $$c_{i,j} = a_{i,1} \cdot b_{1,j} + a_{i,2} \cdot b_{2,j} + \cdots + a_{i,p} \cdot b_{p,j}$$

EXERCISE SET 5.2

Find the order of each of the following matrices.

1. $\begin{bmatrix} 1 & 2 \\ 3 & 4 \end{bmatrix}$ **2.** $\begin{bmatrix} 1 & 2 & 3 \\ 4 & 5 & 6 \end{bmatrix}$ **3.** $\begin{bmatrix} 1 & 2 \\ 3 & 4 \\ 5 & 6 \end{bmatrix}$

4. $\begin{bmatrix} 1 & 3 & 5 & 7 \\ 2 & 4 & 6 & 8 \\ 0 & 1 & 2 & 3 \end{bmatrix}$ **5.** $\begin{bmatrix} 1 & 0 & 0 \\ 0 & 1 & 0 \\ 0 & 0 & 1 \\ 1 & 1 & 1 \end{bmatrix}$ **6.** $\begin{bmatrix} 1 & 2 \\ 3 & 4 \\ 5 & 6 \\ 7 & 8 \end{bmatrix}$

Find the sum of the matrices A and B.

7. $A = \begin{bmatrix} 1 & 2 \\ 3 & 4 \end{bmatrix}, B = \begin{bmatrix} 2 & 1 \\ 4 & 6 \end{bmatrix}$

8. $A = \begin{bmatrix} 0.1 & 3 \\ 5 & 0.7 \end{bmatrix}, B = \begin{bmatrix} 2 & 0.4 \\ 6 & 0.8 \end{bmatrix}$

9. $A = \begin{bmatrix} 1 & 2 & 4 \\ 3 & 1 & 0 \end{bmatrix}, B = \begin{bmatrix} 3 & 7 & 5 \\ 6 & 0 & 1 \end{bmatrix}$

10. $A = \begin{bmatrix} 0.4 & 0 & -3 \\ 0.1 & 3 & 0.3 \end{bmatrix}, B = \begin{bmatrix} -0.2 & 1 & 3 \\ 0.5 & 1 & 0.4 \end{bmatrix}$

11. $A = \begin{bmatrix} 1 & 3 & 4 \\ 2 & 0 & 7 \\ 5 & 6 & 9 \end{bmatrix}, B = \begin{bmatrix} 10 & 2 & 0 \\ 7 & 1 & 3 \\ 4 & 5 & 6 \end{bmatrix}$

12. $A = \begin{bmatrix} 1 & 2 & -1 & 0 \\ 4 & 0 & 2 & -1 \\ 2 & -5 & 1 & 2 \end{bmatrix}, B = \begin{bmatrix} 3 & -4 & 1 & 2 \\ 1 & 5 & 0 & -3 \\ 2 & -2 & 3 & -1 \end{bmatrix}$

Which of the following are true?

13. $\begin{bmatrix} 1 & 2 \\ 3 & 4 \end{bmatrix} = \begin{bmatrix} 1 & 2 \\ 3 & 4 \end{bmatrix}$

14. $\begin{bmatrix} 1 & 2 \\ 3 & 4 \end{bmatrix} = \begin{bmatrix} 2 & 1 \\ 4 & 3 \end{bmatrix}$

15. $\begin{bmatrix} 1 & 2 \\ 1 & 0 \end{bmatrix} + \begin{bmatrix} 2 & 0 \\ 3 & 1 \end{bmatrix} = \begin{bmatrix} 2 & 0 \\ 3 & 1 \end{bmatrix} + \begin{bmatrix} 1 & 2 \\ 1 & 0 \end{bmatrix}$

16. $\begin{bmatrix} 0.1 & 2 \\ 1 & 0 \end{bmatrix} + \begin{bmatrix} 1 \\ 1 \end{bmatrix} = \begin{bmatrix} 1.1 & 3 \\ 2 & 1 \end{bmatrix}$

Find the results of the following scalar multiplications.

17. $2\begin{bmatrix} 1 & 2 \\ 0.3 & -4 \end{bmatrix}$

18. $-3\begin{bmatrix} 0.1 & 0.02 \\ 3 & 0.4 \end{bmatrix}$

19. $3\begin{bmatrix} 1 & 2 & 4 \\ 0.3 & 1 & 0.5 \end{bmatrix}$

20. $-0.2\begin{bmatrix} 3 & 0.7 & -5 \\ 6 & 0 & 1 \end{bmatrix}$

21. $-1\begin{bmatrix} 1 & 3 & 4 \\ 2 & 0 & 7 \\ 5 & 6 & 9 \end{bmatrix}$

22. $0.3\begin{bmatrix} 1 & 2 & -1 & 0 \\ 4 & 0 & 2 & 1 \\ 2 & -5 & 1 & -2 \end{bmatrix}$

Find the indicated differences.

23. $\begin{bmatrix} 1 & 2 \\ 3 & 4 \end{bmatrix} - \begin{bmatrix} 2 & 1 \\ 4 & 6 \end{bmatrix}$

24. $\begin{bmatrix} 1 & 3 \\ 0.5 & 7 \end{bmatrix} - \begin{bmatrix} 0.2 & 4 \\ 0.6 & -8 \end{bmatrix}$

25. $\begin{bmatrix} 1 & 2 & 4 \\ 3 & 1 & 0 \end{bmatrix} - \begin{bmatrix} 3 & 7 & 5 \\ 6 & 0 & 1 \end{bmatrix}$

26. $\begin{bmatrix} -2 & 0.1 & 3 \\ 0.5 & 1 & 4 \end{bmatrix} - \begin{bmatrix} 4 & 0.3 & 2 \\ 0.1 & 3 & 3 \end{bmatrix}$

27. $\begin{bmatrix} 10 & 2 & 0 \\ 7 & 1 & 3 \\ 4 & 5 & 6 \end{bmatrix} - \begin{bmatrix} 1 & 3 & 4 \\ 2 & 0 & 7 \\ 5 & 6 & 9 \end{bmatrix}$

28. $\begin{bmatrix} 1 & 2 & -1 & 0 \\ 4 & 0 & 2 & 1 \\ 2 & -5 & 1 & 2 \end{bmatrix} - \begin{bmatrix} 3 & -4 & 1 & 2 \\ 1 & 5 & 0 & 3 \\ 2 & -2 & 3 & -1 \end{bmatrix}$

Perform the indicated operations.

29. $3\begin{bmatrix} 1 & 2 \\ 2 & 4 \end{bmatrix} - 2\begin{bmatrix} 2 & 1 \\ 4 & 6 \end{bmatrix}$

30. $-2\begin{bmatrix} 1 & 3 \\ 5 & 7 \end{bmatrix} + 3\begin{bmatrix} 2 & 4 \\ 6 & 8 \end{bmatrix}$

31. $0.4\begin{bmatrix} 1 & 2 & 4 \\ 3 & 1 & 0 \end{bmatrix} - 0.1\begin{bmatrix} 4 & 0 & 2 \\ 1 & 3 & 3 \end{bmatrix}$

32. $2\begin{bmatrix} 1 & 2 & -1 & 0 \\ 4 & 0 & 2 & 1 \\ 2 & -5 & 1 & 2 \end{bmatrix} - 3\begin{bmatrix} 3 & -4 & 1 & 2 \\ 1 & 5 & 0 & 3 \\ 2 & -2 & 3 & -1 \end{bmatrix}$

Find the products AB and BA (when they exist).

33. $A = [1 \quad 2 \quad 4]$, $B = \begin{bmatrix} 7 \\ 0 \\ 2 \end{bmatrix}$

34. $A = \begin{bmatrix} 1 & 3 & 4 \\ 2 & 0 & 7 \\ 5 & 6 & 9 \end{bmatrix}, B = \begin{bmatrix} 2 \\ 1 \\ 3 \end{bmatrix}$

35. $A = \begin{bmatrix} 1 & 3 & 4 \\ 2 & 0 & 7 \\ 5 & 6 & 9 \end{bmatrix}, B = \begin{bmatrix} 10 & 2 & 0 \\ 7 & 1 & 3 \\ 4 & 5 & 6 \end{bmatrix}$

36. $A = \begin{bmatrix} 1 & 2 & -1 & 0 \\ 4 & 0 & 2 & 1 \\ 2 & -5 & 1 & 2 \end{bmatrix}, B = \begin{bmatrix} 1 & 0 & 1 \\ 2 & 3 & -1 \end{bmatrix}$

37. $A = \begin{bmatrix} 1 & 3 & 4 \\ 2 & 0 & 7 \\ 5 & 6 & 9 \end{bmatrix}, B = \begin{bmatrix} 1 & 0 & 0 \\ -1 & 1 & 0 \\ -1 & 1 & 1 \end{bmatrix}$

38. $A = \begin{bmatrix} 1 & 3 & 4 \\ 2 & 0 & 7 \\ 5 & 6 & 9 \end{bmatrix}, B = \begin{bmatrix} 1 & 0 & 0 \\ 0 & 0 & 1 \\ 0 & 1 & 0 \end{bmatrix}$

Using the matrices

$$0 = \begin{bmatrix} 0 & 0 \\ 0 & 0 \end{bmatrix}, A = \begin{bmatrix} 1 & 2 \\ 3 & 4 \end{bmatrix}, B = \begin{bmatrix} 2 & -1 \\ 4 & 3 \end{bmatrix},$$

$$C = \begin{bmatrix} 3 & 1 \\ 7 & 5 \end{bmatrix}, \text{ and } I = \begin{bmatrix} 1 & 0 \\ 0 & 1 \end{bmatrix},$$

verify that the following are true. (All matrices are assumed to have the proper dimensions for the statements to make sense.)

39. $A + 0 = 0 + A$ (additive identity for matrices)

40. $0 + C = C + 0$ (additive identity for matrices)

41. $A + B = B + A$ (commutative property of matrix addition)

42. $(A + B) + C = A + (B + C)$ (associative property of matrix addition)

43. $2(A + C) = 2A + 2C$ (distributive property of scalar multiplication)

44. $(2 + 3)B = 2B + 3B$ (distributive property of scalar multiplication)

45. $A(BC) = (AB)C$ (associative property of matrix multiplication)

46. $(A + B)C = AC + BC$ (distributive property of matrix multiplication)

47. $C(A + B) = CA + CB$ (distributive property of matrix multiplication)

48. $IC = CI$ (multiplicative identity for matrices)

49. $2(AB) = (2A)B = A(2B)$

50. Which of the properties in Problems 39–49 do you think hold in general for all matrices?

51. A survey on car pooling showed the following results:

150 people with incomes under \$20,000 car pooled
250 people with incomes between \$20,000 and \$40,000 car pooled
100 people with incomes of more than \$40,000 car pooled
450 people with incomes under \$20,000 did not car pool
400 people with incomes between \$20,000 and \$40,000 did not car pool
650 people with incomes of more than \$40,000 did not car pool

Arrange these data in a matrix.

CHAPTER REVIEW

IMPORTANT TERMS AND SYMBOLS

5.1 Vectors and Vector Operations

scalar A real number.

vector Ordered set of numbers or elements; represented by boldface lowercase letter (such as **a**, **b**).
Dimension of a vector: number of elements of the vector.
Subscript: refers to a particular element

vector operations Let **a** and **b** be two vectors of the same dimension and let s be a scalar.
Addition: two vectors of the same dimension are added by adding corresponding elements; represented by $\mathbf{a} + \mathbf{b}$.
Dot product: $\mathbf{a} \cdot \mathbf{b} = \mathbf{a}_1 \cdot \mathbf{b}_1 + \mathbf{a}_2 \cdot \mathbf{b}_2 + \cdots + \mathbf{a}_n \cdot \mathbf{b}_n$.
Equality: two vectors of the same dimension are equal if their corresponding elements are equal; represented by $\mathbf{a} = \mathbf{b}$.
Multiplication by a scalar: multiply each element of the vector by the scalar; represented by $s\mathbf{a}$.
Subtraction: two vectors of the same dimension are subtracted by subtracting corresponding elements; represented by $\mathbf{a} - \mathbf{b}$.

5.2 Matrices and Matrix Operations

column vector An $m \times 1$ matrix.

matrix A rectangular array (table) of numbers or elements; represented by capital letters such as A, B.
Dimension (order) of a matrix: its number of rows and number of columns.
Identity matrix: square matrix with 1's on main diagonal and 0's elsewhere, represented by I.

Rectangular matrix: matrix that isn't necessarily square.
Square matrix: matrix that has the same number of rows and columns.
Zero matrix: matrix whose elements are all zeros.

matrix operations Let A and B be two matrices of the same dimension and let s be a scalar.
Addition: two matrices of the same dimension are added by adding corresponding elements; represented by $A + B$.
Equality: two matrices of the same dimension are equal if their corresponding elements are equal; represented by $A = B$.
Matrix multiplication: $C = AB$ is defined by

$$c_{i,j} = a_{i,1} \cdot b_{1,j} + a_{i,2} \cdot b_{2,j} + \cdots + a_{i,p} \cdot b_{p,j},$$

where the number of *columns* of A = number of *rows* of B.
Multiplication by a scalar: multiply each element of the matrix by the scalar; represented by sA.
Subtraction: two matrices are subtracted by subtracting corresponding elements; represented by $A - B$.

row vector A $1 \times n$ matrix.

transpose of matrix Found by switching the rows and columns; represented by A^T.

REVIEW PROBLEMS

What are the dimensions of the following vectors?

1. $(1, 3)$ **2.** $(3, -5, 7, 0.9)$ **3.** $(2, -6, 3)$

Find the indicated sums.

4. $(1, 2) + (-5, -4)$
5. $(7, 0.9) + (-3, 0.2)$
6. $(1, 0.1, 1) + (3, 0.5, -1)$
7. $(0.4, 8, 1) + (-0.2, 4, 6)$
8. $(2, 4, 0.4, 0.2) + (0.1, 1, -1, -0.5)$
9. $(0.1, 2, 0.03, 4) + (0.4, 3, -0.3, 4)$

Which of the following are true?

10. $(1, 2) = (1, 2)$
11. $(3, 5) = (-3, 5)$
12. $(-0.3, -0.5) = -0.1(3, 5)$
13. $(1, 1, 1) = (1, 1)$

Perform the indicated operations.

14. $2(3, -4)$
15. $-2(-2, 0.3)$
16. $-3(1, -1, 2)$
17. $4(1, -1, -0.2, 2)$
18. $(3, -5) - (-3, 5)$
19. $(0.1, 2) - 2(-0.3, 4)$
20. $(9, 8, 7) - (-2)(1, 2, 3)$
21. $(4, 7, 3) - 2(1, 0, -2)$

What are the dimensions (orders) of the following matrices?

22. $\begin{bmatrix} 1 & 2 \\ 4 & 3 \end{bmatrix}$ **23.** $\begin{bmatrix} 1 & 2 & 2 \\ 3 & 2 & 1 \end{bmatrix}$ **24.** $\begin{bmatrix} 6 & 5 \\ 7 & 9 \\ 8 & 4 \end{bmatrix}$

Perform the indicated operations.

25. $\begin{bmatrix} 1 & 2 \\ 4 & 3 \end{bmatrix} + \begin{bmatrix} -2 & 3 \\ 4 & 4 \end{bmatrix}$

26. $\begin{bmatrix} 1 & 0.2 \\ 2 & 1 \end{bmatrix} + \begin{bmatrix} 1 & 0.3 \\ 5 & 7 \end{bmatrix}$

27. $\begin{bmatrix} 0.5 & 7 \\ 9 & 0.6 \end{bmatrix} - \begin{bmatrix} 0.1 & 2 \\ 2 & 0.4 \end{bmatrix}$

28. $0.3\begin{bmatrix} 5 & 7 \\ 9 & 6 \end{bmatrix} + 0.4\begin{bmatrix} 1 & 2 \\ 2 & 4 \end{bmatrix}$

29. $\begin{bmatrix} 1 & 3 & 0 \\ 2 & 4 & 6 \end{bmatrix} + \begin{bmatrix} 3 & -2 & 0 \\ 1 & 1 & -1 \end{bmatrix}$

30. $-0.3\begin{bmatrix} 1 & 3 & 0 \\ 2 & 4 & 6 \end{bmatrix} + 0.2\begin{bmatrix} 3 & -2 & 0 \\ 1 & 1 & -1 \end{bmatrix}$

Which of the following are true?

31. $\begin{bmatrix} 1 & 3 \\ 4 & 4 \end{bmatrix} = \begin{bmatrix} 1 & 3 \\ 4 & 4 \end{bmatrix} + \begin{bmatrix} 0 & 0 \\ 0 & 0 \end{bmatrix}$

32. $\begin{bmatrix} 1 & 3 \\ 4 & 4 \end{bmatrix} + 0 = \begin{bmatrix} 1 & 3 \\ 4 & 4 \end{bmatrix}$

33. $\begin{bmatrix} 1 & 3 \\ 4 & 4 \end{bmatrix}\begin{bmatrix} 1 & 0 \\ 0 & 1 \end{bmatrix} = \begin{bmatrix} 1 & 3 \\ 4 & 4 \end{bmatrix}$

34. $\begin{bmatrix} 1 & 3 \\ 4 & 4 \end{bmatrix}\begin{bmatrix} 1 & 1 \\ 1 & 1 \end{bmatrix} = \begin{bmatrix} 4 & 4 \\ 8 & 8 \end{bmatrix}$

35. $\begin{bmatrix} 1 & 1 \\ 1 & 1 \end{bmatrix}\begin{bmatrix} 1 & 3 \\ 4 & 4 \end{bmatrix} = \begin{bmatrix} 4 & 4 \\ 8 & 8 \end{bmatrix}$

36. $\begin{bmatrix} 3 & 4 & 4 \\ 1 & 1 & -1 \end{bmatrix}\begin{bmatrix} 1 & 0 \\ 0 & 1 \\ 1 & 0 \end{bmatrix} = \begin{bmatrix} 7 & 4 \\ 0 & 1 \end{bmatrix}$

37. Tom and Ann went grocery shopping. Tom bought two quarts of milk, one pound of salami, one pound of cheese, and one loaf of bread. Ann bought one quart of milk, one pound of salami, three pounds of cheese, and two loaves of bread.

a) Represent each set of purchases as a vector.

b) Assume that milk is \$0.98 per quart, salami is \$4.09 per pound, cheese is \$3.25 per pound, and bread is \$0.59 per loaf. How much did Tom spend? How much did Ann spend? Who spent more? How much did they spend together?

38. Y'All Haul rents trailers. Last month 150 small trailers, 200 medium trailers, and 100 large trailers were rented. This month 100 small trailers. 250 medium trailers, and 150 large trailers were rented. The projections for next month are calculated as follows:

$$0.9 \times \text{this month's rentals} + 0.1 \times \text{last month's rentals}$$

a) What are the actual numbers for next months' projections?

b) Suppose small trailers rent for \$15 per day, medium trailers for \$20 per day, and large trailers for \$25 per day. Compute the rental fees for each month.

c) Show that total rental fees can be computed in two ways. One way is to find the fees for each month and then add them up. The other way is to find the total number of trailers rented over the three-month time span and then multiply by the daily fees.

6 SOLVING SYSTEMS OF LINEAR EQUATIONS

a) The Kleins would like to invest $25,000 into a mutual stock fund (which pays 15% annual interest but is somewhat risky) and bonds (which pay 9% annual interest with very little risk). They would like the investment to pay $250 per month. How much should they invest in each alternative? (To be discussed below and solved in Example 8, Section 6.2.)

b) A dentist employs three assistants in her practice; her office has two rooms with chairs and equipment (operatories). The dentist provides both pedodontic and general dentistry services. Experience indicates that an average pedodontic service requires 0.75 hour of operatory time, 1.50 hours of an assistant's time, and 0.25 hour of dentist's time, while an average general service requires 0.75 hour of operatory time, 1.00 hour of an assistant's time, and 0.50 hour of dentist's time. Given the dentist's three assistants and two operatories are each to work 8 hours per day, how many patients can be scheduled during an 8-hour day? (To be solved in Example 26, Section 6.3.)

PROBLEM FROM LIFE

INTRODUCTION

In Chapter 3 you saw a variety of problems that could be modeled and solved using a linear equation. In this chapter we'll discuss situations that require more than one linear equation to model, and how we can solve such problems. For example, consider part (a) of the Problem from Life. Let's see how we can model this problem.

As you know, the first step in solving a problem is to read it carefully. In the case of the Kleins, you should note that they have a fixed sum of money ($25,000) to invest in two investment opportunities (a mutual stock fund and bonds), and they require a fixed return per month from their investment ($250). Let's start by defining variables. Let

s = amount of money to invest in the stock fund,

b = amount of money to invest in bonds.

Now, since they wish to invest $25,000, we can write

$$s + b = 25{,}000.$$

Let's think about the other requirement. They want to earn $250 per month. Moreover, the stock fund returns 15% interest and the bonds 9%. However, these are *yearly* rates. The monthly rates are one-twelfth of those rates: 1.25% for the stock fund and 0.75% for the bonds. Hence each month the Kleins should earn $0.0125s + 0.0075b$. Thus the second equation is

$$0.0125s + 0.0075b = 250.$$

To summarize, the Kleins can solve their problem by finding values of s and b that satisfy both of the linear equations

$$\begin{aligned} s + \qquad\quad b &= 25{,}000, \\ 0.0125s + 0.0075b &= 250. \end{aligned}$$

We'll actually solve this problem in Example 8.

6.1 SYSTEMS OF LINEAR EQUATIONS

Many problems require the solution of a number of equations. We refer to a set (or collection) of equations as a **system of equations.** In this chapter we'll study several methods for solving systems in which all the equations are linear. As you could guess, such systems are called **systems of linear equations** or simply **linear systems.**

In Chapter 3 we learned that

$$4x + 3y = 12$$

is a linear equation with two variables and

$$4x - 2y + 6z = 24$$

is a linear equation with three variables. In general, we can write a linear (or first-degree) equation with n variables (or unknowns)

$$a_1x_1 + a_2x_2 + \cdots + a_nx_n = k,$$

where $a_1, a_2, \ldots, a_n$ and k are all real numbers. An example of a system of two linear equations and two variables is

$$\begin{aligned} 2x - 3y &= 6, \\ x + 3y &= 12. \end{aligned}$$

For simplicity, we'll limit our discussion to linear systems involving two or three variables. However, the solution techniques we'll study can be used to solve much larger systems. We will also stop using the term *linear* explicitly, because we are studying only linear systems. Thus when we refer to a system of equations, we really mean a system of linear equations.

In Chapter 3 you learned that solving a single equation meant finding its solution set. The same thing is true for a system of equations: We wish to find the solution set of the system. In fact, the solution set of a system of equations is actually the *intersection* of the solution sets of the individual equations. Using this fact will help us describe the possible types of solutions.

In Chapter 3 you also learned the graph of a linear equation with two variables is a straight line. By studying Fig. 6.1, we can see there are exactly three possibilities for the solution set of a system of two equations and two variables:

1. The two lines intersect at a single point. The coordinates of this point tell us the values of the variables that solve the system. (See Fig. 6.1a.)
2. The two lines are parallel, so that they do not intersect. Here there are no values of the variables that satisfy the system, so the solution set is ∅ (the empty set). Systems of equations whose solution set is empty are called **inconsistent.** (See Fig. 6.1b.)
3. The two lines coincide; that is, they are identical. In this case the solution set for the system contains all the points in the solution set of the equation. (The two given equations are just different forms of the same equation.) Therefore the solution set actually contains an infinite number of points. We say that systems of equations with infinite solution sets are **dependent**. (See Fig. 6.1c.)

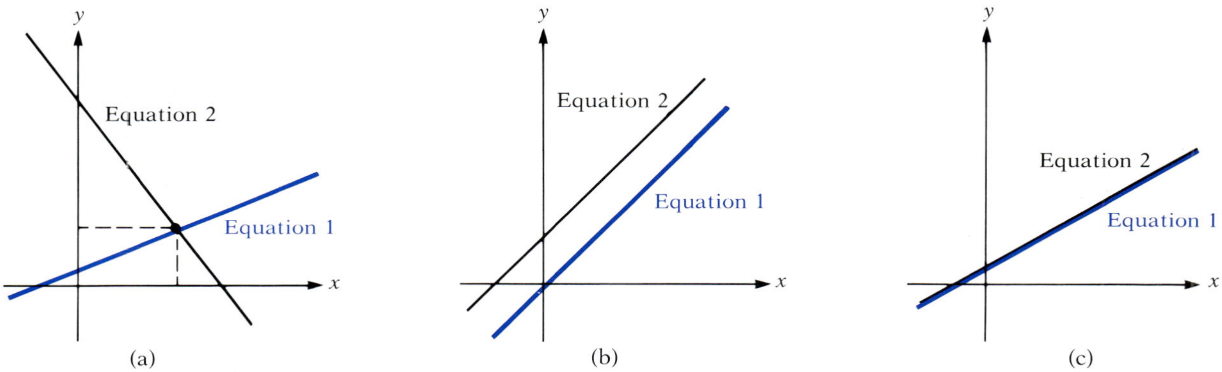

Figure 6.1 **Possible solutions of a system of two equations and two variables. (a) A unique solution. (b) Inconsistent equations have no solution. (c) Dependent equations have an unlimited number of solutions.**

In fact, it can be shown that these three cases describe the possible solutions of any system of (linear) equations. In other words, there will either be one unique solution, no solution (the solution set is empty so the system is inconsistent), or an infinite number of solutions (the system is dependent).

EXAMPLE 1

Determine graphically the type of solution of the following system:

$$2x - 3y = 6,$$
$$x + 3y = 12.$$

SOLUTION

From the graph shown in Fig. 6.2, we see there is one solution: the point (6, 2). In other words, $x = 6$ and $y = 2$ is the only solution.

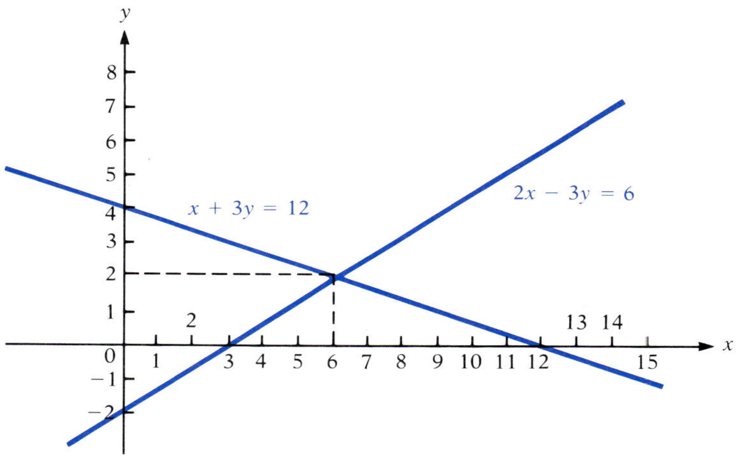

Figure 6.2 **Solution of the system $2x - 3y = 6$ and $x + 3y = 12$.** □

EXAMPLE 2

Determine graphically the type of solution of the system

$$2x - 3y = 6,$$
$$-4x + 6y = 6.$$

SOLUTION

From the graph shown in Fig. 6.3, we see the lines are parallel. Thus, the system is inconsistent, which means there is no solution.

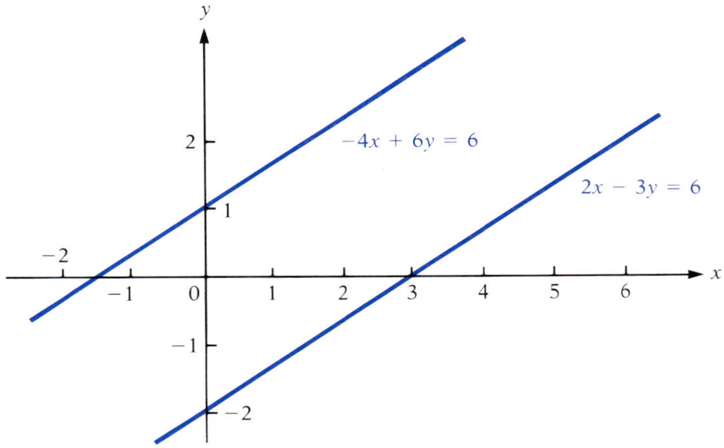

Figure 6.3 **The system $2x - 3y = 6$ and $-4x + 6y = 6$ has no solution.** □

EXAMPLE 3

Determine graphically the type of solution of the system

$$3x + 2y = 6,$$
$$6x + 4y = 12.$$

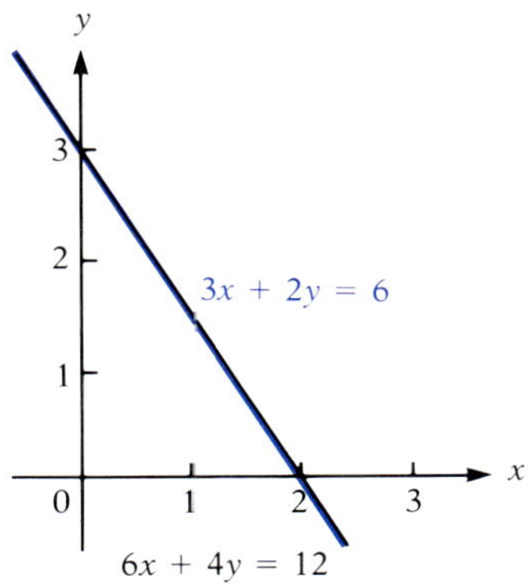

Figure 6.4 **The system $3x + 2y = 6$ and $6x + 4y = 12$ has infinitely many solutions.**

SOLUTION

From the graph shown in Fig. 6-4, we see that both equations generate the same line. In other words, the lines coincide. Thus the system is dependent and the solution set is given by $\{(x, y) \mid 3x + 2y = 6\}$, or equivalently, $\{(x, y) \mid 6x + 4y = 12\}$. In practical terms we can generate different solutions by choosing x to be any real number and by solving $3x + 2y = 6$ for y to yield $y = -1.5x + 3$. For instance, if we let $x = 0$, then $y = -1.5(0) + 3 = 3$, and if $x = 1$ then, $y = -1.5(1) + 3 = 1.5$. [We could also choose y to be any real number and then solve for x, which gives us $x = -0.667y + 2$. Thus the value of one of the variables is completely arbitrary (it can be any real number).] Since there are infinitely many real numbers, there must be an infinite number of solutions to the system $3x + 2y = 6$ and $6x + 4y = 12$. □

By now you may be asking yourself why we can't solve systems of equations with two variables by drawing the graph. In fact, we can. However, there is a catch. It may be very difficult to draw the graph accurately enough to determine the answer (solution set), particularly when the answer is a unique point. *There is no reason to expect that the solution set for a system of equations will have all integer values.* However, you should always draw the graph whenever possible to give yourself a visual ballpark check on your algebraic solution. (You should get in the habit of performing ballpark checks on every calculation you make, whether in this course or not. A simple check of your results can often save you embarrassment and money.)

Practice Problem

What type of solution does each of the following systems have?

a) $\begin{aligned} 3x - 2y &= 10 \\ -6x + 4y &= 15 \end{aligned}$ **b)** $\begin{aligned} x - 2y &= 8 \\ 2x + 2y &= 13 \end{aligned}$

c) $\begin{aligned} -2x + 12y &= 18 \\ x - 6y &= -9 \end{aligned}$

Answers: **a)** No answer: the system is inconsistent.
b) Unique: the solution is $x = 7, y = -\frac{1}{2}$.
c) Infinitely many answers: the system is dependent. The solution set can be written $\{(x, y) \mid x = -9 + 6y\}$ or $\{(x, y) \mid y = (x + 9)/6\}$.

6.2 SOLUTION BY SUBSTITUTION

Equal Numbers of Equations and Variables

There are several approaches we can use to determine the solution set of a system of equations. The simplest is called **substitution.** It is particularly effective when dealing with systems that contain two variables. Let's apply this approach to the three systems we graphed in Examples 1–3.

EXAMPLE 4

Solve the system

$$\begin{aligned} 2x - 3y &= 6, \\ x + 3y &= 12. \end{aligned}$$

SOLUTION

Notice that we can solve the second equation for x very easily. Doing so gives us

$$x + 3y = 12,$$

or

$$x = 12 - 3y.$$

Suppose we replace x by the expression $12 - 3y$ in the first equation. We would then have an equation in which the only variable would be y.

$$
\begin{aligned}
2x - 3y &= 6 && \textbf{But } x = 12 - 3y. \\
\text{so} \quad 2(12 - 3y) - 3y &= 6 && \textbf{Replace } x \textbf{ by } 12 - 3y. \\
24 - 6y - 3y &= 6 && \textbf{Solve for } y. \\
24 - 9y &= 6 \\
-9y &= -18 \\
y &= 2
\end{aligned}
$$

Now substitute this value ($y = 2$) back into the equation for x:

$$
\begin{aligned}
x &= 12 - 3y, && \textbf{Replace } y \textbf{ by 2.} \\
\text{or} \quad x &= 12 - 3 \cdot 2, \\
x &= 6.
\end{aligned}
$$

Thus there is a unique solution at the point (6, 2). □

Let's think about what we just did. The basic idea is to solve one equation for one variable in terms of the other. We then substitute this solution for the variable into the remaining equation, which reduces it to one equation with one variable. We solve this equation for the value of the variable and plug it into the other equation, giving us the value of the first variable.

EXAMPLE 5

Solve the system

$$
\begin{aligned}
2x - 3y &= 6, \\
-4x + 6y &= 6.
\end{aligned}
$$

SOLUTION

Neither variable appears with a coefficient of 1, so there is no obvious choice of variable for which to solve. Perhaps the easiest approach is to solve the first equation for x.

$$
\begin{aligned}
\text{Start with} \quad 2x - 3y &= 6. \\
\text{Then} \quad 2x &= 6 + 3y, \\
\text{so} \quad x &= 0.5(6 + 3y). \\
\text{Since} \quad -4x + 6y &= 6, \\
\text{then} \quad -4[0.5(6 + 3y)] + 6y &= 6, \\
-2(6 + 3y) + 6y &= 6, \\
-12 - 6y + 6y &= 6, \\
-12 &= 6.
\end{aligned}
$$

Substitute this expression for x into the other equation and solve it for y (or, at least attempt to do so).

This condition is obviously impossible, which means that the system is inconsistent. Make careful note of this: *Deriving an impossible condition tells us the original system must be inconsistent.* □

EXAMPLE 6

Solve the system

$$3x + 2y = 6,$$
$$6x + 4y = 12.$$

SOLUTION

Again, there is no obvious place to start, so let's solve the first equation for y (just to show it doesn't matter for which variable you solve). This results in

$$3x + 2y = 6,$$

or
$$2y = 6 - 3x,$$

so
$$y = 0.5(6 - 3x).$$

Then
$$6x + 4y = 12,$$

or
$$6x + 4[0.5(6 - 3x)] = 12,$$
$$6x + 2(6 - 3x) = 12,$$
$$6x + 12 - 6x = 12,$$
$$12 = 12.$$

Substitute the expression for y into the other equation.

This condition is an identity (an equation that is always true). The meaning is that any values of x and y such that

$$y = 0.5(6 - 3x)$$

will satisfy the system. In other words, the system is dependent. Had we solved for x originally, we would have obtained

$$x = (6 - 2y)/3.$$

Again, take note of this result. *Whenever substitution (or any other algebraic procedure) yields an identity (an equation that is always true) then the original system is dependent.* Furthermore, the substitution used describes the solution set.

Compare these results with those obtained graphically in Examples 1–3. □

Practice Problem

Solve the following systems of equations by substitution.

a) $2x - y = 4$
$4x - 2y = 6$

b) $3x + y = 12$
$x - y = -2$

c) $$\begin{aligned} 4x - 2y &= 10 \\ -2x + y &= -5 \end{aligned}$$

Answers: **a)** Inconsistent system: no solution; **b)** $x = 2.5$, $y = 4.5$; **c)** Infinitely many solutions: $y = 2x - 5$ or $x = 0.5(y + 5)$

We can use substitution to analyze systems of three equations and three variables, as the following example shows. However, we do *not* recommend the use of substitution for systems with more than two variables because the algebra can become very messy. Moreover, the methods discussed in the next two sections are more efficient.

EXAMPLE 7

Solve the system of equations

$$\begin{aligned} x - y - z &= -4, \\ 3x - y + 2z &= -3, \\ 8x - 4y + 3z &= 6. \end{aligned}$$

SOLUTION

First, let's solve the top equation for x in terms of y and z.

$$x - y - z = -4,$$

or

$$x = -4 + y + z.$$

Next, we can substitute this expression into the remaining two equations, so we'll have two equations in two variables (y and z). The middle equation becomes

$$3x - y + 2z = -3,$$

or

$$3(-4 + y + z) - y + 2z = -3, \quad \textbf{Substitute for } \boldsymbol{x}.$$

$$-12 + 3y + 3z - y + 2z = -3,$$

$$2y + 5z = 9.$$

The bottom equation becomes

$$8x - 4y + 3z = 6,$$

or

$$8(-4 + y + z) - 4y + 3z = 6, \quad \textbf{Substitute for } \boldsymbol{x}.$$

$$-32 + 8y + 8z - 4y + 3z = 6,$$

$$4y + 11z = 38.$$

We now have reduced the original system to the equations

$$\begin{aligned} x &= -4 + y + z, \\ 2y + 5z &= 9, \\ 4y + 11z &= 38. \end{aligned}$$

Let's solve the middle equation for y, yielding

$$2y = 9 - 5z,$$

or

$$y = 0.5(9 - 5z).$$

Substituting this result into the third equation gives us

$$4[0.5(9 - 5z)] + 11z = 38, \quad \textbf{Substitute for } y.$$
$$2(9 - 5z) + 11z = 38,$$
$$18 - 10z + 11z = 38,$$

so

$$z = 20.$$

We now substitute this into the expression for y:

$$y = 0.5(9 - 5z)$$
$$y = 0.5(9 - 5 \cdot 20) \quad \textbf{Substitute the value of } z.$$
$$y = 0.5(-91)$$

so

$$y = -45.5$$

Finally, we can substitute $y = -45.5$ and $z = 20$ into the expression

$$x = -4 + y + z,$$

so

$$x = -4 + (-45.5) + 20, \quad \textbf{Substitute the values of } y \textbf{ and } z.$$

or

$$x = -29.5.$$

Therefore the solution is $(-29.5, -45.5, 20)$. □

Now we've learned enough to solve the Kleins' investment problem (Problem from Life, part a).

EXAMPLE 8

Problem from Life, part (a) Recall that we formulated this problem in the introduction to this chapter:

$$s + b = 25{,}000$$
$$0.0125s + 0.0075b = 250$$

SOLUTION

To solve the Klein's problem, we need only solve the preceding system. We can solve the top equation easily for either s or b, so let's arbitrarily solve it for s:

$$s = 25{,}000 - b.$$

Substituting this result into the second equation we have

$$0.0125(25{,}000 - b) + 0.0075b = 250.$$

Before proceeding any farther, let's multiply both sides of this equation by 10,000 (to remove the decimal fractions):

$$125(25{,}000 - b) + 75b = 2{,}500{,}000.$$

Now let's divide both sides by 25 to reduce the size of the numbers.

$$\begin{aligned} 5(25{,}000 - b) + 3b &= 100{,}000 \\ 125{,}000 - 5b + 3b &= 100{,}000 \\ -2b &= -25{,}000 \\ b &= 12{,}500 \end{aligned}$$

Substituting this result into the expression for s gives us

$$s = 25{,}000 - b,$$

so

$$s = 25{,}000 - 12{,}500,$$

$$s = 12{,}500.$$

Therefore the Kleins should invest \$12,500 in the mutual stock fund and \$12,500 in bonds. □

Unequal Numbers of Variables and Equations

At this point you may have the impression that we are concerned only with systems of equations that have one equation for each variable, for example, two equations and two variables. Actually, this is not the case. It is entirely possible for a system to have more variables than equations, the same number of variables and equations, or fewer variables than equations. Since we have discussed in some detail the possibilities when the number of variables and the number of equations are equal, let's analyze the other two situations (by examples).

EXAMPLE 9

More Variables than Equations Solve the system

$$\begin{aligned} x + 2y + 3z &= 9, \\ 3x + 7y + 2z &= 12. \end{aligned}$$

SOLUTION

Here there are three variables and only two equations. Solving the top equation for x gives us

$$x + 2y + 3z = 9,$$

or

$$x = 9 - 2y - 3z.$$

We can substitute this into the second equation to yield

$$3x + 7y - 2z = 12,$$

or

$$3(9 - 2y - 3z) + 7y + 2z = 12, \quad \textbf{Substitute for } x.$$

$$27 - 6y - 9z + 7y + 2z = 12,$$

$$y - 7z = -15,$$

so

$$y = -15 + 7z.$$

Now we substitute the expression for y back into the expression for x:

$$x = 9 - 2y - 3z,$$

$$x = 9 - 2(-15 + 7z) - 3z, \quad \textbf{Substitute for } y.$$

$$x = 9 + 30 - 14z - 3z,$$

so

$$x = 39 - 17z.$$

Now we're finished. We've managed to express both x and y in terms of z, but what is the value of z? The answer is any real number. Thus we have a dependent system so that the solution set contains an infinite number of possibilities. We can write the solution set as

$\{(x, y, z) \mid x = 39 - 17z$ and $y = -15 + 7z$, where z is any real number$\}$. □

It is also possible for a system of equations with more variables than equations to be inconsistent (have no solution), as the next example shows.

EXAMPLE 10

More Variables than Equations Solve the system

$$x + 2y + 3z = 9,$$

$$2x + 4y + 6z = 12.$$

SOLUTION

As in the preceding example,

$$x = 9 - 2y - 3z.$$

Now, however, see what happens when we substitute this expression into the second equation:

$$2x + 4y + 6z = 12,$$

$$2(9 - 2y - 3z) + 4y + 6z = 12, \quad \textbf{Substitute for } x.$$

$$18 - 4y - 6z + 4y + 6z = 12,$$

$$18 = 12.$$

This is obviously impossible, so we conclude the system is inconsistent.

To summarize: Systems of equations with more variables than equations are either dependent or inconsistent. □

EXAMPLE 11

Fewer Variables than Equations Solve the system

$$x + 2y = 6,$$
$$2x + 5y = 14,$$
$$x + y = 4.$$

SOLUTION

The first equation tells us

$$x + 2y = 6,$$

so

$$x = 6 - 2y.$$

We can substitute this result into the second equation as follows:

$$2x + 5y = 14,$$
$$2(6 - 2y) + 5y = 14, \qquad \textbf{Substitute for } x.$$
$$12 - 4y + 5y = 14,$$

so

$$y = 2.$$

Substituting this equation into the expression for x we have

$$x = 6 - 2y,$$
$$x = 6 - 2 \cdot 2 \qquad \textbf{Substitute for } y.$$
$$x = 2.$$

Thus $x = 2$ and $y = 2$. What about the extra equation? Simply substitute these values into it and check to see if it is satisfied. Here we have

$$x + y = 4, \qquad ?$$
$$2 + 2 = 4. \qquad \text{Yes!}$$

Therefore this system has the unique solution (2, 2). □

EXAMPLE 12

Fewer Variables than Equations Solve the system

$$x + 2y = 6,$$
$$2x + 5y = 14,$$
$$x + y = 3.$$

SOLUTION

Since the first two equations are the same as in Example 11, $x = 2$ and $y = 2$. However, upon substituting these values into the last equation we have

$$\begin{aligned} x + y &= 3, && ? \\ 2 + 2 &= 3. && \text{No!} \end{aligned}$$

Therefore this system is inconsistent and has no sclution. □

EXAMPLE 13

Fewer Variables than Equations Solve the system

$$\begin{aligned} x + 2y &= 5, \\ 3x + 6y &= 15, \\ 5x + 10y &= 25. \end{aligned}$$

SOLUTION

The first equation yields the expression

$$x = 5 - 2y.$$

Substituting this into the second equation results in

$$\begin{aligned} 3x + 6y &= 15, \\ 3(5 - 2y) + 6y &= 15, && \textbf{Substitute for } \boldsymbol{x}. \\ 15 - 6y + 6y &= 15, \\ 15 &= 15, && \text{an identity!} \end{aligned}$$

Let's try substituting the expression for x into the last equation:

$$\begin{aligned} 5x + 10y &= 25, \\ 5(5 - 2y) + 10y &= 25, && \textbf{Substitute for } \boldsymbol{x}. \\ 25 - 10y + 10y &= 25, \\ 25 &= 25, && \text{another identity!} \end{aligned}$$

We must conclude that the system is dependent; the solution set is $\{(x, y) \mid x = 5 - 2y, \text{ where } y \text{ is any real number}\}$. □

Systems with fewer variables than equations can have unique solutions, be inconsistent, or be dependent.

These possibilities are summarized in Table 6.1.

TABLE 6.1 Possible outcomes when solving systems of linear equations

	When the system has		
	More variables than equations	Equal number of variables and equations	Fewer variables than equations
The system can	Be inconsistent or dependent	Have a unique solution, be inconsistent, or dependent	Have a unique solution, be inconsistent, or dependent

EXERCISE SET 6.2

Solve the following systems of equations. Be sure to characterize the system as having a unique solution, being inconsistent, or being dependent.

1. $x + 2y = 4$
$3x + 4y = 10$

2. $x - 2y = 6$
$2x + 2y = 6$

3. $x - 3y = 4$
$5x + 3y = 8$

4. $x - 5y = 5$
$3x + 10y = 10$

5. $x - y = 5$
$7x + 3y = 15$

6. $2x - 4y = 8$
$x + 4y = 13$

7. $2x + 2y = 6$
$3x + 5y = 15$

8. $3x - 6y = 6$
$7x + 8y = 15$

9. $x - 3y = 8$
$3x - 9y = 24$

10. $x + 2y = 4$
$-2x + 4y = -8$

11. $x + 3y = 7$
$6x + 18y = 35$

12. $2x - 5y = 1$
$6x - 15y = 10$

13. $x + 3y + z = 4$
$-x + 5y + z = 6$

14. $x + 2y = 6$
$5x - y = 8$
$x - 2y = -2$

15. $4x + y = 13$
$2x + 5y = 11$
$x + y = 4$

16. $4x + y = 7$
$2x + 5y = 17$
$x + y = 4$

17. $4x + 5y = 13$
$2x + 5y = 17$
$x + y = 4$

18. $x + 3y + 2z = 6$
$-x - 3y + z = 3$

19. $2x - y + 3z = 12$
$-4x + 2y - 6z = 15$

20. $x + y + z = 4$
$2x + 5y - 2z = 3$

21. The amount or quantity of an item that people are willing to buy (at any given time) is generally a function of its price. Suppose that the demand for gas in some city can be described by the function

$$d(p) = 10{,}000 - 2500p, \qquad \$1.00 \le p \le \$4.00,$$

where p is the price per gallon and $d(p)$ is the number of gallons demanded. We can write a similar expression to describe the available supply, which is the amount of the item the suppliers are willing to offer at a given price. Let's assume the supply relationship can be written

$$s(p) = -1250 + 5000p, \qquad \$1.00 \leq p \leq \$4.00,$$

where $s(p)$ is the number of gallons supplied.

Now, suppose p is \$1.00 per gallon. Then the amount demanded will be $d(1.00) = 7500$ gallons, but the amount supplied will be only $s(1.00) = 3750$ gallons. However, when p is \$3.00 per gallon, the amount demanded will be only $d(3.00) = 2500$ gallons, while the amount supplied will be $s(3.00) = 13{,}750$ gallons. As you might guess, when demand exceeds supply (for example, when p is \$1.00) the price will tend to increase; when supply exceeds demand the price will tend to decrease. However, there is a price, called the **equilibrium price,** at which supply and demand will be equal.

a) Find the equilibrium price by setting $d(p) = s(p)$ and finding the value of p. What is the equilibrium quantity of gasoline supplied (and demanded)? Check your solution by graphing the demand and supply functions.

b) Suppose the supply function becomes

$$s(p) = -1250 + 6000p, \qquad \$1.00 \leq p \leq \$4.00.$$

What are the new equilibrium price and quantity?

22. In the same city the demand and supply equations for coffee (in pounds) are given by

$$\begin{aligned} d(p) &= 1800 - 600p \\ s(p) &= -900 + 600p \end{aligned} \qquad \$2.00 \leq p \leq = \$3.00.$$

Determine the equilibrium price and quantity. Check your answer graphically.

23. The biology labs at State University must combine two food mixes in certain proportions to constitute a diet for the laboratory mice. The diet must contain exactly 22.5 grams of protein and 6 grams of fat. The characteristics of the mixes are as follows.

Mix	% Protein	% Fat
A	20	6
B	25	6

How much of each mix should be used?

24. In another experiment at State University, the technician has been told that a diet containing 22 grams of protein, 7.5 grams of fat, and 6.6 grams of fiber must be constituted from the three food mixes described as follows.

Mix	% Protein	% Fat	% Fiber
A	20	6	6
B	25	6	9
C	20	9	6

How much of each mix should be used?

* **25.** Firenze Leather makes purses and briefcases from leather. Each purse requires 3 hours of labor and 3 square feet of leather, while each briefcase requires 4 hours of labor and 6 square feet of leather. There are 72 hours of labor and 96 square feet of leather available next week. How many purses and briefcases should Firenze make?

26. Advanced Computer Systems manufactures computers. The relevant data are as follows.

Model	Labor hours	Test hours	Packaging hours
System 1	15	5	1
System 2	20	10	1.5
System 3	25	5	2

There are exactly 400 hours of labor, 120 hours of testing, and 30 hours of packaging available next week. How many of each system should be manufactured?

6.3 SOLUTION BY GAUSS–JORDAN ELIMINATION

Algebraic Elimination

In the previous section you learned one of the simplest methods of solving a system of equations: substitution. However, as Example 7 showed, substitution becomes awkward for systems with only three variables and essentially hopeless for larger systems. Fortunately, a more efficient approach is available for such systems: Gauss–Jordan elimination.

Before we formally discuss Gauss–Jordan elimination, we'll analyze a system first solved in Example 4 using algebraic elimination.

* Reprinted with permission from Wu and Coppins, *Linear Programming and Extensions* (McGraw-Hill, New York), 1981.

EXAMPLE 14

Solve the following system using algebraic elimination:

$$\begin{aligned} 2x - 3y &= 6, \\ x + 3y &= 12. \end{aligned}$$

SOLUTION

Let's begin by switching the order of the equations, which will place x with a coefficient of 1 in the top equation. (The reason for this will become apparent shortly.)

$$\begin{aligned} x + 3y &= 12 \\ 2x - 3y &= 6 \end{aligned}$$

Next, let's eliminate x from the bottom row. We can do this by multiplying the first equation by -2 and then adding it to the second equation. This is valid because we know we can multiply both sides of an equation by a constant. Then, when we add the result to the second equation, we are adding equal amounts to both sides of the second equation. The new second equation will be

$$\begin{array}{rl} 2x - 3y = 6 & \text{old second equation} \\ +(-2x - 6y = -24) & +\ (-2)\text{first equation} \\ \hline -9y = -18 & \text{new second equation} \end{array}$$

so

Multiplying both sides by $-\frac{1}{9}$ gives us

$$y = 2.$$

Now our system of equations is

$$\begin{aligned} x + 3y &= 12, \\ y &= 2. \end{aligned}$$

Next, let's eliminate y from the first equation. We can do this by multiplying the second equation by -3 and adding it to the first equation.

$$\begin{array}{rl} x + 3y = 12 & \text{old first equation} \\ +\ \ (-3y = -6) & +\ (-3)\text{second equation} \\ \hline x \qquad = 6 & \text{new first equation} \end{array}$$

Now our system has become

$$\begin{aligned} x \qquad &= 6, \\ y &= 2. \end{aligned}$$

This system of equations can be solved by inspection because each equation contains only one variable (and with a coefficient of 1). Thus, by manipulating the original system, we obtained a new system that could be solved by inspection. The key here is that the manipulations produced an **equivalent system:** one that has the same solution set as the original system. □

Augmented Matrices and Row Operations

Now let's see how we can formalize this approach (algebraic elimination) just demonstrated. The first step is to organize the problem data by putting them into an **augmented matrix.** Thus the system

$$\begin{aligned} 2x - 3y &= 6, \\ x + 3y &= 12, \end{aligned}$$

is represented by the augmented matrix

$$\left[\begin{array}{cc|c} 2 & -3 & 6 \\ 1 & 3 & 12 \end{array}\right].$$

Notice that each row of this matrix represents an equation. We've simplified matters by not writing the variables and using a vertical line rather than equality signs.

Let's think about the algebraic manipulations we used to solve Example 14:

1. We switched the order of the equations.
2. We multiplied both sides of an equation by a (nonzero) constant.
3. We multiplied an equation by a constant and added it to another equation.

We want to be able to perform the equivalent operations on the augmented matrix that represents the system of equations. We can do this by using the three **elementary row operations,** which can be performed on a matrix:

Elementary Row Operations

1. Two rows can be interchanged.
2. A row can be multiplied by a nonzero scalar.
3. A row can be multiplied by a nonzero scalar and added to another row.

Let's rework Example 14 using elementary row operations.

EXAMPLE 15 Solve the system

$$\begin{aligned} 2x - 3y &= 6, \\ x + 3y &= 12. \end{aligned}$$

SOLUTION

First, let's write the system as an augmented matrix:

$$\left[\begin{array}{cc|c} 2 & -3 & 6 \\ 1 & 3 & 12 \end{array}\right].$$

In Example 14 we eliminated x from the second equation after first switching the order of the equations. Here let's switch the two rows in the augmented matrix because this will place a 1 in the first row and first column.

$$R1 \leftrightarrow R2 \quad \left[\begin{array}{cc|c} 1 & 3 & 12 \\ 2 & -3 & 6 \end{array}\right]$$

Notice the notation used to indicate the switch of the rows. Next, let's eliminate the 2 from the second row by multiplying the top row by -2 and adding it to the bottom row.

$$R2 + (-2)R1 \quad \left[\begin{array}{cc|c} 1 & -3 & 12 \\ 0 & -9 & -18 \end{array}\right].$$

In Example 14 we eliminated y from the top equation. We can start by multiplying the bottom row by $-\frac{1}{9}$.

$$(-\tfrac{1}{9})R2 \quad \left[\begin{array}{cc|c} 1 & 3 & 12 \\ 0 & 1 & 2 \end{array}\right]$$

Next, multiply the bottom row by -3 and add it to the top row to eliminate the 3.

$$R1 + (-3)R2 \quad \left[\begin{array}{cc|c} 1 & 0 & 6 \\ 0 & 1 & 2 \end{array}\right]$$

We're finished. The augmented matrix preceding, which contains an identity matrix on the left of the line, represents the system

$$\begin{aligned} x \phantom{{}+y} &= 6, \\ y &= 2. \end{aligned}$$

This means the solution to the original system is (6, 2). Because we used elementary row operations, the system resulting from our operations is equivalent to the original system of equations. □

Let's review the procedure we utilized in Example 15. First, we used an augmented matrix to represent the original system of equations. Next, we systematically performed elementary row operations until the left side of the augmented matrix became an identity matrix. At this point we were able to read the solution to the original system quite easily. This approach is called **Gauss–Jordan elimination.** Because it is so systematic, it can be programmed so that computers can solve systems of equations with hundreds or even thousands of equations and variables.

We can summarize the procedure for systems with an equal number of variables and equations as follows.

Gauss–Jordan Elimination: Systems with Equal Number of Variables and Equations

1. Write the augmented matrix representation of the given system of equations.
2. Get a 1 in the first column, first row, by multiplying the first row by the multiplicative inverse of the number that appears there. (If the number is 0, switch the first row with another row that has a non-zero entry in the first column.) Next, take appropriate multiples of the first row and add them to the other rows to convert all the other elements in the first column to 0's.
3. Get a 1 in the second column, second row, by multiplying the second row by the inverse of the (nonzero) number that appears there. Take appropriate multiples of the second row and add them to the other rows to convert all other elements in the second column to 0's.
4. Continue as necessary.

EXAMPLE 16

Solve the following system using Gauss–Jordan elimination:

$$2x - 3y = 6,$$
$$-4x + 6y = 6.$$

SOLUTION

Step 1 The augmented matrix that represents the original system is

$$\left[\begin{array}{cc|c} 2 & -3 & 6 \\ -4 & 6 & 6 \end{array}\right].$$

Step 2 Multiply the first row by 0.5 to obtain

$$0.5R1 \quad \left[\begin{array}{cc|c} 1 & -1.5 & 3 \\ -4 & 6 & 6 \end{array}\right].$$

Next, multiply the top row by 4 and add it to the bottom row (to eliminate the -4 from the second row).

$$R2 + 4R1 \quad \left[\begin{array}{cc|c} 1 & -1.5 & 3 \\ 0 & 0 & 18 \end{array}\right]$$

The bottom row is equivalent to the equation

$$0x + 0y = 18,$$

which is impossible. Thus this system must be inconsistent. □

EXAMPLE 17

Solve the following system using Gauss–Jordan elimination:

$$3x + 2y = 6,$$
$$6x + 4y = 12.$$

SOLUTION

Step 1 The augmented matrix is

$$\left[\begin{array}{cc|c} 3 & 2 & 6 \\ 6 & 4 & 12 \end{array}\right].$$

Step 2 Multiply the first row by $\frac{1}{3}$:

$$(\tfrac{1}{3})R1 \quad \left[\begin{array}{cc|c} 1 & \frac{2}{3} & 2 \\ 6 & 4 & 12 \end{array}\right].$$

Next, multiply the top row by -6 and add it to the bottom row.

$$R2 + (-6)R1 \quad \left[\begin{array}{cc|c} 1 & \frac{2}{3} & 2 \\ 0 & 0 & 0 \end{array}\right]$$

The bottom row is equivalent to the equation

$$0x + 0y = 0,$$

which is an identity. This result indicates that the system of equations is dependent. We solve either equation for one variable in terms of the other to describe the solution set. For example,

$$y = 3 - 1.5x,$$

or

$$x = 2 - \tfrac{2}{3}y. \quad \square$$

Practice Problem

Solve the following systems of equations by Gauss–Jordan elimination.

a) $4x - 2y = 10$
$2x - y = 5$

b) $2x - y = 8$
$4x - 2y = 6$

c) $3x + y = 12$
$2x - 2y = -4$

Answers: **a)** infinitely many solutions: $y = 2x - 5$ or $x = 0.5(y + 5)$; **b)** inconsistent system: no solution; **c)** $x = 2.5$, $y = 4.5$

Let's see how this works with a larger problem. Consider the following system (originally solved by substitution as Example 7).

EXAMPLE 18 Solve the following system using Gauss–Jordan elimination:

$$\begin{aligned} x - y - z &= -4, \\ 3x - y + 2z &= -3, \\ 8x - 4y + 3z &= 6. \end{aligned}$$

SOLUTION

Step 1 Form the augmented matrix:

$$\left[\begin{array}{ccc|c} 1 & -1 & -1 & -4 \\ 3 & -1 & 2 & -3 \\ 8 & -4 & 3 & 6 \end{array}\right].$$

Step 2 Eliminate the 3 from the second row and the 8 from the third row, in the first column. We need not transform the top row because there is a 1 in the first position. Therefore, multiply the top row by -3 and add it to the second row.

$$R2 + (-3)R1 \quad \left[\begin{array}{ccc|c} 1 & -1 & -1 & -4 \\ 0 & 2 & 5 & 9 \\ 8 & -4 & 3 & 6 \end{array}\right]$$

Next, multiply the top row by -8 and add it to the bottom row.

$$R3 + (-8)R1 \quad \left[\begin{array}{ccc|c} 1 & -1 & -1 & -4 \\ 0 & 2 & 5 & 9 \\ 0 & 4 & 11 & 38 \end{array}\right]$$

Step 3 Put a 1 in the middle of the second column and eliminate the -1 and 4 in the second column. Begin by multiplying the middle row by 0.5.

$$0.5R2 \quad \left[\begin{array}{ccc|c} 1 & -1 & -1 & -4 \\ 0 & 1 & 2.5 & 4.5 \\ 0 & 4 & 11 & 38 \end{array}\right]$$

Add the middle row to the top row.

$$R1 + R2 \quad \left[\begin{array}{ccc|c} 1 & 0 & 1.5 & 0.5 \\ 0 & 1 & 2.5 & 4.5 \\ 0 & 4 & 11 & 38 \end{array}\right]$$

Multiply the middle row by -4 and add it to the bottom row.

$$R3 + (-4)R1 \quad \left[\begin{array}{ccc|c} 1 & 0 & 1.5 & 0.5 \\ 0 & 1 & 2.5 & 4.5 \\ 0 & 0 & 1 & 20 \end{array}\right]$$

Step 4 Eliminate the 1.5 and 2.5 from the third column. Multiply the bottom row by -1.5 and add it to the top row.

$$R1 + (-1.5)R3 \quad \left[\begin{array}{ccc|c} 1 & 0 & 0 & -29.5 \\ 0 & 1 & 2.5 & 4.5 \\ 0 & 0 & 1 & 20 \end{array}\right]$$

Multiply the bottom row by -2.5 and add it to the middle row.

$$R2 + (-2.5)R3 \quad \left[\begin{array}{ccc|c} 1 & 0 & 0 & -29.5 \\ 0 & 1 & 0 & -45.5 \\ 0 & 0 & 1 & 20 \end{array}\right]$$

We're done! The answer can be read from the final matrix:

$$(-29.5, -45.5, 20).$$

(Compare this approach with that shown in Example 7.) □

Practice Problem

What are the solutions of the following systems?

a) $\begin{aligned} x - 2y &= 8 \\ 2x + 2y &= 13 \end{aligned}$

b) $\begin{aligned} 3x - 2y &= 10 \\ -6x + y &= 16 \end{aligned}$

c) $\begin{aligned} 3x - 2y &= 18 \\ -6x + 4y &= 10 \end{aligned}$

d) $\begin{aligned} 4x - 2y &= 12 \\ -x + y &= -6 \end{aligned}$

Answers: **a)** $x = 7, y = -0.5$; **b)** $x = -4.667, y = -12$; **c)** inconsistent system; **d)** $x = 0, y = -6$

At this point you should be able to solve any system with the same number of equations and variables by Gauss–Jordan elimination. However, you may be wondering what happens when the number of equations and the number of variables are not equal. (In these cases we certainly can't expect to obtain an identity matrix on the left-hand portion of the augmented matrix because it isn't square.) Let's see.

EXAMPLE 19

Fewer Equations than Variables Solve the following system by Gauss–Jordan elimination:

$$\begin{aligned} x + 2y + 3z &= 17, \\ 3x - 4y + 6z &= 13. \end{aligned}$$

SOLUTION

We know this system is either dependent or inconsistent, depending on the numbers in the equations since there are three variables and only two equations. We can try to solve for x and y in terms of z.

Step 1 Form the augmented matrix

$$\left[\begin{array}{ccc|c} 1 & 2 & 3 & 17 \\ 3 & -4 & 6 & 13 \end{array}\right].$$

Step 2 Eliminate the 3 from the first column. Multiply the first row by -3 and add it to the bottom row.

$$R2 + (-3)R1 \quad \left[\begin{array}{ccc|c} 1 & 2 & 3 & 17 \\ 0 & -10 & -3 & -38 \end{array}\right]$$

Step 3 Eliminate the 2 from the second column, top row. Begin by multiplying the bottom row by -0.1.

$$-0.1R2 \quad \left[\begin{array}{ccc|c} 1 & 2 & 3 & 17 \\ 0 & 1 & 0.3 & 3.8 \end{array}\right]$$

Next, multiply the bottom row by -2 and add it to the top row.

$$R1 + (-2)R2 \quad \left[\begin{array}{ccc|c} 1 & 0 & 2.4 & 9.4 \\ 0 & 1 & 0.3 & 3.8 \end{array}\right]$$

We cannot proceed any further. The final matrix represents the system

$$\begin{aligned} x \qquad + 2.4z &= 9.4, \\ y + 0.3z &= 3.8, \end{aligned}$$

or

$$\begin{aligned} x &= 9.4 - 2.4z, \\ y &= 3.8 - 0.3z, \\ z &= \text{any real number}. \end{aligned}$$

Thus, the original system is dependent. □

When there are more variables than equations, eliminate variables (as we did in Example 19) until you reach a situation in which the *first* nonzero number in each row is a 1 and the 1's form a zig-zag pattern (or stair-step) down and to the right. Moreover, the first 1 in a row should be the *only* nonzero number in the column. This is called **reduced row echelon form.** (See Fig. 6.5.) Once the augmented matrix is in reduced row echelon form, you can determine the dependent solution.

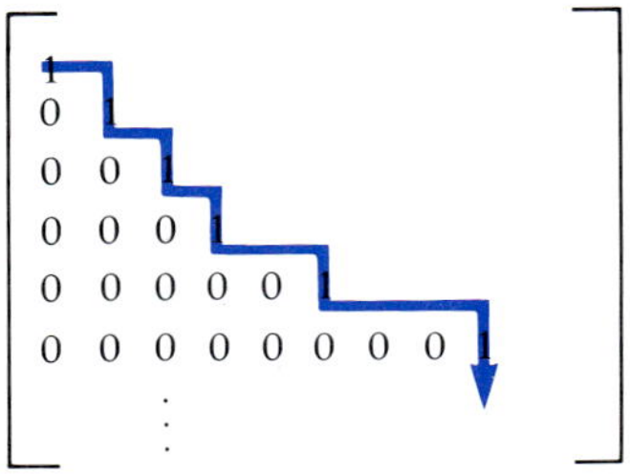

Figure 6.5 In reduced row echelon form, the 1's form a zig-zag.

EXAMPLE 20

The following matrices are in reduced row echelon form.

$$\left[\begin{array}{cc|c} 1 & 0 & 2.4 \\ 0 & 1 & 0.3 \end{array}\right] \qquad \left[\begin{array}{ccc|c} 1 & -1 & 0 & 1.5 \\ 0 & 0 & 1 & -2.5 \end{array}\right]$$

$$\left[\begin{array}{ccccc|c} 1 & 2 & 0 & 3 & 0 & 4 \\ 0 & 0 & 1 & 0 & 0 & 3 \\ 0 & 0 & 0 & 0 & 1 & -2 \end{array}\right]$$

The following matrices are *not* in reduced row echelon form.

$$\left[\begin{array}{ccc|c} 1 & 0 & 2 & 4 \\ 0 & -2 & 3 & 2 \end{array}\right]$$ The first entry on the second row is not a 1.

$$\left[\begin{array}{ccc|c} 1 & -1 & 3 & 6 \\ 0 & 1 & 4 & 9 \end{array}\right]$$ There is a nonzero number (-1) in the second column above the 1. □

Practice Problem

Which of the following matrices are in reduced row echelon form?

a) $\left[\begin{array}{ccc|c} 1 & 0 & -2 & 6 \\ 0 & 3 & 4 & 8 \end{array}\right]$ b) $\left[\begin{array}{ccc|c} 1 & 0 & -2 & 6 \\ 0 & 1 & 6 & 5 \end{array}\right]$

c) $\left[\begin{array}{cccc|c} 1 & 0 & 2 & -1 & 5 \\ 0 & 0 & 1 & 0 & 3 \end{array}\right]$ d) $\left[\begin{array}{cccc|c} 1 & 0 & 0 & 6 & 8 \\ 0 & 1 & 0 & 1 & 7 \\ 0 & 0 & 1 & -2 & 5 \end{array}\right]$

e) $\left[\begin{array}{cccc|c} 1 & 0 & 0 & 2 & -1 \\ 0 & 2 & 0 & 3 & 6 \\ 0 & 0 & 1 & -2 & 3 \end{array}\right]$ f) $\left[\begin{array}{ccc|c} 1 & 0 & 0 & 2 \\ 0 & 1 & 0 & 3 \\ 0 & 0 & 1 & 6 \end{array}\right]$

Answer: Matrices b, d, and f are in reduced row echelon form.

Now we can give a more general form for Gauss–Jordan elimination.

Gauss–Jordan Elimination: General Approach

1. Write the augmented matrix representation of the given system of equations.
2. Get a 1 in the first column of the first row that doesn't contain all zeros by multiplying the first row by the multiplicative inverse of the number that appears there. (If the number is 0, switch the first row with another row that has a non-zero entry in the first column.) Next, take appropriate multiples of the first row and add them to the other rows to convert all the other elements in the first column to 0's.
3. Get a 1 in the next column on the next row that doesn't contain all 0's by multiplying the row by the multiplicative inverse of the first nonzero number in the row. (If necessary, switch the row with one below it to eliminate the 0.) Take appropriate multiples of the row and add them to the other rows to convert all other elements in this column to 0.
4. Continue as necessary until the matrix is in reduced row echelon form.

EXAMPLE 21

Solve the following system using Gauss–Jordan elimination:

$$x + 2y + 3z = 9,$$
$$3x + 7y + 2z = 12.$$

SOLUTION

Step 1 The augmented matrix is

$$\left[\begin{array}{ccc|c} 1 & 2 & 3 & 9 \\ 3 & 7 & 2 & 12 \end{array}\right].$$

Step 2 Eliminate the 3 from the first column. Multiply the top row by -3 and add it to the bottom row.

$$R2 + (-3)R1 \quad \left[\begin{array}{ccc|c} 1 & 2 & 3 & 9 \\ 0 & 1 & -7 & -15 \end{array}\right]$$

Step 3 Eliminate the 2 from the second column. Multiply the bottom row by -2 and add it to the top row.

$$R1 + (-2)R2 \quad \left[\begin{array}{ccc|c} 1 & 0 & 17 & 39 \\ 0 & 1 & -7 & -15 \end{array}\right]$$

We're done. Translating the augmented matrix back into equation form, we have

$$\begin{aligned} x \qquad + 17z &= 39, \\ y - \ \ 7z &= -15. \end{aligned}$$

so

$$\begin{aligned} x &= 39 - 17z, \\ y &= -15 + 7z, \\ z &= \text{any real number}. \end{aligned}$$

Compare this to Example 9. □

EXAMPLE 22 Solve the following system using Gauss–Jordan elimination:

$$\begin{aligned} 2w + 4x - \ \ 4y + 6z &= 20, \\ 3w - 6x + 18y - 3z &= 18. \end{aligned}$$

SOLUTION

Here we have four variables and only two equations. Thus we can get w and x in terms of y and z.

Step 1 The augmented matrix is

$$\left[\begin{array}{cccc|c} 2 & 4 & -4 & 6 & 20 \\ 3 & -6 & 18 & -3 & 18 \end{array}\right].$$

Step 2 Eliminate the 3 from the first column. First, multiply the top row by 0.5:

$$0.5R1 \quad \left[\begin{array}{cccc|c} 1 & 2 & -2 & 3 & 10 \\ 3 & -6 & 18 & -3 & 18 \end{array}\right].$$

Multiply the top row by -3 and add it to the bottom row.

$$R2 + (-3)R1 \quad \left[\begin{array}{cccc|c} 1 & 2 & -2 & 3 & 10 \\ 0 & -12 & 24 & -12 & -12 \end{array}\right]$$

Step 3 Eliminate the 2 from the second column. Begin by multiplying the bottom row by $-\frac{1}{12}$ (the decimal equivalent would have to be rounded, so we'll use the fraction).

$$(-\tfrac{1}{12})R2 \quad \left[\begin{array}{cccc|c} 1 & 2 & -2 & 3 & 10 \\ 0 & 1 & -2 & 1 & 1 \end{array}\right]$$

Multiply the bottom row by -2 and add it to the top row.

$$R1 + (-2)R2 \quad \left[\begin{array}{cccc|c} 1 & 0 & 2 & 1 & 8 \\ 0 & 1 & -2 & 1 & 1 \end{array}\right]$$

This is reduced row echelon form so we're finished. We have

$$\begin{aligned} w \qquad + 2y + z &= 8, \\ x - 2y + z &= 1, \end{aligned}$$

or

$$\begin{aligned} w &= 8 - 2y - z, \\ x &= 1 + 2y - z, \text{ where } y \text{ and } z \text{ are any real numbers.} \end{aligned}$$ □

The following examples (first solved as Examples 11–13) show how to proceed when there are fewer variables than equations.

EXAMPLE 23

Fewer Variables Than Equations Solve the following system using Gauss–Jordan elimination:

$$\begin{aligned} x + 2y &= 6, \\ 2x + 5y &= 14, \\ x + y &= 4. \end{aligned}$$

SOLUTION

Step 1 The augmented matrix is

$$\left[\begin{array}{cc|c} 1 & 2 & 6 \\ 2 & 5 & 14 \\ 1 & 1 & 4 \end{array}\right].$$

Step 2 Eliminate the 2 and (bottom) 1 from the first column. Multiply the first row by -2 and add it to the second row.

$$R2 + (-2)R1 \quad \left[\begin{array}{cc|c} 1 & 2 & 6 \\ 0 & 1 & 2 \\ 1 & 1 & 4 \end{array}\right]$$

Multiply the top row by -1 and add it to the bottom row.

$$R3 + (-1)R1 \quad \left[\begin{array}{cc|c} 1 & 2 & 6 \\ 0 & 1 & 2 \\ 0 & -1 & -2 \end{array}\right]$$

Step 3 Eliminate the 2 and -1 from the second column. Multiply the second row by -2 and add it to the top row.

$$R1 + (-2)R2 \quad \left[\begin{array}{cc|c} 1 & 0 & 2 \\ 0 & 1 & 2 \\ 0 & -1 & -2 \end{array}\right]$$

Add the second row to the bottom row.

$$R3 + R2 \quad \left[\begin{array}{cc|c} 1 & 0 & 2 \\ 0 & 1 & 2 \\ 0 & 0 & 0 \end{array}\right]$$

The solution is (2, 2). The bottom row consists of $0x + 0y = 0$, which is an identity. □

EXAMPLE 24

Solve the following system using Gauss–Jordan elimination.

$$\begin{aligned} x + 2y &= 6, \\ 2x + 5y &= 14, \\ x + y &= 3. \end{aligned}$$

SOLUTION

Perform the same set of row operations as in Example 23. This will result in

$$\left[\begin{array}{cc|c} 1 & 0 & 2 \\ 0 & 1 & 2 \\ 0 & 0 & -1 \end{array}\right].$$

Here the bottom row tells us

$$0x + 0y = -1,$$

which is impossible. Therefore the given system is inconsistent. □

EXAMPLE 25

Solve the following system using Gauss–Jordan elimination:

$$\begin{aligned} x + 2y &= 5, \\ 3x + 6y &= 15, \\ 5x + 10y &= 25. \end{aligned}$$

SOLUTION

Step 1 The augmented matrix is

$$\left[\begin{array}{cc|c} 1 & 2 & 5 \\ 3 & 6 & 15 \\ 5 & 10 & 25 \end{array}\right]$$

Step 2 Eliminate the 3 and 5 from the first column. Multiply the top row by -3 and add it to the second row.

$$R2 + (-3)R1 \quad \left[\begin{array}{cc|c} 1 & 2 & 5 \\ 0 & 0 & 0 \\ 5 & 10 & 25 \end{array}\right]$$

Multiply the top row by -5 and add it to the third row.

$$R3 + (-5)R1 \quad \left[\begin{array}{cc|c} 1 & 2 & 5 \\ 0 & 0 & 0 \\ 0 & 0 & 0 \end{array}\right]$$

Here we have

$$x + 2y = 5$$

and the identity $0x + 0y = 0$ (twice). Thus we have the dependent solution $\{(x, y) \mid x = 5 - 2y$, where y is any real number$\}$, or $\{(x, y) \mid y = 2.5 - 0.5x$, where x is any real number$\}$. □

Now we can solve part (b) of the Problem from Life.

EXAMPLE 26

Problem from Life, part (b) A dentist employs three assistants in her practice; her office has two rooms with chairs and equipment (operatories). The dentist provides both pedodontic and general dentistry services. Experience indicates that an average pedodontic service requires 0.75 hour of operatory time, 1.50 hours of an assistant's time, and 0.25 hour of dentist's time, while an average general service requires 0.75 hour of operatory time, 1.00 hour of an assistant's time, and 0.50 hour of dentist's time. Given the dentist's three assistants and two operatories are each to work 8 hours, how many patients can be scheduled during an 8-hour day?

SOLUTION

We can see from the last sentence of the problem description that we must determine how many of each type of patient to schedule each day. Let

p = number of pedodontic patients,

g = number of general dentistry patients.

Since there are three resources (the dentist, the assistants, and the operatories), we will have three equations to ensure that each resource is used appropriately.

$$\begin{aligned} &\text{Dentist:} && 0.25p + 0.50g = 8, \\ &\text{Assistants:} && 1.50p + 1.00g = 24, \\ &\text{Operatories:} && 0.75p + 0.75g = 16. \end{aligned}$$

Note that the right-hand side of the second equation is 24 because each of the three assistants works 8 hours; the right-hand side of the last equation is 16 because each operatory is available 8 hours.

Step 1 Form the augmented matrix

$$\left[\begin{array}{cc|c} 0.25 & 0.50 & 8 \\ 1.50 & 1.00 & 24 \\ 0.75 & 0.75 & 16 \end{array}\right].$$

Step 2 Eliminate the 1.5 and 0.75 from the first column. Begin by multiplying the first row by 4.

$$4R1 \quad \left[\begin{array}{cc|c} 1 & 2 & 32 \\ 1.50 & 1.00 & 24 \\ 0.75 & 0.75 & 16 \end{array}\right]$$

Next, multiply the top row by -1.5 and add it to the second row.

$$R2 + (-1.5)R1 \quad \left[\begin{array}{cc|c} 1 & 2 & 32 \\ 0 & -2 & -24 \\ 0.75 & 0.75 & 16 \end{array}\right]$$

Multiply the top row by -0.75 and add it to the third row.

$$R3 + (-0.75)R1 \quad \left[\begin{array}{cc|c} 1 & 2 & 32 \\ 0 & -2 & -24 \\ 0 & -0.75 & -8 \end{array}\right]$$

Step 3 Eliminate the 2 and -0.75 from the second column. Multiply the second row by -0.5.

$$-0.5R2 \quad \left[\begin{array}{cc|c} 1 & 2 & 32 \\ 0 & 1 & 12 \\ 0 & -0.75 & -8 \end{array}\right]$$

Multiply the second row by -2 and add it to the top row.

$$R1 + (-2)R2 \quad \left[\begin{array}{cc|c} 1 & 0 & 8 \\ 0 & 1 & 12 \\ 0 & -0.75 & -8 \end{array}\right]$$

Multiply the second row by 0.75 and add it to the third row.

$$R3 + 0.75R2 \quad \left[\begin{array}{cc|c} 1 & 0 & 8 \\ 0 & 1 & 12 \\ 0 & 0 & 1 \end{array}\right]$$

We're done. The third row tells us

$$0p + 0g = 1,$$

which means that the system is inconsistent. In other words, there is no way to schedule patients so as to completely use all of the time each resource is available. Thus, based upon the amount of time required by each pedodontic and each general dentistry patient, there are unequal amounts of the resources provided. □

We'll take another look at this problem in Chapter 8 (Linear Programming by the Simplex Method) to study how the dentist could maximize her profits.

EXERCISE SET 6.3

Convert each of the following systems of equations into an augmented matrix. (Do *not* attempt to solve.)

1. $3x - 5y = 10$
$4x - 2y = 17$

2. $2x + 9y = 8$
$3x - 2y = 6$

3. $6x - 5y = 2$
$2x - y = 4$

4. $7x - 2y + z = 6$
$3x - 5y - 6z = -15$

5. $2x + 9y - 2z = 27$
$3x - y + 10z = 32$

6. $x + 4y + 6z = 4$
$3x + 2y + 8z = 2$
$5x + y + 10z = 16$

7. $w - 3x + 2y + z = 15$
$3w - 8x + 5z = 29$
$6w - 16x + 2y - 4z = 36$
$x - 3y - 18z = 44$

Solve each of the following systems of equations by Gauss–Jordan elimination:

8. $3x - 5y = 10$
$4x + 2y = 12$

9. $2x + 9y = 8$
$3x - 2y = 2$

10. $6x - 5y = 12$
$2x - y = 4$

11. $\dfrac{x}{2} + \dfrac{y}{4} = 2$
$\dfrac{x}{3} - \dfrac{y}{6} = 3$

12. $\dfrac{2x}{5} - \dfrac{y}{7} = 1$
$-\dfrac{x}{10} + \dfrac{2y}{3} = 2$

13. $2x - 3y = -2$
$5x + 6y = 18$

14. $2x + 9y - 2z = 27$
$3x - y + 10z = 32$

15. $x - 3y + 6z = 15$
$2x + y - 3z = 8$

16. $x + 4y + 6z = 8$
$3x - 5y - 8z = 2$
$5x + y - z = 16$

17. $3x + 2y + z = 6$
$2x - y + 4z = -8$
$2x + 2y - 4z = 10$

18.
$$\begin{aligned} x + y + z &= 6 \\ -3x + 2y + z &= -1 \\ -2x + 3y - 4z &= -13 \end{aligned}$$

19.
$$\begin{aligned} 2x + 4y + 4z &= 22 \\ -x + y + z &= 4 \\ 2x + 5y + 9z &= 39 \end{aligned}$$

20.
$$\begin{aligned} w - 3x + 4y + z &= 8 \\ 3w - 2x + 5z &= 16 \\ 6w + 2y - 4z &= 12 \end{aligned}$$

21.
$$\begin{aligned} 2w - 4x + 3y + z &= 8 \\ w + x + y + z &= 12 \\ -2w + 3x - y + z &= 6 \end{aligned}$$

22. Ann Jones has a total income of \$1000 per year from her two investments. Fund A pays 8% annual interest and fund B pays 12% annual interest. (Thus the income from fund A is 8% of the amount invested in the fund.) She has twice as much invested in fund B as in fund A. How much does she have in each fund?

23. Solve Problem 22 assuming Ms. Jones has equal amounts invested in each fund.

24. Jim Kopetsky, the head tennis pro at the racquet club, earned \$480 giving lessons. He is paid \$12.00 for 30-minute lessons and \$18.00 for 45-minute lessons. How many of each type did he give, given that the number of 30-minute lessons was one half the number of 45-minute lessons?

25. The Department of Management Sciences must assign a total of 208 students to three different classes. Class section 001 is in the largest room. Class section 002 is in a small room that contains only one third the number of seats as the room for section 001. Section 003 is in a room that contains exactly four fifths as many seats as the room for section 001. Formulate as a problem with three equations and three variables and solve by Gauss–Jordan elimination to find the number of students in each class.

6.4 SOLUTION BY MATRIX INVERSION

There is one more approach to solving a system of equations with which you should be familiar. To employ it, we must represent the system as a matrix equation. We will then be able to solve the system by using the inverse of the matrix. This approach is particularly useful when you must solve a number of systems that are the same except for the constants on their right-hand sides.

The Inverse of a Square Matrix

Suppose A is $n \times n$ (in other words, square). We say that B is an **inverse** of A if

$$AB = BA = I.$$

EXAMPLE 27

Let $A = \begin{bmatrix} 1 & 2 \\ 3 & 4 \end{bmatrix}$. Then $B = \begin{bmatrix} -2 & 1 \\ 1.5 & -0.5 \end{bmatrix}$ is an inverse of A, because

$$\begin{aligned} AB &= \begin{bmatrix} 1 & 2 \\ 3 & 4 \end{bmatrix}\begin{bmatrix} -2 & 1 \\ 1.5 & -0.5 \end{bmatrix} \\ &= \begin{bmatrix} 1(-2) + 2 \cdot 1.5 & 1 \cdot 1 + 2(-0.5) \\ 3(-2) + 4 \cdot 1.5 & 3 \cdot 1 + 4(-0.5) \end{bmatrix} \\ &= \begin{bmatrix} 1 & 0 \\ 0 & 1 \end{bmatrix} = I \end{aligned}$$

and

$$\begin{aligned} BA &= \begin{bmatrix} -2 & 1 \\ 1.5 & -0.5 \end{bmatrix}\begin{bmatrix} 1 & 2 \\ 3 & 4 \end{bmatrix} \\ &= \begin{bmatrix} -2 \cdot 1 + 1 \cdot 3 & -2 \cdot 2 + 1 \cdot 4 \\ 1.5 \cdot 1 + (-0.5)3 & 1.5 \cdot 2 + (-0.5)4 \end{bmatrix} \\ &= \begin{bmatrix} 1 & 0 \\ 0 & 1 \end{bmatrix} = I. \quad \square \end{aligned}$$

Not every square matrix has an inverse.

EXAMPLE 28

Let $C = \begin{bmatrix} 2 & 4 \\ 1 & 2 \end{bmatrix}$. Does C have an inverse?

SOLUTION

Suppose $D = \begin{bmatrix} a & b \\ c & d \end{bmatrix}$ is an inverse of C. Then it must be true that

$$CD = I$$

or

$$\begin{bmatrix} 2 & 4 \\ 1 & 2 \end{bmatrix}\begin{bmatrix} a & b \\ c & d \end{bmatrix} = \begin{bmatrix} 1 & 0 \\ 0 & 1 \end{bmatrix}.$$

Multiplying the matrices gives

$$\begin{bmatrix} 2 \cdot a + 4 \cdot c & 2 \cdot b + 4 \cdot d \\ 1 \cdot a + 2 \cdot c & 1 \cdot b + 2 \cdot d \end{bmatrix} = \begin{bmatrix} 1 & 0 \\ 0 & 1 \end{bmatrix}.$$

We can equate elements to obtain

$$\begin{aligned} 2a + 4c &= 1, \\ 1a + 2c &= 0; \end{aligned}$$

$$\begin{aligned} 2b + 4d &= 0, \\ 1b + 2d &= 1. \end{aligned}$$

The first pair of equations can be rewritten to read

$$\begin{aligned} 2(a + 2c) &= 1, \\ a + 2c &= 0. \end{aligned}$$

It should be apparent that these two equations are inconsistent. In other words, there are no values of a and c that can possibly satisfy them. The same is true for the second pair:

$$\begin{aligned} 2(b + 2d) &= 0, \\ b + 2d &= 1. \end{aligned}$$

Therefore, the given matrix (C) has no inverse. □

A (square) matrix that has an inverse is said to be **nonsingular,** while a matrix that does not have an inverse is said to be **singular.**

Practice Problem

How would you describe the matrices A and C of Examples 27 and 28?

Answer: nonsingular, singular

The following statements are true for a nonsingular matrix.

1. The inverse of an $n \times n$ matrix must be $n \times n$. (Why?)
2. The inverse of a matrix is unique.

If we have a matrix A, we represent its inverse by A^{-1}, so that

$$AA^{-1} = A^{-1}A = I.$$

Practice Problem

What is the inverse of an identity matrix of size $n \times n$?

Answer: itself, since $I_n I_n = I_n$

We can use an augmented matrix and row operations to find the inverse of any nonsingular matrix. Here's how. Suppose A is nonsingular. Form the augmented matrix $[A \mid I]$, where I is (of course) the same size as A. Then perform row operations on the augmented matrix until you obtain $[I \mid B]$. The matrix B is the inverse of A. In other words, the procedure

$$[A \mid I] \xrightarrow[\text{operations}]{\text{row}} [I \mid A^{-1}]$$

enables us to **invert** (find the inverse of) any nonsingular matrix.

EXAMPLE 29

Find the inverse of

$$A = \begin{bmatrix} 1 & 2 \\ 3 & 4 \end{bmatrix}$$

using an augmented matrix and row operations.

SOLUTION

Step 1 Form the augmented matrix

$$\left[\begin{array}{cc|cc} 1 & 2 & 1 & 0 \\ 3 & 4 & 0 & 1 \end{array}\right].$$

Step 2 We must transform (by row operations) the matrix so that the left-hand portion becomes an identity matrix. We'll do this column by column, as we did in Gauss–Jordan elimination.

There is a 1 in the upper-left corner, so we don't need to modify the first row. To eliminate the 3 in the first column, multiply the first row by -3 and add it to the second row.

$$R2 + (-3)R1 \quad \left[\begin{array}{cc|cc} 1 & 2 & 1 & 0 \\ 0 & -2 & -3 & 1 \end{array}\right]$$

Step 3 There should be a 1 in the second column on the second row, so we multiply the second row by -0.5:

$$-0.5R2 \quad \left[\begin{array}{cc|cc} 1 & 2 & 1 & 0 \\ 0 & 1 & 1.5 & -0.5 \end{array}\right]$$

Multiply the second row by -2 and add it to the first row:

$$R1 + (-2)R2 \quad \left[\begin{array}{cc|cc} 1 & 0 & -2 & 1 \\ 0 & 1 & 1.5 & -0.5 \end{array}\right].$$

Therefore

$$A^{-1} = \begin{bmatrix} -2 & 1 \\ 1.5 & -0.5 \end{bmatrix}.$$

(See Example 27.) □

Take careful note of the strategy utilized in the preceding example. We used a **column-by-column** approach to invert the given matrix. We can summarize the procedure as follows.

Inverting A Square Matrix With Row Operations

1. Augment the matrix with an identity matrix of the same size.
2. Get a 1 in the first column, first row, by multiplying the first row by the multiplicative inverse of the number that appears there. (If the number is 0, switch the first row with another row that has a non-zero entry in the first column.) Next, take appropriate multiples of the first row and add them to the other rows to convert all the other elements in the first column to 0's.
3. Get a 1 in the second column, second row, by multiplying the second row by the multiplicative inverse of the (nonzero) number that appears there. Take appropriate multiples of the second row and add them to the other rows to convert all other elements in the second column to 0's.
4. Continue until you have obtained the identity matrix on the left side.

As you may have noticed, this procedure is similar to Gauss–Jordan elimination with the exception that the augmented matrix contains an identity matrix rather than the right-hand sides of a system of equations.

EXAMPLE 30

Find the inverse of the 3×3 matrix

$$B = \begin{bmatrix} 4 & 5 & 6 \\ 1 & 2 & 3 \\ 2 & 4 & 7 \end{bmatrix}.$$

SOLUTION

Step 1 First we form the augmented matrix

$$\left[\begin{array}{ccc|ccc} 4 & 5 & 6 & 1 & 0 & 0 \\ 1 & 2 & 3 & 0 & 1 & 0 \\ 2 & 4 & 7 & 0 & 0 & 1 \end{array}\right].$$

Step 2 Next, we must work on the first column. We could multiply the first row by 0.25. However, note the 1 in the first column on the second row. Let's just switch the first and second rows:

$$R1 \leftrightarrow R2 \quad \left[\begin{array}{ccc|ccc} 1 & 2 & 3 & 0 & 1 & 0 \\ 4 & 5 & 6 & 1 & 0 & 0 \\ 2 & 4 & 7 & 0 & 0 & 1 \end{array}\right].$$

(Switching rows in this manner avoids some fractions, which is always desirable). We can eliminate the 4 in the first column by multiplying the top row by -4 and adding it to the second row:

$$R2 + (-4)R1 \qquad \left[\begin{array}{rrr|rrr} 1 & 2 & 3 & 0 & 1 & 0 \\ 0 & -3 & -6 & 1 & -4 & 0 \\ 2 & 4 & 7 & 0 & 0 & 1 \end{array}\right].$$

Eliminate the 2 in the first column by multiplying the top row by -2 and adding it to the third row:

$$R3 + (-2)R1 \qquad \left[\begin{array}{rrr|rrr} 1 & 2 & 3 & 0 & 1 & 0 \\ 0 & -3 & -6 & 1 & -4 & 0 \\ 0 & 0 & 1 & 0 & -2 & 1 \end{array}\right].$$

Step 3 Now we must work on the second column. Begin by multiplying the second row by $-\frac{1}{3}$ (note that we can't do anything sneaky here like switching rows):

$$(-\tfrac{1}{3})R2 \qquad \left[\begin{array}{rrr|rrr} 1 & 2 & 3 & 0 & 1 & 0 \\ 0 & 1 & 2 & -\frac{1}{3} & \frac{4}{3} & 0 \\ 0 & 0 & 1 & 0 & -2 & 1 \end{array}\right].$$

Eliminate the 2 in the second column by multiplying the second row by -2 and adding it to the top row:

$$R1 + (-2)R2 \qquad \left[\begin{array}{rrr|rrr} 1 & 0 & -1 & \frac{2}{3} & -\frac{5}{3} & 0 \\ 0 & 1 & 2 & -\frac{1}{3} & \frac{4}{3} & 0 \\ 0 & 0 & 1 & 0 & -2 & 1 \end{array}\right].$$

The bottom row has a 0 in the second column, so we need not do anything to it.

Step 4 Finally, we must attack the third column. There's already a 1 on the third row, so we don't modify it. We can eliminate the -1 in the top row by adding the third row to the top row:

$$R1 + R3 \qquad \left[\begin{array}{rrr|rrr} 1 & 0 & 0 & \frac{2}{3} & -\frac{11}{3} & 1 \\ 0 & 1 & 2 & -\frac{1}{3} & \frac{4}{3} & 0 \\ 0 & 0 & 1 & 0 & -2 & 1 \end{array}\right].$$

Eliminate the 2 by multiplying the bottom row by -2 and adding it to the second row:

$$R2 + (-2)R3 \qquad \left[\begin{array}{rrr|rrr} 1 & 0 & 0 & \frac{2}{3} & -\frac{11}{3} & 1 \\ 0 & 1 & 0 & -\frac{1}{3} & \frac{16}{3} & -2 \\ 0 & 0 & 1 & 0 & -2 & 1 \end{array}\right].$$

Thus the inverse is

$$B^{-1} = \begin{bmatrix} \frac{2}{3} & -\frac{11}{3} & 1 \\ -\frac{1}{3} & \frac{16}{3} & -2 \\ 0 & -2 & 1 \end{bmatrix}.$$

Prove it by showing that $BB^{-1} = B^{-1}B = I_3$. □

We can also use this approach to determine whether a matrix is singular or nonsingular. If it's nonsingular, we know we'll obtain its inverse. If it's singular, we'll discover we cannot obtain an identity matrix on the left of the augmented matrix because there will be a row of zeros at some stage of the process.

EXAMPLE 31

Show that $A = \begin{bmatrix} 2 & 4 \\ 1 & 2 \end{bmatrix}$ is singular.

SOLUTION

Step 1 Form the augmented matrix

$$\left[\begin{array}{cc|cc} 2 & 4 & 1 & 0 \\ 1 & 2 & 0 & 1 \end{array}\right].$$

Step 2 Next, switch the top and bottom rows:

$$R1 \leftrightarrow R2 \qquad \left[\begin{array}{cc|cc} 1 & 2 & 0 & 1 \\ 2 & 4 & 1 & 0 \end{array}\right].$$

Eliminate the 2 in the first column by multiplying the top row by -2 and adding it to the bottom row:

$$R2 + (-2)R1 \qquad \left[\begin{array}{cc|cc} 1 & 2 & 0 & 1 \\ 0 & 0 & 1 & -2 \end{array}\right].$$

Since there is a row of zeros, we cannot continue. We conclude that A is singular. (See Example 28.) □

This procedure works so well that it can be used to find the inverse of 40 × 40 (or even larger) matrices in just a few seconds on a computer.

Solving Systems of Equations

Let's begin by learning how to represent a system of equations as a matrix equation.

Consider the system

$$2x + 3y = 10,$$
$$5x + 8y = 26.$$

Let

$$A = \begin{bmatrix} 2 & 3 \\ 5 & 8 \end{bmatrix}.$$

We call A the **coefficient matrix** of the system. Next, define

$$X = \begin{bmatrix} x \\ y \end{bmatrix} \quad \text{and} \quad B = \begin{bmatrix} 10 \\ 26 \end{bmatrix}.$$

Now we can rewrite the original system as $AX = B$. Note that $AX = B$ is equal to

$$\begin{bmatrix} 2 & 3 \\ 5 & 8 \end{bmatrix}\begin{bmatrix} x \\ y \end{bmatrix} = \begin{bmatrix} 10 \\ 26 \end{bmatrix}.$$

The entries on the left sides of these equations equal the entries on the right sides precisely when the original equations are satisfied.

Now, let's suppose we have a system of equations that we have represented by the matrix equation

$$AX = B$$

and further suppose that A^{-1} exists. Then we can multiply both sides of the equation by A^{-1} to get

$$A^{-1}AX = A^{-1}B$$

but

$$A^{-1}A = I$$

so we have

$$IX = A^{-1}B$$

Also, $IX = X$, which means

$$X = A^{-1}B.$$

In other words, the solution to the matrix equation is given by the matrix product $A^{-1}B$. This approach is called the **matrix inversion technique**.

EXAMPLE 32

a) Solve the system we just converted to matrix equation form by matrix inversion:

$$2x + 3y = 10,$$
$$5x + 8y = 26.$$

b) Solve the system

$$2x + 3y = 14,$$
$$5x + 8y = 36.$$

SOLUTION

a) We know that

$$A = \begin{bmatrix} 2 & 3 \\ 5 & 8 \end{bmatrix}, \quad X = \begin{bmatrix} x \\ y \end{bmatrix}, \text{ and } B = \begin{bmatrix} 10 \\ 26 \end{bmatrix},$$

so we need to find A^{-1}.

Step 1 Augment A by I to obtain

$$\left[\begin{array}{cc|cc} 2 & 3 & 1 & 0 \\ 5 & 8 & 0 & 1 \end{array}\right]$$

Step 2 Now we'll proceed by using row operations in order to transform the left-hand side to I (which will give us A^{-1} on the right). First, multiply the top row by 0.5.

$$0.5R1 \quad \left[\begin{array}{cc|cc} 1 & 1.5 & 0.5 & 0 \\ 5 & 8 & 0 & 1 \end{array}\right]$$

Multiply the first row by -5 and add it to the bottom row.

$$R2 + (-5)R1 \quad \left[\begin{array}{cc|cc} 1 & 1.5 & 0.5 & 0 \\ 0 & 0.5 & -2.5 & 1 \end{array}\right]$$

Step 3 Multiply the bottom row by 2.

$$2R2 \quad \left[\begin{array}{cc|cc} 1 & 1.5 & 0.5 & 0 \\ 0 & 1 & -5 & 2 \end{array}\right]$$

Multiply the bottom row by -1.5 and add it to the top row.

$$R1 + (-1.5)R2 \quad \left[\begin{array}{cc|cc} 1 & 0 & 8 & -3 \\ 0 & 1 & -5 & 2 \end{array}\right]$$

Thus, A^{-1} is given by $\begin{bmatrix} 8 & -3 \\ -5 & 2 \end{bmatrix}$, which means

$$X = A^{-1}B$$

or

$$X = \begin{bmatrix} 8 & -3 \\ -5 & 2 \end{bmatrix}\begin{bmatrix} 10 \\ 26 \end{bmatrix} = \begin{bmatrix} 80 - 78 \\ -50 + 52 \end{bmatrix} = \begin{bmatrix} 2 \\ 2 \end{bmatrix}.$$

In other words, the solution is (2, 2). (You can confirm this by substituting these values into the system and showing that they do indeed satisfy both equations.)

b) Once again $A = \begin{bmatrix} 2 & 3 \\ 5 & 8 \end{bmatrix}$ and $X = \begin{bmatrix} x \\ y \end{bmatrix}$, while $B = \begin{bmatrix} 14 \\ 36 \end{bmatrix}$. Because A is the same as it was in part (a), A^{-1} is still $\begin{bmatrix} 8 & -3 \\ -5 & 2 \end{bmatrix}$, so

$$X = A^{-1}B$$

or

$$X = \begin{bmatrix} 8 & -3 \\ -5 & 2 \end{bmatrix}\begin{bmatrix} 14 \\ 36 \end{bmatrix} = \begin{bmatrix} 4 \\ 2 \end{bmatrix}.$$

(Confirm this!) □

We were able to solve the second system quite easily because we had already found A^{-1}. Thus we needed only to perform a simple matrix multiplication.

We hope it's obvious that this method will only work when A has an inverse. Because only a square matrix can have an inverse, the method will never work for systems in which the number of variables is different from the number of equations. In addition, we know that not every square matrix has an inverse. If you attempt to invert a coefficient matrix and cannot, you can conclude the system is dependent or inconsistent. In this case, use Gauss–Jordan elimination.

Practice Problem

Find the solutions to the following systems by finding the inverse of the coefficient matrix.

a) $x + 3y = 16$
$3x + y = 8$

b) $x + 3y = 9$
$3x + y = 11$

c) $2x + y = 15$
$-3x + y = 5$

d) $2x + y = 6$
$-3x + y = -14$

Answers: For **a)** and **b)**, $A^{-1} = \begin{bmatrix} -0.125 & 0.375 \\ 0.375 & -0.125 \end{bmatrix}$.

For **a)**,

$X = \begin{bmatrix} 1 \\ 5 \end{bmatrix}$ and for **b)**, $X = \begin{bmatrix} 3 \\ 2 \end{bmatrix}$. For **c)** and **d)**,

$A^{-1} = \begin{bmatrix} 0.2 & -0.2 \\ 0.6 & 0.4 \end{bmatrix}$. For **c)**, $X = \begin{bmatrix} 2 \\ 11 \end{bmatrix}$ and for **d)**, $X = \begin{bmatrix} 4 \\ -2 \end{bmatrix}$.

EXERCISE SET 6.4

Find the inverse of the following matrices.

1. $\begin{bmatrix} 1 & 3 \\ 2 & 4 \end{bmatrix}$ **2.** $\begin{bmatrix} 3 & 2 \\ 2 & 4 \end{bmatrix}$ **3.** $\begin{bmatrix} 3 & 2 \\ 2 & 1 \end{bmatrix}$

4. $\begin{bmatrix} 1 & 2 & 3 \\ 4 & 5 & 6 \\ 1 & 2 & 4 \end{bmatrix}$ **5.** $\begin{bmatrix} 2 & 2 & 3 \\ 3 & 4 & 4 \\ 2 & 2 & 4 \end{bmatrix}$ **6.** $\begin{bmatrix} 2 & 1 & 0 \\ 2 & 3 & 5 \\ 1 & 0 & 1 \end{bmatrix}$

7. $\begin{bmatrix} 1 & 3 & 5 & 4 \\ 2 & 0 & 1 & 3 \\ 0 & 1 & 1 & 0 \\ 1 & 0 & 0 & 1 \end{bmatrix}$ **8.** $\begin{bmatrix} 2 & 6 & 10 & 8 \\ 4 & 0 & 2 & 6 \\ 0 & 2 & 2 & 0 \\ 2 & 0 & 0 & 2 \end{bmatrix}$

Show that the following matrices are singular.

9. $\begin{bmatrix} 1 & 2 \\ 1 & 2 \end{bmatrix}$ **10.** $\begin{bmatrix} 1 & 2 \\ 3 & 6 \end{bmatrix}$ **11.** $\begin{bmatrix} 3 & 2 \\ 6 & 4 \end{bmatrix}$

12. $\begin{bmatrix} 1 & 2 & 3 \\ 3 & 6 & 8 \\ 4 & 8 & 11 \end{bmatrix}$ **13.** $\begin{bmatrix} 2 & 2 & 3 \\ 3 & 4 & 4 \\ 4 & 4 & 6 \end{bmatrix}$

14. $\begin{bmatrix} 1 & 3 & 5 & 4 \\ 2 & 0 & 1 & 3 \\ 0 & 1 & 1 & 0 \\ 3 & 4 & 7 & 7 \end{bmatrix}$ **15.** $\begin{bmatrix} 2 & 0 & 4 & 1 \\ 3 & 2 & 2 & 1 \\ 0 & 1 & 1 & 0 \\ 5 & 3 & 7 & 2 \end{bmatrix}$

For the following matrices, either find the inverse or show that the matrix is singular.

16. $\begin{bmatrix} 4 & 5 \\ 16 & 20 \end{bmatrix}$ **17.** $\begin{bmatrix} 3 & 4 \\ -3 & 4 \end{bmatrix}$

18. $\begin{bmatrix} 1 & 2 & 2 \\ 3 & 6 & -6 \\ 1 & 1 & 1 \end{bmatrix}$ **19.** $\begin{bmatrix} 1 & 2 & 2 \\ -3 & -6 & -6 \\ 1 & 1 & 1 \end{bmatrix}$

20. $\begin{bmatrix} 1 & 1 & 0 & 0 \\ 0 & 0 & 1 & 1 \\ 1 & 0 & 1 & 0 \\ 0 & 1 & 0 & 1 \end{bmatrix}$ **21.** $\begin{bmatrix} 1 & 1 & 0 & 0 \\ 0 & 1 & 1 & 0 \\ 0 & 0 & 1 & 1 \\ 0 & 0 & 0 & 1 \end{bmatrix}$

Solve the following systems by matrix inversion.

22. $\begin{aligned} x + 2y &= 4 \\ 3x + 4y &= 10 \end{aligned}$ **23.** $\begin{aligned} x - 2y &= 6 \\ 2x + 2y &= 6 \end{aligned}$

24. $\begin{aligned} x - 3y &= 4 \\ 5x + 3y &= 8 \end{aligned}$ **25.** $\begin{aligned} x - 5y &= 5 \\ 3x + 10y &= 10 \end{aligned}$

26. $-x + y = -5$
$7x + 3y = 15$

27. $2x - 4y = 8$
$x + 4y = 13$

28. $2x + 2y = 6$ and $2x + 2y = 12$
$3x + 5y = 15$ $\quad$ $3x + 5y = 6$

29. $3x - 6y = 6$ and $3x - 6y = -6$
$7x + 8y = 15$ $\quad$ $7x + 8y = 24$

30. $x - 3y = 8$
$3x - 9y = 24$

31. $2x - 5y = 1$
$-6x + 15y = -10$

32. $x + 4y + 6z = 4$ and $x + 4y + 6z = 42$
$3x + 2y + 8z = 2$ $\quad$ $3x + 2y + 8z = 56$
$5x + y + 10z = 16$ $\quad$ $5x + y + 10z = 69$

33. $w - 3x + 2y + z = 15$ and $w - 3x + 2y + z = 0$
$3w - 8x + 2y + 5z = 29$ $\quad$ $3w - 8x + 2y + 5z = -7$
$6w - 16x + 2y - 4z = 36$ $\quad$ $6w - 16x + 2y - 4z = -41$
$x - 3y - 18z = -44$ $\quad$ $x - 3y - 18z = -41$

34. Advanced Computer Systems manufactures computers. The relevant data are as follows:

Model	Labor hours	Test hours	Packaging hours
System 1	15	5	1
System 2	20	10	1.5
System 3	25	5	2

a) Represent these data by a matrix and find the inverse of this matrix.
b) Suppose there are exactly 400 hours of labor, 120 hours of testing, and 30 hours of packaging available next week. How many of each system should be manufactured?
c) Now suppose there will be 315 hours of labor, 105 hours of testing, and 23.5 hours of packaging available in two weeks. How many of each system should be manufactured?

* **35.** Firenze Leather makes purses and briefcases from leather. Each purse requires 3 hours of labor and 3 square feet of leather, while each briefcase requires 4 hours of labor and 6 square feet of leather.

a) Suppose there are 72 hours of labor and 96 square feet of leather available next week. How many purses and briefcases should Firenze make?
b) Now suppose a mistake was made and there will be 72 hours of labor but only 84 square feet of leather available next week. What should Firenze do?

* Reprinted with permission from Wu and Coppins, *Linear Programming and Extensions*(McGraw-Hill, New York), 1981.

6.5 APPLICATION: INPUT-OUTPUT ANALYSIS

Input-output analysis (also referred to as interindustry analysis) is a means of analyzing an economy that stresses the interdependence of economic activities. The model is intended to allow economists to predict the various outputs that would be required to meet future demands for goods and services. Developed in the 1930s by Wassily Leontief (who received a Nobel prize in 1973), input-output analysis has been used around the world at both the national and regional levels, by highly developed and by underdeveloped nations, by free economies and by centrally planned economies.

The basis of input-output analysis is the input-output table, which shows how the output of each industry is distributed among the various industries to be used as inputs. Let's look at a simplified economy which has only two industries, food and energy. (By way of contrast, Leontief's original model divided the economy of the United States into 500 interacting sectors.) (See Table 6.2.)

To read the table, note that each *row* details how the output of each industry is used, while each *column* denotes the inputs required by an industry. The figures are measured in suitable units (for example, billions of dollars). Thus reading across the row for food, we see that the food output was used as follows: 76 units for food, 99 units for energy, and 25 units for final demand. Reading down the column for energy, we see that energy used 99 units of food output and 150 units of energy output. The final demand category includes exports, government purchases, inventory accumulation, and payments to households. (Because this sector is outside the production sector in the sense that final products are "exported" to the inhabitants of the final demand sector, our model is said to be **open.** Variations in which labor is treated as a production commodity, that is, included as a producing industry, are said to be **closed.**) The total output column is simply the sum across each row of the outputs.

To summarize, we can say that to meet the final demands of 25 units of food and 30 units of energy, there were 200 units of food produced and 300 units of energy produced. Now, suppose that in two years we expect the final demands to be 28 units of food and 32 units of energy. How much should each industry produce?

Let's call the new production levels of food and energy x and y, respectively. By studying Table 6.2, we claim the following relationships must be true:

$$\text{Output of food} = \begin{matrix}\text{units required}\\ \text{by food}\end{matrix} + \begin{matrix}\text{units required}\\ \text{by energy}\end{matrix} + \begin{matrix}\text{final demand}\\ \text{for food}\end{matrix}$$

TABLE 6.2 **An input-output table for an economy with two industries**

Outputs \ Inputs	Food	Energy	Final demand	Total output
Food	76	99	25	200
Energy	120	150	30	300

and

$$\text{Output of energy} = \begin{matrix}\text{units required}\\ \text{by food}\end{matrix} + \begin{matrix}\text{units required}\\ \text{by energy}\end{matrix} + \begin{matrix}\text{final demand}\\ \text{for energy.}\end{matrix}$$

(Think about it!) In addition, the food industry used 76 units of its own output and 120 units of energy output to produce 200 total units of output. To produce one unit of output, we could then expect the food industry to require $\frac{76}{200}$ units of food output and $\frac{120}{200}$ units of energy output. In a similar fashion, energy required 99 units of food and 150 units of its own output to produce 300 total units of output. To produce one unit of output, we could then expect energy to require $\frac{99}{300}$ units of food and $\frac{150}{300}$ units of energy. Therefore the output of the food industry must satisfy the relationship

$$x = \frac{76}{200}x + \frac{99}{300}y + 28$$

(recall 28 is our new estimate of final demand for food). And energy's output must satisfy the relationship

$$y = \frac{120}{200}x + \frac{150}{300}y + 32.$$

(Remember: We are analyzing where the outputs go.) In matrix form we have

$$\begin{bmatrix} x \\ y \end{bmatrix} = \begin{bmatrix} \frac{76}{200} & \frac{99}{300} \\ \frac{120}{200} & \frac{150}{300} \end{bmatrix} \begin{bmatrix} x \\ y \end{bmatrix} + \begin{bmatrix} 28 \\ 32 \end{bmatrix},$$

which we can write as

$$X = AX + D$$

where

$$X = \begin{bmatrix} x \\ y \end{bmatrix}, \quad A = \begin{bmatrix} \frac{76}{200} & \frac{99}{300} \\ \frac{120}{200} & \frac{150}{300} \end{bmatrix}, \quad \text{and} \quad D = \begin{bmatrix} 28 \\ 32 \end{bmatrix}.$$

We refer to X as the **output matrix,** to D as the **demand matrix,** and to A as the **input-output** matrix. The elements of A are called the **input-output coefficients.** The equation $X = AX + D$ is called the **input-output equation.**

Now, since

$$X = AX + D$$

we have

$$X - AX = D.$$

Then

$$(I - A)X = D$$

(for the appropriate sized I) and

$$X = (I - A)^{-1}D$$

assuming the inverse exists.

In this case, with

$$A = \begin{bmatrix} \frac{76}{200} & \frac{99}{300} \\ \frac{120}{200} & \frac{150}{300} \end{bmatrix} = \begin{bmatrix} 0.38 & 0.33 \\ 0.60 & 0.50 \end{bmatrix},$$

$$(I - A) = \begin{bmatrix} 1 & 0 \\ 0 & 1 \end{bmatrix} - \begin{bmatrix} 0.38 & 0.33 \\ 0.60 & 0.50 \end{bmatrix} = \begin{bmatrix} 0.62 & -0.33 \\ -0.60 & 0.50 \end{bmatrix}.$$

Thus

$$(I - A)^{-1} = \begin{bmatrix} 4.464 & 2.946 \\ 5.357 & 5.537 \end{bmatrix} \quad \text{(to three decimal places).}$$

(This inverse was computed by hand using the procedure described in the previous section.) In addition,

$$X = (I - A)^{-1}D,$$

so

$$X = \begin{bmatrix} 4.464 & 2.946 \\ 5.357 & 5.537 \end{bmatrix}\begin{bmatrix} 28 \\ 32 \end{bmatrix} = \begin{bmatrix} 219.3 \\ 327.3 \end{bmatrix}$$

In other words, the food industry must produce (approximately) 219.3 units and energy must produce (approximately) 327.2 units to meet the projected demands two years from now.

EXAMPLE 33

Consider an economy with two sectors, food and clothing, and the input-output table shown in Table 6.3.

a) Find the input-output coefficients.

b) Find the production levels for final demands of 300 and 150 units for food and clothing, respectively.

TABLE 6.3 An input-output table for Example 33

Outputs \ Inputs	Food	Clothing	Final demand	Total output
Food	240	640	320	1200
Clothing	600	220	180	1000

SOLUTION

a) To find the input-output coefficients, divide the first column of Table 6.3 by 1200 (food's output) and divide the second column of Table 6.3 by 1000 (clothing's output).

This gives us

$$A = \begin{bmatrix} \frac{240}{1200} & \frac{640}{1000} \\ \frac{600}{1200} & \frac{220}{1000} \end{bmatrix} = \begin{bmatrix} 0.20 & 0.64 \\ 0.50 & 0.22 \end{bmatrix}.$$

b) Start by finding $(I - A)^{-1}$. We have

$$(I - A) = \begin{bmatrix} 1 & 0 \\ 0 & 1 \end{bmatrix} - \begin{bmatrix} 0.20 & 0.64 \\ 0.50 & 0.22 \end{bmatrix},$$

$$(I - A) = \begin{bmatrix} 0.80 & -0.64 \\ -0.50 & 0.78 \end{bmatrix},$$

and

$$(I - A)^{-1} = \begin{bmatrix} 2.566 & 2.106 \\ 1.645 & 2.632 \end{bmatrix} \quad \text{(to three places).}$$

Since

$$X = (I - A)^{-1}D$$

$$X = \begin{bmatrix} 2.566 & 2.106 \\ 1.645 & 2.632 \end{bmatrix} \begin{bmatrix} 300 \\ 150 \end{bmatrix} = \begin{bmatrix} 1086 \\ 888 \end{bmatrix} \quad \text{(rounded to the nearest unit).}$$

Thus the food industry should produce 1086 units and the clothing industry should produce 888 units. □

One word of caution. We have assumed the input-output matrix remains constant over time. Over the long run, technological advances will occur, which means that long-term predictions will probably not be reliable. (A dynamic model, which can overcome this problem, is available.)

EXERCISE SET 6.5

1. Table 6.4 gives the input-output table of a hypothetical economy.

a) Find the input-output matrix A.

b) Suppose the final demands change to 110 units for food and 85 units for energy. What should the industries produce?

TABLE 6.4 **An input-output table for Problem 1**

Outputs	Inputs: Food	Inputs: Energy	Final demand	Total output
Food	20	56	24	100
Energy	50	8	22	80

2. Table 6.5 gives the input-output table of a hypothetical economy.

a) Find the input-output matrix A.

b) Suppose the final demands change to 190 units for industry S and 130 units for industry T. What should they produce?

TABLE 6.5 **An input-output table for Problem 2**

Outputs	Inputs: S	Inputs: T	Final demand	Total output
S	40	110	50	200
T	60	20	21	121

3. Consider an economy with three industries with the following input-output matrix:

$$A = \begin{bmatrix} 0.05 & 0.25 & 0.10 \\ 0.15 & 0.20 & 0.15 \\ 0.15 & 0.10 & 0.20 \end{bmatrix}.$$

How much should industry produce to meet final demands of 20 units, 50 units, and 40 units, respectively?

CHAPTER REVIEW

IMPORTANT TERMS AND SYMBOLS

6.1 Systems of Linear Equations

graphical solution Draw the graphs of the individual equations.
solution set of a system of equations Intersection of the solution sets of the individual equations.
system of equations A set of two or more equations.
Dependent system: has an infinite number of solutions.
Inconsistent system: solution set is empty.

6.2 Solution by Substitution

possible solutions for systems of linear equations:

	When the system has		
	More variables than equations	Equal number of variables and equations	Fewer variables than equations
The system can	Be inconsistent or dependent	Have a unique solution, be inconsistent, or dependent	Have a unique solution, be inconsistent, or dependent

substitution (works well for systems of equations with two variables) Solve an equation for one variable in terms of the other variable and substitute this expression into the remaining equation.

6.3 Solution by Gauss–Jordan Elimination

algebraic elimination Eliminate variables by adding appropriate multiples of equations.
augmented matrix A matrix that includes an additional column or columns.
elementary row operations (for matrices)

1. Interchange two rows.
2. Multiply a row by a nonzero scalar.
3. Multiply a row by a nonzero scalar and add it to another row.

equivalent systems Two systems of equations that have the same solution set.
Gauss–Jordan elimination Systematic procedure for solving a system of linear equations by utilizing an augmented matrix and row operations.

1. Write the augmented matrix representation of the system.

2. Get a 1 in the first column of the first row that doesn't contain all 0's by multiplying the first row by the multiplicative inverse of the (nonzero) number that appears there (or switch rows if a 0 is there). Take appropriate multiples of the first row and add them to the other rows to convert all other elements in the column to 0's.
3. Get a 1 in the next column on the second row that doesn't contain 0's by multiplying the second row by the multiplicative inverse of the (nonzero) number that appears there. (If necessary, switch the second row with one below it to eliminate the 0.) Take appropriate multiples of the second row and add them to the other rows to convert all other elements of the second column to 0's.
4. Continue as necessary.

reduced row echelon form of a matrix The first nonzero element in each row must be a 1 and the 1's must form a zig-zag pattern down and to the right, so that the first 1 in a row should be the only nonzero number in its column.

6.4 Solution by Matrix Inverse

inverse of a square matrix A square matrix of the same size as the original matrix which has the property that the product of the original matrix and its inverse in either order is an identity matrix of the same size; represented by A^{-1}, where A is the original matrix. Finding the inverse using row operations:

1. Augment the original matrix with an identity matrix.
2. Use row operations to transform the original matrix into an identity matrix.
3. The original identity matrix is transformed into the inverse.

nonsingular matrix A square matrix that has an inverse.
singular matrix A square matrix that does not have an inverse.
using the inverse of a matrix to solve a system of equations Represent the system as $AX = B$, in which case

$$X = A^{-1}B.$$

6.5 Application: Input-Output Analysis

demand matrix Set of final demands; represented by D.
final demands Includes uses of final products "outside' the production sector.
input-output analysis A means of analyzing an economy that stresses the interdependence of economic activities.
input-output equation $X = AX + D$.
input-output matrix Matrix of coefficients that tell what fraction of each industry's output is required as input by a given industry; represented by A.
input-output table A table in which rows show industry outputs and columns show the inputs required by each industry.
output matrix The set of outputs required to satisfy the final demands; represented by X.

REVIEW PROBLEMS

Solve the following problems graphically, by Gauss–Jordan elimination, and by matrix inversion (where possible). Classify the system as to whether it has a unique solution, is inconsistent, or is dependent.

1. $\begin{aligned} x + 2y &= 8 \\ 2x - 3y &= -2 \end{aligned}$

2. $\begin{aligned} 3x + 2y &= 9 \\ 2x + 8y &= 12 \end{aligned}$

3. $\begin{aligned} -5x + 4y &= -8 \\ 2x + y &= 7 \end{aligned}$

4. $\begin{aligned} 5x - 2y &= 10 \\ -10x + 4y &= 20 \end{aligned}$

5. $\begin{aligned} -x + 3y &= 5 \\ 2x - y &= 6 \end{aligned}$

6. $\begin{aligned} -x + 3y &= 5 \\ 2x - 6y &= -10 \end{aligned}$

7. $\begin{aligned} 2x + y &= 3 \\ x - 2y &= 5 \\ 3x - y &= 8 \end{aligned}$

8. $\begin{aligned} x + 2y &= 6 \\ 3x + 3y &= 9 \\ 5x + 7y &= 12 \end{aligned}$

Solve the following problems by Gauss–Jordan elimination and by matrix inversion (where possible). Classify the system as to whether it has a unique solution, is inconsistent, or is dependent.

9. $\begin{aligned} 2x + 3y + z &= 6 \\ x - 2y + 3z &= 5 \end{aligned}$

10. $\begin{aligned} x + 3y + z &= 12 \\ 2x + y - 2z &= 4 \end{aligned}$

11. $\begin{aligned} x + y + z &= 10 \\ 2x - y + 3z &= 12 \\ -x + 3y + z &= 6 \end{aligned}$

12. $\begin{aligned} x + y + z &= 18 \\ 2x - y + 3z &= 6 \\ -x + 3y + z &= 10 \end{aligned}$

13. $\begin{aligned} 4x - y + z &= 8 \\ 2x + 2y - z &= 2 \\ 6x + y \phantom{{}+z} &= 10 \end{aligned}$

14. $\begin{aligned} x - y + z &= 5 \\ 3y - 2z &= -4 \\ 2x \phantom{{}-y} + 6z &= 12 \end{aligned}$

15. $\begin{aligned} 4x - y + z &= 4 \\ -2x - 3y + 2z &= 9 \\ 2x - 3y + 2z &= 3 \end{aligned}$

16. $\begin{aligned} w + 2x - y + z &= 8 \\ -2x + 3y + 4z &= 6 \\ 3w \phantom{{}+2x} - 5y + 6z &= 10 \end{aligned}$

17. An economy has the input-output table shown in Table 6.6.

a) Find the input-output matrix A.

b) Suppose the final demands become 22 units of necessities and 26 units of luxuries. What should each industry produce?

c) Suppose the final demands become 25 units of necessities and 30 units of luxuries. What should each industry produce?

TABLE 6.6 An input-output table for Problem 17

Outputs	Inputs: Necessities	Inputs: Luxuries	Final demand	Total output
Necessities	25	54	21	100
Luxuries	48	36	24	108

7 LINEAR PROGRAMMING: GEOMETRIC APPROACH

Ann and Fred have opened a small shop in which they make and sell tables and chairs. Each table requires 5 board-feet* of oak, 2 board-feet of pine, and 4 labor hours. Each chair requires 2 board-feet of oak, 3 board-feet of pine, and 3 labor hours. There are 122 board-feet of oak, 100 board-feet of pine, and 120 labor hours available for the upcoming week. Given that each table makes a profit of $12 and each chair makes a profit of $8, how many tables and chairs should Ann and Fred make during the next week so as to maximize their profits? (To be discussed throughout the chapter.)†

* A board-foot is a unit of lumber equal to the volume of a board that is 12 × 12 × 1 inches.

† Reprinted with permission from Wu and Coppins, *Linear Programming and Extensions* (McGraw-Hill, New York), 1981.

PROBLEM FROM LIFE

INTRODUCTION

In this chapter you will learn about a rather recent development called linear programming. We will discuss just what linear programming is and present a geometric interpretation and approach to solving linear programming problems. We will also discuss the importance of linear programming and how it can be used to help solve certain problems. In the next chapter we will present an analytical approach to solving linear programming problems. Much of the material in these chapters relies upon your ability to graph inequalities in two variables, which we discussed in Section 3.4, and upon your ability to solve systems of simultaneous equations, which we discussed in Chapter 6. You may want to review these topics before you continue with this chapter.

7.1

DEFINITION OF LINEAR PROGRAMMING

Let's take another look at the Problem from Life that we described at the beginning of this chapter. The information contained in the problem is presented in Table 7.1.

The problem facing Ann and Fred is a typical **resource allocation** problem. In this case, two products (tables and chairs) are competing for three scarce resources (oak, pine, and labor). The final production should reflect an optimal (the best possible) allocation of the scarce resources (in this case, so as to maximize profits).

Before we model this problem and show how to solve it, let's discuss it informally. We hope it's obvious that the more product (tables and chairs) that Ann and Fred produce, the more money they'll make. This, in turn, suggests that we need to worry about the limited resources that are available. After all, if resources were available in unlimited quantity, Ann and Fred could make unlimited numbers of tables and chairs and therefore unlimited amounts of money. One possible plan would be to make all tables. But how many tables can be produced? If the production of tables were limited by oak, we could make $24\frac{2}{5}$ tables (122 board-feet ÷ 5 board feet); if it were limited by pine we could make 50 (100 board-feet ÷ 2 board-feet); if it were limited by labor we could make 30 (120 board-feet ÷ 4 board-feet). Another plan would be to make all chairs, but again we must determine how many chairs could be produced. Here if we were limited by the oak, we could make 61 chairs; if we were limited by pine, $33\frac{1}{3}$; if we were limited by labor, 40 chairs. Finally, Ann and Fred could make both tables and chairs. However, there are many possible combinations of table and chair production, so it's

TABLE 7.1 Furniture manufacturing data

	Resources required per product unit			
	Oak, bd-ft	**Pine, bd-ft**	**Labor hours**	**Profit, $ per unit**
Tables	5	2	4	12
Chairs	2	3	3	8
Available resources	122	100	120	

not at all apparent which combinations we should examine. In fact, this problem seems to be rather complex, doesn't it?

Actually, we can model and solve this problem without much difficulty. First, let's see how we can formulate this problem using a linear programming model. The first step (in this or any other linear programming problem) is to define the **decision variables.** These are variables whose values we need to determine in order to solve the problem. (The term *decision variables* arises because we must make a series of decisions that require us to determine the values of one or more variables.) In this case, we must decide how many tables and chairs to produce during the coming week. Let

$$t = \text{number of tables to produce next week},$$
$$c = \text{number of chairs to produce next week}.$$

Given these variables, we can express the profits to be made from the manufacture of tables and chairs as

$$P = 12t + 8c.$$

The function P of the decision variables is called the **objective function** because it describes our objective in solving the problem: In this case, to maximize profits, we wish to

$$\text{Maximize} \quad P = 12t + 8c.$$

A little thought should convince you that there must be some restrictions on the values of t and c that we can choose. Otherwise, a solution to the problem of maximizing profits would be to say, let's make several million tables and chairs, because that would result in millions of dollars of profits. Thus our next step is to determine the restrictions on the values of the decision variables. These restrictions are typically called **constraints.**

Here there are three constraints: one for each scarce resource (oak, pine, and labor hours). The constraints are a set of three inequalities:

$$\begin{aligned} &\text{Oak:} && 5t + 2c \le 122 \\ &\text{Pine:} && 2t + 3c \le 100 \\ &\text{Labor:} && 4t + 3c \le 120 \end{aligned}$$

Let's see why these constraints are correct.

First, consider oak. Since each table requires 5 board-feet, the total amount of oak necessary to produce t tables must be $5t$. Similarly, since each chair requires 2 board-feet, the total amount of oak necessary to produce c chairs must be $2c$. Thus, to produce t tables and c chairs, a total of

$$5t + 2c$$

board-feet of oak is required. Recall that only 122 board-feet of oak are available. Thus, total usage of oak must be at most 122 board-feet. But our total usage is $5t + 2c$. Putting it all together we have

$$5t + 2c \leq 122.$$

Similar analysis can be used to generate the constraints for pine and for labor.

Notice that we wrote the constraints as inequalities. We did this because the optimal (best) solution may not require all of each of the three scarce resources. In fact, in this case, there is no solution (optimal or otherwise) that uses up every resource. Constraints should be written as inequalities unless the problem language specifically states otherwise.

In addition to the scarce resource constraints, there are two more constraints that we should include:

$$t \geq 0, \qquad c \geq 0.$$

These inequalities are called the **nonnegativity constraints.** They are invariably included in a linear programming problem because negative values for the decision variables would have no meaning. (For example, how could you produce -3 tables?)

Here is the summary of the linear programming model of Ann and Fred's problem:

$$\begin{array}{lll}
\text{Maximize} & P = 12t + 8c & \\
\text{Subject to} & 5t + 2c \leq 122 & \text{(oak)} \\
& 2t + 3c \leq 100 & \text{(pine)} \\
& 4t + 3c \leq 120 & \text{(labor)} \\
& \left.\begin{array}{r} t \geq 0 \\ c \geq 0 \end{array}\right\} & \text{(nonegativity)}
\end{array}$$

Now we can make a definition. A **linear programming problem** is a problem that is concerned with optimizing (either maximizing or minimizing) a linear function of the decision variables (called the objective function) over a system of linear inequalities (called the constraints). [A linear function of the variables $x_1, x_2, \ldots, x_n$ has the form

$$f(x_1, x_2, \ldots, x_n) = a_1x_1 + a_2x_2 + \cdots + a_nx_n,$$

where $a_1, a_2, \ldots, a_n$ are constants. A linear inequality has one of the forms

$$a_1x_1 + a_2x_2 + \cdots + a_nx_n \leq b \quad \text{or} \quad a_1x_1 + a_2x_2 + \cdots + a_nx_n \geq b,$$

where $a_1, a_2, \ldots, a_n$, and b are constants.] We assume in general that the decision variables are nonnegative.

Let's study some examples of problems that can be modeled as linear programming problems.

EXAMPLE 1

Diet Mix Problem A pet-store owner has decided to make and sell dog food. He has decided that the mix must contain at least 10 units of protein, 5 units of carbohydrates, and 12 units of vitamins. There are two ingredients that he can use to obtain the desired mix. Each pound of ingredient A contains 2 units of protein, 2 units of carbohydrates, and 2 units of vitamins. Each pound of ingredient B contains 3 units of protein, 1 unit of carbohydrates, and 4 units of vitamins. Ingredient A costs \$0.25 per pound and ingredient B costs \$0.30 per pound. How should the mix be made so that it has sufficient protein, carbohydrates, and vitamins and its cost is as low as possible?

SOLUTION

First, let's put the problem data into a table. (See Table 7.2.)

In verbal terms, the pet-shop owner's problem is to determine the amount of each ingredient (A and B) so that the resulting mix has sufficient protein, carbohydrates, and vitamins and its cost is as low as possible. Let

$$a = \text{number of pounds of ingredient A to use,}$$
$$b = \text{number of pounds of ingredient B to use.}$$

Then the objective function is $C = 0.25a + 0.30b$, which gives the cost of the mix. The resulting model is as follows.

$$\begin{array}{lrl} \text{Minimize} & C = 0.25a + 0.30b & \\ \text{Subject to} & 2a + 3b \geq 10 & \text{(protein requirement)} \\ & 2a + b \geq 5 & \text{(carbohydrate requirement)} \\ & 2a + 4b \geq 12 & \text{(vitamin requirement)} \\ & \left.\begin{array}{r} \mathrm{a} \geq 0 \\ a \geq 0 \end{array}\right\} & \text{(nonnegativity)} \end{array}$$

This is a simplified version of one of the earliest linear programming problems. The diet mix problem was originally formulated by economist George Stigler in the early 1940s.

TABLE 7.2 Diet mix problem data

	Units per pound		Minimum requirements
	A	**B**	
Protein	2	3	10 units
Carbohydrates	2	1	5 units
Vitamins	2	4	12 units
Cost per pound	\$0.25	\$0.30	

□

EXAMPLE 2

Media planning A local college has decided to put on a concert with students to raise money. The planning committee has a limited advertising budget of \$500. It must decide whether to advertise in the local newspaper at \$50 per ad or on the local radio station at \$75 per ad. Each newspaper ad will reach 5000 adults and 2500 teenagers, while each radio ad will reach 3500 adults and 7000 teenagers. The committee believes the advertising campaign must reach at least 25,000 adults and 28,000 teenagers to be successful. What should the committee do to achieve its goals at minimum cost?

SOLUTION

The data are summarized in Table 7.3. Let

$$n = \text{number of newspaper ads to place,}$$
$$r = \text{number of radio ads to place.}$$

The committee's objective function is $C = 50n + 75r$, which gives the cost of the advertising campaign (to be minimized). The model is

$$\begin{array}{lrl} \text{Minimize} & C = 50n + 75r & \\ \text{Subject to} & 5000n + 3500r \geq 25{,}000 & \text{(adults to reach)} \\ & 2500n + 7000r \geq 28{,}000 & \text{(teenagers to reach)} \\ & \left.\begin{array}{r} n \geq 0 \\ r \geq 0 \end{array}\right\} & \text{(nonnegativity)} \end{array}$$

The use of linear programming in media planning dates back to the early 1960s.

TABLE 7.3 Media planning problem data

	People reached per ad		Minimum requirement
	Newspaper	**Radio**	
Adults	5000	3500	25,000
Teenagers	2500	7000	28,000
Cost per ad	\$50	\$75	

□

EXAMPLE 3

Transportation Problem Y'All Haul, a truck rental firm, has an excess of trucks in Atlanta and Richmond, while there is a shortage of trucks in Boston and New York. Given the cost data shown in Table 7.4, how can Y'All Haul reallocate its trucks so as to spend as little money as possible?

SOLUTION

Let

a_b = number of trucks shipped from Atlanta to Boston,
a_n = number of trucks shipped from Atlanta to New York,
r_b = number of trucks shipped from Richmond to Boston,
r_n = number of trucks shipped from Richmond to New York.

(Yes, there are four variables in this problem because there are four possible ways to reallocate trucks. Note that Table 7.4 has, indeed, four different cost figures in it.) The model is

$$\begin{aligned}
&\text{Minimize} && C = 75a_b + 60a_n + 70r_b + 60r_n \\
&\text{Subject to} && a_b + a_n \le 300 && \text{(truck supply in Atlanta)} \\
& && r_b + r_n \le 200 && \text{(truck supply in Richmond)} \\
& && a_b + r_b \ge 225 && \text{(trucks needed at Boston)} \\
& && a_n + r_n \ge 275 && \text{(trucks needed at New York)} \\
& && \left.\begin{aligned} a_b &\ge 0 \\ a_n &\ge 0 \\ r_b &\ge 0 \\ r_n &\ge 0 \end{aligned}\right\} && \text{(nonnegativity)}
\end{aligned}$$

This model is a bit more complicated than the others we've shown you, because we didn't want to give the impression that all linear programming models have only two variables. Notice here we have four variables and four constraints: one variable for each possible route and one constraint for each supply point (source of trucks) and one constraint for each demand point.

The transportation model was first worked on by L. V. Kantorovich in 1939. F. L. Hitchcock in 1941 and T. C. Koopmans in 1947 also did important early research on this problem.

TABLE 7.4 Shipping costs, supplies, and demands for Example 3

From	To: Boston	To: New York	Excess trucks
Atlanta	\$75.00	\$60.00	300
Richmond	\$70.00	\$60.00	200
Trucks required	225	275	500

□

Now that we've seen some examples of linear programming problems, you will find it useful to follow these steps when formulating a linear programming problem.

Steps in Formulating a Linear Programming Problem

1. Read the problem carefully.
2. Define the decision variables (those variables whose values you need in order to solve the problem). Sometimes it helps to make believe you are in charge and must tell your subordinates exactly what to do. This, in turn, requires that you know certain values that can only be determined by solving the desired linear programming problem.
3. Determine the objective function in verbal terms. Express the objective function in terms of the decision variables.
4. Write the constraints in terms of the decision variables.

EXERCISE SET 7.1

Formulate the following as linear programming problems. Do *not* attempt to solve them.

1. A manufacturing company makes two types of skis, the downhill and the cross-country. The necessary data are shown in the accompanying table. Formulate the problem that will tell the company how many pairs of skis of each type to make each week so as to maximize profit.

	Labor hours per pair		
Department	**Downhill**	**Cross-country**	**Maximum labor hours available per week**
Cutting	6	5	600
Finishing	1	2	400
Profit per pair	\$65	\$55	

2. A local lens maker has begun grinding lenses for astronomers. She offers two models: the amateur and the pro. Given the data shown in the accompanying table, how many of each lens should she make each month to maximize her profit?

	Labor hours per lens		
Operation	**Amateur**	**Pro**	**Maximum labor hours available per month**
Grinding	15	25	320
Finishing	10	20	160
Profit per lens	\$70	\$125	

3. An investor has $50,000 to invest. Option A yields a return of 15% per year and has a risk factor of 25 (the higher the risk factor, the higher the possibility of a loss). Option B yields a return of 10% per year and has a risk factor of 15. How much should the investor invest in each option if
   a) he wants to maximize return?
   b) he wants to achieve at least 13% return and minimize risk?
4. Phyllis Wentworth operates a local shop that sells coffee. She has decided to introduce two new blends: custom and premium. Each pound of custom contains 25% Blue Mountain and 15% Kona, while each pound of premium contains 30% Blue Mountain and 10% Kona. (The remainder of each blend is made of cheaper coffees which are in plentiful supply.) Given the cost and other relevant data shown in the accompanying table, how much of each blend should Phyllis sell to
   a) maximize her weekly profits?
   b) maximize her total weekly production?

Available supply		Profit per pound	
Blue Mountain	**Kona**	**Custom**	**Premium**
250 lb	100 lb	$0.85	$0.95

5. Sam and Dave have decided to advertise their record store. They have a budget of $1500 for the next quarter. They can buy radio ads for $250 each or television ads for $525 each. Each radio ad reaches 5000 people, with an estimated 50% of the total being 15 years of age or below. Each television ad reaches 8000 people, with an estimated 40% of the total being 15 years old or below. Sam and Dave wish to maximize the number of listeners age 15 or below, given their budget restriction and the fact they also wish to reach at least 25,000 total people. How many ads of each type should Sam and Dave purchase?
6. A dietician in a hospital is trying to arrange a special restricted diet using only two foods. The relevant data are shown in the accompanying table. How many ounces of each food should be used to minimize cholesterol intake while attaining the minimum requirements?

	Units per ounce		
Nutrient	**Food A**	**Food B**	**Minimum requirement**
Iron	20	20	250 units
Vitamin A	15	25	375 units
Vitamin C	18	12	200 units
Cholesterol	8	6	

7. Joe Armstrong has 250 acres of land on which to plant corn and soybeans. Each acre of corn requires 4 units of storage space and 2 hours of labor.

Each acre of soybeans requires 5 units of storage space and 9 hours of labor. Corn returns \$40 profit per acre and soybeans return \$50 profit per acre. Given that there are 1500 labor hours and 600 units of storage space, how many acres should be devoted to each crop?

8. Nuts to You Company has 120 pounds of cashews and 150 pounds of peanuts to use to make 5 pound boxes of mixed nuts. A box of Standard Mix requires 3 pounds of peanuts and 2 pounds of cashews and sells for \$9.00. A box of Premium Mix requires 2 pounds of peanuts and 3 pounds of cashews and sells for \$13.00.

   a) How many boxes of each mix must be made (and sold) to maximize sales revenues?

   b) Suppose the number of boxes of standard mix must be no more than three times the number of boxes of premium mix. What should be done to maximize revenue?

9. Roberta Cannon of Scientific Computer Systems is placing an order with the manufacturer for disk drives to go into her company's products. The manufacturer requires at least 100 drives per order. Single-sided drives cost \$60, double-sided drives cost \$80, and fixed drives cost \$500. Fixed drives must make up at least 15% of the total number of drives ordered. How many drives of each type should Roberta order, given that she wishes to minimize the cost of the order?

10. Custom Bootery specializes in made-to-order shoes. A pair of dress shoes requires 2 hours of skilled labor and 1 hour of unskilled labor. A pair of boots requires 3 hours of skilled labor and 2.5 hours of unskilled labor. There are 200 hours of skilled labor and 250 hours of unskilled labor available each week. The profit on shoes is \$20 and the profit on boots is \$35.

   a) How many pairs of each type should be made each week to maximize profits?

   b) Suppose that the number of pairs of boots should be at least one half the number of pairs of dress shoes. What should be done to maximize profits?

7.2

A GEOMETRIC INTERPRETATION OF LINEAR PROGRAMMING

In the preceding section we learned what a linear programming problem is and we also saw some examples. In this section we will present a geometric interpretation of linear programming. This will enable us to solve small problems, as well as setting the stage for the analytic approach to solution (the simplex method) which we will study in the next chapter.

Before we begin our discussion of the geometric approach to solving linear programming problems, let's use a pair of examples to review how to graph inequalities in two variables.

EXAMPLE 4

Graph the solution set for the inequality $3x + 2y \leq 12$.

SOLUTION

We will follow the approach described in Section 3.4. First, we graph the given inequality as if it were an equation. Here, we graph the equation $3x + 2y = 12$. Since the sense of the inequality is $\leq$, we use a solid line. (See Fig. 7.1.) Next, we must choose a test point that does not lie on the graph of the equation. Since the line doesn't pass through the origin, we will use (0, 0) as our test point. This gives us

$$3(0) + 2(0) = 0 \leq 12$$

(in other words, the origin satisfies the original inequality). The solution set is the half-plane containing the origin.

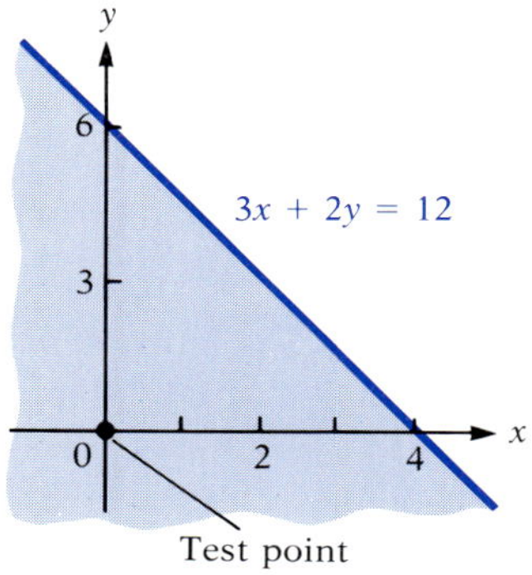

Figure 7.1 The solution set for $3x + 2y \leq 12$. □

EXAMPLE 5

Graph the solution set for the inequalities

$$\begin{aligned} 3x + 2y &\leq 12, \\ x &\geq 0, \\ y &\geq 0. \end{aligned}$$

SOLUTION

The solution set for a system of inequalities is the set of points that satisfies all of the inequalities simultaneously. It is formed from the intersection of the solution sets of the individual inequalities. The easiest way to find the intersection of the solution sets is to do it

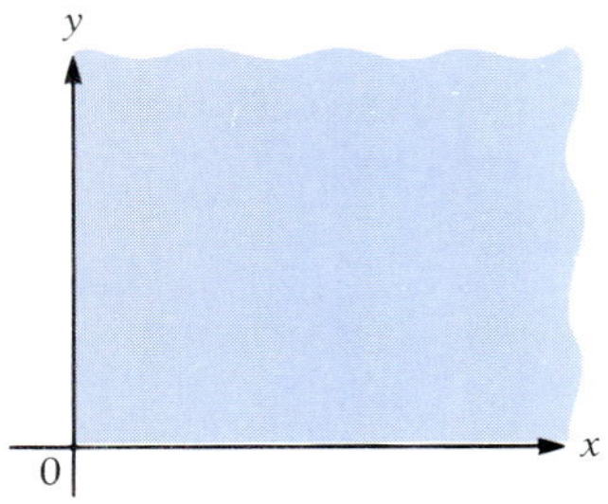

Figure 7.2 **The solution set for $x \geq 0$ and $y \geq 0$.**

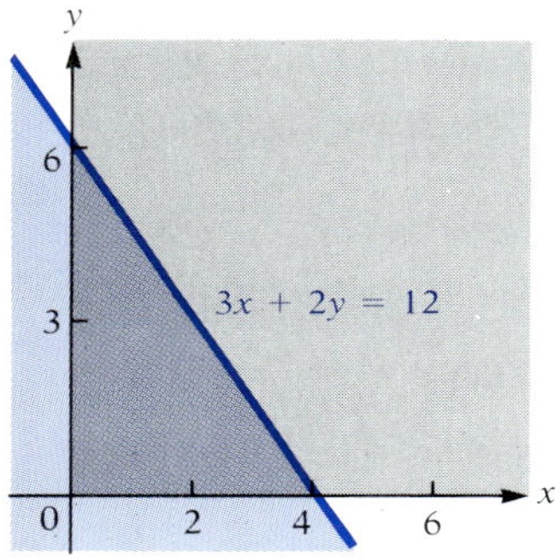

Figure 7.3 **The solution set for the three inequalities $3x + 2y \leq 12$, $x \geq 0$, and $y \geq 0$.**

graphically. Now, we just studied the first inequality in Example 4. The solution set for the other inequalities is shown in Fig. 7.2.

To produce the intersection of these solution sets, we simply draw them all on the same graph with different shadings. The points that are shaded for every inequality (cross-hatched) are the points in the intersection. (See Fig. 7.3.)

Notice that all the points in the solution set in Fig. 7.3 are in the first quadrant. This is because of the inequalities $x \geq 0$ and $y \geq 0$. But these inequalities are equivalent to the statement that both x and y should be nonnegative. Since all the linear programming problems we will encounter have nonnegative variables, we can conclude that the solution sets to all the problems we will solve graphically occur in the first quadrant. In general, the solution sets to linear programming problems have nonnegative variables. □

Practice Problem

Graph the solution set for the inequalities

$$5x + 2y \leq 10, \qquad x \geq 0, \qquad y \geq 0.$$

Answer:

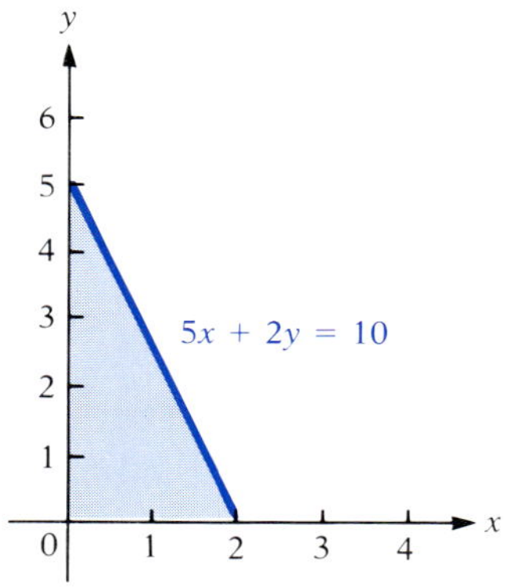

A Maximization Problem

Now let's determine the solution to the linear programming faced by Ann and Fred (the Problem from Life). Their problem is

$$\begin{array}{lrll} \text{Maximize} & P = & 12t + 8c & \\ \text{Subject to} & & 5t + 2c \leq 122 & \text{(oak)} \\ & & 2t + 3c \leq 100 & \text{(pine)} \\ & & 4t + 3c \leq 120 & \text{(labor)} \\ & & \left.\begin{array}{r} t \geq 0 \\ c \geq 0 \end{array}\right\} & \text{(nonnegativity)} \end{array}$$

where t and c are the number of tables and chairs, respectively, to manufacture during the next week.

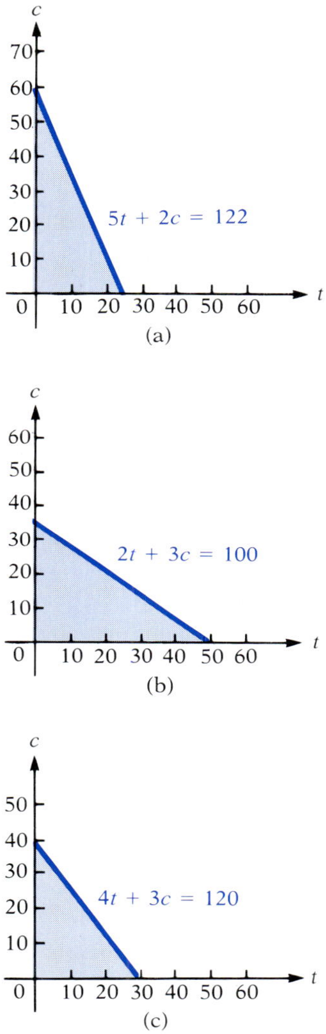

Figure 7.4 **(a) The oak constraint, (b) the pine constraint, (c) the labor constraint.**

Our first step is to examine the set of constraints in order to determine the solution set. We know (from the previous example) the nonnegativity restrictions ($t \geq 0$ and $c \geq 0$) ensure that the solution set will occur in the first quadrant, so let's consider the other constraints. The oak constraint is shown in Fig. 7.4(a), the pine constraint is shown in Fig. 7.4(b), and the labor constraint is shown in Fig. 7.4(c). Placing all three constraints on a common set of axes gives us the solution set to the constraints as shown in Fig. 7.5. In linear programming terms the set of all possible solutions (the solution set of the constraints) is called the **feasible region.**

Take a good look at Fig. 7.5 because it illustrates the important features of a feasible region in a linear programming problem. First, the feasible region is contained in the first quadrant (as we have previously noted). Second, the sides of the feasible region are always straight-line segments, never curved lines. The line segments run between points where two of the constraints cross. These crossing points are called **corner points** or **extreme points.** (Notice, however, that not all crossing points are in the feasible region. For example, the oak and pine constraints cross outside the feasible region. Their crossing point is *not* a corner point.) Finally, notice that the feasible region has no "holes" or indentations in it. In fact, take any two points in the feasible region and connect them with a straight line. You will find that all of the points on the connecting line are also part of the feasible region. Regions where this is true are called **convex regions.** Thus the feasible region in a linear programming

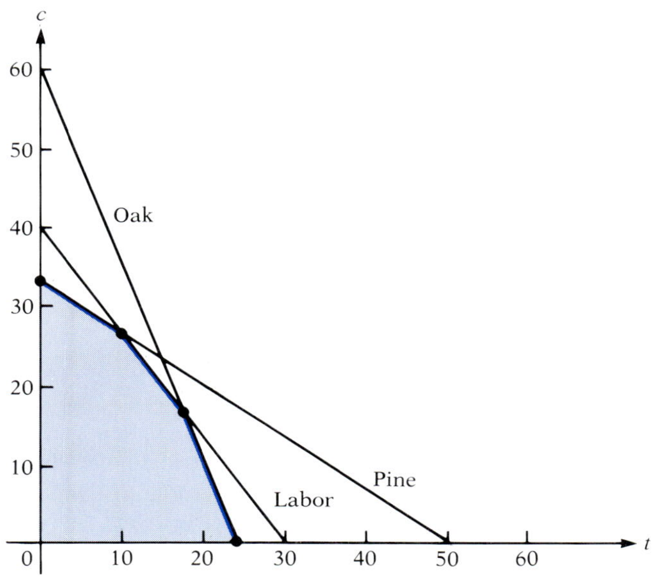

Figure 7.5 **The feasible region for Ann and Fred's problem.**

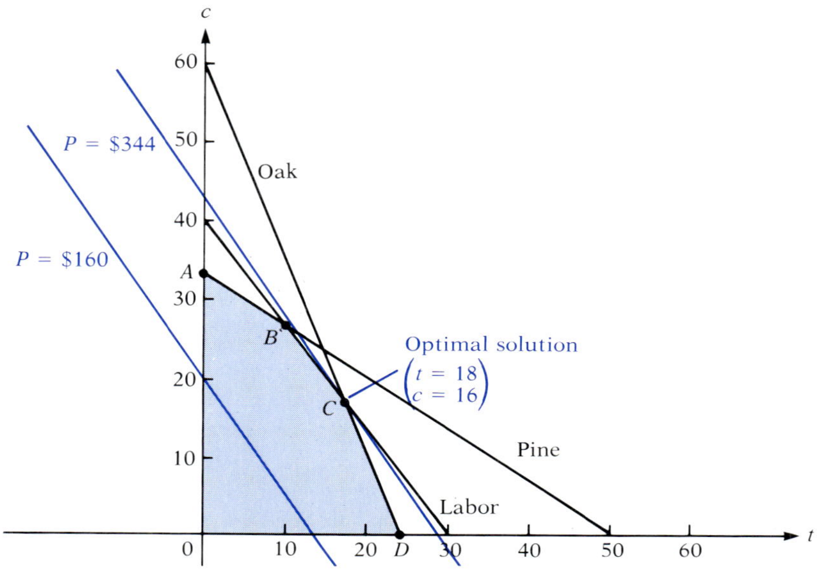

Figure 7.6 Finding the optimal solution for Ann and Fred's problem.

problem is always convex. Finally, note that the feasible region is determined *only by the constraints:* The objective function has no role in determining the feasible region of a linear programming problem.

Given that we've drawn the feasible region for the problem of Ann and Fred, we must now determine the best or **optimal** solution. In other words, we must choose the point (or points) in the feasible region that make the profit function $P = 12t + 8c$ as large as possible.

Let's draw several **isoprofit** (constant-profit) lines on a graph that also shows the feasible region. We do this by assigning P several different values and then drawing the graphs of the resulting equations. (See Fig. 7.6.) Each line represents all the combinations of tables and chairs that will produce a particular profit. For example, all points on the line $P = \$160$ will produce \$160 profit. Note that all the constant-profit lines are parallel, because they all have the same slope. We can show this by writing the function $P = 12t + 8c$ in slope-intercept form (recognizing that P is a constant):

$$c = -\left(\frac{12}{8}\right)t + \left(\frac{P}{8}\right).$$

Thus all the constant-profit lines have a slope of $-\frac{12}{8} = -1.5$ (since we've chosen to plot t on the horizontal axis and c on the vertical axis). Note also that as P increases, the lines move farther out from the origin. This suggests that to find the point(s) of maximum profit we simply look for the point(s) in the feasible region on the constant-profit line that is the farthest away from the origin. As we can see in Fig. 7.6, this occurs at the point (18, 16). In other words, if Ann and

Fred make 18 tables and 16 chairs each week, they will maximize their weekly profits at $P = 12(18) + 8(16) = \$344$.

What do you think would happen if we had a different objective function? We would simply draw constant-profit lines as before and again look for the point in the feasible region on the constant-profit line which is farthest away from the origin. Of course, these lines would probably have a different slope from the ones that we drew originally. However, a little experimentation with a ruler and Fig. 7.6 should convince you that no matter what the exact value of the slope, in general the maximum profit will occur at a corner point of the feasible region. In fact, the following statement is true (although we won't prove it):

> If a linear programming problem has a finite solution (a meaningful solution), then the optimal value of the objective function will occur at a corner point of the feasible region.

This suggests that to solve a linear programming problem graphically, simply determine the corner points of the feasible region and compute the value of the objective function at each corner point. The optimal solution can then be identified easily.

How can we determine all of the corner points? Draw the graph of the feasible region. This will give us a visual determination of the corner points. It will also identify, for each point, precisely which constraints are crossing. Once the constraints have been identified, solve the appropriate pair of simultaneous linear equations to determine the coordinates of the corner point. For example, in Fig. 7.6 the corner points are labeled O, A, B, C, and D. Now we know that O is the point $(0, 0)$. We also know that A is $(0, 33\frac{1}{3})$ (because it is the c-intercept for the pine equation) and that D is $(24\frac{2}{5}, 0)$ (because it is the t-intercept for the oak equation). To determine B and C requires more work, however. More specifically, at B the pine and labor equations cross. Therefore B is the solution of

$$\begin{aligned} 2t + 3c &= 100, \\ 4t + 3c &= 120, \end{aligned}$$

so $B = (10, 26\frac{2}{3})$. (Check this result yourself.) Similarly, C is the point at which the oak and labor equations cross, so that C is the solution of

$$\begin{aligned} 5t + 2c &= 122, \\ 4t + 3c &= 120, \end{aligned}$$

so $C = (18, 16)$. We summarize this information in Table 7.5. Not surprisingly, point C is indeed the best corner point solution and therefore the optimal solution to Ann and Fred's problem.

TABLE 7.5 Corner points for Ann and Fred's problem

Corner point	Coordinates	Profit = $12t + 8c$
O	(0, 0)	\$ 0.
A	$(0, 33\frac{1}{3})$	266.67
B	$(10, 26\frac{2}{3})$	333.33
C	(18, 16)	344.00
D	$(24\frac{2}{5}, 0)$	292.80

We can summarize this discussion in the following procedure for the geometric (graphical) solution of linear programming problems with two variables. (Actually, it could be used to solve problems with three variables, but drawing the necessary graphs would be very difficult.)

Steps for Geometric Solution of a Linear Programming Problem

1. Formulate the problem in mathematical form.
2. Draw the graph of the feasible region by graphing the constraints.
3. Determine the exact coordinates of the corner points of the feasible region.
4. Compute the value of the objective function at each corner point. Choose the appropriate point(s).

Note: Not all linear programming problems have solutions. See Examples 7 and 8.

A Minimization Problem

The problem of Ann and Fred was a maximization problem. Let's take a look at a minimization problem to see what differences there are in the solution procedure.

EXAMPLE 6

Diet Mix Problem, revisited Solve the diet mix problem of Example 1, in which a pet store owner was attempting to determine the cheapest mix for his dog food.

SOLUTION

We formulated the problem as follows. Let

$$a = \text{number of pounds of ingredient A to use,}$$
$$b = \text{number of pounds of ingredient B to use.}$$

$$
\begin{array}{lll}
\text{Minimize} & C = 0.25a + 0.30b & \\
\text{Subject to} & 2a + 3b \geq 10 & \text{(protein requirement)} \\
 & 2a + b \geq 5 & \text{(carbohydrate requirement)} \\
 & 2a + 4b \geq 12 & \text{(vitamin requirement)} \\
 & \left.\begin{array}{r} a \geq 0 \\ b \geq 0 \end{array}\right\} & \text{(nonnegativity)}
\end{array}
$$

We've already formulated the problem, so our next step is to graph the constraints in order to determine the feasible region. The protein requirement is shown in Fig. 7.7(a). Notice the solution set for this constraint does not include the origin because the constraint is $\geq$. The carbohydrate requirement and vitamin requirement constraints are shown in Fig. 7.7(b) and 7.7(c), respectively. The feasible region is shown in Fig. 7.8. This feasible region is different from the one that we studied earlier (in Ann and Fred's problem) because it is not bounded. (A region in a plane is **bounded** if we can draw a circle around it.) In other words, the feasible region in Fig. 7.8 goes out in the first quadrant without limit. Although an unbounded feasible region can cause problems (as we will see in Example 7), here it does not bother us.

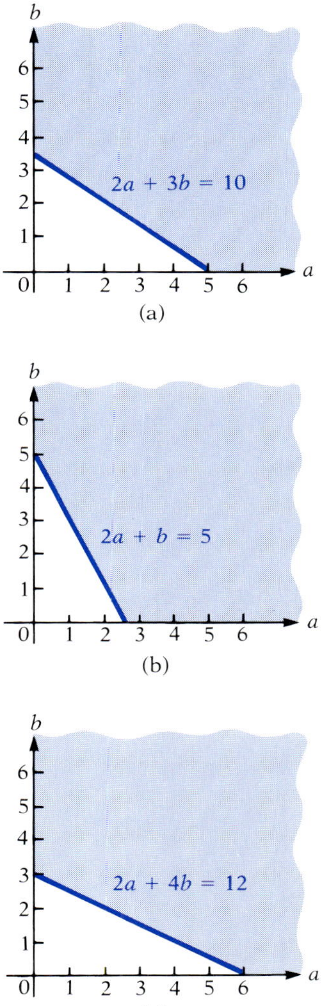

Figure 7.7 **(a) Protein constraint, (b) carbohydrate constraint, (c) vitamin constraint.**

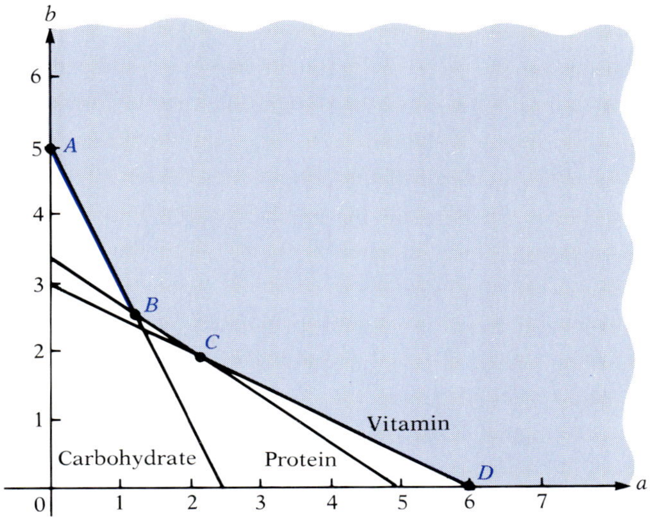

Figure 7.8 **The feasible region for the diet mix problem.**

TABLE 7.6 Corner points for the diet mix problem

Corner point	Coordinates	Cost = 0.25*a* + 0.30*b*
A	(0, 5)	\$1.50
B	(1.25, 2.5)	\$1.0625
C	(2, 2)	\$1.10
D	(6, 0)	\$1.50

Once we have drawn the feasible region, the next step is to locate the corner points. As shown in Fig. 7.8, there are four corner points, labeled *A*, *B*, *C*, and *D*. The coordinates of point *A* are (0, 5) while those of point *D* are (6, 0). The other corner points are not as easy to locate. Point *B* is the intersection of the protein and carbohydrate equations. Thus we can find its coordinates by solving the equations

$$2a + 3b = 10,$$
$$2a + b = 5.$$

The solution is $a = 1.25$ and $b = 2.5$. Point *C* is the intersection of the protein and vitamin equations, so that we must solve the equations

$$2a + 3b = 10,$$
$$2a + 4b = 12.$$

The solution here is $a = 2$ and $b = 2$. We summarize these data in Table 7.6.

As we can see from Table 7.6, the least-cost mix occurs at point *B*. Thus the optimal solution is to use 1.25 pounds of ingredient *A* and 2.5 pounds of ingredient *B*, at a cost of \$1.0625. □

Practice Problem

Solve the following linear programming problems:

a) Maximize $P = 3x + 2y$
Subject to
$$\begin{aligned} x &\le 700 \\ y &\le 900 \\ 3x + y &\le 2400 \\ x + 2y &\le 1600 \\ x &\ge 0 \\ y &\ge 0 \end{aligned}$$

b) Minimize $C = 3m + 4n$
Subject to
$$\begin{aligned} m &\ge 4 \\ n &\ge 6 \\ 4m + 3n &\ge 40 \\ m &\ge 0 \\ n &\ge 0 \end{aligned}$$

Answers: **a)** $x = 640$, $y = 480$, $P = 2880$; **b)** $m = 5.5$, $n = 6$, $C = 40.5$

As you can see, we used exactly the same approach in Example 6 as in the maximization problem of Ann and Fred. Thus we solve maximization and minimization problems exactly the same way.

However, because the diet mix problem is unbounded, we must be careful. As we were attempting to minimize the objective function, the unboundedness of the feasible region is not significant. After all, although points such as (1 million, 1 million) are feasible, they would hardly be cheap. If we were maximizing, though, such points would obviously be attractive. However, a simple listing of the corner points would not indicate that such solutions were possible. Therefore, when the feasible region is unbounded, you must consider the usefulness of points with extremely large coordinates in addition to the corner points. The next example shows what could happen with an unbounded feasible region.

EXAMPLE 7

Unbounded Solution Solve the following problem:

$$\begin{aligned} \text{Maximize} \quad & P = 5x + 4y \\ \text{Subject to} \quad & 2x - y \geq 6 \\ & x + y \geq 3 \\ & x \geq 0 \\ & y \geq 0 \end{aligned}$$

SOLUTION

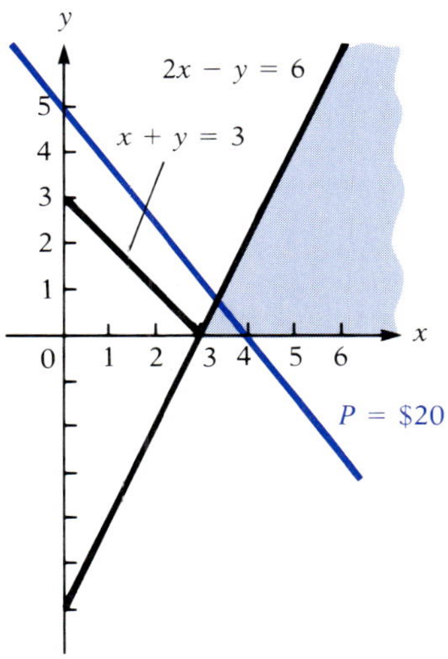

Figure 7.9 The feasible region for Example 7 (unbounded solution).

The feasible region for this problem, as shown in Fig. 7.9, is unbounded. Moreover, here we are maximizing. A little thought should be able to convince you that there is no limit to the value of the objective function. This means that no matter what point you choose in the feasible region, there is always (at least) one point that is even more profitable. We have drawn the isoprofit line $P = \$20$ to help illustrate this. As you can see, there is nothing to keep us from drawing other isoprofit lines as far to the right as we wish (except, perhaps, the boundaries of the page). We say that this problem has an **unbounded solution.**

Some of you might be asking the question, "What's going on? I thought the optimal solution to a linear programming problem always occurs at corner point!" The answer is that you're *almost* correct. Reread our earlier statement; we said that if a linear programming problem has a finite (meaningful) solution, then it occurs at a corner point. The key word here is *finite.* Since an unbounded solution is not finite (meaningful), our statement is still correct. □

Actually, there is one other case in which a linear programming problem will not have an optimal solution at a corner point: the case in which it has no feasible solutions at all.

EXAMPLE 8

Infeasible Problem Solve the linear programming problem

$$\begin{aligned} \text{Maximize} \quad & P = 2x + 4y \\ \text{Subject to} \quad & x + y \leq 3 \\ & x + y \geq 5 \\ & x \geq 0 \\ & y \geq 0 \end{aligned}$$

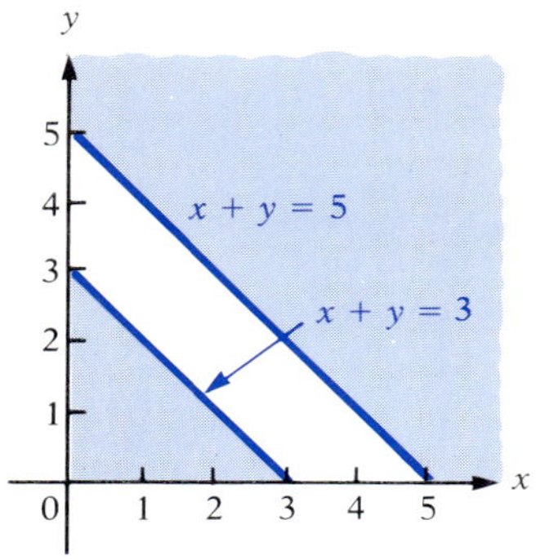

SOLUTION

As Fig. 7.10 illustrates, this problem has no feasible solutions at all. (In other words, the feasible region is the empty set.) This shouldn't be a surprise, since there is no pair of numbers (x, y) such that their sum is both 3 or less and 5 or more. Such problems are said to be **infeasible.**

Figure 7.10 The feasible region for Example 8 is the empty set (infeasible problem). □

Practice Problem

Characterize the following linear programming problems as having an unbounded solution or an infeasible solution:

a) $$\begin{aligned} \text{Maximize} \quad & P = 5x + 6y \\ \text{Subject to} \quad & -x + y \geq 4 \\ & x + y \leq 2 \\ & x \geq 0 \\ & y \geq 0 \end{aligned}$$

b) $$\begin{aligned} \text{Maximize} \quad & P = 3x + y \\ \text{Subject to} \quad & -x + y \leq 2 \\ & x - y \leq 0 \\ & x \geq 0 \\ & y \geq 0 \end{aligned}$$

Answers: **a)** infeasible solution; **b)** unbounded solution

There is one other unusual situation, as the following example shows.

EXAMPLE 9

* **Multiple Optimal Solutions** Solve the linear programming problem

$$\begin{aligned} \text{Maximize} \quad & P = 3x + 2y \\ \text{Subject to} \quad & 6x + 4y \leq 24 \\ & 10x + 5y \leq 35 \\ & x \geq 0 \\ & y \geq 0 \end{aligned}$$

* Reprinted with permission from Wu and Coppins, *Linear Programming and Extensions* (McGraw-Hill, New York), 1981.

SOLUTION

As Fig. 7.11 shows, the feasible region has four corner points: O, A, B, and C. Their coordinates and objective function values are shown in Table 7.7. Notice that both points A and B have the same objective function value, 12. Therefore, this problem has two optimal corner points! When more than one corner point is optimal, we say the problem has **multiple optimal solutions.**

TABLE 7.7 Corner points for Example 9

Corner point	Coordinates	Profit = $3x + 2y$
O	(0, 0)	\$ 0
A	(0, 6)	\$12
B	(2, 3)	\$12
C	(3.5, 0)	\$10.50

The cause of the multiple optimal solutions is that the objective function has the same slope as a constraint that is part of the boundary of the feasible region. In this case, the objective function has exactly the same slope as the first constraint. You can see this by placing a ruler on the isoprofit line $P = \$6$ and moving it out as far as it will go and still remain in the feasible region. In fact, in this problem, not only are there two optimal corner point solutions, there are also an infinite number of non-corner point solutions that are optimal. Can you determine where they are? The other optimal solutions are all the points on the line segment between points A and B.

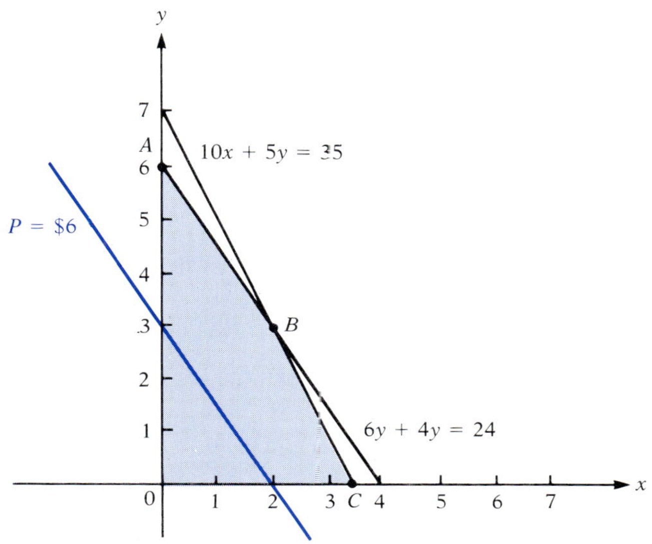

Figure 7.11 **The feasible region for Example 9 (multiple optimal solutions).** □

Practice Problem

Determine, by solving them, which of the following linear programming problems have multiple optimal solutions.

a) Maximize $P = 2x + y$
Subject to
$$8x + 4y \le 32$$
$$5x + 8y \le 40$$
$$x \ge 0$$
$$y \ge 0$$

b) Minimize $C = 6x + 9y$
Subject to
$$6x + 3y \le 18$$
$$4x + 5y \ge 10$$
$$x \ge 0$$
$$y \ge 0$$

Answers: **a)** multiple optimal solutions because both (4, 0) and $(\frac{24}{11}, \frac{40}{11})$ are optimal, with $P = 8$; **b)** only the point (2.5, 0) with $C = 15$ is optimal

EXERCISE SET 7.2

Solve the following linear programming problems.

1. Maximize $P = 4x + 4y$
Subject to $2x + y \le 10$, $x + 2y \le 8$, $x \ge 0$, $y \ge 0$

2. Maximize $P = 2x + 3y$
Subject to $3x + 6y \le 24$, $6x + 3y \le 30$, $x \ge 0$, $y \ge 0$

3. Minimize $C = 8x + 12y$
Subject to $3x + 2y \ge 12$, $2x + 3y \ge 12$, $x \ge 0$, $y \ge 0$

4. Minimize $C = 6x + 7y$
Subject to $5x + 4y \ge 20$, $4x + 7y \ge 28$, $x \ge 0$, $y \ge 0$

5. Maximize $P = 15x + 20y$
Subject to $4x + 6y \le 36$, $9x + 5y \le 45$, $8x + 6y \le 48$, $x \ge 0$, $y \ge 0$

6. Minimize $C = 8x + 12y$
Subject to $x + y \ge 4$, $4x + 2y \ge 8$, $3x + 5y \ge 15$, $x \ge 0$, $y \ge 0$

7. Minimize $C = 5x + 10y$
Subject to $x + y \ge 2$, $8x + 12y \le 72$, $4x + 8y \le 40$, $x \ge 0$, $y \ge 0$

8. Maximize $P = 3x + y$
Subject to $2x - y \ge 6$, $3x + 6y \le 30$, $x \le 9$, $x \ge 0$, $y \ge 0$

9. Maximize $P = 20x + 35y$
Subject to $x + y \ge 2$, $12x + 16y \le 144$, $x - y \ge 1$, $x \ge 0$, $y \ge 0$

10. Minimize $C = 14x + 21y$
Subject to $2x + 3y \le 66$, $3x + 9y \ge 27$, $4x + 11y \ge 44$, $x \ge 0$, $y \ge 0$

11. Minimize $C = 15x + 25y$
Subject to
$$\begin{aligned} x - y &\le 0 \\ 4x + 6y &\ge 12 \\ 6x + 3y &\le 18 \\ x &\ge 0 \\ y &\ge 0 \end{aligned}$$

12. Maximize $P = 3x - 2y$
Subject to
$$\begin{aligned} -x + y &\ge 3 \\ 6x + 4y &\ge 24 \\ 6x + 8y &\le 48 \\ x &\ge 0 \\ y &\ge 0 \end{aligned}$$

13. Kustom Kars does van conversions. The custom conversion requires 15 hours of shop time, 8 hours of painting time, and 1 hour of inspection time. The deluxe conversion requires 12 hours of shop time, 12 hours of painting time, and 1.5 hours of inspection time. There are 90 hours of shop time, 72 hours of painting time, and 10 hours of inspection time available during the coming two weeks. How many conversions of each type should Kustom Kars perform assuming that each custom conversion results in \$175 profit and each deluxe conversion results in \$225 profit? What is the maximum profit? Does your answer change if the amount of shop time available is 108 hours?

14. A trucking company carries chemical waste in sealed containers. Each container for acid waste weighs 1000 pounds and is 75 cubic feet in volume, while each container for caustic waste weighs 1000 pounds and is 45 cubic feet in volume. The truck can carry only 10 tons of waste containers and only 1125 cubic feet of space is available. The charge for carrying acid containers is \$500 and the charge for carrying caustic containers is \$450. What is the maximum possible revenue for the truck? Is there any empty space remaining in the truck?

15. American Scientific Systems manufactures floppy disks and diskettes. Each 100 disks require 6400 square inches of material, 1.5 hours of manufacturing time, and 2.5 hours of packaging time. Each 100 diskettes require 2500 square inches of material, 1.25 hours of manufacturing time, and 1.5 hours of packaging time. There are 125,000 square inches of material, 45 hours of manufacturing time, and 60 hours of packaging time available. If disks sell for \$4.75 and diskettes for \$3.00, how many of each should the company sell to maximize revenue? Does the answer change if the company must sell at least 1500 of each type?

16. The local golf course must buy fertilizer to keep the grass green. The requirements and other relevant information are summarized in the accompanying table. How many cubic yards of each fertilizer should the course buy? What is the cost?

	Pounds of nutrient per cubic yard		
	Mix A	**Mix B**	**Minimum requirements**
Phosphoric acid	20	10	690 lb
Nitrogen	30	30	1440 lb
Potash	5	10	330 lb
Cost per cubic yard	\$30	\$35	

17. An engineering firm is bidding on a certain job. Determine how many junior engineers and how many senior engineers it should use to minimize its personnel costs and still satisfy the job requirements. The relevant data are given in the following table.

	Work-hours per engineer		Minimum work-hours required
	Senior engineer	**Junior engineer**	
Development	1200	800	20,000
Design	600	800	15,000
Test	200	400	8,000
Cost per engineer	\$55,000	\$30,000	

18. Juanita Torres has \$25,000 to invest in stocks. Stock A has a growth rate of 15% per year, pays dividends of 5% per year, and has a risk factor of 2.5. Stock B has a growth rate of 8% per year, pays dividends of 9% per year, and has a risk factor of 2. Juanita has determined that her portfolio must have a composite growth rate of 10% per year and pay at least 7.5% dividends per year. At the moment, each stock sells for \$100 per share. How many shares of each stock should Juanita purchase to minimize her risk?

19. Reconsider Problem 18. Suppose Juanita decides that she would rather maximize the composite growth rate while limiting risk to an average factor of 2.35. How many shares of each stock should she purchase?

CHAPTER REVIEW

IMPORTANT TERMS AND SYMBOLS

7.1 Definition of Linear Programming

linear programming problem Determine the values of the decision variables that optimize the objective function (a linear function of the decision variables) subject to a set of constraints (linear inequalities, which include nonnegativity restrictions).

steps in formulating a linear programming problem

1. Read the problem carefully.
2. Define the decision variables (those variables whose values you need in order to solve the problem). Sometimes it helps to make believe

you are in charge and must tell your subordinates exactly what to do. This, in turn, requires that you know certain values that can only be determined by solving the desired linear programming problem.

3. Determine the objective function in verbal terms. Express the objective function in terms of the decision variables.
4. Write the constraints in terms of the decision variables.

7.2 A Geometric Interpretation of Linear Programming

corner (extreme) points Points on the boundary of the feasible region where two (or more) constraints cross.

feasible region of a linear programming problem The solution set of the constraints.

geometric solution of a linear programming problem

1. Formulate the problem in mathematical form.
2. Draw the graph of the feasible region by graphing the constraints.
3. Determine the exact coordinates of the corner points of the feasible region.
4. Compute the value of the objective function at each corner point. Choose the appropriate point(s).

infeasible problem A problem in which the feasible region is empty.

multiple optimal solutions A situation in which more than one corner point is optimal.

unbounded solution A situation for which the solution does not lie at a corner point because the feasible region is unbounded.

REVIEW PROBLEMS

Formulate the following as linear programming problems. Do not attempt to solve them.

1. An oil company owns and operates two refineries. Refinery A produces (each day) 200 barrels of high-grade oil, 400 barrels of medium-grade oil, and 600 barrels of low-grade oil at a cost of \$20,000. Refinery B produces (each day) 300 barrels of high-grade oil, 150 barrels of medium-grade oil, and 300 barrels of low-grade oil at a cost of \$13,500. An order has been received for 2000 barrels of high-grade oil, 1750 barrels of medium-grade oil, and 3000 barrels of low-grade oil. How many days should each refinery be operated to minimize the cost of filling the order?
2. A truck manufacturer produces small and large pickup trucks. Each small truck requires 8 hours of assembly, 2 hours of painting, and 1 hour of upholstery while yielding a profit of \$1000. Each large truck

requires 18 hours of assembly, 3 hours of painting, and 1 hour of upholstery while yielding a profit of \$1500. Each week there are 1800 hours of assembly, 250 hours of painting, and 200 hours of upholstery available. How many small and large trucks should the manufacturer produce each week to maximize profit?

3. The Nut Shoppe has decided to sell two raisin-nut mixtures. Mixture I contains 12 ounces of peanuts, 2 ounces of raisins, and 8 ounces of cashews and sells for \$1.75. Mixture II contains 12 ounces of peanuts, 3 ounces of raisins, and 2 ounces of cashews and sells for \$1.50. Management has decided to allocate 6000 ounces of peanuts, 1500 ounces of raisins, and 3000 ounces of cashews. How many boxes of each mixture should be sold in order to maximize revenues?

4. Scenic Tours must drive 500 people and 15,000 pounds of luggage to Atlantic City. It can charter buses from two different companies. Company A's buses carry 55 passengers and hold 1750 pounds of luggage at a cost of \$1200 for the trip. Company B's buses carry up to 65 passengers and hold 1800 pounds of luggage at a cost of \$1350 for the trip. How many buses of each type should be chartered to minimize the cost of the trip?

5. A farmer has 500 acres of land on which to plant wheat and corn. Each acre of wheat requires \$100 investment and 8 labor hours. Each acre of corn requires \$125 investment and 6 labor hours. There are \$60,000 available for investment and 3500 labor hours. Given that wheat returns a profit of \$65 per acre and corn returns a profit of \$70 per acre, how many acres should be devoted to each crop to maximize profit?

6. Suppose, in the previous problem, the farmer learns that government regulations prevent him from planting more than 300 acres of wheat. Moreover, suppose he has decided to plant at least three acres of corn for every two acres of wheat. What does the linear programming model for the farmer look like under these conditions?

Solve the following linear programming problems geometrically. Characterize each problem as having a unique solution, an unbounded solution, being infeasible, or having multiple optimal solutions.

7. Maximize $P = 3x + y$

Subject to
$$\begin{aligned} x + 2y &\le 7 \\ 4x + 2y &\le 10 \\ 2x + 3y &\le 12 \\ x &\ge 0 \\ y &\ge 0 \end{aligned}$$

8. Minimize $C = 2x + 5y$

Subject to
$$\begin{aligned} 3x - y &\ge 8 \\ x + 2y &\ge 4 \\ x &\ge 0 \\ y &\ge 0 \end{aligned}$$

9. Minimize $C = x + 2y$

Subject to
$$\begin{aligned} 2x + y &\ge 12 \\ x + 3y &\ge 27 \\ x &\ge 0 \\ y &\ge 0 \end{aligned}$$

10. Maximize $P = 4x + 5y$

Subject to
$$\begin{aligned} 8x + 16y &\le 64 \\ 9x + 6y &\le 54 \\ x + y &\ge 3 \\ x &\ge 0 \\ y &\ge 0 \end{aligned}$$

11. Minimize $C = 7x + 3y$
Subject to
$$\begin{aligned} 3x + 5y &\geq 45 \\ 12x + 4y &\leq 40 \\ 2x + 6y &\leq 36 \\ x &\geq 0 \\ y &\geq 0 \end{aligned}$$

12. Maximize $P = -3x + 2y$
Subject to
$$\begin{aligned} 2x - 3y &\geq 0 \\ x + y &\geq 4 \\ 5x + 3y &\geq 15 \\ x &\geq 0 \\ y &\geq 0 \end{aligned}$$

13. Maximize $P = 4x - 3y$
Subject to
$$\begin{aligned} -3x + 2y &\leq 0 \\ 8x + 6y &\leq 72 \\ 2x + y &\geq 8 \\ x &\geq 0 \\ y &\geq 0 \end{aligned}$$

14. Minimize $C = 5x + 7y$
Subject to
$$\begin{aligned} 3x + y &\geq 9 \\ 12x + 8y &\leq 144 \\ 5x - 7y &\geq 35 \\ x &\geq 0 \\ y &\geq 0 \end{aligned}$$

15. Solve Review Problem 1.

16. Solve Review Problem 2.

17. Solve Review Problem 3.

18. Solve Review Problem 4.

19. Solve Review Problem 5.

20. Solve Review Problem 6.

8 LINEAR PROGRAMMING BY THE SIMPLEX METHOD

Sally Green, owner of Sally's Hamburger Stands, has decided to interview some of her customers in an attempt to improve sales. Her marketing specialist has indicated that the survey design should meet the following requirements: (1) at least 100 families must be contacted, (2) at least 40 families with children must be contacted, and (3) at least 25 families without children must be contacted. It will cost Sally \$5 for each family with children contacted and \$4 for each family without children contacted. How many families of each type should be contacted so as to minimize the cost of the survey and still meet the design requirements? (To be solved in Section 8.3.)*

* Reprinted with permission from Wu and Coppins, *Linear Programming and Extensions* (McGraw-Hill, New York), 1981.

PROBLEM FROM LIFE

INTRODUCTION

In the preceding chapter you learned how to formulate linear programming problems and how to solve them geometrically. In this chapter you will learn how to solve linear programming problems analytically by a process called the simplex method. This process (with some modifications) can be used to solve problems with as many as several thousand constraints and five to ten thousand variables. You will have more practice in formulating linear programming problems. You will also learn about duality and how it relates to linear programming.

We would not be surprised if at this point you were wondering why we're spending so much effort on linear programming. After all, does anybody use the stuff? The answer to this question is a resounding yes. In fact, IBM estimated that in 1970 fully 25% of all scientific computing costs in the world were devoted to solving linear programming problems. As another example of the importance of linear programming, on November 7, 1979, and November 19, 1984, *The New York Times* had articles describing what were thought to be more efficient approaches to solving such problems.

We will rely heavily upon your ability to perform row operations (see Section 6.3) and your ability to solve systems of simultaneous equations using matrices. You may want to review these topics before you continue with this chapter. Be sure that you are familiar with the notation used for row operations.

8.1

MAXIMIZATION WITH ≤ CONSTRAINTS

In this section we will describe the simplex method and show how to use it to solve maximization problems in which all constraints are of the ≤ variety.

The simplex method was developed in 1947 by George B. Dantzig and the Project "Scoop" team for solving linear programming problems. The method itself is an algorithmic (repetitive) procedure in which the optimal solution to a linear programming problem is found after the steps of the algorithm have been repeated a (finite) number of times. (The term *simplex* comes from one of the first applications of the method: a problem in which the corner points formed a geometric shape called a simplex.)

Standard Form for ≤ Constraints

At the beginning of this chapter we asked you to review row operations and how they can be used to solve a system of linear equations. The reason for this is that the simplex method deals with problems in which the inequality constraints have been converted to equalities. A linear programming problem in which all constraints are expressed as equations is said to be in **standard form.**

To begin, let's see how we can convert inequality constraints written with ≤ into equations. We'll assume the right-hand sides of these constraints are nonnegative.

Suppose we're given the constraint

$$5x + 6y + 3z \le 54.$$

To convert this to an equation, we add a slack variable s to the left-hand side, which gives us

$$5x + 6y + 3z + s = 54, \qquad \text{where } s \ge 0.$$

The variable s is called a **slack variable** because it measures the difference (slack) between the right-hand side and the left-hand side of the equation.

A slack variable has no impact on the objective function, so its objective-function coefficient is zero. Let's see why. By simple algebra,

$$s = 54 - (5x + 6y + 3z).$$

Thus s simply measures the difference between the original 54 units on the right-hand side of the constraint and the amount used up by the linear function $(5x + 6y + 3z)$. We don't need an additional variable to measure this difference; we could just as easily perform the subtraction without it.

EXAMPLE 1

Convert the following linear programming problem to standard form.

$$\begin{aligned} \text{Maximize} \quad & P = 4x + 2y \\ \text{Subject to} \quad & 3x + y \le 25 \\ & 2x + 4y \le 35 \\ & x, y \ge 0 \end{aligned}$$

SOLUTION

To convert this to standard form, we must add a slack variable (s) to the first constraint and another slack variable (t) to the second constraint. The problem in standard form is

$$\begin{aligned} \text{Maximize} \quad & P = 4x + 2y + 0s + 0t \\ \text{Subject to} \quad & 3x + y + s + 0t = 25 \\ & 2x + 4y + 0s + t = 35 \\ & x, y, s, t \ge 0 \end{aligned}$$

Notice that we used two slack variables because we had two constraints. Had we started with four constraints, we would have needed four slack variables, etc. □

Practice Problem

Convert the following linear programming problem to standard form.

$$\begin{aligned} \text{Maximize} \quad & P = 3x + 5y \\ \text{Subject to} \quad & 5x + 8y \le 40 \\ & x + 3y \le 12 \\ & x, y \ge 0 \end{aligned}$$

Answer:

$$\begin{aligned} \text{Maximize} \quad & P = 3x + 5y + 0s + 0t \\ \text{Subject to} \quad & 5x + 8y + s + 0t = 40 \\ & x + 3y + 0s + t = 12 \\ & x, y, s, t \ge 0 \end{aligned}$$

Corner Points and Basic Feasible Solutions

In Chapter 7 we stated that if a linear programming problem had a meaningful (finite) solution, then the optimal value of the objective function will occur at a corner point of the feasible region. Let's see how to represent corner points algebraically.

You learned in Chapter 6 that systems of equations with more variables than equations had an infinite number of solutions (were dependent) or had no solution. Consider the problem of Example 1 in standard form. It has four variables (x, y, s, and t) and only two equations, so that it has either an infinite number of solutions or no solution. Let's assume it does indeed have an infinite number of solutions. (Our procedure will detect the case when no solution exists.) We can write these solutions

$$s = 25 - 3x - y,$$
$$t = 35 - 2x - 4y,$$

by doing a little simple algebra. Thus, once we know the values of x and y, we can determine the values of s and t. Now, suppose we make both x and y equal to zero. Then s must be 25 and t must be 35. In fact, if we set any two of the variables equal to zero, then we'll have two variables remaining and two equations, which we can solve for the unique values of the remaining variables. The variables set to zero are called **nonbasic variables,** while the remaining variables are called **basic variables.** Table 8.1 shows the different basic solutions for the constraints of Example 1. We can generate these solutions by choosing any two of the variables to be nonbasic (zero) and then solving for the values of the remaining two basic variables.

Let's generalize this. Suppose we have m equations and n variables, where $n > m$. If we make $(n - m)$ of the variables nonbasic (equal to zero), then we can solve the m equations for the values of the m basic variables (unless the resulting system is inconsistent or dependent). This is called a **basic solution.** Of course, there is no reason to expect that the values of the basic variables will always be nonnegative.

TABLE 8.1 Basic solutions to the system of equations

$$3x + y + s + 0t = 25$$
$$2x + 4y + 0s + t = 35$$

Nonbasic variables	Value of basic variables
(s, t)	$x = 6.5$, $y = 5.5$
(s, x)	$t = -65$, $y = 25$
(s, y)	$t = 18.33$, $x = 8.33$
(t, x)	$s = 16.25$, $y = 8.75$
(t, y)	$s = -27.5$, $x = 17.5$
(x, y)	$s = 25$, $t = 35$

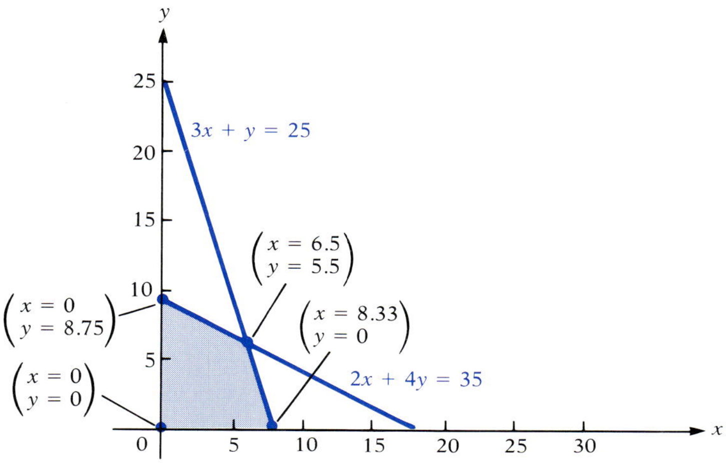

Figure 8.1 **There are four corner points in the feasible region for the constraints**

$$\begin{aligned} 3x + y &\le 25 \\ 2x + 4y &\le 35 \\ x, y &\ge 0 \end{aligned}$$

In a linear programming problem, we are interested in solutions in which the variables are nonnegative. Basic solutions with this property are called **basic feasible solutions.** Now, here's the reason for all these terms:

> In a linear programming problem, the corner points of the feasible region are basic feasible solutions of the constraints in standard form.

In Table 8.1 there are six basic solutions, but only the four solutions in color are basic feasible solutions. In other words, there are only four corner points in the feasible region. (See Fig. 8.1.)

Developing the Simplex Method

Consider the following linear programming problem:

$$\begin{aligned} \text{Maximize} \quad & P = 3x + 5y \\ \text{Subject to} \quad & 5x + 8y \le 40 \\ & x + 2y \le 9 \\ & x, y \ge 0 \end{aligned}$$

We'll use this problem to illustrate the simplex method, which is an efficient way of examining the corner points of the feasible region. In brief, we'll look at a corner point (basic feasible solution) and determine whether it's the best possible (optimal) solution. If it isn't, we move to a new corner point and then determine whether it is optimal. The method stops when an optimal solution is found.

First we must convert the problem to standard form. Since we have two constraints, we'll need two slack variables. The converted problem is

$$\begin{aligned} \text{Maximize} \quad & P = 3x + 5y + 0s + 0t \\ \text{Subject to} \quad & 5x + 8y + s + 0t = 40 \\ & x + 2y + 0s + t = 9 \\ & x, y, s, t \geq 0 \end{aligned}$$

Next we must determine an initial basic feasible solution. Since there are two constraints and four variables, we will have 4 − 2 or two nonbasic variables (which are set to zero). We must also identify two variables to make up the basic feasible solution. The procedure we shall use is as follows.

> Whenever the original problem contains all ≤ inequalities, choose the initial basic feasible solution to consist of the slack variables (assuming the right-hand sides of the inequalities are nonnegative).

In this case, we use the slack variables s and t. Notice that finding their values is trivial. Because both x and y are nonbasic, they must both be equal to zero. Thus we must have $s = 40$ and $t = 9$.

We normally organize the problem data into a table called the **simplex tableau.** As you can see from the portion of the initial simplex tableau shown in Fig. 8.2, the simplex tableau is an augmented matrix. Our choice of the initial set of basic variables, called the **basis,** is shown in the rightmost column. The column to the right of the vertical line, labeled RHS, contains the values of the basic variables.

At this point we've identified an initial basic feasible solution and copied the problem data into an augmented matrix called a simplex tableau. However, we don't know whether our current solution is optimal. To determine this we must look at the objective function. To be consistent with the way we've written the constraints, we should write the objective function as an equation with all the variables on the left and the value of P on the right. We have

$$-3x - 5y - 0s - 0t + P = 0.$$

(Because both x and y are zero in our initial solution, the value of P must be zero.) We enter this row on the bottom of the simplex tableau, as shown in Fig. 8.3.

Figure 8.2 Part of the initial simplex tableau for the maximization example.

x	y	s	t	RHS	Basis
5	8	1	0	40	s
1	2	0	1	9	t

Figure 8.3 The complete initial simplex tableau for the maximization example.

x	y	s	t	RHS	Basis
5	8	1	0	40	s
1	2	0	1	9	t
−3	−5	0	0	0	P

Now we can determine whether it's possible to improve the value of the objective function. We know that variables s and t are currently basic because they appear in the basis column. Thus x and y are the only nonbasic variables. Since both x and y have negative values in the P row, we can improve the value of the objective function by making either x or y positive. (Take a look at the P row, and let x be 1. Then we must make $P = 3$.) Since the coefficient of y (-5) is the most negative, y should be increased. But only basic variables can be positive. Thus, the only way we can increase y is to make it basic. In other words, the variable to be increased (in this case y) "enters" (becomes a member of) the basis. This gives us the following rule.

Rule for Choosing Entering Variable (Maximizing)

The variable to enter the basis (the variable to be increased) is the one with the most negative coefficient on the bottom row.

For y to become positive (basic), either s or t must become zero (nonbasic), because we must always have two basic and two nonbasic variables for this problem. How do we know which of the variables s or t will become zero? The answer is based upon the fact that *each row within the main body of the simplex tableau is an equation.* Thus the first row tells us

$$5x + 8y + s + 0t = 40,$$

or

$$s = 40 - 5x - 8y.$$

Therefore, for every unit y increases, s must decrease by eight units. (The coefficients of the variables in the body of the tableau are sometimes called **substitution coefficients** because they tell us how many units of the basic variable we must give up to substitute one unit of a nonbasic variable.) Similarly, the second row tells

$$x + 2y + 0s + t = 9,$$

or

$$t = 9 - x - 2y$$

so that for every unit y increases, t must decrease by two units. Both s and t must remain nonnegative, so there is a limit to how much we can increase y. Based on the first equation, y cannot exceed $40 \div 8 = 5$. Using the second equation, y cannot exceed $9 \div 2 = 4.5$. In order to satisfy the condition that both s and t are nonnegative, y cannot exceed the smaller of 5 and 4.5, which is 4.5.

We normally do these computations with the aid of a new column headed Ratio in the simplex tableau. The column of the entering variable is marked with an arrow and the word In. We compute the entries in the ratio column one row at a time. For each row, divide the number in the RHS column by the number (on the same row) in the column of the incoming variable, *unless the denominator would be zero or negative.* For the first row, we have

$$\text{ratio}_1 = \frac{40}{8} = 5$$

and for the second row we have

$$\text{ratio}_2 = \frac{9}{2} = 4.5.$$

Given that we have computed all the ratios, we can determine which variable must leave the basis, the *exiting* variable, as follows.

Rule for Choosing the Exiting Variable

The variable that must leave the basis is the one with the minimum ratio. (This rule holds in all cases, regardless of whether you are maximizing or minimizing.) If two or more variables have the same ratio, choose one variable arbitrarily.

Based on this rule, we choose the variable on the second row, t, to leave the basis. Thus we have chosen y to enter the basis and t to leave the basis. Mark the row of the exiting variable with an arrow and the word Out. (See Fig. 8.4.)

You should note that we only compute ratios for rows in which the denominator will be positive. The reason for this is we are concerned only with the possibility that a basic variable could become negative. For example, suppose some row in a simplex tableau had substitution coefficients of 4 and -2, with an RHS of 100. Then,

Figure 8.4 **Initial simplex tableau with ratio column and entering and exiting variables shown.**

x	y	s	t	RHS	Basis	Ratio
5	8	1	0	40	s	5
1	(2)	0	1	9	t	4.5 Out →
−3	−5	0	0	0	P	
	↑ In					

assuming that the basic variable on that row is s, the equivalent equation is

$$s = 100 - (4x - 2y) = 100 - 4x + 2y.$$

Now, if we had chosen y to be the entering variable, notice that increasing its value would *increase* the value of s. Since that can't possibly make s negative, we don't have to worry about this row. Similarly, if a substitution coefficient for the entering variable is zero on some row, then increasing the value of the entering variable won't change the value of the basic variable on that row at all. Thus, when a substitution coefficient is zero or negative, we don't need to compute a ratio for that row.

Now we can generate the new solution by doing a series of row operations called **pivoting.** The entering variable is y and the exiting variable is t. The column of the entering variable is called the **pivot column** and the row of the exiting variable is called the **pivot row.** The substitution coefficient in both the pivot column and the pivot row is called the **pivot element.** Circle the pivot element, as in Fig. 8.4.

To accomplish the pivot, do the following. First, multiply the pivot row by the inverse of the pivot element to obtain a 1. Here we multiply the old t row by 0.5. Next, use row operations to transform the rest of the elements in the pivot column into 0's. (See Fig. 8.5.)

Figure 8.5 **The first pivot operation, maximization example.**

Operation	x	y	s	t	RHS	Basis
	5	8	1	0	40	s
$0.5R_2$	0.5	1	0	0.5	4.5	y
	−3	−5	0	0	0	P
$R_1 + (-8)R_2$	1	0	1	−4	4	s
	0.5	1	0	0.5	4.5	y
	−3	−5	0	0	0	P
	1	0	1	−4	4	s
	0.5	1	0	0.5	4.5	y
$R_3 + 5R_2$	−0.5	0	0	2.5	22.5	P

Notice that we have included both the substitution coefficients and the value of RHS for each row. The new pivot row is the y row because y is replacing t in the basic solution. The new basis consists of variables s (on the first row) and y (on the second row).

Notice that the row operations we performed enable us to read the values of the basic values directly under the RHS column (remember the nonbasic variables are equal to zero). We always perform row operations when pivoting to maintain this property.

That's it! We simply repeat the steps of the simplex method until we can no longer improve the objective function. The current solution is not optimal because x has a negative coefficient on the P row. Thus x will be the new entering variable.

Next, compute the ratios to determine which basic variable should exit:

$$\text{ratio}_1 = \frac{4}{1} = 4, \qquad \text{ratio}_2 = \frac{4.5}{0.5} = 9.$$

The smaller ratio is the first one, which means that s will be removed from the basis and x will enter the basis. (See Fig. 8.6.)

Now we must perform the pivot operations to generate the new tableau. (See Fig. 8.7.)

Since all the coefficients on the P row are zero or positive, the current solution must be optimal. The optimal values of the variables are

$$x = 4, \qquad s = 0,$$
$$y = 2.5, \qquad t = 0,$$

and the optimal value of the objective function is \$24.50.

Because this problem has only two decision variables, we can also solve it graphically. The extreme points that corresponded to the tableaux generated by the simplex method were

First extreme point:	O	(the origin)
Second extreme point:	A	$(x = 0, y = 4.5, P = 22.5)$
Third extreme point:	B	$(x = 4, y = 2.5, P = 24.5)$

(See Fig. 8.8.)

Figure 8.6 Entering and exiting variables, second tableau (maximization example).

x	y	s	t	RHS	Basis	Ratio	
(1)	0	1	−4	4	s	4	Out →
0.5	1	0	0.5	4.5	y	9	
−0.5	0	0	2.5	22.5	P		
↑ In							

Figure 8.7 Second pivot operation, maximization example.

	x	y	s	t	RHS	Basis
	1	0	1	−4	4	x
$R_2 + (-0.5)R_1$	0	1	−0.5	2.5	2.5	y
	−0.5	0	0	2.5	22.5	P

	x	y	s	t	RHS	Basis
	1	0	1	−4	4	x
	0	1	−0.5	2.5	2.5	y
$R_3 + 0.5R_1$	0	0	0.5	0.5	24.5	P

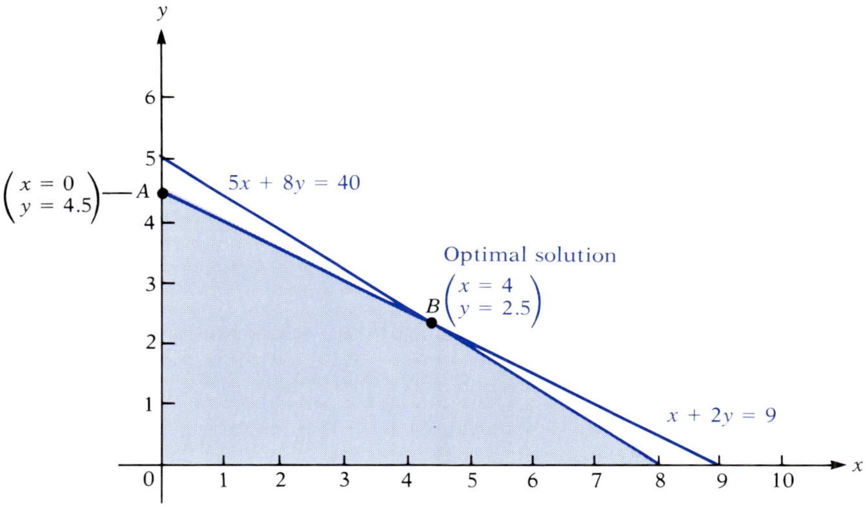

Figure 8.8 Graphical solution for maximization example.

The steps of the simplex method correspond to examining the corner points of the feasible region of a linear programming problem. In general terms, we can describe the method as follows, assuming the problem contains m rows and n variables.

The Simplex Method: Maximization

Step 1 State the linear programming problem in standard form.

Step 2 Select a set of m variables (where m = number of constraints *not* counting the nonnegativity constraints) that form an initial basis that is feasible. All other variables are nonbasic and are set equal to zero. Calculate the value of the objective function.

Step 3 Examine the bottom row to determine which variable(s) would improve the objective function. If no coefficients are negative, the current basic solution is optimal. Otherwise, the variable with the most negative coefficient is chosen to enter the basis.

Step 4 Select the current basic variable to leave the basis so that the nonnegativity constraints are not violated. Let x_* be the entering variable and let a_{i*} be the substitution coefficient of x_* on row i. Compute

$$\text{ratio}_i = \frac{b_i}{a_{i*}} \qquad \text{where } a_{i*} > 0.$$

The exiting variable is the basic variable on the row with the minimum ratio.

Step 5 Generate the new solution by pivoting, using row operations. The new (entering) variable was selected in Step 3; the exiting variable in Step 4. The new basic solution becomes the current solution. Set all nonbasic variables to zero and go to Step 3.

Another Example: The Dentist's Problem

Let's solve the problem of the dentist, introduced in Chapter 6 in the Problem from Life, using the simplex method.

EXAMPLE 2

Let p represent the number of pedodontic patients and g the number of general dentistry patients. Then, assuming the dentist earns a profit of \$8 for each pedodontic patient and \$6 for each general dentistry patient, the problem is

$$\begin{array}{lrl} \text{Maximize} & P = 8p + 6g & \\ \text{Subject to} & 0.25p + 0.50g \le 8 & \text{(dentist's time)} \\ & 1.50p + 1.00g \le 24 & \text{(assistant's time)} \\ & 0.75p + 0.75g \le 16 & \text{(operatory time)} \\ & p, g \ge 0 & \end{array}$$

(See also Example 26 in Chapter 6.)

SOLUTION

Step 1 We must convert the problem to standard form. Since we have three constraints, we'll need three slack variables. Let's use s, t, and u, respectively, in the three constraints. (The letters we use to represent the slack variables are not important as long as we remember which ones we did use.) The converted problem is

$$\begin{array}{ll} \text{Maximize} & P = 8p + 6g + 0s + 0t + 0u \\ \text{Subject to} & 0.25p + 0.50g + s + 0t + 0u = 8 \\ & 1.50p + g + 0s + t + 0u = 24 \\ & 0.75p + 0.75g + 0s + 0t + u = 16 \\ & p, g, s, t, u \geq 0 \end{array}$$

Step 2 Next we must select the intial basic feasible solution. Since there are three constraints, a basic feasible solution must contain three variables. As in the first example, we can use the slack variables s, t, and u. Notice that finding their values is trivial. Since p and g are nonbasic, they must both be equal to zero. By looking at the problem in standard form, we can see that s must be 8, t must be 24, and u must be 16. At this point, we can construct the first simplex tableau, as shown in Fig. 8.9. The value of this solution is zero, because p and g are both zero.

Step 3 We must examine the nonbasic variables (those not currently in the solution) to see if any can improve the value of the objective function. As before, we do this by looking at their coefficients in the P row. Since both p and g have negative coefficients, the current solution is not optimal. In addition, since p has the more negative coefficient (-8), it should enter the solution.

Step 4 Given that p should enter the solution, we must determine the exiting variable. Computing the ratios gives us

$$\text{ratio}_1 = \frac{8}{0.25} = 32,$$

$$\text{ratio}_2 = \frac{24}{1.5} = 16,$$

$$\text{ratio}_3 = \frac{16}{0.75} = 21.333.$$

The minimum ratio occurs on row two, so that slack variable t should leave the solution. The entering and exiting variables and the ratios are also shown in Fig. 8.9.

p	g	s	t	u	RHS	Basis	Ratio	
0.25	0.5	1	0	0	8	s	32	
(1.5)	1	0	1	0	24	t	16	Out →
0.75	0.75	0	0	1	16	u	21.333	
−8	−6	0	0	0	0	P		
↑ In								

Figure 8.9 **Initial simplex tableau for dentist's problem.**

	p	g	s	t	u	RHS	Basis
	0.25	0.5	1	0	0	8	s
$0.667R_2$	1	0.667	0	0.667	0	16	p
	0.75	0.75	0	0	1	16	u
	−8	−6	0	0	0	0	P

	p	g	s	t	u	RHS	Basis
$R_1 + (-0.25)R_2$	0	0.333	1	−0.167	0	4	s
	1	0.667	0	0.667	0	16	p
	0.75	0.75	0	0	1	16	u
	−8	−6	0	0	0	0	P

	p	g	s	t	u	RHS	Basis
	0	0.333	1	−0.167	0	4	s
	1	0.667	0	0.667	0	16	p
$R_3 + (-0.75)R_2$	0	0.25	0	−0.5	1	4	u
	−8	−6	0	0	0	0	P

	p	g	s	t	u	RHS	Basis
	0	0.333	1	−0.167	0	4	s
	1	0.667	0	0.667	0	16	p
	0	0.25	0	−0.5	1	4	u
$R_4 + 8R_2$	0	−0.664	0	5.336	0	128	P

Figure 8.10 **Initial pivot operation for dentist's problem.**

Step 5 Perform the row operations to generate the new solution. The entering variable is p and the exiting variable is t. (See Fig. 8.10.)

Notice that we followed the same procedure we did in our first example, even though this problem has three rows and not two. The size of the problem does not affect the calculations!

Because we just obtained a new solution, we must now return to Step 3.

Step 3 Determine if the current solution is optimal. Again, we must examine the coefficients of the P row. Since g has a negative coefficient, the current solution is not optimal; g should enter the solution. Notice that the value of P has increased to 128. This means the current solution generates a profit of \$128.

Step 4 We must determine the exiting variable by computing the ratios. We have

$$\text{ratio}_1 = \frac{4}{0.333} = 12.01, \qquad \text{ratio}_2 = \frac{16}{0.667} = 23.988, \qquad \text{ratio}_3 = \frac{4}{0.25} = 16.$$

p	g	s	t	u	RHS	Basis	Ratio	
0	(0.333)	1	−0.167	0	4	s	12.01	Out →
1	0.667	0	0.667	0	16	p	23.988	
0	0.25	0	−0.5	1	4	u	16	
0	−0.664	0	5.336	0	128	P		
	↑ In							

Figure 8.11 **Second simplex tableau, dentist's problem, entering and exiting variables.**

The minimum ratio is ratio$_1$, so variable s should leave the solution. The entering and exiting variables and the ratios are shown in Fig. 8.11.

Step 5 Generate the new solution. The entering variable is g and the exiting variable is s. (See Fig. 8.12.)

	p	g	s	t	u	RHS	Basis
$3.003R_1$	0	1	3.003	−0.502	0	12.01	g
	1	0.667	0	0.667	0	16	p
	0	0.25	0	−0.5	1	4	u
	0	−0.664	0	5.336	0	128	P
	0	1	3.003	−0.502	0	12.01	g
$R_2 + (-0.664)R_1$	1	0	−2.002	1.002	0	7.989	p
	0	0.25	0	−0.5	1	4	u
	0	−0.664	0	5.336	0	128	P
	0	1	3.003	−0.502	0	12.01	g
	1	0	−2.002	1.002	0	2.989	p
$R_3 + (-0.25)R_1$	0	0	−0.751	−0.374	1	0.998	u
	0	−0.664	0	5.336	0	128	P
	0	1	3.003	−0.502	0	12.01	g
	1	0	−2.002	1.002	0	7.989	p
	0	0	−0.751	−0.374	1	0.998	u
$R_4 + 0.664R_1$	0	0	1.994	5.003	0	135.975	P

Figure 8.12 **Second pivot operation, dentist's problem.**

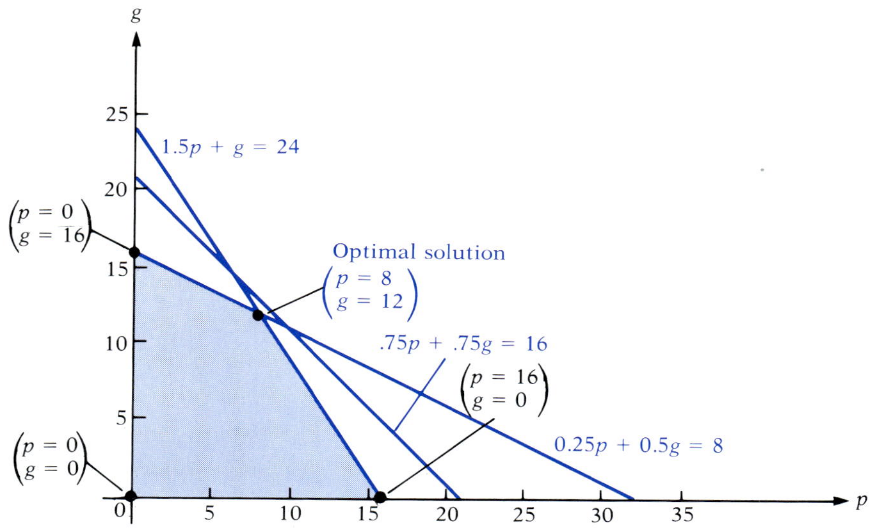

Figure 8.13 **Graphical solution to the dentist's problem.**

As before, since we generated a new solution, we go from Step 5 to Step 3.

Step 3 Determine if the current solution is optimal by checking the coefficients on the P row. Since all coefficients are nonnegative, the current solution is optimal. The optimal values of the variables are (rounding off to the nearest integer)

$$g = 12, \quad p = 8, \quad s = 0, \quad t = 0, \quad u = 1,$$

and the optimal value of the objective function is \$136. Compare this to the graphical solution shown in Fig. 8.13. □

EXERCISE SET 8.1

Solve the following linear programming problems using the simplex method.

1. Maximize $P = 4x + 4y$
Subject to
$$2x + y \le 10$$
$$x + 2y \le 8$$
$$x, y \ge 0$$

2. Maximize $P = 2x + 3y$
Subject to
$$3x + 6y \le 24$$
$$6x + 3y \le 30$$
$$x, y \ge 0$$

3. Maximize $P = 2x + y$
Subject to
$$8x + 4y \le 32$$
$$5x + 8y \le 40$$
$$x, y \ge 0$$

4. Maximize $P = 5x + 3y$
Subject to
$$3x + 6y \le 18$$
$$8x + 4y \le 32$$
$$x, y \ge 0$$

5. Maximize $P = 15x + 20y$
Subject to
$$\begin{aligned} 4x + 6y &\le 36 \\ 9x + 5y &\le 45 \\ 8x + 6y &\le 48 \\ x, y &\ge 0 \end{aligned}$$

6. Maximize $P = 12x + 8y$
Subject to
$$\begin{aligned} x &\le 700 \\ y &\le 900 \\ 3x + y &\le 2400 \\ x + 2y &\le 1600 \\ x, y &\ge 0 \end{aligned}$$

7. Maximize $P = 3x + y$
Subject to
$$\begin{aligned} 3x + 6y &\le 30 \\ 3x - 3y &\le 9 \\ x &\le 8 \\ x, y &\ge 0 \end{aligned}$$

8. Maximize $P = 3x - 2y$
Subject to
$$\begin{aligned} 6x + 3y &\le 24 \\ x - y &\le 2 \\ y &\le 4 \\ x, y &\ge 0 \end{aligned}$$

Formulate the following as linear programming problems and solve using the simplex method.

9. A manufacturing company makes two types of skis, the downhill and the cross-country. The necessary data are shown in the following table. How many pairs of each type of ski should the company make so as to maximize profit?

Department	Labor hours per pair: Downhill	Labor hours per pair: Cross-country	Maximum labor hours available per week
Cutting	6	5	900
Finishing	2	2	600
Profit per pair	\$65	\$55	

10. An investor has \$100,000 to invest. Option A offers a return of 12% per year and has a risk factor of 20. Option B offers a return of 15% per year and has a risk factor of 25. The investor has also decided the amount invested in option B must be no more than \$50,000. How much should the investor invest in each option if she wants to maximize return while limiting the average risk of her investments to no more than 22?

11. Phyllis Wentworth operates a local shop that sells coffee. She has decided to introduce two new blends: deluxe and ne plus ultra. Each pound of deluxe contains 30% Blue Mountain and 20% Kona, while each pound of ne plus ultra contains 25% Blue Mountain and 25% Kona. (The remainder of each blend is made of cheaper coffees, which are in plentiful supply.) Given the cost and other relevant data shown in the table, how much of each blend should Phyllis sell to maximize her weekly profit?

Available supply: Blue Mountain	Available supply: Kona	Profit per pound: Deluxe	Profit per pound: Ne Plus Ultra
275 lb	200 lb	\$0.95	\$1.00

8.2

MAXIMIZATION WITH ≤ AND ≥ CONSTRAINTS

Standard Form for ≥ Constraints

In Section 8.1 you learned to convert ≤ constraints to standard form by adding a slack variable. Now we'll discuss the procedure for dealing with ≥ constraints.

Suppose we are given the constraint

$$4x + 3y + 2z \geq 15.$$

To convert this to an equation, we subtract a **surplus variable,** s, from the left-hand side, which gives us

$$4x + 3y + 2z - s = 15, \qquad \text{where } s \geq 0.$$

Here the variable s is called a surplus variable because it measures the amount (surplus) by which the left-hand side exceeds the right-hand side. As with slack variables, a surplus variable has no impact on the objective function so its objective-function coefficient is zero.

EXAMPLE 3

Convert the following linear programming problem to standard form.

$$\begin{aligned} \text{Maximize} \quad & P = 4x + 3y \\ \text{Subject to} \quad & -x + 2y \geq 4 \\ & 6x - y \geq 12 \\ & x, y \geq 0 \end{aligned}$$

SOLUTION

To convert this problem to standard form, we must subtract a surplus variable, s, from the first constraint and another surplus variable, t, from the second constraint. This gives us

$$\begin{aligned} \text{Maximize} \quad & P = 4x + 3y + 0s + 0t \\ \text{Subject to} \quad & -x + 2y - s + 0t = 4 \\ & 6x - y + 0s - t = 12 \\ & x, y, s, t \geq 0 \end{aligned}$$ □

EXAMPLE 4

Convert the following linear programming problem to standard form.

$$\begin{aligned} \text{Maximize} \quad & P = 3x + 2y \\ \text{Subject to} \quad & 5x + 8y \leq 59 \\ & x - y \geq 4 \\ & x, y \geq 0 \end{aligned}$$

SOLUTION

To convert this to standard form, we must add a slack variable, s, to the first constraint and subtract a surplus variable, t, from the second constraint. The problem in standard form is

$$\begin{aligned} \text{Maximize} \quad & P = 3x + 2y + 0s + 0t \\ \text{Subject to} \quad & 5x + 8y + s + 0t = 59 \\ & x - y + 0s - t = 4 \\ & x, y, s, t \geq 0 \end{aligned}$$ □

Practice Problem

Convert the following problem to standard form.

$$\begin{aligned} \text{Maximize} \quad & P = 7x + 3y \\ \text{Subject to} \quad & 8x + 12y \leq 96 \\ & 3x + 2y \geq 12 \\ & x, y \geq 0 \end{aligned}$$

Answer:

$$\begin{aligned} \text{Maximize} \quad & P = 7x + 3y + 0s + 0t \\ \text{Subject to} \quad & 8x + 12y + s + 0t = 96 \\ & 3x + 2y + 0s - t = 12 \\ & x, y, s, t \geq 0 \end{aligned}$$

A Simple Example

Let's solve the following linear programming problem.

$$\begin{aligned} \text{Maximize} \quad & P = 4x + 5y \\ \text{Subject to} \quad & 4x + 2y \leq 16 \\ & -x + y \geq 4 \\ & x, y \geq 0 \end{aligned}$$

Step 1 Convert the problem to standard form. Adding the slack variable, s, to the first inequality and subtracting a surplus variable, t, from the second inequality results in

$$\begin{aligned} \text{Maximize} \quad & P = 4x + 5y + 0s + 0t \\ \text{Subject to} \quad & 4x + 2y + s + 0t = 16 \\ & -x + y + 0s - t = 4 \\ & x, y, s, t \geq 0 \end{aligned}$$

Step 2 Our next task is to select an initial basic feasible solution. We can use the slack variable, s, for the first equation (since letting $s = 16$ and $x = y = 0$ will satisfy it), but what do we do for the second equation? We can't use t, because its value would be -4, and we can't use x or y, because we want them to be nonbasic (and therefore zero). The answer is that we introduce a new variable, a, called an **artificial variable** into the second equation. The new second equation is

$$-x + y + 0s - t + a = 4, \qquad \text{where } a \geq 0.$$

This will work, since we can let $a = 4$. Thus our initial basic feasible solution will contain s (with a value of 16) and a (with a value of 4).

Why can we pull a variable out of thin air (so to speak) and just arbitrarily put it into an equation? Doesn't this change the problem? The answer is Yes. The variable a didn't appear in our original problem, so it had better not appear in the optimal solution. Our extra variable is an artifice (hence the name *artificial variable*) that we hope to be able to use just to get the simplex method going (like pulling ourselves up by our bootstraps). Later on, we must make sure that a becomes zero. We can do this by giving it a very large negative objective-function coefficient. This means that any solution in which a is positive will be very undesirable with respect to the objective function and therefore the simplex method itself will force a to become zero during the course of the solution procedure.

To summarize: Consider a linear programming problem in standard form. Add an artificial variable to each constraint that contains a surplus variable. The initial basic feasible solution will consist of any slack variables and the artificial variables. Give each artificial variable a very large negative objective-function coefficient (since we are maximizing).

Returning to our example, the revised problem becomes

$$\begin{array}{ll} \text{Maximize} & P = 4x + 5y + 0s + 0t - 1000a \\ \text{Subject to} & 4x + 2y + s + 0t + 0a = 16 \\ & -x + y + 0s - t + a = 4 \\ & \qquad\qquad x, y, s, t, a \geq 0 \end{array}$$

Before we can write down the initial simplex tableau, we must perform some algebra. We begin with the equation

$$-4x - 5y + 0s + 0t + 1000a + P = 0.$$

However, we want the variable a to have a coefficient of 0, because all basic variables have such coefficients, and a will be part of the initial basis. Thus we can use the equation in which a appears,

$$-x + y + 0s - t + a = 4,$$

and subtract 1000 times this equation from the P equation:

$$\begin{array}{r} -4x - 5y + 0s + 0t + 1000a + P = 0, \\ -1000(-x + y + 0s - t + a \qquad = 4), \end{array}$$

x	y	s	t	a	RHS	Basis	Ratio	
4	2	1	0	0	16	s	8	
−1	(1)	0	−1	1	4	a	4	Out →
996	−1005	0	1000	0	−4000	P		
	↑ In							

Figure 8.14 **Initial simplex tableau, example of Section 8.2.**

which yields

$$996x - 1005y + 0s + 1000t + 0a + P = -4000.$$

The new P equation can now be used in the initial tableau. The value of P, -4000, simply means we have a solution in which an artificial variable is positive. (See Fig. 8.14.)

Step 3 To see which of the nonbasic variables can improve the objective function we must evaluate them by examining their coefficients on the P row. Since y has a negative coefficient, the current solution is not optimal and y should enter the solution.

Step 4 To choose the current basic variable to exit the basis, we compute the ratios:

$$\text{ratio}_1 = \frac{16}{2} = 8, \qquad \text{ratio}_2 = \frac{4}{1} = 4.$$

Ratio$_2$ is the smaller, so a exits from the basis. (See Fig. 8.14.)

Step 5 Generate the new solution by performing the appropriate series of row operations. The entering variable is y and the exiting variable is a, so the pivot element is 1. The pivot operation is shown in Fig. 8.15. After performing the pivot, we return to Step 3.

	x	y	s	t	a	RHS	Basis
$R_1 + (-2)R_2$	6	0	1	2	−2	8	s
	−1	1	0	−1	1	4	y
	996	−1005	0	1000	0	−4000	P
	6	0	1	2	−2	8	s
	−1	1	0	−1	1	4	y
$R_3 + 1005R_2$	−9	0	0	−5	1005	20	P

Figure 8.15 **First pivot operation, maximization example with ≥ constraint.**

x	y	s	t	a	RHS	Basis	Ratio	
(6)	0	1	2	−2	8	s	1.333	Out →
−1	1	0	−1	1	4	y	—	
−9	0	0	−5	1005	20	P		
↑ In								

Figure 8.16 **Second simplex tableau with entering and exiting variables, maximization example with ≥ constraint.**

Step 3 Once again, we must evaluate the nonbasic variables by examining the P row. Both x and t have negative coefficients, so the current solution is not optimal. Since the coefficient of x (-9) is more negative than the coefficient of t (-5), variable x is chosen to enter the solution.

Step 4 Now we must choose the variable to exit from the basis, by computing the ratios. Since the coefficient in the x column on the second row is negative, only one ratio can be computed:

$$\text{ratio}_1 = \frac{8}{6} = 1.333.$$

Thus the exiting variable is s. (See Fig. 8.16.)

Step 5 The row operations to generate the new solution must be performed. Variable x will enter and variable s will exit. The pivot element is 6. The pivot operation is shown in Fig. 8.17. As before, after completing the pivot, we return to Step 3.

	x	y	s	t	a	RHS	Basis
$0.167R_1$	(1)	0	0.167	0.334	−0.334	1.333	x
	−1	1	0	−1	1	4	y
	−9	0	0	1005	1005	20	P
	1	0	0.167	0.334	−0.334	1.333	x
$R_2 + R_1$	0	1	0.167	−0.666	0.666	5.333	y
	−9	0	0	−5	1005	20	P
	1	0	0.167	0.334	−0.334	1.333	x
	0	1	0.167	−0.666	0.666	5.333	y
$R_3 + 9R_1$	0	0	1.503	−1.994	1001.994	31.997	P

Figure 8.17 **Second pivot operation, maximization example with ≥ constraint.**

x	y	s	t	a	RHS	Basis	Ratio	
1	0	0.167	(0.334)	−0.334	1.333	x	3.991	Out ⟶
0	1	0.167	−0.666	0.666	5.333	y		
0	0	1.503	−1.994	1001.994	31.997	P		
			↑ In					

Figure 8.18 **Third simplex tableau with entering and exiting variables, maximization with ≥ constraint.**

Step 3 Studying the P row, we see that the current basic solution is not optimal because t has a negative coefficient. Thus t should enter the basis.

Step 4 Once again, we need compute only one ratio because a_{24} is negative. Thus x must be the exiting variable. (See Fig. 8.18.)

Step 5 The entering variable is t and the exiting variable is x. The pivot element is 0.334. The pivot operation is shown in Fig. 8.19. Again, after performing the pivot, we return to Step 3.

Step 3 Since all the coefficients in the P row are nonnegative, the current solution is optimal. Thus we have (rounding off to the nearest integer)

$$x = 0, \quad s = 0, \quad a = 0,$$
$$y = 8, \quad t = 4,$$

and the optimal value of the objective function is $40.

	x	y	s	t	a	RHS	Basis
$2.994R_1$	2.994	0	0.5	1	−1	3.991	t
	0	1	0.167	−0.666	0.666	5.333	y
	0	0	1.503	−1.994	1001.944	31.997	P
	2.994	0	0.5	1	−1	3.991	t
$R_2 + 0.666R_1$	1.997	1	0.5	0	0	7.991	y
	0	0	1.503	−1.994	1001.944	31.997	P
	2.994	0	0.5	1	−1	3.991	t
	1.997	1	0.5	0	0	7.991	y
$R_3 + 1.994R_1$	6	0	2.5	0	1000	39.955	P

Figure 8.19 **Third pivot operation, maximization example with ≥ constraint.**

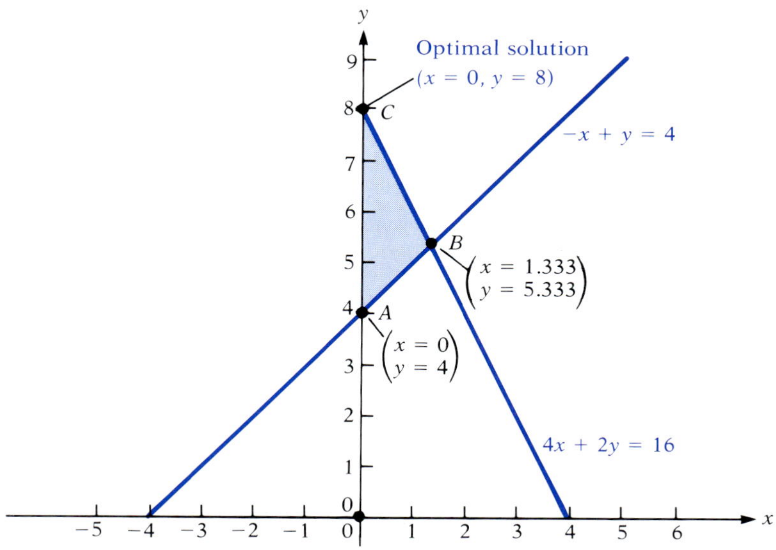

Figure 8.20 **Geometric solution of the maximization example with a ≥ constraint.**

You will find it useful to consider the geometric solution to this problem shown in Fig. 8.20. The extreme points that corresponded to the tableaux generated by the simplex method were (after rounding)

First extreme point:	O	(the origin)
Second extreme point:	A	$(x = 0, y = 4, P = 20)$
Third extreme point:	B	$(x = 1.333, y = 5.333, P = 32)$
Fourth extreme point:	C	$(x = 0, y = 8, P = 40)$

Note especially that the origin is *not* feasible in this problem. Recall that our initial solution had the artificial variable (a) equal to 4, and recall also that we stated the artificial variable had to be 0 for any solution to be feasible. This is true in general. Any solution to a linear programming problem in which one (or more) artificial variables is positive is infeasible with respect to the original statement of the problem. Thus we must always attempt to make all artificial variables 0 by giving them the appropriate objective function coefficients.

Dealing with Equality Constraints

Now that you've been introduced to artificial variables, you can solve maximization problems that include equality constraints. Study the following example.

EXAMPLE 5

Determine an initial basic feasible solution for the following linear programming problem.

$$\begin{aligned} \text{Maximize} \quad & P = 5x + 8y \\ \text{Subject to} \quad & x \leq 400 \\ & y \geq 200 \\ & x + y = 500 \\ & x, y \geq 0 \end{aligned}$$

SOLUTION

First, let's convert this to standard form. Use a slack variable (s) for the first constraint and a surplus variable (t) for the second. The third constraint is an equality, so we need not alter it. This results in

$$\begin{aligned} \text{Maximize} \quad & P = 5x + 8y + 0s + 0t \\ \text{Subject to} \quad & x + s + 0t = 400 \\ & y + 0s - t = 200 \\ & x + y + 0s + 0t = 500 \\ & x, y, s, t \geq 0 \end{aligned}$$

To begin the simplex method, we must identify an initial basic solution. We know that we can use the slack variable s from the first equation, and we know it is necessary to introduce an artificial variable into the second equation. What about the third equation? Since this equation has no slack variable either, we simply use another artificial variable! Letting the artificial variables be a and b, respectively, our problem becomes

$$\begin{aligned} \text{Maximize} \quad & P = 5x + 8y + 0s + 0t - 1000a - 1000b \\ \text{Subject to} \quad & x + s + 0t + 0a + 0b = 400 \\ & y + 0s - t + a + 0b = 200 \\ & x + y + 0s + 0t + 0a + b = 500 \\ & x, y, s, t, a, b \geq 0 \end{aligned}$$

The initial basic feasible solution for this problem is

$$s = 400, \qquad a = 200, \qquad b = 500. \quad \square$$

Practice Problem

Convert the following linear programming problem into a form that contains an initial basic feasible solution. Identify the variables in the initial basis and give their values.

$$\begin{aligned} \text{Maximize} \quad & P = 5x + 3y \\ \text{Subject to} \quad & 6x + 2y \leq 54 \\ & 5x + 5y \geq 40 \\ & 2x + 4y = 28 \\ & x, y \geq 0 \end{aligned}$$

Answer: The converted problem is

$$\begin{array}{ll} \text{Maximize} & P = 5x + 3y + 0s + 0t - 1000a - 1000b \\ \text{Subject to} & 6x + 2y + s + 0t + 0a + 0b = 54 \\ & 5x + 5y + 0s - t + a + 0b = 40 \\ & 2x + 4y + 0s + 0t + 0a + b = 28 \\ & x, y, s, t, a, b \geq 0 \end{array}$$

The initial basic feasible solution is

$$s = 54, \qquad a = 40, \qquad b = 28.$$

To summarize: When dealing with a constraint that involves $\geq$ or $=$, an artificial variable must be used to create an initial basis.

EXERCISE SET 8.2

Solve the following linear programming problems using the simplex method.

1. Maximize $P = x + y$
Subject to
$$\begin{aligned} x + y &\geq 1 \\ 5x + 10y &\leq 50 \\ y &\leq 4 \\ x, y &\geq 0 \end{aligned}$$

2. Maximize $P = 4x + 3y$
Subject to
$$\begin{aligned} x + y &\leq 10 \\ 2x + 2y &\geq 40 \\ x, y &\geq 0 \end{aligned}$$

3. Maximize $P = 2x + y$
Subject to
$$\begin{aligned} 11x + 3y &\geq 33 \\ 8x + 5y &\leq 40 \\ 7x + 10y &\leq 70 \\ x, y &\geq 0 \end{aligned}$$

4. Maximize $P = 5x + 10y$
Subject to
$$\begin{aligned} x + y &\geq 2 \\ 8x + 12y &\leq 72 \\ 4x + 8y &\leq 40 \\ x, y &\geq 0 \end{aligned}$$

5. Maximize $P = 3x + y$
Subject to
$$\begin{aligned} 2x - y &\geq 6 \\ 3x + 6y &\leq 30 \\ x &\leq 9 \\ x, y &\geq 0 \end{aligned}$$

6. Maximize $P = 20x + 35y$
Subject to
$$\begin{aligned} x + y &\geq 2 \\ 12x + 16y &\leq 144 \\ x - y &\geq 1 \\ x, y &\geq 0 \end{aligned}$$

7. Maximize $P = 3x - 2y$
Subject to
$$\begin{aligned} -x + y &\geq 3 \\ 6x + 4y &\geq 24 \\ 6x + 8y &\leq 48 \\ x, y &\geq 0 \end{aligned}$$

8. Maximize $P = 2x - y + z$
Subject to
$$\begin{aligned} x + y - 2z &\leq 8 \\ 4x - y + z &\geq 2 \\ 4x + 6y - 2z &\geq 8 \\ x, y, z &\geq 0 \end{aligned}$$

9. Maximize $P = 8x + 5y$
Subject to
$$\begin{aligned} x &\geq 300 \\ y &\leq 300 \\ x + y &= 450 \\ x, y &\geq 0 \end{aligned}$$

10. Maximize $P = 8x + 10y$
Subject to
$$\begin{aligned} -2x + 2y &\geq 8 \\ 2x + y &\leq 8 \\ x + y &= 6 \\ x, y &\geq 0 \end{aligned}$$

* 11. Surplus-to-You purchases surplus products of all descriptions which it then sells to the general public. It buys surplus fabric from two large textile mills, S and B. Surplus-to-You then sells the fabric through fabric stores, discount stores, and catalog mail-order houses. The fabric from the two mills is not marketed identically. Of the fabric from mill S, 40% is sold in fabric stores, 15% in discount stores, and 25% by mail. Of the fabric from mill B, 30% is sold in fabric stores, 25% in discount stores, and 25% by mail. Regardless of the source, 20% of the fabric is unusable. Analysis has revealed that the company makes \$6 profit for every bolt of fabric purchased from mill S and \$8 profit for every bolt purchased from mill B. Sales forecasts indicate that at most 6000 bolts can be sold in fabric stores, 5000 bolts in discount stores, and 5000 bolts through direct mail within the next year. Given that Surplus-to-You has contracted to buy at least 1500 bolts from mill S, how many bolts of fabric should it buy from each of the mills in order to maximize profit?

12. The management of Awesome Autos must determine its advertising media selections for the next quarter. The available budget is \$10,000, which can be used to buy radio ads for \$250 each or television ads for \$525 each. Each radio ad reaches 5000 people, with an estimated 50% of the total being 18 years of age or below. Each television ad reaches 8000 people, with an estimated 40% of the total being 18 years of age or below. Management wishes to maximize the number of listeners older than 18 years, given the budget restriction and the fact that it also wishes to reach at least 150,000 total people. How many ads of each type should the management of Awesome Autos purchase?

8.3 MINIMIZATION

We've discussed how to solve maximization problems. Here we will show you how to deal with the other type of linear programming problem: the minimization problem. In fact, you already know almost everything necessary.

The simplest way to deal with minimization problems is to convert them to maximization problems by modifying the objective function. All other aspects of the problem remain the same. Thus the steps of the simplex method for minimization problems are exactly as we showed earlier with the single exception of Step 1. Here is the

* Reprinted with permission from Wu and Coppins, *Linear Programming and Extensions* (McGraw-Hill, New York), 1981.

new version of Step 1:

> **Step 1** State the problem in standard form. If the problem is one of maximization, simply proceed. If the problem specifies minimization, multiply *all* objective function coefficients by -1.

The only difference between solving maximization and minimization problems is that we multiply the objective function of minimization problems by -1. *All the other steps are exactly the same.*

This procedure works because when we multiply the objective function of a minimization problem by -1, we convert it to a maximization problem! Although we won't prove it, we are relying on a theorem that states the following. Let problem P be

$$\begin{array}{ll} \text{Minimize} & f = c_1x_1 + c_2x_2 + \cdots + c_nx_n \\ \text{Subject to} & \text{a given set of linear constraints} \end{array}$$

and let problem P1 be

$$\begin{array}{ll} \text{Maximize} & (-f) = -c_1x_1 - c_2x_2 - \cdots - c_nx_n \\ \text{Subject to} & \text{a given set of linear constraints} \end{array}$$

(where both problem P and P1 have the same set of constraints). Then problems P and P1 have exactly the same optimal solution. (In other words, problems P and P1 are optimized by the same extreme point.) However, since we multiplied the objective function of problem P by -1 to get the objective function of problem P1, the values of the optimal solutions will be the negative of each other.

EXAMPLE 6

Convert the following linear programming problems to maximization problems in standard form:

a) $$\begin{array}{ll} \text{Minimize} & C = 3x + 2y \\ \text{Subject to} & x + y \geq 12 \\ & 3x + 4y \geq 36 \\ & x, y \geq 0 \end{array}$$

b) $$\begin{array}{ll} \text{Minimize} & C = 2x + 5y \\ \text{Subject to} & 5x + 6y \geq 30 \\ & 20x + 45y \leq 90 \\ & x, y \geq 0 \end{array}$$

SOLUTION

All we need do is multiply each objective function coefficient by -1. *We must not alter the constraints.* The converted problems are

a) Maximize $-C = -3x - 2y + 0s + 0t$
Subject to
$$x + y - s + 0t = 12$$
$$3x + 4y + 0s - t = 36$$
$$x, y, s, t \geq 0$$

b) Maximize $-C = -2x - 5y + 0s + 0t$
Subject to
$$5x + 6y - s + 0t = 30$$
$$20x + 45y + 0s + t = 90$$
$$x, y, s, t \geq 0 \quad \square$$

Practice Problem

Convert the following linear programming problems to maximization problems in standard form.

a) Minimize $C = x + y$
Subject to
$$x - y \geq 2$$
$$2x + 3y \geq 12$$
$$x, y \geq 0$$

b) Minimize $C = 3x + 9y$
Subject to
$$-x + 2y \leq 5$$
$$4x + 3y \geq 12$$
$$x, y \geq 0$$

Answer: **a)** Maximize $-C = -x - y + 0s + 0t$
Subject to
$$x - y - s + 0t = 2$$
$$2x + 3y + 0s - t = 12$$
$$x, y, s, t \geq 0$$

b) Maximize $-C = -3x - 9y + 0s + 0t$
Subject to
$$-x + 2y + s + 0t = 5$$
$$4x + 3y + 0s - t = 12$$
$$x, y, s, t \geq 0$$

Let's solve one minimization problem just to prove that only the objective function need be altered. Consider the Problem from Life that opened this chapter. We can model this problem as follows. Let

c = number of families with children to contact,
n = number of families without children to contact.

Then we wish to

Minimize $C = 5c + 4n$
Subject to
$$c + n \geq 100$$
$$c \geq 40$$
$$n \geq 25$$
$$c, n \geq 0$$

Step 1 Stating the problem in standard form and converting the objective function to maximization gives us

$$\begin{aligned}
\text{Maximize} \quad & -C = -5c - 4n + 0s + 0t + 0u \\
\text{Subject to} \quad & c + n - s + 0t + 0u = 100 \\
& c + 0s - t + 0u = 40 \\
& n + 0s + 0t - u = 25 \\
& c, n, s, t, u \geq 0
\end{aligned}$$

The variables s, t, and u are surplus variables.

Step 2 We must select an initial basic feasible solution. Since there are no slack variables present, we must introduce an artificial variable into each constraint. Let's use w, x, and y to denote these new variables. Since we are now maximizing, we give each of the variables w, x, and y an objective function coefficient of -1000. Including the artificial variables, the problem is

$$\begin{aligned}
\text{Maximize} \quad & -C = -5c - 4n + 0s + 0t + 0u - 1000w - 1000x - 1000y \\
\text{Subject to} \quad & c + n - s + 0t + 0u + w + 0x + 0y = 100 \\
& c + 0s - t + 0u + 0w + x + 0y = 40 \\
& n + 0s + 0t - u + 0w + 0x + y = 25 \\
& c, n, s, t, u, w, x, y \geq 0
\end{aligned}$$

Because there are artificial variables present, we must perform some algebra on the C equation so that we can continue with the simplex method. Actually, we must perform three row operations in order to make the C row coefficient of *each* artificial variable zero. We begin with the equation

$$5c + 4n + 0s + 0t + 0u + 1000w + 1000x + 1000y + C = 0.$$

First, let's make the coefficient of w equal to zero. We multiply the equation in which w appears by -1000 and add it to the C equation:

$$\begin{aligned}
& 5c + 4n + 0s + 0t + 0u + 1000w + 1000x + 1000y + C = 0, \\
+ (-1000)(& c + n - s + 0t + 0u + w + 0x + 0y = 100),
\end{aligned}$$

results in

$$-995c - 996n + 1000s + 0t + 0u + 0w + 1000x + 1000y + C = -100{,}000,$$

Next, make the coefficient of x equal to zero. Multiply the equation containing x by -1000 and add it to the current C equation:

$$\begin{aligned}
& -995c - 996n + 1000s + 0t + 0u + 0w + 1000x + 1000y + C = -100{,}000, \\
+(-1000)(& c + 0s - t + 0u + 0w + x + 0y = 40),
\end{aligned}$$

which yields

$$-1995c - 996n + 1000s + 1000t + 0u + 0w + 0x + 1000y + C = -140{,}000.$$

c	n	s	t	u	w	x	y	RHS	Basis	Ratio	
1	1	−1	0	0	1	0	0	100	w	100	
1	0	0	−1	0	0	1	0	40	x	—	
0	(1)	0	0	−1	0	0	1	25	y	25	Out →
−1995	−1996	1000	1000	1000	0	0	0	−165,000	C		
	↑ In										

Figure 8.21 **The initial tableau, Sally Green's problem.**

Finally, make the coefficient of y equal to zero. Multiply the equation containing y by -1000 and add it to the current C row:

$$\begin{aligned} &-1995c - 996n + 1000s + 1000t + 0u + 0w + 0x + 1000y + C = -140{,}000,\\ &\quad +(-1000)(n + \quad 0s + \quad 0t - \quad u + 0w + 0x + \qquad y \qquad = 25), \end{aligned}$$

which results in

$$-1995c - 1996n + 1000s + 1000t + 1000u + 0w + 0x + 0y + C = -165{,}000.$$

Now all the coefficients of the artificial variables have been converted to zero, so we can insert the C row into the initial tableau. (See Fig. 8.21.)

As you can see from Fig. 8.21, this problem requires the use of six extra variables: three surplus variables and three artificial variables. However, this presents no new difficulties; it simply means that each tableau will be a bit larger than those we have worked with previously.

Since there is nothing new from this point on, we'll simply show the remaining tableaux required to solve the problem. (See Figs. 8.22–8.24.) As you can see, the optimal solution has $c = 40$ and $n = 25$, with $C = 440$. In practical terms, this means that Sally's interviewers should interview 40 families with children and 25 families without children, at a projected cost of \$440.

c	n	s	t	u	w	x	y	RHS	Basis	Ratio	
1	0	−1	0	1	1	0	−1	75	w	75	
(1)	0	0	−1	0	0	1	0	40	x	40	Out →
0	1	0	0	−1	0	0	1	25	n	—	
−1995	0	1000	1000	−996	0	0	1996	−115,100	C		
↑ In											

Figure 8.22 **Second tableau, Sally Green's problem.**

c	n	s	t	u	w	x	y	RHS	Basis	Ratio	
0	0	−1	1	(1)	1	−1	−1	35	w	35	Out →
1	0	0	−1	0	0	1	0	40	c	—	
0	1	0	0	−1	0	0	1	25	n	—	
0	0	1000	−995	−996	0	1995	1996	−35,300	C		
				↑ In							

Figure 8.23 **Third tableau, Sally Green's problem.**

c	n	s	t	u	w	x	y	RHS	Basis
0	0	−1	1	1	1	−1	−1	35	u
1	0	0	−1	0	0	1	0	40	c
0	1	−1	1	0	1	−1	0	25	n
0	0	4	1	0	996	999	1000	−440	C

Figure 8.24 **Fourth tableau, Sally Green's problem.**

EXERCISE SET 8.3

Solve the following linear programming problems using the simplex method.

1. Minimize $C = 8x + 12y$
Subject to $3x + 2y \geq 12$
$2x + 3y \geq 12$
$x, y \geq 0$

2. Minimize $C = 6x + 7y$
Subject to $5x + 4y \geq 20$
$4x + 7y \geq 28$
$x, y \geq 0$

3. Minimize $C = 8x + 12y$
Subject to $x + y \geq 4$
$4x + 2y \geq 8$
$3x + 5y \geq 15$
$x, y \geq 0$

4. Minimize $C = 5x + 10y$
Subject to $x + y \geq 2$
$8x + 12y \leq 72$
$4x + 8y \leq 40$
$x, y \geq 0$

5. Minimize $C = 15x + 25y$
Subject to $x - y \leq 1$
$4x + 6y \geq 12$
$6x + 3y \leq 18$
$x, y \geq 0$

6. Minimize $C = 14x + 21y$
Subject to $2x + 3y \leq 66$
$3x + 9y \geq 27$
$4x + 11y \geq 44$
$x, y \geq 0$

7. Minimize $C = 6x + 10y + 2z$
Subject to $2x + 2y + 2z \geq 40$
$2y + 4z \geq 20$
$x, y, z \geq 0$

8. Minimize $C = 3a + 4b + c + 6d$
Subject to
$$\begin{aligned} 5a - 2b + c - 3d &\geq 2 \\ 6a + b - 5c - 3d &\geq 5 \\ -a + 4b + 3c + 7d &\geq 6 \\ a, b, c, d &\geq 0 \end{aligned}$$

9. An investor has \$10,000 to invest in stocks. Stock A has a growth rate of 10% per year, pays dividends of 7% per year, and has a risk factor of 2.5. Stock B has a growth rate of 8% per year, pays dividends of 12% per year, and has a risk factor of 2.8. Stock C has a growth rate of 8% per year, pays dividends of 9% per year, and has a risk factor of 2. The investor has determined that her portfolio must have a composite growth rate of 8.5% per year and must pay at least 8.2% dividends per year. At the moment, stock A sells for \$100 per share, stock B sells for \$50 per share, and stock C sells for \$50 per share. How many shares of each stock should the investor purchase to minimize her risk?

10. A consulting firm is bidding on a contract. The firm must determine how many junior analysts and how many senior analysts it should use to minimize its personnel cost and still satisfy the job requirements. The relevant data are given in the table.

	Work-hours per analyst		Minimum work-hours required
	Senior analyst	Junior analyst	
Problem analysis	1200	800	10,000
Program design	700	800	7,500
Program test	100	400	4,000
Cost per analyst	\$55,000	\$35,000	

* 11. John Snell has decided to go into business for himself with a company called No-Fail Exterminators. He plans to manufacture a pesticide called No-roach. A variety of ingredients can be used to manufacture the pesticide. Because of federal regulations and supply limitations, there are restrictions on four ingredients as shown in the table:

Ingredient	Restriction
I	No more than 30 units/day
II	No more than 25 units/day
III	No more than 18 units/day
IV	No more than 38 units/day

The product can be manufactured by three different methods, as de-

* Reprinted with permission from Wu and Coppins, *Linear Programming and Extensions* (McGraw-Hill, New York), 1981.

scribed in the following table.

	Units of ingredients per gallon			
Method	**I**	**II**	**III**	**IV**
A	4	2	3	3
B	2	3	1	4
C	2	2	2	4
Cost per unit	\$2.00	\$3.00	\$1.50	\$1.75

Mr. Snell needs at least 10 gallons of the pesticide. How should he manufacture it so as to minimize his costs?

8.4 DUALITY

So far we have studied how to formulate and solve linear programming problems. Next we will study how a related problem, called the *dual,* can be associated with a linear programming problem and its economic interpretation. In addition, knowledge of the dual can be used to study the effects upon the optimal solution of changing the problem parameters. This area of study, called post-optimality analysis, is extremely important in any practical application of linear programming (although it is beyond the scope of this text).

The Furniture Problem of Ann and Fred Revisited

Let's consider once more the furniture manufacturing problem of Ann and Fred (first formulated and solved in Chapter 7). Ann and Fred can use oak, pine, and labor to manufacture tables and chairs. Their problem is to maximize their profits, given limited availabilities for the three inputs. In mathematical form, the problem is to

$$\begin{array}{lll} \text{Maximize} & P = 12t + 8c & \\ \text{Subject to} & 5t + 2c \le 122 & \text{(oak)} \\ & 2t + 3c \le 100 & \text{(pine)} \\ & 4t + 3c \le 120 & \text{(labor)} \\ & t, c \ge 0 & \end{array}$$

This formulation is called the **primal** problem because it is a direct representation of the given problem. The initial and final simplex tableaux are shown in Figs. 8.25 and 8.26. Variables q, r, and s are

	t	c	q	r	s	RHS	Basis	Ratio	
	(5)	2	1	0	0	122	q	24.4	Out →
	2	3	0	1	0	100	r	50	
	4	3	0	0	1	120	s	30	
	−12	−8	0	0	0	0	P		
	↑ In								

Figure 8.25 Initial simplex tableau, Ann and Fred's problem.

t	c	q	r	s	RHS	Basis
1	0	0.428	0	−0.286	18	t
0	0	0.856	1	−1.571	16	r
0	1	−0.571	0	0.714	16	c
0	0	0.573	0	2.285	344	P

Figure 8.26 Final simplex tableau, Ann and Fred's problem.

slack variables that represent unused oak, unused pine, and unused labor hours, respectively. (Check our calculations by solving the problem yourself.)

The optimal solution is to produce 18 tables and 16 chairs for a profit of $344. Note that 16 board-feet of pine are left over by this solution, while all of the oak and labor hours are used.

Suppose Ann and Fred were able to obtain more of the scarce resources oak and labor hours. A little thought should convince you that they would be able to make more profit because the additional resources would enable them to make more tables and chairs. Moreover, additional board-feet of pine would be useless because there is currently an excess (so that obtaining more pine would simply lead to a larger excess quantity of pine). The key consideration is how much additional profit would each board-foot of oak and each labor hour produce. These values are called **shadow prices** (or the **marginal values**) of oak and labor hours. We can determine these shadow prices by solving a related linear programming problem called the dual.

The **dual** of Ann and Fred's problem is formulated as follows. In order to produce tables and chairs, two types of wood and labor are required. Each of these resources has a shadow price (marginal value). Let

w = shadow price of 1 board-foot of oak,

x = shadow price of 1 board-foot of pine,

y = shadow price of 1 labor hour.

Then the **dual problem** is

$$\begin{aligned} \text{Minimize} \quad & C = 122w + 100x + 120y \\ \text{Subject to} \quad & 5w + 2x + 4y \geq 12 \\ & 2w + 3x + 3y \geq 8 \\ & w, x, y, \geq 0 \end{aligned}$$

Why? First, note that the objective function must give the total value of the scarce resources used, since the variables w, x, and y give the shadow price (value) of unit amounts and we started with 122 board-feet of oak, 100 board-feet of pine, and 120 labor hours. Next, we have one constraint for each product of the primal (original) problem. The first constraint states that the marginal value of the resources used to produce one table must be at least as large as the marginal profit gained from selling one table. (This derives from economic theory, which says produce units of output until the marginal cost of production exceeds the marginal profit. If you haven't studied this yet, just accept it for now.) Similarly, the second constraint states that the marginal value of the resources used to produce one chair must be at least as large as the marginal profit gained from selling one chair.

The initial and final simplex tableaux for the dual problem are shown in Figs. 8.27 and 8.28, respectively. We used surplus variables a and b and artificial variables c and d to define the initial basic feasible solution. The optimal solution is w = \$0.57, x = \$0, and y = \$2.28 with the optimal value of the objective function equal to \$344 (rounded). In other words, the shadow prices of oak, pine, and labor are \$0.57 per board-foot, \$0, and \$2.28 per hour, respectively.

w	x	y	a	b	c	d	RHS	Basis	Ratio	
5	2	4	−1	0	1	0	12	c	3	Out
2	3	(3)	0	−1	0	1	8	d	2.667	⟶
−6878	−4900	−6880 ↑ In	1000	1000	0	0	−20,000	C		

Figure 8.27 **Initial tableau, dual of Ann and Fred's problem.**

w	x	y	a	b	c	d	RHS	Basis
1	−0.858	0	−0.429	0.571	0.429	−0.571	0.571	w
0	1.572	1	0.286	−0.714	−0.286	0.714	2.286	y
0	16.004	0	18.002	16.002	981.998	983.998	−343.998	C

Figure 8.28 **Final tableau, dual of Ann and Fred's problem.**

This means that if one more board-foot of oak were available, Ann and Fred could make an additional \$0.57. (This would require changing the product mix. Without proof, the new solution would be to produce 18.428 tables and 15.429 chairs for a profit of \$348.568, an increase of approximately \$0.57.)

It is important for you to realize that the shadow prices of oak, pine, and labor in Ann and Fred's problem are determined strictly by the fact they can be used to make tables and chairs which can be sold for a profit. In other words, the shadow prices are determined by the problem parameters and *not* by the market costs of the resources. For example, if the profits for tables and chairs were \$10 and \$5, respectively, the shadow prices would be \$1.43 per board-foot for oak, \$0 for pine, and \$0.71 per hour for labor (which you can prove by re-solving the dual with the new right-hand sides).

Formulation of the dual

Now let's see how we can write the dual of a linear programming problem. To do so, we must write the primal in **canonical form.** For maximization problems, this means all constraints are $\leq$ relations, while for minimization problems, all constraints must be $\geq$ relations. (It is not necessary that all right-hand sides be nonnegative in canonical form.)

EXAMPLE 7

Convert the following linear programming problem to canonical form.

$$\begin{array}{ll} \text{Maximize} & P = 3x + 4y + 7z \\ \text{Subject to} & 4x - 12y + 3z \geq 18 \\ & 2x + y + z \leq 50 \\ & -14x + 25y + 10z \geq 25 \\ & x, y, z \geq 0 \end{array}$$

SOLUTION

We must convert the first and third constraints to $\leq$ form. This is done by multiplying them by -1, which gives us

$$\begin{array}{ll} \text{Maximize} & P = 3x + 4y + 7z \\ \text{Subject to} & -4x + 12y - 3z \leq -18 \\ & 2x + y + z \leq 50 \\ & 14x - 25y - 10z \leq -25 \\ & x, y, z \geq 0 \quad \square \end{array}$$

Practice Problem

Convert the following problem to canonical form.

$$\begin{array}{ll} \text{Minimize} & C = 4a + 3b \\ \text{Subject to} & 2a + 4b \geq 24 \\ & 12a + 30b \leq 150 \\ & a, b \geq 0 \end{array}$$

Answer:

$$\begin{array}{ll} \text{Minimize} & C = 4a + 3b \\ \text{Subject to} & 2a + 4b \geq 24 \\ & -12a - 30b \geq -150 \\ & a, b \geq 0 \end{array}$$

Once the primal has been written in canonical form, the following procedure can be used to generate the dual.

Generating the Dual of a Linear Programming Problem

1. There is one dual variable for each constraint of the primal.
2. The objective-function coefficients of the dual are the right-hand-side constants of the primal.
3. The right-hand-side constants of the dual are the objective-function coefficients of the primal.
4. There is one dual constraint for each primal variable.
5. The constraints of the dual are $\geq$ when the constraints of the primal are $\leq$ and $\leq$ when the primal constraints are $\geq$.
6. The row coefficients of the dual constraints are the column coefficients of the primal. Symbolically, a_{ij} in the dual $= a_{ji}$ in the primal.
7. The dual is a minimization problem if the primal objective is maximization and vice versa.

EXAMPLE 8

Apply these rules for generating the dual of a linear programming problem to the canonical form we developed in Example 7.

$$\begin{array}{ll} \text{Maximize} & P = 3x + 4y + 7z \\ \text{Subject to} & -4x + 12y - 3z \leq -18 \\ & 2x + y + z \leq 50 \\ & 14x - 25y - 10z \leq -25 \\ & x, y, z \geq 0 \end{array}$$

SOLUTION

Step 1 Since there are three primal constraints, there will be three dual variables. Let's call them a, b, and c, respectively.

Step 2 The objective-function coefficients of the dual are the right-hand-side constants of the primal: -18, 50, and -25.

Step 3 The right-hand-side constants of the dual are the objective-function coefficients of the primal: 3, 4, and 7.

Step 4 Since there are three primal variables, there will be three dual constraints.

Step 5 Since the primal constraints are $\leq$, the dual constraints will be $\geq$.

Step 6 The row coefficients of the dual constraints are the column coefficients of the primal. This gives us

$$\text{First primal column} \rightarrow \text{First dual row}$$
$$\begin{matrix} -4 \\ 2 \\ 14 \end{matrix} \qquad \begin{matrix} -4 & 2 & 14 \end{matrix}$$

$$\text{Second primal column} \rightarrow \text{Second dual row}$$
$$\begin{matrix} 12 \\ 1 \\ -25 \end{matrix} \qquad \begin{matrix} 12 & 1 & -25 \end{matrix}$$

$$\text{Third primal column} \rightarrow \text{Third dual row}$$
$$\begin{matrix} -3 \\ 1 \\ -10 \end{matrix} \qquad \begin{matrix} -3 & 1 & -10 \end{matrix}$$

Combining this with the right-hand-side constants from Step 3 we have the following dual constraints.

$$\begin{aligned} -4a + 2b + 14c &\geq 3 \\ 12a + b - 25c &\geq 4 \\ -3a + b - 10c &\geq 7 \end{aligned}$$

Step 7 Since the primal is a maximization problem, the dual is a minimization problem. Combining this with the objective-function coefficients of Step 2 we have the dual objective function:

$$\text{Minimize} \quad C = -18a + 50b - 25c.$$

Putting everything together we have

$$\begin{array}{ll} \text{Minimize} & C = -18a + 50b - 25a \\ \text{Subject to} & -4a + 2b + 14c \geq 3 \\ & 12a + b - 25c \geq 4 \\ & -3a + b - 10c \geq 7 \\ & a, b, c \geq 0 \quad \square \end{array}$$

Using vector notation, we can summarize the rules for generating the dual of a linear programming problem quite compactly.

1. Suppose the primal in canonical form is

$$\begin{array}{ll} \text{Maximize} & P = \mathbf{c} \cdot \mathbf{x} \\ \text{Subject to} & A\mathbf{x} \leq \mathbf{b} \\ & \mathbf{x} \text{ nonnegative} \end{array}$$

where $\mathbf{c} = (c_1, c_2, \ldots, c_n)$ is the row vector of objective-function coefficients;

$$\mathbf{x} = \begin{pmatrix} x_1 \\ x_2 \\ \cdot \\ \cdot \\ \cdot \\ x_n \end{pmatrix} \text{ is the column vector of variables;}$$

$A = [a_{i,j}]$ is the $m \times n$ matrix of substitution coefficients;

$$\mathbf{b} = \begin{pmatrix} b_1 \\ b_2 \\ \cdot \\ \cdot \\ \cdot \\ b_m \end{pmatrix} \text{ is the column vector of right-hand sides.}$$

Then the dual is

$$\begin{array}{ll} \text{Minimize} & C = \mathbf{b}^T \cdot \mathbf{y} \\ \text{Subject to} & A^T\mathbf{y} \geq \mathbf{c}^T \\ & \mathbf{y} \text{ nonnegative} \end{array}$$

where A^T is the transpose of A (see Section 5.2) and $\mathbf{b}^T$ is a row vector (the transpose of a column vector is a row vector and vice versa) and $\mathbf{c}^T$ is a column vector.

2. Similarly, when the primal is

$$\begin{array}{ll} \text{Minimize} & C = \mathbf{c} \cdot \mathbf{x} \\ \text{Subject to} & A\mathbf{x} \geq \mathbf{b} \\ & \mathbf{x} \text{ nonnegative} \end{array}$$

Then the dual is

$$\begin{array}{ll} \text{Maximize} & P = \mathbf{b}^T \cdot \mathbf{y} \\ \text{Subject to} & A^T\mathbf{y} \le \mathbf{c}^T \\ & \quad \mathbf{y} \text{ nonnegative} \end{array}$$

EXAMPLE 9

Given the following primal problem, find its dual:

$$\begin{array}{ll} \text{Minimize} & C = 6x + 8y \\ \text{Subject to} & 3x + y \ge 4 \\ & 5x + 2y \le 10 \\ & 12x + y \le 25 \\ & x, y \ge 0 \end{array}$$

SOLUTION

The primal in canonical form is

$$\begin{array}{ll} \text{Minimize} & C = 6x + 8y \\ \text{Subject to} & 3x + y \ge 4 \\ & -5x - 2y \ge -10 \\ & -12x - y \ge -25 \\ & x, y \ge 0 \end{array}$$

so we have

$$\mathbf{b} = \begin{pmatrix} 4 \\ -10 \\ -25 \end{pmatrix}, \qquad \mathbf{c} = (6, 8) \quad \text{and} \quad A = \begin{bmatrix} 3 & 1 \\ -5 & -2 \\ -12 & -1 \end{bmatrix}$$

Then

$$A^T = \begin{bmatrix} 3 & -5 & -12 \\ 1 & -2 & -1 \end{bmatrix}$$

and the dual is

$$\begin{array}{ll} \text{Maximize} & P = 4a - 10b - 25c \\ \text{Subject to} & 3a - 5b - 12c \le 6 \\ & a - 2b - c \le 8 \\ & a, b, c \ge 0 \end{array}$$ □

Practice Problem

Find the dual of the following linear programming problem.

$$\begin{array}{ll} \text{Maximize} & P = 4x + 3y + 2z \\ \text{Subject to} & 2x + 3y + z \le 40 \\ & 10x + 9y + 8z \ge 70 \\ & x, y, z \ge 0 \end{array}$$

Answer: Minimize $C = 40a - 70b$
Subject to
$$\begin{aligned} 2a - 10b &\geq 4 \\ 3a - 9b &\geq 3 \\ a - 8b &\geq 2 \\ a, b &\geq 0 \end{aligned}$$

EXERCISE SET 8.4

Find the dual of the following linear programming problems.

1. Maximize $P = 4x + 4y$
Subject to
$$\begin{aligned} 2x + y &\leq 10 \\ x + 2y &\leq 8 \\ x, y &\geq 0 \end{aligned}$$

2. Minimize $C = 8x + 12y$
Subject to
$$\begin{aligned} 3x + 2y &\geq 12 \\ 2x + 3y &\geq 12 \\ x, y &\geq 0 \end{aligned}$$

3. Maximize $P = 2x + y$
Subject to
$$\begin{aligned} 8x + 4y &\leq 32 \\ 5x + 8y &\leq 40 \\ x, y &\geq 0 \end{aligned}$$

4. Minimize $C = 6x + 7y$
Subject to
$$\begin{aligned} 5x + 4y &\geq 20 \\ 4x + 7y &\geq 28 \\ x, y &\geq 0 \end{aligned}$$

5. Maximize $P = 3x - 2y$
Subject to
$$\begin{aligned} 6x + 3y &\leq 24 \\ x - y &\leq 2 \\ y &\leq 4 \\ x, y &\geq 0 \end{aligned}$$

6. Minimize $C = 8x + 12y$
Subject to
$$\begin{aligned} x + y &\geq 4 \\ 4x + 2y &\geq 8 \\ 3x + 5y &\geq 15 \\ x, y &\geq 0 \end{aligned}$$

7. Maximize $P = x + y$
Subject to
$$\begin{aligned} x + y &\geq 1 \\ 5x + 10y &\leq 50 \\ y &\leq 4 \\ x, y &\geq 0 \end{aligned}$$

8. Minimize $C = 15x + 25y$
Subject to
$$\begin{aligned} x - y &\leq 1 \\ 4x + 6y &\geq 12 \\ 6x + 3y &\leq 18 \\ x, y &\geq 0 \end{aligned}$$

9. Maximize $P = 2x - y + z$
Subject to
$$\begin{aligned} x + y - 2z &\leq 8 \\ 4x - y + z &\geq 2 \\ 4x + 6y - 2z &\geq 8 \\ x, y, z &\geq 0 \end{aligned}$$

10. Minimize $C = 3a + 4b + c + 6d$
Subject to
$$\begin{aligned} 5a - 2b + c - 3d &\geq 2 \\ 6a + b - 5c - 3d &\geq 5 \\ a - 4b - 3c - 7d &\leq 25 \\ a, b, c, d &\geq 0 \end{aligned}$$

11. Consider Problem 9 of Exercise Set 8.1. How much more profit could be made if one additional hour of cutting time were available? If one additional hour of finishing time were available?

12. Consider Problem 11 of Exercise Set 8.1. How much more profit could Phyllis Wentworth make if she had an additional pound of Blue Mountain? an additional pound of Kona?

13. Consider Problem 11 of Exercise Set 8.3. How much more profit could John Snell make if he had an additional pound of ingredient I? of ingredient II? of ingredient III? of ingredient IV?

CHAPTER REVIEW

IMPORTANT TERMS AND SYMBOLS

8.1 Maximization with ≤ Constraints

basic feasible solution A basic solution in which all variables are nonnegative.

basic solution Given a set of m equations in n variables, where $n > m$, set $n - m$ of the variables equal to zero and solve uniquely for the remaining m variables; corner points of the feasible region are equivalent to the basic feasible solutions of the constraint equations in standard form.

basic variable One of the m variables for which a value is to be found in a basic solution.

nonbasic variable One of the $n - m$ variables set equal to zero in a basic solution.

slack variable Added to a ≤ constraint to convert it to an equation; has an objective-function coefficient of zero.

standard form A linear programming problem in which slack variables were added as needed to convert the original inequality constraints to equations.

simplex method For solving linear programming problems:

1. State the problem in standard form.
2. Select a set of m variables to form an initial basic feasible solution. Set all nonbasic variables to zero.
3. Evaluate the nonbasic variables by examining their bottom row coefficients. If all coefficients are nonnegative, stop; the current solution is optimal. Otherwise, select the nonbasic variable with the most negative coefficient to enter the solution.
4. Select the current basic variable to exit the solution by computing the ratios b_i/a_{i*} for $a_{i*} > 0$, where the entering variable is x_*. The variable with the smallest ratio is chosen to exit the basis.
5. Generate the new solution by pivoting, using row operations.

The simplex method is generally performed using the simplex tableau.

8.2 Maximization with ≤ and ≥ Constraints

artificial variable Introduced into a constraint that was originally ≥ or =, in order to be part of the initial basic feasible solution; has large negative objective function coefficient.

surplus variable Subtracted from a ≥ constraint to convert it to an equation; has objective-function coefficient of zero.

8.3 Minimization

minimization with simplex method Multiply all objective-function coefficients by −1 in Step 1 and then proceed as in maximization.

8.4 Duality

canonical form For maximization problems, all constraints are $\leq$; for minimization problems, all constraints are $\geq$.

dual problem Linear programming problem generated from statement of primal which addresses the value of scarce resources.
Generating the dual of a linear programming problem:

1. There is one dual variable for each constraint of the primal.
2. The objective-function coefficients of the dual are the right-hand-side constants of the primal.
3. The right-hand-side constants of the dual are the objective-function coefficients of the primal.
4. There is one dual constraint for each primal variable.
5. The constraints of the dual are $\geq$ when the constraints of the primal are $\leq$ and $\leq$ when the primal constraints are $\geq$.
6. The row coefficients of the dual constraints are the column coefficients of the primal. Symbolically, a_{ij} in the dual $= a_{ji}$ in the primal.
7. The dual is a minimization problem if the primal objective is maximization and vice versa.

primal problem The original (given) linear programming problem.

shadow price The value of one additional unit of a scarce resource = its marginal value.

REVIEW PROBLEMS

Solve the following linear programming problems.

1. Maximize $P = 4x + 5y$
Subject to
$$3x + 6y \leq 24$$
$$5x + 4y \leq 20$$
$$x, y \geq 0$$

2. Minimize $C = 3x + 2y$
Subject to
$$2x + 6y \geq 18$$
$$4x + 4y \geq 16$$
$$x, y \geq 0$$

3. Maximize $P = 16x + 32y$
Subject to
$$x + y \geq 1$$
$$x + y \leq 20$$
$$x - 2y \geq 2$$
$$x, y \geq 0$$

4. Minimize $C = 10a + 12b$
Subject to
$$a + b \leq 100$$
$$a + b \geq 2$$
$$-a + 3b \leq 3$$
$$a, b \geq 0$$

5. A manufacturer of shampoos is planning production levels for its fashion shampoos, type O (for oily hair) and type N (for normal hair). Sufficient ingredients are available to make 50,000 bottles of each shampoo but there are only 75,000 bottles available. The production facility can make 1000 bottles of type O in 4 hours and 1000 bottles of type N in 3.5 hours; there are 210 production hours available. The manufacturer makes 8 cents profit per bottle of type O and 6 cents profit per bottle of type N. How many bottles of each type should the manufacturer produce?

6. The Quality Feed Company has received an order for 100 pounds of high-protein feed. Ellen Armstrong, the manager in charge, has noted that the customer's order can be filled using an appropriate combination of three feeds. The end product must contain at least 60 pounds of feed A and no more than 50 pounds of feeds B and C combined. The warehouse has 80 pounds of feed A, 40 pounds of feed B, and 40 pounds of feed C which can be used. Feed A costs Quality \$1.50 per pound, while feed B costs \$1.20 per pound and feed C costs \$1.00 per pound. How should Ellen Armstrong combine the feeds to minimize the cost of the mix and satisfy the customer's requirements?

Formulate the dual of the following linear programming problems.

7. Maximize $P = 4x + 6y$
Subject to
$$\begin{aligned} -4x + 5y &\geq 20 \\ 11x - 4y &\geq 0 \\ 5x + 4y &\leq 40 \\ x, y &\geq 0 \end{aligned}$$

8. Minimize $C = 4u - 6v$
Subject to
$$\begin{aligned} -3u + 4v &\leq 10 \\ -5u + 6v &\leq 8 \\ u, v &\geq 0 \end{aligned}$$

9. Maximize $P = 2a + 4b - 6c$
Subject to
$$\begin{aligned} 2a + 4b - 4c &\leq 16 \\ 8a - 2b + 6c &\leq 14 \\ -6a + 4b + 8c &\leq 8 \\ a, b, c &\geq 0 \end{aligned}$$

10. Perfect Paints specializes in the production of industrial paints and varnish. The company has the capability of making 200 gallons per hour of high gloss varnish and 100 gallons per hour of its flat varnish. The production facility works an eight-hour day. Given that daily demand is at least 400 gallons for the flat varnish and that profit levels are \$1.00 per gallon for high gloss and \$1.50 for flat varnish, answer the following questions.

    a) How much of each varnish should Perfect Paints produce?
    b) How much is an additional hour of production time worth?

9 EXPONENTIAL AND LOGARITHMIC FUNCTIONS

a) A rock group has been measured at a sound level of 120 dB. How much more intense is this sound than that of a factory with a sound level of 95dB? (To be solved in Example 18, Section 9.2.)

b) A company producing a product has found that the first unit required 100 hours while the tenth unit required only 50 hours. Determine the rate of learning. How many hours will be required to produce the twentieth unit? the hundredth unit? the thousandth unit? (To be solved in Example 21, Section 9.2.)

PROBLEM FROM LIFE

INTRODUCTION

In this chapter we'll discuss exponential and logarithmic functions and a variety of applications of them. For example, part (a) of the Problem from Life deals with the intensity of sound, as measured in decibels (dB). This scale is logarithmic (which means nothing to you now but will shortly). There are a number of other logarithmic scales in common use, including the Richter scale (for measuring the intensity of earthquakes), the pH scale (for measuring the acidity or alkalinity of a substance) and the scale of magnitudes (brightness) of stars.

Part (b) of the Problem from Life deals with a phenomenon called learning, which describes a situation in which repeated performances of a task require less and less time.

In addition to the applications we've just mentioned, we'll present other uses of exponential and logarithmic functions. Moreover, these functions will be used in the next chapter (which discusses the mathematics of finance) and in various places throughout the book.

The material in this chapter requires the use of a business or scientific calculator that provides natural logarithms (a key labeled [ln x]) and exponential functions (with key [e^x] or [INV] [ln x]), powers of 10 (with key [10^x] or [INV] [log x]), and a key labeled [y^x]. Because such a calculator can be purchased for less than \$20, we have eliminated discussing the use of logarithm and exponential tables. Instead, we will continue to use the text to aid you in learning to use your calculator. You will find that doing the mathematics on a calculator is faster, easier, and quite a bit more accurate! (*Note:* As we did earlier, we assume throughout our discussion that your calculator employs algebraic notation. If your calculator employs reverse Polish notation (RPN), please refer to your owner's manual.)

9.1 EXPONENTIAL FUNCTIONS

Basic Properties

So far in this book we've discussed a variety of functions that all have one thing in common: they were **algebraic functions.** Such functions are those that can be defined by using the algebraic operations (addition, subtraction, multiplication, division, powers, and roots). However, we have not considered functions in which the variable has appeared in an exponent. In this chapter we'll discuss functions that have variables appearing in exponents.

Let's begin by defining a new function. An **exponential function** has the form

$$f(x) = b^x, \qquad b > 0 \quad \text{and} \quad b \neq 1.$$

Here b is a constant, called the **base.** The variable x can be any real number, so that the domain of f is the set of all real numbers. Because b is positive, the range of f includes only positive numbers.

EXAMPLE 1

The functions $f(x) = 2^x$ and $g(x) = 2^{-x} = 1/2^x = 0.5^x$ are both exponential functions. The base of $f(x)$ is 2, while the base of $g(x)$ is 0.5. □

EXAMPLE 2

The functions $h(x) = x^2$ and $k(x) = x^{-2} = 1/x^2$ are not exponential functions; they are simple algebraic functions. □

EXAMPLE 3

The function $m(x) = (-3)^x$ is not a valid exponential function because its base is -3. Can you see why we don't allow a negative base? Suppose, for example, that $x = 0.5$. Then $m(0.5) = (-3)^{0.5} = \sqrt{-3}$, which is not a real number. (Obviously, the function is defined for some values of x, such as $x = 2$. However, because it doesn't make sense for all values of x, we don't allow a negative base. In addition, as we pointed out in Chapter 2, your calculator won't allow you to use a negative base.) □

EXAMPLE 4

The function $p(x) = 1^x$ is not an exponential function because the base is 1. The reason that this is not allowed should be obvious. □

WARNING

DON'T CONFUSE -3^x AND $(-3)^x$

The placement of the parentheses is important. The first form, -3^x, means: raise 3 to the power x and *then* multiply the result by -1. This is a valid function. The second form, $(-3)^x$, means raise -3 to the power x. This is not a valid exponential function.

Actually we have created a problem with our definition of an exponential function because we allow the exponent, x, to be any real number. Can you guess what the problem is?

Suppose we consider the exponential function $f(x) = 2^x$. We know how to evaluate $f(x)$ whenever x is an integer. For example, $f(1) = 2^1 = 2$, while $f(5) = 2^5 = 32$. In fact, we can evaluate $f(x)$ whenever x is rational. For example, $f(\frac{2}{3}) = 2^{2/3} = (2^2)^{1/3} = (2^{1/3})^2 =$ (approximately) 1.587401. (On your calculator, enter 2 [y^x] 1.5 [$1/x$] [=].) However, we do not know how to evaluate $f(x)$ when x is irrational. For example, how can we evaluate $2^{\sqrt{3}}$ or 2^{π}? (Recall that an irrational number is a number that requires an infinite number of decimal places to express.) Actually, we cannot answer this question completely without the use of calculus. Fortunately the result is easy to explain. We can approximate 2^x (or any function b^x) for irrational x as closely as we want by using rational numbers to approximate x. In other words, since $\pi = 3.1415926535\ldots$, we can approximate 2^{π} by the sequence

$$2^3,\ 2^{3.1},\ 2^{3.14},\ 2^{3.141},\ \ldots.$$

The better the approximation of x, the better the approximation of 2^x (or of b^x).

Now we can plot the functions $f(x) = 2^x$ and $g(x) = 2^{-x} = 0.5^x$. (See Fig. 9.1.) We actually choose convenient (in other words, integer) values of x and then connect the points with a smooth curve. You should take careful note of the shapes of the curves in Fig. 9.1. The graph of $f(x)$ increases without limit as x increases (as we move from left to right), while the graph of $g(x)$ decreases asymptotically to the x-axis as x increases. Every exponential function must have one of these two forms: increasing without limit or decreasing to an asymptote. An exponential function of the form of $f(x)$ (increasing) is said to be a **growth** function, while a function of the form $g(x)$ (decreasing asymptotically) is said to be a **negative growth** or **decay** function.

Practice Problem

Characterize the following exponential functions as growth or decay functions: $f(x) = 0.25^x$, $g(x) = 3^x$. (*Hint:* Draw their graphs.)

Answers: $f(x)$ is a decay function; $g(x)$ is a growth function

x	$f(x) = 2^x$
−4	0.0625
−3	0.125
−2	0.25
−1	0.50
0	1.
1	2.
2	4.
3	8.
4	16.

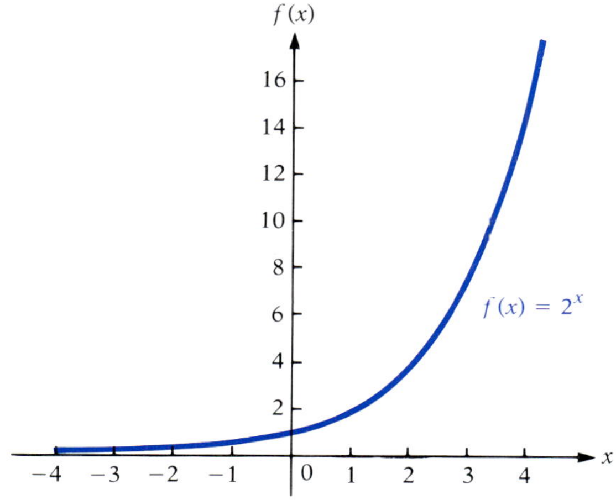

a) The function $f(x) = 2^x$

x	$g(x) = 0.5^x$
−4	16.
−3	8.
−2	4.
−1	2.
0	1.
1	0.5
2	0.25
3	0.125
4	0.0625

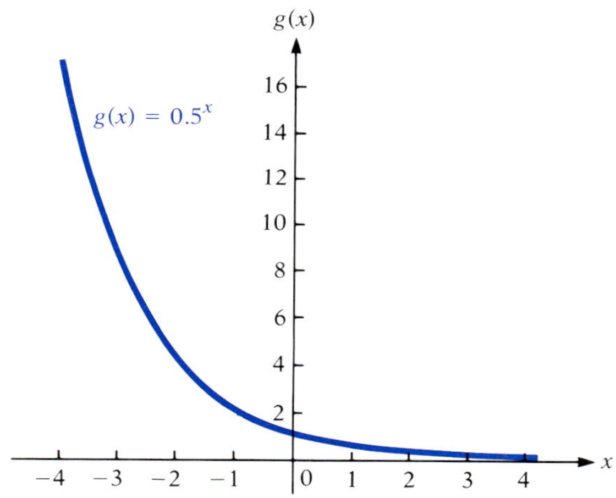

b) The function $g(x) = 2^{-x} = 0.5^x$

Figure 9.1 **Exponential functions can grow without limit or decrease asymptotically.**

Let's look at some examples of growth functions.

EXAMPLE 5

A single bacterium of a certain type divides every hour to produce two complete bacteria. Given that a colony of bacteria starts with 1000 bacteria, after t hours the colony will have

$$B(t) = 1000 \cdot 2^t$$

bacteria (assuming an adequate food supply).

t	2^t	$B(t) = 1000 \cdot 2^t$
0	1	1000
1	2	2000
2	4	4000
3	8	8000
4	16	16,000
5	32	32,000

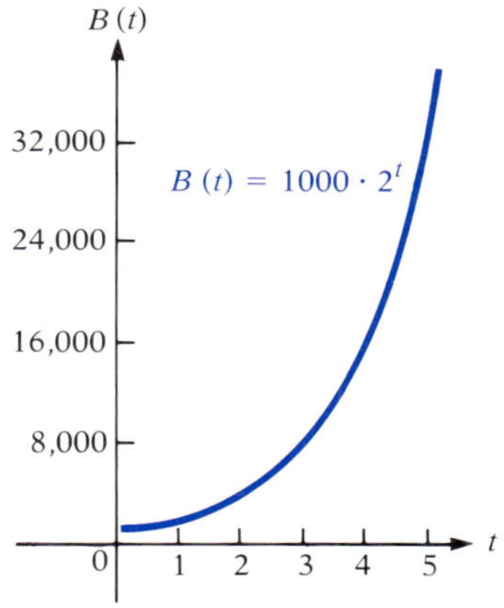

Figure 9.2 **Bacteria growing exponentially.**

The number of bacteria for integer values of t, $0 \le t \le 5$ is given above; the graph is shown in Fig. 9.2. This is an example of a growth function. (In actuality, bacteria will not grow in this fashion for very long because they will not be able to receive adequate food nor remove waste properly.) □

EXAMPLE 6

Money deposited in a savings account (and now in many checking accounts) earns **interest,** which is a fee paid by the bank to the depositor for allowing the bank to use the money. When interest can be earned on interest, we say the account is paying **compound interest** (we'll discuss this in far more detail in Chapter 10). An amount of money P deposited in an account paying 8% compound interest per year will be worth $A(t)$ after t years as defined by

$$A(t) = P(1.08)^t, \qquad 0 \le t.$$

Thus an initial amount of \$1000 will be worth

$$A(1) = 1000(1.08)^1 = \$1080 \text{ after 1 year,}$$
$$A(5) = 1000(1.08)^5 = \$1469.32 \text{ after 5 years,}$$

and

$$A(10) = 1000(1.08)^{10} = \$2158.92 \text{ after 10 years.}$$

(We rounded down in all cases, which banks are allowed to do.) □

Practice Problem

How much would a deposit of \$2000 at 8% compound interest be worth after **a)** 1 year **b)** 10 years **c)** 20 years?

Answers: **a)** \$2160; **b)** \$4317.84; **c)** \$9321.91

The Exponential Function

Recall that our definition of an exponential function allows the base to be any positive constant (other than 1). There is an irrational number, denoted by e, that is used very frequently as a base. (The name e was chosen to honor the mathematician Leonhard Euler.) In fact, the function $f(x) = e^x$ is often referred to as *the* exponential function. Although e is irrational, we can approximate it to any desired accuracy by using the expression $(1 + 1/n)^n$ for large enough integer values of n; thus we find that e is approximately 2.71828. . . . We can also approximate e^x as accurately as desired by evaluating the expression $(1 + 1/n)^{nx}$ for large enough integer values of n. (We cannot go into more detail without the use of calculus.) Calculations involving e were rather tedious (to say the least) before the advent of computers and calculators. Now, we can evaluate e^x with an electronic calculator by simply entering x and then the key [e^x]. (On some calculators you may have to enter x and then [INV] [ln x].)

The function $f(x) = e^x$ is a special case of an exponential function (with a base of e), so its domain is the set of all real numbers and its range is the set of all positive real numbers. Now let's look at some examples that involve the exponential function.

EXAMPLE 7

Evaluate and graph the functions $f(x) = e^x$, $g(x) = e^{-x}$, and $h(x) = e^{-x^2}$ for $-3 \le x \le 3$. [The function $h(x)$ is of great importance in statistics.]

SOLUTION

We know how to evaluate e^x. To evaluate e^{-x}, enter x [+/−] [e^x]. (Or, enter x [+/−] [INV] [ln x].) Finally, to evaluate e^{-x^2}, enter x [x^2] [+/−] [e^x]. (Or, enter x [x^2] [+/−] [INV] [ln x].) The data are shown in the following table; the graphs are shown in Figs. 9.3, 9.4, and 9.5.

x	$f(x) = e^x$	$g(x) = e^{-x}$	$h(x) = e^{-x^2}$
−3	0.049787	20.085537	0.000123
−2	0.135335	7.389056	0.018316
−1	0.367879	2.718282	0.367879
0	1.	1.	1.
1	2.718282	0.367879	0.367879
2	7.389056	0.135335	0.018316
3	20.085537	0.049787	0.000123

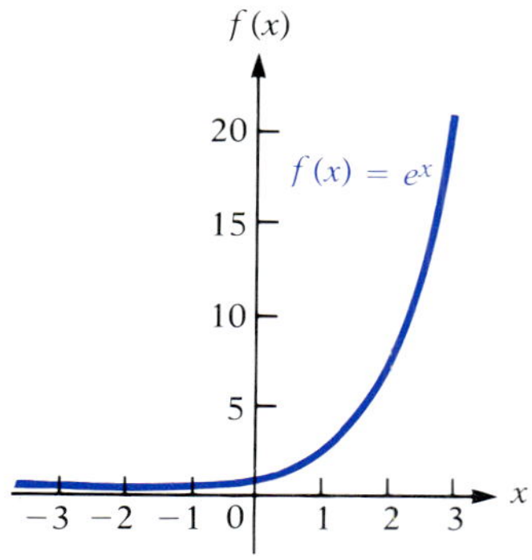

Figure 9.3 **The function $f(x) = e^x$, $-3 \le x \le 3$.**

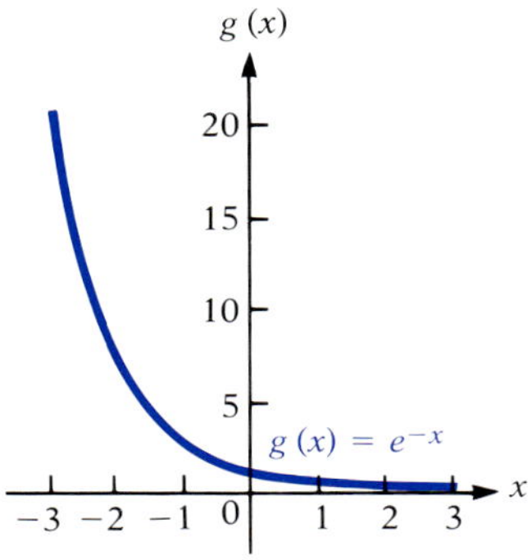

Figure 9.4 The function $g(x) = e^{-x}$, $-3 \le x \le 3$.

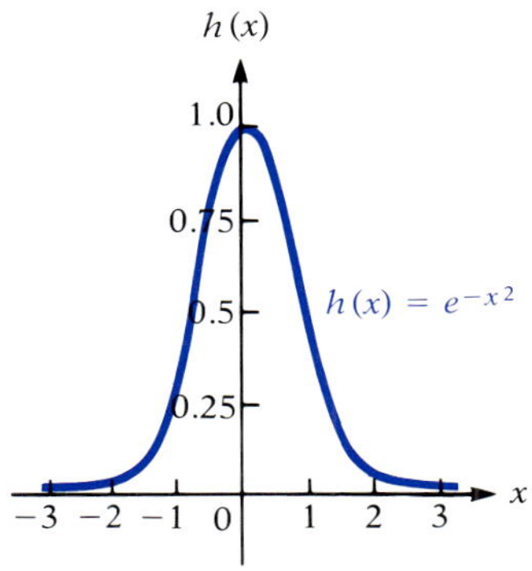

Figure 9.5 The function $h(x) = e^{-x^2}$, $-3 \le x \le 3$. □

EXAMPLE 8

The population of a certain city during the years 1950 to 2000 can be modeled by the function

$$P(t) = 10{,}000e^{0.05t}, \qquad 0 \le t \le 50,$$

where t is the number of years elapsed after 1950. The size of the population in 1960 is given by $P(10)$, since 1960 is 10 years after 1950. We have

$$P(10) = 10{,}000e^{0.05 \cdot 10} = 10{,}000e^{0.5} \cong 10{,}000 \cdot 1.64872$$
$$\cong 16{,}487 \qquad \text{(rounding to the nearest integer).}$$

Similarly, in 1975 the population is given by

$$P(25) = 10{,}000e^{0.05 \cdot 25} = 10{,}000e^{1.25} \cong 34{,}903$$

and in 1990 the population is given by

$$P(40) = 10{,}000e^{0.05 \cdot 40} = 10{,}000e^{2} \cong 73{,}890.$$

In this case, we've used an exponential function to model the growth because we can't assume that people will produce children at fixed intervals (as the bacteria in Example 5 did). This is actually a case of **continuous compounding.** □

EXAMPLE 9

Another example of continuous compounding is provided by some financial institutions, which pay compound interest continuously rather than once per period. An amount of money P invested at an interest rate of r per year will result in an amount $A(t)$ after t years given by

$$A(t) = Pe^{rt}, \qquad 0 \le t.$$

For example, suppose an account pays 8% per year compounded continuously. For an initial deposit of $1000, the amounts of money after 1 year, 5 years, and 10 years are given by (respectively)

$$A(1) = 1000e^{0.08 \cdot 1} = 1000e^{0.08} \cong \$1083.28,$$
$$A(5) = 1000e^{0.08 \cdot 5} = 1000e^{0.40} \cong \$1491.82,$$
$$A(10) = 1000e^{0.08 \cdot 10} = 1000e^{0.80} \cong \$2225.54.$$

(Compare these amounts to those shown in Example 6.) □

Practice Problem

For an initial deposit of $1000, compute the amounts of money after 1, 5, and 10 years when interest is paid at a rate of 6% per year, compounded continuously.

Answers: after 1 year, $A(1)$ = $1061.83; after 5 years, $A(5)$ = $1349.85; after 10 years, $A(10)$ = $1822.11

EXERCISE SET 9.1

Graph the following functions for $-3 \leq x \leq 3$.

1. $f(x = 3^x$
2. $g(x) = 3^{-x} = (\frac{1}{3})^x$
3. $f(x) = 3^{0.5x}$
4. $f(x) = 3^{-0.5x} = (\frac{1}{3})^{0.5x}$
5. $g(x) = 4^{-x}$
6. $f(x) = 4^{-0.25x}$
7. $f(x) = 9^{-x/3}$
8. $g(x) = 9^{x/3}$
9. $g(x) = e^{0.5x}$
10. $f(x) = e^{-0.5x}$
11. $h(x) = 2e^x$
12. $g(x) = e^{2x}$
13. $f(x) = 10e^{-2x}$
14. $h(x) = 10e^{2x}$
15. $g(x) = 100 - 100e^{-x}$
16. $f(x) = 100 - 100e^{-2x}$
17. Manhattan was supposedly purchased for $24 in 1620. Assume this amount of money was placed in a bank at 6% compound interest.
    a) Assuming the money was compounded once per year, the value of the money after t years would be given by $A(t) = 24(1.06^t)$. How much would the money be worth today?
    b) Assuming the money was compounded continuously, the value of the money after t years would be given by $A(t) = 24(e^{1.06t})$. How much would the money be worth today?
18. Suppose a colony of bacteria doubles its size every 30 minutes. Assuming that there were 10 bacteria originally, how many bacteria are present
    a) after 30 minutes? b) after 12 hours? c) after t hours?
19. Suppose that the amount of a certain radioactive substance present after t months is given by $a(t) = 10e^{-3t}$, where $a(t)$ is expressed in pounds. How much of the substance is present
    a) initially (at time $t = 0$)? b) after six months? c) after one year?

20. The number of units of a new product that can be produced by a certain process is given by the relationship $p(x) = 1000 - 800e^{-x/6}$, where x represents the number of months of production and $p(x)$ is the monthly production rate.

a) How many units per month can be produced initially?

b) How many units per month can be produced after one year? after two years?

c) Graph the function $p(x)$ for $0 \le x \le 24$.

d) Suppose that production costs for the process average $10,000 per month. What happens to the unit cost of production? (The unit cost of production is given by the relationship unit cost = monthly cost ÷ monthly production.) Compute the unit costs initially, after one year and after two years.

9.2 LOGARITHMS

Definition and Basic Properties

Logarithms enable us to work certain problems easily, especially when combined with the use of an electronic calculator. In addition, logarithms are used often in scales of measure. For example, logarithmic scales (called log scales) are often used when making graphs in which one variable (or both) has a very broad range of values. (See, for example, Fig. 9.6.)

It wasn't until 1628 that a table of logarithms (base 10) of all the integers from 1 to 100,000 was published. Today such tables are no longer necessary, because electronic calculators can easily compute (almost) any desired logarithm.

Logarithms are actually another name for exponents. For example, since

$100 = 10^2$,	the logarithm of 100 (base 10) is 2,
$81 = 3^4$,	the logarithm of 81 (base 3) is 4,
$64 = 2^6$,	the logarithm of 64 (base 2) is 6.

Let's be more formal and actually define a **logarithm.** We say

$$f(x) = \textbf{logarithm}_b\, x = a \text{ when } b^a = x, \qquad \text{where } b \text{ is the base.}$$

In other words, a is the logarithm of x, base b, when x is equal to b raised to the power a. We commonly abbreviate logarithm by "$\log_b$." As with exponential functions, we require $b > 0$ and also $b \neq 1$. Here

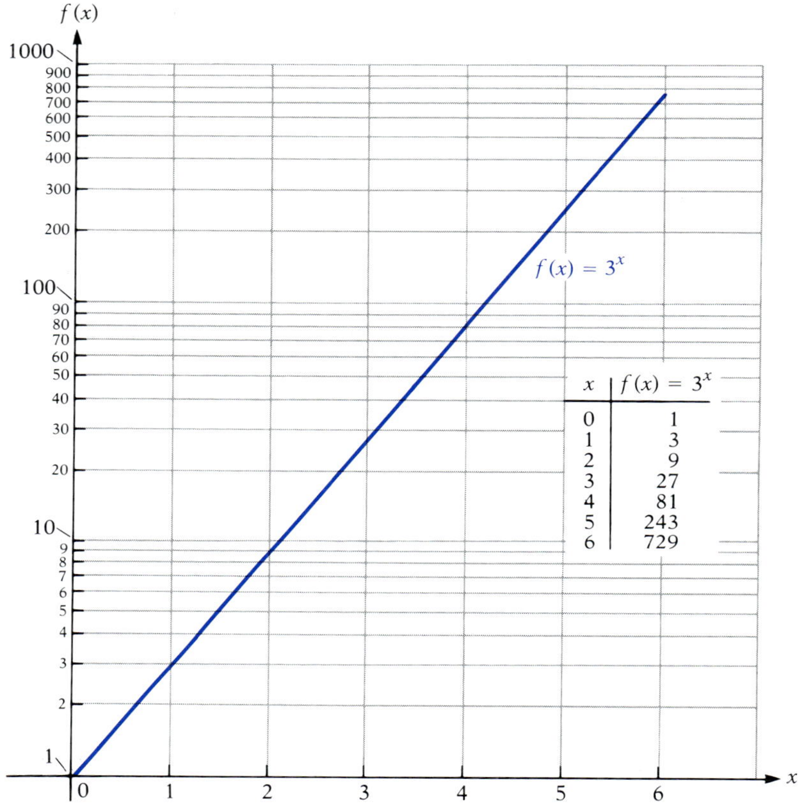

Figure 9.6 The graph of an exponential function is a straight line with a vertical logarithmic scale.

x	$f(x) = \log_2 x$
0.125	−3
0.25	−2
0.5	−1
1	0
2	1
4	2
8	3
16	4
32	5

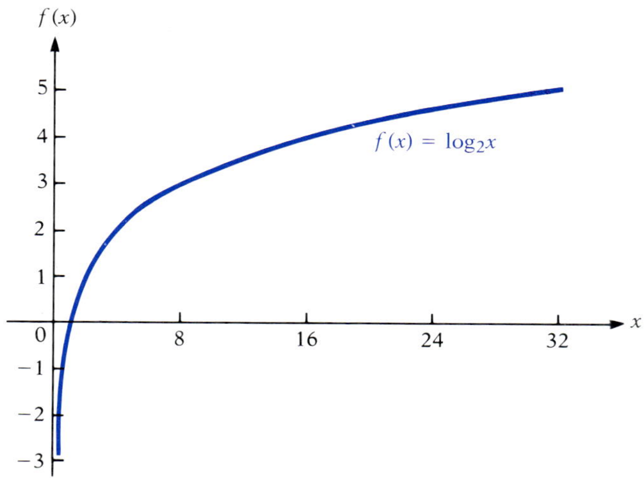

Figure 9.7 The function $f(x) = \log_2 x$.

we must restrict the domain to positive real numbers only. (Think about this. Can b^a ever be negative when $b > 0$?) The range of the logarithm function is the set of all real numbers.

In Fig. 9.7 we show the graph of the function $f(x) = \log_2 x$. As with the exponential functions, we've drawn the graph by plotting some known values and then drawing a smooth curve through them. For comparison, see the graph of $f(x) = 2^x$ in Fig. 9.1(a).

When working with logarithms, it is very helpful to think of $\log_b x = a$ as follows: a is the exponent to which we must raise b to get x. In other words, when you see $\log_b x = a$, think $b^a = x$

EXAMPLE 10

a) $\log_5 25 = 2$ because $x = 25$, $b = 5$, and $25 = 5^2$.

b) $\log_7 49 = 2$ because $x = 49$, $b = 7$, and $49 = 7^2$.

c) $\log_{10} 1000 = 3$ because $x = 1000$, $b = 10$, and $1000 = 10^3$.

d) $\log_9 27 = \frac{3}{2}$ because $x = 27$, $b = 9$, and $27 = 9^{3/2}$.

e) $\log_8 4 = \frac{2}{3}$ because $x = 4$, $b = 8$, and $4 = 8^{2/3}$. □

Practice Problem

Find the logarithms.

a) $\log_4 16$ **b)** $\log_5 125$ **c)** $\log_{16} 8$

Answers: **a** 2; **b)** 3; **c)** $\frac{3}{4}$

EXAMPLE 11

Find the number x: $\log_8 x = 2$.

SOLUTION

Here the log (2) and b (8) are known. From the definition of a logarithm, $x = b^{\log}$, so $x = 8^2 = 64$. □

EXAMPLE 12

Find the base b: $\log_b 27 = \frac{3}{2}$.

SOLUTION

Here we know the log ($\frac{3}{2}$) and x (27). From the definition, $x = b^{\log}$, so we have $27 = b^{3/2}$. To solve this, raise both sides to the $1 \div (\frac{3}{2}) = \frac{2}{3}$ power:

$$27 = (b^{3/2}),$$

$$27^{2/3} = (b^{3/2})^{2/3}$$

so

$$9 = b. \quad \square$$

EXAMPLE 13 Find the base b: $\log_b 64 = \frac{3}{5}$.

SOLUTION

Again, we know the log ($\frac{3}{5}$) and x (64). Using the definition, $x = b^{\log}$, so $64 = b^{3/5}$.

$$64 = b^{3/5}$$

$$64^{5/3} = (b^{3/5})^{5/3} \qquad \text{Raise to the power } 1 \div \tfrac{3}{5} = \tfrac{5}{3}$$

so

$$1024 = b. \quad \square$$

Practice Problem

Find either x or b.

a) $\log_7 x = 3$ **b)** $\log_b 625 = \frac{4}{3}$ **c)** $\log_b 243 = \frac{5}{6}$

Answers: **a)** 343; **b)** 125; **c)** 729

There are three important properties of logarithms with which you should be familiar.

Properties of Logarithms

For x and y real nonnegative numbers, and m a real number, and $b > 0$ and $b \neq 1$, we have

1. $\log_b (x \cdot y) = \log_b x + \log_b y$
2. $\log_b \left(\frac{x}{y}\right) = \log_b x - \log_b y$
3. $\log_b x^m = m \log_b x$

Let's see why these properties are true. Remember that $\log_b x = a$ means that $b^a = x$. Now, consider Property 1. Suppose we represent the logarithm (base b) of x by a, and the logarithm (base b) of y by c. This means $x = b^a$ and $y = b^c$. Then $x \cdot y = b^a \cdot b^c = b^{a+c}$, which says that $\log_b (x \cdot y) = a + c$ or $\log_b (x \cdot y) = \log_b x + \log_b y$. *In other words, the logarithm of the product of two numbers is the sum of their logarithms.*

Next consider Property 2. Let $\log_b x = a$ and $\log_b y = c$, as before. Then $(x/y) = b^a \div b^c = b^{a-c}$, so that $\log_b (x/y) = a - c$ or $\log_b (x/y) = \log_b x - \log_b y$. *The logarithm of a quotient of two numbers is the difference of their logarithms.*

Finally, let's look at Property 3. Again, let $\log_b x = a$. Then $x^m = (b^a)^m = b^{a \cdot m}$ so $\log_b x = m \cdot a$ or $\log_b x = m \log_b x$. *The logarithm of a number raised to a power is the power times the logarithm of the number.*

You should note that all of this discussion is really nothing more than a restating of some of the properties of exponents. Recall that the product of x^a and $x^c = x^{a+c}$. Property 1 simply is another way of expressing this. The same is true for the other properties.

Let's demonstrate these properties with some examples.

EXAMPLE 14

(*Note:* To get the base 10 log of a number, use the [log x] key.) Show that $\log_{10}(150 \cdot 35) = \log_{10} 150 + \log_{10} 35$.

SOLUTION

(This is true by Property 1.) First, enter 150 [×] 35 [=] 5250. Next, hit [log x], which results in 3.720159 (to six places). Next, enter 150 [log x] (which is 2.176091) [+] 35 [log x] (which is 1.544068) [=] 3.720159, as expected. (*Note:* You can convert from a base 10 logarithm back to the original number, called the **antilogarithm,** by keying [10^x] or [INV] [log x]. Thus, with 3.720159 displayed, keying [10^x] gives 5249.996, which approximates the product of 150 and 35. Had we carried out the logarithms to more decimal places, keying [10^x] would have resulted in a more accurate approximation.) □

EXAMPLE 15

Show that $\log_{10}(\frac{85}{5}) = \log_{10} 85 - \log_{10} 5$.

SOLUTION

(This is true by Property 2.) First, let's compute $\frac{85}{5}$ and find its logarithm. Enter 85 [÷] 5 [=] 17. Key [log x] to get 1.230449. Next, enter 85 [log x] [−] 5 [log x] [=], which results in 1.230449. □

EXAMPLE 16

Find 1.06^{10} using Property 3.

SOLUTION

By Property 3, $\log_{10}(1.06^{10}) = 10 \cdot \log_{10} 1.06$. Enter 10 [×] 1.06 [log x] [=], which gives (approximately) 0.2530586. Now, key [10^x], which gives us 1.790848. In other words, $1.06^{10} = 1.790848$. □

We can also use Property 3 to find roots. Suppose we want $16^{1/4}$. We can use Property 3 to find the logarithm and then take the antilogarithm. We have $\log_{10} 16^{1/4} = (\frac{1}{4}) \log_{10} 16$, so enter 0.25 [×] 16 [log x] [=], which gives 0.30103 (approximately). Now key [10^x], which results in 2. In other words, 2 is the fourth root of 16. (You can check this by raising 2 to the fourth power.)

Practice Problem

Using logarithms, find $500^{1/5}$ and $800^{1/10}$.

Answers: 3.465724, 1.951232

There are two commonly used bases for logarithms. **Common logarithms** have a base of 10 and **natural logarithms** have a base of e (the irrational number 2.7182818 . . .). These are denoted as follows.

- common logarithms: "log" (When no base is shown, assume 10.)
- natural logarithm: "ln" (*Not* In)

Of course, all of the properties of logarithms hold for both common and natural logarithms. You can convince yourself of this by reworking Examples 14–16 with natural logarithms instead of common logarithms. (Replace [log x] by [ln x] and [10^x] by [e^x].)

An important use of logarithms is solving equations such as $2 = 1.05^n$. The following example illustrates the technique.

EXAMPLE 17

Solve the equation $2 = 1.05^n$.

SOLUTION

Begin by taking the logarithm of both sides (either common or natural; we will use natural logarithms to help familiarize you with them).

$$\ln 2 = \ln (1.05^n)$$

which by Property 3 gives us

$$\ln 2 = n \ln 1.05.$$

Solving for n yields

$$n = \frac{\ln 2}{\ln 1.05}$$

so

$$n = \frac{0.6931472}{0.0487902} = 14.206699. \quad \square$$

WARNING

DON'T CONFUSE $\ln \frac{2}{1.05}$ and $\frac{\ln 2}{\ln 1.05}$.

Practice Problem

Find n so that $2 = 1.02^n$.

Answer: $n = 35.002789$

Suppose $\ln x = a$. What is e^a? By definition, $\ln x = a$ means $e^a = x$. We can write this another way. Since $a = \ln x$, it must be true that $e^a = e^{\ln x} = x$. Similarly, $10^{\log x} = x$. In other words, taking natural logarithms and raising to a power of e are the inverse of one another (as are taking common logarithms and raising to a power of 10). Therefore, when you must solve an equation in which the variable occurs in an exponent, take the logarithm (either natural or common) of both sides of the equation, and then use the properties of logarithms. (See Example 17.)

Because taking logarithms and raising to a power are the inverse of one another, some calculators don't have keys for e^x and 10^x. Instead, you use INV $\ln x$ and INV $\log x$, respectively.

Applications

Sound Intensity

EXAMPLE 18

Problem from Life, part (a) A rock group has been measured at a sound level of 120 dB. How much more intense is this sound than that at a factory with a sound level of 95 dB?

SOLUTION

Within recent years there has been more concern over the noise levels to which people are exposed. Sound intensity is measured in bels. However, this is a fairly large unit so that the commonly used unit is the **decibel.** The formula for measuring sound intensity in decibels, abbreviated dB, is

$$L = 10 \log \frac{I}{I_0},$$

where L is the level of sound in decibels, I is the observed intensity, and I_0 is an arbitrarily chosen standard. For example, suppose the noise level in an office is 3000 times as intense as the base intensity I_0. Then the value of L would be

$$10 \log \frac{3000\, I_0}{I_0} = 10 \log 3000 = 34.77 \text{ dB} \qquad \text{(approximately).}$$

Now let's consider the Problem from Life, part (a), as described. Let

r = sound intensity of the rock group and
f = sound intensity of the factory

Then we can form the difference

$$120 \text{ dB} - 95 \text{ dB} = 10 \log \left(\frac{r}{I_0}\right) - 10 \log \left(\frac{f}{I_0}\right)$$

by using the definition of L. Subtracting 95 from 120 on the left and then dividing both sides by 10 gives us

$$2.5 = \log \left(\frac{r}{I_0}\right) - \log \left(\frac{f}{I_0}\right)$$

Now raise 10 to the numbers on both sides of the equation:

$$10^{2.5} = \frac{r/I_0}{f/I_0}$$

or

$$316.22777 = r/f.$$

In other words, the rock group is approximately 316.23 times as loud as the factory. □

Practice Problem

How much louder is an office with a noise level of 50 dB than an office with a noise level of 45 dB?

Answer: approximately 3.162 times as loud

Earthquake Intensity. The intensity of earthquakes is commonly measured on a logarithmic scale called the **Richter scale.** The formula for determining the magnitude of an earthquake is

$$R = \log \frac{I}{I_0},$$

where R is the Richter scale intensity, I is the observed intensity, and I_0 is a fixed intensity which defines the base of the scale. For example, an earthquake with an intensity 1000 times greater than the base intensity would have a Richter scale value of

$$R = \log \frac{1000\, I_0}{I_0} = \log 1000 = 3.$$

EXAMPLE 19

What is the intensity of an earthquake that measures 5.8 on the Richter scale?

SOLUTION

Let E = the intensity of the earthquake. Then

$$5.8 = \log \frac{E}{I_0}.$$

By Property 2, the right side is the difference of the logs, so

$$5.8 = \log E - \log I_0$$

or

$$5.8 + \log I_0 = \log E.$$

Now raise 10 to the numbers shown on both sides, using Property 3 on the left side.

$$10^{5.8} I_0 = E$$

or

$$630{,}957 I_0 = E \qquad \text{(approximately).} \quad \square$$

EXAMPLE 20

How much more intense is an earthquake with a Richter scale value of 7.5 than one with a Richter value of 5?

SOLUTION

We can solve this problem in the same manner we used to solve Example 18. Let

$$x = \text{intensity of first earthquake,}$$
$$y = \text{intensity of second earthquake.}$$

Then we have the difference

$$7.5 - 5 = \log\left(\frac{x}{I_0}\right) - \log\left(\frac{y}{I_0}\right)$$

by using the definition of Richter scale intensity. This is equivalent to

$$2.5 = \log\left(\frac{x}{I_0}\right) - \log\left(\frac{y}{I_0}\right).$$

Raise 10 to the numbers shown on both sides of the equation:

$$10^{2.5} = \frac{x/I_0}{y/I_0}$$

or

$$316.22777 = x/y.$$

The first earthquake is approximately 316.23 times as intense as the second earthquake. $\square$

Practice Problem

How much more intense is an earthquake with a Richter scale intensity of 8 than one with a Richter intensity of 4?

Answer: 10,000 times as intense

Learning. In certain settings it has been found that people can become more efficient at accomplishing a task (take less time) the more they perform the task. This phenomenon has been called **learning.** When learning occurs, the time to do a repetitive task drops by a given percentage each time the total number of repetitions doubles. For example the time to accomplish a task might drop from 1 hour after 10 repetitions to 0.85 hour after 20 repetitions. Doubling the number of repetitions to 40 would drop the time for the last repetition to 0.85 (0.85 hour) = 0.7225 hour. In this case we would say there is an 85% rate of learning. (See Fig. 9.8.) In Fig. 9.9 we've plotted the 85% learning curve on a log-log scale. Notice that the plot is now a straight line with a slope of −0.85. We can model the learning phenomenon with a function of the form

$$X = X_0\left(\frac{P}{P_0}\right)^{-b},$$

where

$$X = \text{time to accomplish a task after } P \text{ repetitions},$$
$$X_0 = \text{time to accomplish the task after } P_0 \text{ repetitions},$$
$$b = \text{a constant}, 0 < b < 1.$$

First, let's see what constant value of b is associated with an 85% learning curve. From the definition of 85% learning, when we double the number of repetitions, the time per repetition is reduced by a

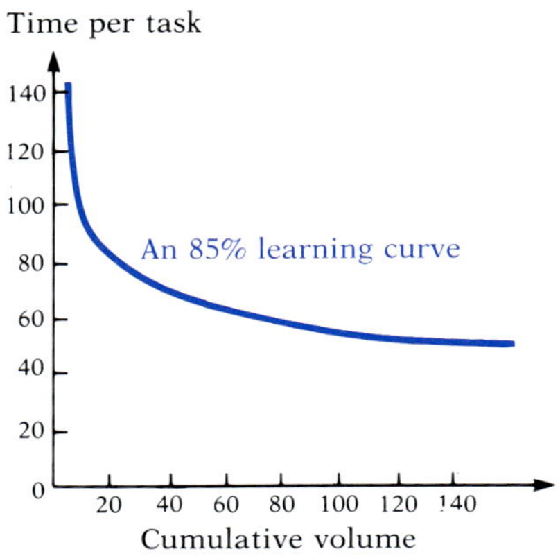

Figure 9.8 **An 85% learning curve**

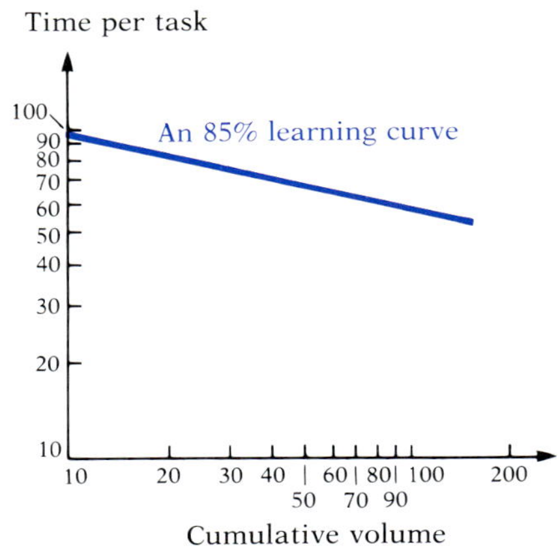

Figure 9.9 **An 85% learning curve on a log-log scale**

factor of 0.85. Therefore we have

$$X = 0.85X_0 \qquad \text{and} \qquad P = 2P_0.$$

Then

$$0.85X_0 = X_0 \cdot (2P_0/P_0)^{-b}$$

or

$$0.85 = 2^{-b}.$$

We must solve this for b by using the same approach we used in Example 17: Take the natural logarithm of both sides and apply Property 3 of logarithms. We have

$$\begin{aligned} \ln 0.85 &= \ln(2^{-b}), \\ \ln 0.85 &= -b(\ln 2), \\ b &= -\frac{\ln 0.85}{\ln 2}, \\ b &= -\frac{-0.1625189}{0.6931472} = 0.234465. \end{aligned}$$

Now we can solve part (b) of the Problem from Life.

EXAMPLE 21

Problem from Life, part (b) A company producing a product has found that the first unit of a product required 100 hours while the tenth unit required only 50 hours. How many hours will be required to produce the twentieth unit? the hundredth unit? the thousandth unit?

SOLUTION

To solve this problem we can begin by determining the constant b and the rate of learning. We have

$$X = 50 \text{ hours}, \qquad X_0 = 100 \text{ hours}, \qquad P = 10, \qquad P_0 = 1,$$

so

$$\begin{aligned} 50 &= 100 \cdot \left(\frac{10}{1}\right)^{-b}, \\ 0.5 &= 10^{-b}. \end{aligned}$$

Since the right side involves a power of 10, let's take common logs on both sides of the equation:

$$\begin{aligned} \log 0.5 &= \log(10^{-b}), \\ -0.30103 &= -b, \\ b &= 0.30103. \end{aligned}$$

Remember that $\log 10^x = x$.

Now let's determine the number of hours required to produce the twentieth unit, which will also enable us to determine the rate of learning. We have

$$X_0 = 50, \qquad P = 20, \qquad P_0 = 10, \qquad b = 0.30103.$$

Therefore

$$X = 50 \cdot \left(\frac{20}{10}\right)^{-0.30103}$$
$$X = 50 \cdot 2^{-0.30103},$$
$$X \cong 50 \cdot 0.811673,$$
$$X = 40.584 \text{ hours} \qquad \text{(approximately).}$$

In addition, since we multiplied the time for the tenth unit by 0.811673 to derive the time for the twentieth unit, the rate of learning must be 81.1673%.

The time for the hundredth unit is given by

$$X = 50\left(\frac{100}{10}\right)^{-0.30103}$$
$$X = 50 \cdot 0.50,$$
$$X = 25 \text{ hours},$$

and the time for the thousandth unit is

$$X = 50\left(\frac{1000}{10}\right)^{-0.30103}$$
$$X = 50 \cdot 0.25,$$
$$X = 12.5 \text{ hours}.$$

As you can see, the effect of learning can be quite significant. □

Practice Problem

A person takes 0.2 hour to perform a task. Given that the rate of learning is 90%, how long will it take the person to perform the task again?

Answer: 0.18 hour

Practice Problem

After having produced 100 parts, the time to perform a certain operation has been measured at 1.5 hours. After another 100 parts were produced, the time has dropped to 1.2 hours.

a) What is the rate of learning?

b) What is the constant b?

c) How long will it take to produce the four-hundredth unit?

Answers: **a)** 80%; **b)** 0.321928; **c)** 0.96 hour

EXERCISE SET 9.2

Find the appropriate logarithm.

1. $\log_3 9$
2. $\log_2 16$
3. $\log_2 0.25$
4. $\log_5 \sqrt{5}$
5. $\log_7 49$
6. $\log_9 81$
7. $\log_3 27$
8. $\log_{100} 10$
9. $\log_4 64$
10. $\log_5 125$

Find the number x given its logarithm.

11. $\log_{32} x = \frac{1}{5}$
12. $\log_{10} x = 2$
13. $\log_7 x = 3$
14. $\log_{9.5} x = 0$
15. $\log_5 x = -1$
16. $\log_{0.25} x = 2$
17. $\log_3 x = 0$
18. $\log_{0.5} x = 3$
19. $\log_3 x = 3$
20. $\log_4 x = \frac{3}{2}$

Use the appropriate property of logarithms to find the answer.

21. $285 \cdot 62$
22. $482 \cdot 96$
23. $782 \cdot 3$
24. $982 \div 85$
25. $742 \div 843$
26. $325 \div 7842$
27. 602^2
28. 1.02^{15}
29. $82^{1.5}$
30. $100^{1/9}$

Evaluate the following by using logarithms.

31. $\dfrac{2 \cdot 4}{3}$
32. $\dfrac{85 \cdot 63}{15}$
33. $18 \cdot 3$
34. $190/850$
35. $10 \cdot (1.025^{10})$
36. $1500 \cdot (1.02^{20})$
37. $\dfrac{\sqrt{14}}{16}$
38. $9000 \cdot (1.05^{30})$
39. $35{,}000 \cdot (1.07^{20})$
40. $1 - (1.02^{-3})$
41. The San Francisco earthquake of 1906 measured approximately 8.24 on the Richter scale. What was its intensity (compared to the base level)?
42. How much more intense is an earthquake of Richter scale 8.2 compared to one of Richter scale 5.8?
43. Chemists measure the acidity or alkalinity of a solution by calculating its pH factor. The formula is

$$\text{pH} = -\log\,[\text{H}^+],$$

where pH is the relative acidity or alkalinity and $[\text{H}^+]$ is the quantity of hydrogen ions (in moles per liter) present in the solution. A pH of 7 indicates a neutral (neither acidic nor alkaline) solution. For example, suppose a hair shampoo has a hydrogen ion concentration of 1.1×10^{-8}

moles per liter. Its pH is given by pH $= -\log(1.1 \times 10^{-8}) = 7.96$. Suppose a certain brand of apple cider has a pH of 5. What is the hydrogen ion concentration?

44. Find the pH of a solution with a hydrogen ion concentration of 5.5×10^{-8}. Is this solution acidic or alkaline? (A pH of less than 7 indicates an acid solution; a pH of more than 7 an alkaline solution.)

45. Suppose a solution has a pH of 9 and the concentration of hydrogen ions is doubled. What is the new pH? What would the pH be if the concentration of hydrogen ions were halved?

46. Suppose an office has a sound level of 65 dB and a new piece of equipment is purchased that doubles the noise intensity. What is the new sound level in decibels?

47. A rock group has been measured at a sound level of 115 dB. How much more intense is the sound of a jet plane with a sound level of 140 dB?

48. Suppose it has been determined that the time to perform the first repetition of a task is 25 hours and $b = -0.415$. How many hours will it take to produce the second unit? The tenth unit? What is the rate of learning?

49. A student is preparing for a test. She estimates that she will have only 10 minutes to work a certain type of problem. The first time she tried a similar problem it took her 50 minutes, while the second one took her 35 minutes. How many practice problems should she work so that her time will be down to 9 minutes (to give her a little margin for error)?

CHAPTER REVIEW

IMPORTANT TERMS AND SYMBOLS

9.1 Exponential Functions

exponential function Has form $f(x) = b^x$, $b > 0$ and $b \neq 1$, where b is the base; domain includes all real numbers and range includes only positive real numbers.

the exponential function Has form $f(x) = e^x$, where e is the irrational number 2.71828. . . .

growth function Exponential function that increases without limit.

negative growth (decay) function Decreases asymptotically to zero.

9.2 Logarithmic Functions

decibel Measure of sound intensity, abbreviated dB; defined by

$$L = 10 \log \frac{I}{I_0},$$

where L is the sound intensity in decibels, I is the observed intensity, and I_0 is an arbitrarily chosen standard.

learning Situation in which repetitions of a task result in decreased time per repetition; modeled by

$$X = X_0\left(\frac{P}{P_0}\right)^{-b},$$

where X is the time to accomplish a task after P repetitions, X_0 is the time to accomplish the task after P_0 repetitions, and b is a constant, $0 < b < 1$.

logarithm Log of x to the base b is a when $b^a = x$, where $b > 0$ and $b \neq 1$: has domain of positive real numbers and range of all real numbers. Common logarithms: have base 10; use [log x] or [INV] [10^x] keys. Natural logarithms: have base e; use [ln x] or [INV] [e^x] keys. Properties of logarithms:

1. $\log_b (x \cdot y) = \log_b x + \log_b y$
2. $\log_b \left(\frac{x}{y}\right) = \log_b x - \log_b y$
3. $\log_b x^m = m \log_b x$

rate of learning Gives the percentage decrease in the time per repetition after the number of repetitions has doubled.

Richter scale (for earthquakes) Defined by

$$R = \log \frac{I}{I_0},$$

where R is the Richter scale intensity, I is the observed intensity, and I_0 is the fixed base intensity.

REVIEW PROBLEMS

1. Graph the functions $4^{0.5x}$ and $4^{-0.5x}$ for $-3 \leq x \leq 3$.
2. Find $\log_4 32$ and $\log_{125} 25$.

Find the missing quantity.

3. $\log_7 x = \frac{3}{2}$
4. $\log_5 x = \frac{2}{3}$
5. $\log_b 32 = \frac{1}{3}$
6. $\log_b 81 = \frac{4}{3}$
7. $\log_{49} 7 = a$
8. $\log_{625} 125 = a$
9. Graph e^{-2x} for $-3 \leq x \leq 3$.
10. Graph $\ln x^2$ for $0 < x \leq 6$.
11. Graph $\ln x$ for $0 < x \leq 6$.
12. How much more severe is an earthquake with a Richter scale value of 8.5 than one with a value of 6.5?
13. A company has determined that a certain operation took 5 hours on the first unit manufactured and only 4 hours on the one-hundredth unit. What is the rate of learning? How long will it take to perform the operation on the five-hundredth unit?

10 MATHEMATICS OF FINANCE

1986 OLDSMOBILE CIERA
FOR ONLY
$9753 or
$189.68/month

Selling price $9753. Down payment $2400 cash or trade-in. Amount to be financed $7353 for 48 months. Finance charge $1751.64, annual percentage rate 10.9%. Taxes, license, and title fee not included.

a) How was the payment of $189.68 per month determined? (To be solved in Example 24, Section 10.3.)
b) What would the monthly payment be if you wanted to pay off the loan in 36 months rather than 48 months? (To be solved in Example 25, Section 10.3.)
c) What would the monthly payment be if you had $2000 cash or trade-in instead of the $2400? (To be solved in Example 26, Section 10.3.)
d) How much cash or trade-in value is required to limit the payment to $150 per month? (To be solved in Example 27, Section 10.3.)

PROBLEM FROM LIFE

INTRODUCTION

Some people believe that the mathematics of finance is of interest only to those who study business. We believe nothing could be further from the truth. Anyone who expects to buy an item "on time" (for example, a car, a house, or a stereo), anyone who expects to borrow money or even anyone who plans to invest money should be interested in the material in this chapter.

10.1 SIMPLE INTEREST

Basic Concepts

There are two general methods for computing interest. (Interest is a fee you pay when you borrow money from someone, or, alternatively, it is the money you earn when you allow someone the use of your money.) Simple interest is the method used in most cases for short-term loans (those made for less than 1 year). For example, ads for six-month certificates of deposit express the yield in simple interest.

Let's define **simple interest** symbolically. The basic formula is

Simple Interest

$$I = Prt, \tag{10.1}$$

where
I = amount of interest earned,
P = principal (amount invested or loaned),
r = interest rate per year,
t = number of years.

(*Note:* Because there are many equations and formulas in this chapter, we will number them in order to make referring to them easier.)

EXAMPLE 1

Suppose you lend a friend \$100 at 10% simple interest per year. How much interest will you earn after 1 year? after 3 years?

SOLUTION

We have $P = \$100,$
$r = 10\%$ or $0.10,$ **Convert to a decimal fraction.**

and for one year, $t = 1$, so

$$I = Prt,$$
$$I = \$100 \cdot 0.10 \cdot 1 = \$10.$$

Your friend must pay you \$10 interest. For a loan of 3 years, $t = 3$, so

$$I = Prt,$$
$$I = \$100 \cdot 0.10 \cdot 3 = \$30.$$

Here your friend must pay you \$30 interest. □

Practice Problem

In Example 1, how much interest would you earn if

a) your friend borrowed the money for 2.5 years at 10% per year?

b) your friend borrowed the money for 3 years at 15% per year?

Answers: **a)** \$25; **b)** \$45

There are two more formulas that can help us study situations that involve simple interest. The first enables us to compute the total amount of money (principal plus interest):

Total Amount (Simple Interest)

$$A = P(1 + rt), \tag{10.2}$$

where
A = total amount of money,
P = principal,
r = rate per year,
t = number of years.

(This formula is derived from the simple interest formula by noting that $A = P + I$ and then substituting $I = Prt$ into it.)

Another useful formula is

Principal (Simple Interest)

$$P = \frac{A}{1 + rt}. \tag{10.3}$$

Formulas (10.1) and (10.2) tell us the key fact concerning simple interest: *the principal never changes*. Thus, if you invest money at simple interest for 2 years, you earn exactly twice what you would earn for a 1-year investment.

EXAMPLE 2

Suppose you decide to lend money to a friend. She wants to borrow \$50 for 2 years and you agree upon a rate of 10% per year simple interest. How much total money (interest plus principal) must she repay you after 2 years?

SOLUTION

Here

$$P = \$50,$$
$$r = 10\%, \quad \text{or} \quad 0.10,$$
$$t = 2.$$

Using formula (10.2) we have

$$A = P(1 + rt),$$
$$A = 50(1 + 0.10 \cdot 2),$$
$$A = 50(1.2),$$
$$A = \$60.$$

Your friend must repay you \$60, which consists of \$50 original principal and \$10 interest. Check this with formula (10.1). □

EXAMPLE 3

A bond will be worth \$1000 nine months from today. Assuming the (simple) interest rate is 16%, how much is it worth today?

SOLUTION

Here

$$A = \$1000,$$
$$r = 16\%, \quad \text{or} \quad 0.16,$$
$$t = 9 \text{ months}, \quad \text{or} \quad 0.75 \text{ year},$$

so

$$P = \frac{A}{1 + rt},$$
$$P = \frac{1000}{1 + 0.16 \cdot 0.75},$$
$$P = \frac{1000}{1.12},$$
$$P = \$892.86.$$

The bond is worth \$892.86 today. □

Practice Problem

A bond will be worth \$1000 eight months from today. The interest rate is 15% per year. What is it worth today?

Answer: \$909.09

There are two slightly different methods used to convert 1 year into days in financial situations that involve interest. For the sake of simplicity we will use a 360-day year unless otherwise stated. The other method is the obvious 365-day year. (Financial institutions use both.)

EXAMPLE 4

Find the simple interest on a loan of \$100 for 60 days at 12% per year.

SOLUTION

Here $P = \$100$, $r = 0.12$, and $t = 60 \div 360$, so

$$I = Prt,$$

$$I = 100 \cdot 0.12 \cdot \frac{60}{360},$$

$$I = \$2. \quad \square$$

EXAMPLE 5

What interest rate must be charged to earn \$1.00 interest on a principal of \$200 for 30 days, assuming a 365-day year?

SOLUTION

Here $I = \$1$, $P = \$200$, $t = 30 \div 365$, and r is unknown. We can solve formula (10.1) for r as follows:

$$I = Prt,$$

so

$$r = \frac{I}{Pt}.$$

Substituting the appropriate values

$$r = \frac{1}{200 \cdot (30/365)},$$

$$r = 0.06083 \quad \text{or} \quad 6.083\% \text{ per year.}$$

(Interest rates are generally defined on a yearly basis.) $\square$

EXAMPLE 6

Find the time necessary for \$500 invested at 10% simple interest to become \$504.

SOLUTION

Here $A = \$504$, $P = \$500$, and $r = 0.10$. We can find the time, t, by solving formula (10.2) for it:

$$A = P(1 + rt),$$

so $$t = \frac{A - P}{Pr},$$ **Check the algebra!**

$$t = \frac{504 - 500}{500 \cdot 0.1},$$

$$t = 0.08 \text{ years}, \quad \text{or} \quad 28.8 \text{ days} \qquad \text{(based on 360 days per year)} \quad \square$$

EXAMPLE 7

Find the principal invested at 15% simple interest that earns \$9 in 55 days.

SOLUTION

We have $I = \$9$, $r = 0.15$, and $t = 55 \div 360$. We can find the principal, P, by solving formula (10.1):

$$I = Prt,$$

so $$P = \frac{I}{rt},$$

$$P = \frac{9}{0.15 \cdot (55/360)},$$

$$P = \$392.73. \quad \square$$

Practice Problem

a) What interest rate must be charged to earn \$30 interest in 60 days on a principal of \$2500?

b) Find the time necessary for \$5000 invested at 16% simple interest to become \$5050.

c) Find the principal invested at 12% simple interest that earns \$25 in 180 days.

Answers: **a)** 7.3% per year; **b)** 0.0625 year or 22.5 days; **c)** \$416.67

Banker's Discount

Banker's discount (sometimes called *simple discount*) is a form of simple interest used by many banks for short-term loans. The interest is computed in the same manner as we've been doing for simple interest, but the interest is then *subtracted* from the principal *before* the person receives the loan. Thus if you wanted to borrow \$1000 (approximately) for 60 days and the bank charged 18% simple interest, the interest would be computed using (10.1):

$$I = Prt,$$
$$I = 1000 \cdot 0.18 \cdot (60/360),$$
$$I = \$30.$$

The banker's discount would be to subtract the \$30 interest from the \$1000, which is \$970. Thus the bank would actually give you \$970 for 60 days, at which time you would pay back \$1000. (With simple interest you would receive \$1000 and pay back \$1030.) This means the bank is actually earning *more* than 18%!

Let's derive the **effective** or actual interest rate for banker's discount. We know A (the amount you pay back to the bank), P (the amount you receive), and t (the time you hold the loan). Rather than use r, let's use the symbol d (for discount). We can solve formula (10.3) for d (using d rather than r):

$$A = P(1 + dt),$$

so

$$d = \frac{A - P}{Pt},$$

Check the algebra!

where

d = effective rate,
A = amount owed to bank,
P = amount received,
t = length of loan.

In this case

$$d = \frac{1000 - 970}{970 \cdot (60/360)},$$
$$d = 18.5567\%.$$

The bank actually earns 18.5567% rather than the stated (or nominal) rate of 18%.

We can derive a simpler formula for computing the effective rate, d. Let's agree that r is the nominal rate of the loan. Now, the numerator in the calculation, $A - P$, is simply the amount of interest paid to the bank. But this must be given by $I = Art$. Moreover, $P = A - I = A - Art = A(1 - rt)$. Thus we can rewrite the formula for the effective rate as

$$d = \frac{A - P}{Pt},$$

$$d = \frac{Art}{A(1 - rt)t} = \frac{r}{1 - rt}.$$

Banker's Discount

$$d = \frac{r}{1 - rt}, \tag{10.4}$$

where d = effective rate,
r = nominal rate,
t = length of loan.

EXAMPLE 8

Sam and Bertha Jones borrow $5000, discounted, from a bank for 90 days at a nominal rate of 16%. What is the effective rate?

SOLUTION

Using formula (10.4) with $r = 0.16$ and $t = (90/360) = 0.25$, we have

$$d = \frac{0.16}{1 - 0.16 \cdot 0.25},$$
$$d = 16.6667\%.$$

The effective rate is 16.6667%, as compared to the nominal rate of 16%. □

Practice Problem

You borrow $2500 for 30 days, discounted, at a nominal rate of 15%. How much do you actually receive? What is the effective rate?

Answers: You receive $2468.75. The effective rate is 15.1899%

To summarize, with banker's discount you pay more interest than you might expect based upon the nominal rate. Table 10.1 shows the effective rates given a nominal rate of 12% and various lengths of time. Note that as the length of time increases, the effective rate increases.

Time Payments*

Suppose you have a charge account that bills you interest each month of 1.5% of the balance. Let's see how we might compute a

* This section requires knowledge of Section 4.3, Sequences and Series.

TABLE 10.1 Impact of length of loan on effective rate, banker's discount

Nominal rate	Length of time (days)	Effective rate
0.12	30	0.121212
0.12	60	0.122449
0.12	90	0.123711
0.12	180	0.127660
0.12	360	0.136364

sequence of monthly payments that will pay off the balance in a specified period of time.

EXAMPLE 9

Suppose you put \$1200 on your charge card and decide to pay off the amount in 2 years. Each month you are charged interest of 1.5% of the unpaid balance. What should the monthly payments be? What is the total amount of interest paid?

SOLUTION

We can determine the monthly payment necessary to pay off the principal: \$1200 ÷ 24 months = \$50 per month. Now let's compute the interest payments. The first month's interest should be 1.5% of \$1200, or $0.015 \cdot 1200 = \$18$. Thus your first monthly payment is $\$50 + \$18 = \$68$.

After the first month, you owe only \$1150 (because you paid \$50 of the principal). Thus the second month's interest is $0.015 \cdot 1150 = \$17.25$, and the second monthly payment totals \$67.25.

At this point, we hope you agree that it is obvious that the balance is dropping by \$50 each month (the rest of each payment is interest, which does not reduce the balance). Thus the interest payment should drop by 1.5% of \$50 or \$0.75 each month. The last (24th) payment will be $\$68 - 23 \times 0.75 = \$50.75$. This means we have the payment scheme shown in Fig. 10.1.

Month	1	2	3	⋯	24
Payment	\$68.00	\$67.25	\$66.50	⋯	\$50.75

Figure 10.1 **The sequence of monthly payments, Example 9.**

Notice that the monthly payments form an arithmetic sequence with a first term of \$68.00 and a last term of \$50.75. Using the formula for the sum of an arithmetic sequence (see Section 4.3), we have

$$S_n = \left(\frac{n}{2}\right)(a_n + a_1),$$

where

$$n = 24,$$
$$a_{24} = \$50.75,$$
$$a_1 = \$68.00,$$

so

$$S_{24} = \left(\frac{24}{2}\right)(\$50.75 + \$68.00),$$
$$S_{24} = 12 \cdot \$118.75,$$
$$S_{24} = \$1425.$$

You will repay a total of \$1425. Since you borrowed only \$1200, the difference of $\$1425 - \$1200 = \$225$ must be the total amount of interest paid. □

EXAMPLE 10

Let's perform the analysis of Example 9 again, using a monthly interest rate of 2% of the unpaid balance.

SOLUTION

Once again, the basic loan payments are \$50 per month. Now, the first month's interest is 2% of $\$1200 = 0.02 \cdot \$1200 = \$24$. Moreover, since the balance decreases by \$50 each month, the interest payment must decrease by $0.02 \cdot \$50 = \$1.00$. Therefore the sequence of payments is as shown in Fig. 10.2.

We have (again) an arithmetic sequence of 24 payments, with $a_1 = \$74$ and $a_{24} = \$51$. Thus

$$S_{24} = \left(\frac{24}{2}\right)(\$51 + \$74),$$
$$S_{24} = 12 \cdot \$125,$$
$$S_{24} = \$1500.$$

Month	1	2	3	...	24
Payment	\$74.00	\$73.00	\$72.00	...	\$51.00

Figure 10.2 The sequence of monthly payments, Example 10.

Now you will repay a total of \$1500, of which \$300 must be interest, because the original balance was \$1200. □

EXAMPLE 11

Finally, let's analyze the problem of Examples 9 and 10 with 12 monthly payments and an interest charge of 1.75% per month.

SOLUTION

Now, the basic payment is \$1200 ÷ 12 = \$100 per month. The first monthly payment includes interest of \$21 (1.75% of \$1200), for a total of \$121. Since the balance each month will drop by \$100, the interest payments must decrease by 0.0175 · \$100 = \$1.75 each month. The loan repayment scheme is shown in Figure 10.3. This is an arithmetic sequence of 12 payments, with a_1 = \$121.00 and a_{12} = \$101.75. Hence

$$S_{12} = \left(\frac{12}{2}\right)(\$101.75 + \$121.00),$$
$$S_{12} = 6 \cdot \$222.75,$$
$$S_{12} = \$1336.50.$$

Therefore, you repay a total of \$1336.50, of which \$136.50 is interest.

Month	1	2	3	...	12
Payment	\$121.00	\$119.25	\$117.50	...	\$101.75

Figure 10.3 **The sequence of monthly payments, Example 11.** □

Practice Problem

Suppose you want to pay off a balance of \$3600 in 36 months. How much total money and total interest will you pay if the interest charge per month is

a) 1.5%? **b)** 1.75%? **c)** 2%?

Answers: **a)** total money = \$4599, total interest = \$999; **b)** total money = \$4765.50, total interest = \$1165.50; **c)** total money = \$4932.00, total interest = \$1332.00

Practice Problem

Suppose you want to pay off a balance of \$2400 with a monthly interest charge of 2%. How much total money and total interest will you pay if you pay off the loan in

a) 12 months? **b)** 24 months? **c)** 48 months?

Answers: **a)** total money = \$2712, total interest = \$312;
b) total money = \$3000, total interest = \$600;
c) total money = \$3576, total interest = \$1176

EXERCISE SET 10.1

Using formulas (10.1), (10.2), and (10.3), compute the missing values (assume a 360-day year).

1.	P = \$800,	r = 0.10,	t = 3 years.	I = ?	A = ?
2.	P = \$2000,	r = 0.12,	t = 2.5 years.	I = ?	A = ?
3.	P = \$8000,	r = 0.15,	I = \$100.	t = ?	A = ?
4.	P = \$8000,	r = 0.12,	I = \$120.	t = ?	A = ?
5.	P = \$150,	t = 2 years,	I = \$15.	r = ?	A = ?
6.	P = \$1000,	t = 1 month,	I = \$12.	r = ?	A = ?
7.	A = \$220,	r = 0.08,	t = 4 months.	P = ?	I = ?
8.	A = \$2500,	r = 0.10,	t = 15 days.	P = ?	I = ?
9.	A = \$8000,	r = 0.09,	t = 84 days.	P = ?	I = ?
10.	A = \$10,000,	r = 0.11,	t = 5 years.	P = ?	I = ?

11. What principal will earn \$4 interest in 8 months at 12.5% simple interest?

12. Ms. Jones had to repay a total of \$2462 after 8 months on a sum borrowed at 9% simple interest. How much did she borrow? How much interest did she pay?

13. A city borrowed \$2,500,000 at 11.5% for 10 years to build a new high school. How much simple interest will be paid the first year?

14. What principal will become a total of \$50,000 in 5 years at 12%?

15. Paul and Paula buy a stereo system for \$900 and charge it. They must pay 1.75% interest on the unpaid balance each month. How much total interest will they pay if the account is paid off in 1 year?

16. Frankie and Johnny buy furniture for \$3000 and charge it. They must pay 2% interest on the unpaid balance each month. How much total interest will they pay if the charge is paid off in 3 years?

17. Jim borrowed \$1500 for one year at 15% with banker's discount. How much money was he actually given? How much interest did he pay? What was the effective rate?

18. Sally borrowed \$5000 for 90 days at 13.2% with banker's discount. How much money was she actually given? How much interest did she pay? What was the effective rate?
19. What principal will earn \$80 interest in 9 months at 9% simple interest?
20. If you pay \$900 for a note that will pay \$1000 in 6 months, what simple interest will you earn?
21. How much should Kathy Miller pay for a note that will be worth \$2500 in 8 months if she wants to earn 12.5% simple interest?
22. Sam and Susan borrowed \$4800 and have agreed to repay the loan by paying \$200 per month plus 1.5% of the unpaid balance each month. What is the total cost of the loan over 24 months? How much total interest will they pay?
23. Suppose in Problem 22 that Sam and Susan are offered the option of paying \$160 per month plus 1.5% of the unpaid balance each month for a period of 30 months. What is the total cost of the loan? How much total interest will they pay?
24. Carmen and Juan borrowed \$10,000 for 6 months at 13% with banker's discount. How much money did they actually receive? How much interest did they pay? What was the effective rate?
25. John and Nancy borrowed a sum of money with banker's discount. They received \$7000. The term of the loan was 90 days, and the nominal rate was 13.75%. How much money did they borrow? What was the effective rate?

10.2 COMPOUND INTEREST

Basic Concepts

If you have a savings account, your money is probably earning compound interest. Moreover, when you borrow money or begin a series of deposits (say into an IRA), the basic idea used in the analysis is that of compound interest. In addition, there are nonfinancial situations (such as radioactive decay) that also involve the idea of compounding.

The key concept of **compound interest** is that the principal *changes* each period because interest earned during the period is added to the principal. Contrast this to simple interest, in which the principal never changes.

Suppose you deposit \$1000 into a savings account that earns 10% interest, compounded annually. Figure 10.4 illustrates the key concept of compounding.

Time	Now	Year 1	Year 2	Year 3
Principal	\$1000	\$1000	\$1100	\$1210
Interest		+ 100	+ 110	+ 121
		\$1100	\$1210	\$1331

Figure 10.4 The growth of principal with compound interest.

At the end of the first year, you have earned interest of

$$I = Prt = \$1000 \cdot 0.10 \cdot 1 = \$100.$$

This is added to the original principal so that your new principal is now \$1000 + \$100 = \$1100. Your money has been **compounded.**

At the end of the second year, you have earned interest of

$$I = Prt = \$1100 \cdot 0.10 \cdot 1 = \$110.$$

Your principal is now \$1100 + \$110 = \$1210.

At the end of the third year, you have earned interest of

$$I = Prt = \$1210 \cdot 0.10 \cdot 1 = \$121.$$

Your principal is now \$1210 + \$121 = \$1331.

Thus in 3 years your money has grown from \$1000 to \$1331. Had you earned simple interest for 3 years, your account would be worth only

$$A = P(1 + rt) = \$1000 \cdot (1 + 0.1 \cdot 3) = \$1300.$$

Thus the compounding of interest has earned you an extra \$31.

The basic formulas for compound interest are

Total Amount (Compound Interest)

$$A = P(1 + r)^t, \tag{10.5}$$

Present Value (Compound Interest)

$$P = \frac{A}{(1 + r)^t} = A(1 + r)^{-t}, \tag{10.6}$$

where
A = total amount after t periods,
P = principal (present value),
r = rate per period,
t = number of periods.

Formula (10.5) tells us that money (or anything else) grows exponentially when it is compounded. Formula (10.6) tells us how much money must be present originally, P, to be worth A after t periods.

We can justify (10.5) as follows. Suppose you deposit P dollars into an account earning interest at the rate r each period. After one period the interest is $I = Pr$. The new principal (after compounding) is $P + Pr = P(1 + r)$. After two periods the interest is $P(1 + r)r$ and the new principal is

$$\begin{aligned} P(1 + r) + P(1 + r)r &= P(1 + r)(1 + r), \qquad \textbf{Factor out } \boldsymbol{P(1 + r)}. \\ &= P(1 + r)^2. \end{aligned}$$

After three periods the interest is $P(1 + r)^2r$ and the new principal is

$$\begin{aligned} P(1 + r)^2 + P(1 + r)^2r &= P(1 + r)^2(1 + r), \qquad \textbf{Factor out } \boldsymbol{P(1 + r)^2}. \\ &= P(1 + r)^3. \end{aligned}$$

Notice the growth of the amount of money (principal):

after 0 periods:	P
after 1 period:	$P(1 + r)$
after 2 periods:	$P(1 + r)^2$
after 3 periods:	$P(1 + r)^3$
⋮	
after t periods:	$P(1 + r)^t$

Formula (10.6) is easily obtained from formula (10.5) by solving it for P. We can interpret (10.6) as defining the original investment that will become A after t periods at a rate of r per period. The original investment is usually referred to as the **present value** of the amount A.

WARNING

BE CAREFUL WITH *r* AND *t*

The values of r and t in these formulas (and all others involving interest) must be chosen carefully to reflect the rate at which the investment is compounded. For example, suppose an investment earns 6% compounded annually. Then $r = 0.06$ and t is the number of years. When the compounding is semiannual (twice a year), then $r = 0.03$ and t is the number of half-years (twice the number of years). When the compounding is monthly, then r is given by $0.06/12 = 0.005$ and t is the number of months.

EXAMPLE 12

Suppose a sum of \$100 is invested at 10% compound interest for 3 years. What is the amount of money and the amount of interest paid when the money is compounded

a) annually? **b)** semiannually? **c)** monthly? **d)** daily?

SOLUTION

In all cases we will use formula (10.5) with $P = \$100$.

a) $r = 0.10$ and $t = 3$ so

$$A = 100(1 + 0.10)^3.$$

This is calculated as follows. Enter

1.10 [y^x] 3 [=] [×] 100 [=],

which results in \$133.10 (the value of A).

b) $r = 0.10/2 = 0.05$ and $t = 2 \cdot 3 = 6$ so $A = 100 \cdot (1 + 0.05)^6$.

We have

1.05 [y^x] 6 [=] [×] 100 [=],

which gives us $A = \$134.01$.

c) $r = 0.10/12$ which is a repeating decimal. Rather than round off, we can enter it as shown below to avoid errors. Also, $t = 12 \cdot 3 = 36$. Enter

0.1 [÷] 12 [=] [+] 1
[=] [y^x] 36 [=] [×] 100 [=]

which gives us $A = \$134.82$ (approximately).

d) $r = 0.10/360$, another repeating decimal, and $t = 360 \cdot 3 = 1080$. We have

0.1 [÷] 360 [=] [+] 1
[=] [y^x] 1080 [=] [×] 100 [=]

which gives us $A = \$134.98$.

Note that as the rate of compounding increases, so does the growth of the money. The results are summarized in Table 10.2.

TABLE 10.2 The growth of an investment of \$100 for 3 years at 10%

Method of interest	A	Interest earned
Simple[a]	\$130.00	\$30.00
Compounded annually	\$133.10	\$33.10
Compounded semiannually	\$134.01	\$34.01
Compounded monthly	\$134.82	\$34.82
Compounded daily	\$134.98	\$34.98

[a] See Example 1.

(Since A is the amount of money including both the original principal and the interest earned, we can compute the interest by subtracting the original investment from A.) □

Practice Problem

In Example 12, what would the amount of money and interest earned be if the money were compounded

a) quarterly? **b)** weekly?

Answers: **a)** A = \$134.49, so interest = \$34.49; **b)** A = \$134.95, so interest = \$34.95

EXAMPLE 13

Compute the present value of \$1500 compounded annually at 15% for 30 years.

SOLUTION

Here we can use formula (10.6) to find P. We have A = \$1500, $r = 0.15$, and $t = 30$. Since $P = A(1 + r)^{-t}$, enter

1.15 [y^x] 30 [+/−] [=] [×] 1500 [=]

which gives us P = \$22.65 (approximately). In other words, the amount \$22.65 invested at 15% for 30 years will be worth almost \$1500. (Check it!) □

Practice Problem

What is the present value of \$5000 compounded monthly at 12% annually for 10 years?

Answer: \$1514.97

We can also use formula (10.5) to solve for r (or for t), when the other parameters are known.

EXAMPLE 14

An original investment of \$1500 becomes \$2250 after 5 years. What is the interest rate, assuming yearly compounding?

SOLUTION

We have P = \$1500, A = \$2250, and $t = 5$. Solving formula (10.5) for r we have

$$A = P(1 + r)^t,$$

so

$$A/P = (1 + r)^t,$$
$$(1 + r)^t = A/P,$$
$$(1 + r) = (A/P)^{1/t} \qquad \textbf{Raise both sides to } 1/t.$$
$$r = (A/P)^{1/t} - 1.$$

On a calculator this becomes

2250 [÷] 1500 [=] [y^x] 5 [1/x] [=] [−] 1 [=],

which gives $r = 0.0844718$, or 8.44718% per year. □

EXAMPLE 15

An original investment of \$25,000 becomes \$50,000 at a rate of 12% compounded annually. How many years are required?

SOLUTION

We have $P = \$25{,}000$, $A = \$50{,}000$, and $r = 0.12$. Here we must solve formula (10.5) for t.

$$A = P(1 + r)^t,$$

so

$$A/P = (1 + r)^t,$$
$$(1 + r)^t = A/P,$$
$$t \ln(1 + r) = \ln(A/P), \qquad \textbf{Take ln of both sides.}$$

so

$$t = \frac{\ln(A/P)}{\ln(1 + r)}.$$

On a calculator enter

50000 [÷] 25000 [=] [ln x] [÷] 1.12 [ln x] [=],

which gives $t = 6.116255$ years (or 6 years and 41.8518 days, assuming a 360-day year)

Note: A crude but useful rule of thumb for this problem is as follows. With compound interest, money will double in a period of time such that the number of years times the interest rate (in percents) is approximately 70. Why does this work? When we want to compute the time necessary for an investment to double, we're saying

$$A = 2P.$$

Hence

$$t = \frac{\ln(2P/P)}{\ln(1 + r)},$$

or

$$t = \frac{\ln 2}{\ln(1 + r)}.$$

Thus $\quad t \ln (1 + r) = \ln 2.$

Now, $$\ln 2 = 0.69315 \qquad \text{(to five decimal places)}$$
$$\cong 0.70.$$

In addition, it can be shown that for small values of r (such as the interest rates in our problems)

$$\ln (1 + r) = r.$$

Putting these facts together,

$$t \cdot r = 0.70.$$

To express r in percents, multiply both sides by 100 to obtain

$$t \cdot r \text{ (in percents)} = 70. \quad \square$$

Practice Problem

a) At what rate compounded monthly will \$4000 grow to \$6000 in 5 years?

b) How long does it take for \$1000 to double at 6% compounded annually?

Answers: **a)** 6.7806% per year; **b)** 11.8957 years. The rule of thumb gives $70 \div 6 \cong 12$ years.

Continuous Compounding

You may have noticed in Example 12 and the following Practice Problem that as we compounded more often the amount of money, A, became larger at a very gradual rate. Moreover, as the rate of compounding increased, the increase in the amount of money became less and less. It would seem that no matter how often we compounded, A can't get any larger than (about) \$135. Is this the case? And can we actually find the exact amount?

There is a method of compounding, called **continuous compounding,** that provides the answer to both of these questions. Think about the values in Table 10.2 from Example 12.

Here we start with a yearly interest rate (r) of 10% and a period of 3 years (t). Next we decrease the length of the period but increase the number of periods so that their product is 0.3 ($r \cdot t$) in every case. Let N be the number of periods in 1 year, so that $3N$ will be the number of periods in 3 years' time and $0.1/N$ will be the interest rate per period. Then

$$A = P(1 + r)^t$$

will be given by

$$A = P\left(1 + \frac{0.1}{N}\right)^{3N}.$$

You may recall from Chapter 9 that we can approximate e^x for irrational values of x by using

$$\left(1 + \frac{1}{n}\right)^{ny},$$

where y is a good rational approximation of x and n is large. Suppose we have a very large value of N (such as we would have if we compounded every hour). Then the expression

$$P\left(1 + \frac{0.1}{N}\right)^{3N}$$

will be a very good approximation of

$$Pe^{0.1t}, \qquad \text{where } t = 3N = \text{number of years.}$$

In fact, we define **continuous compounding** as follows:

Continuous Compounding

$$A = Pe^{rt}, \tag{10.7}$$

where r = rate per year,
t = number of years.

Continuous compounding actually pays the most interest for a given annual rate, r. Some banks and savings and loans advertise continuous compounding (either directly or by the phrase "the most"). For the data of Example 12, with $P = \$100$, $r = 0.10$, and $t = 3$ years, continuous compounding gives us

$$A = 100e^{0.10 \cdot 3},$$
$$A = 100e^{0.3},$$
$$A \cong \$134.9858808,$$

so A is approximately \$134.98. This number is the limit we referred to earlier. (There is very little difference between compounding daily and continuous compounding in terms of results. However, the computations are easier with continuous compounding.)

EXAMPLE 16

A sum of \$500 is invested for 5 years at 12% compounded continuously. What is the final amount of money?

SOLUTION

We have $P = \$500$, $r = 0.12$, and $t = 5$, so

$$A = 500e^{0.12 \cdot 5},$$
$$A = 500e^{0.6},$$
$$A = \$911.06 \qquad \text{(approximately)}. \qquad \square$$

Practice Problem

What is the final value of \$4000 invested for 8 years at 8% compounded continuously?

Answer: \$7585.92

A word of caution. Banks often use a 360-day year for business purposes (as we've been doing). However, some banks will correct for the fact that a calendar year has 365 days by multiplying the interest rate by the factor 365/360. Thus for banks that do this, in Example 16 we would have

$$r = 0.12 \cdot \left(\frac{365}{360}\right),$$

and

$$A = 500e^{0.12 \cdot \left(\frac{365}{360}\right) \cdot 5},$$
$$A = \$918.68 \qquad \text{after 5 years.}$$

Effective Rate

A common method for comparing investment opportunities is by computing an **effective rate of interest.** This is the annual simple interest rate that would give the same interest as the stated or nominal rate of interest. The easiest way to compute the effective rate is to set $P = \$1$ and then compute A. The difference between A and \$1 will tell you the effective rate.

EXAMPLE 17

Compute the effective rates for money invested at 10% compounded semiannually, monthly, daily, and continuously.

SOLUTION

Semiannually: $r = 0.05$ and $t = 2$ so

$$A = 1(1 + 0.05)^2,$$
$$A = 1.1025,$$

so $A - 1 = 0.1025$. The effective rate is 10.25%.

Monthly: $r = 0.10/12$ and $t = 12$.

$$A = 1(1 + 0.10/12)^{12},$$
$$A = 1.104713,$$

so $A - 1 = 0.104713$. The effective rate is 10.4713%.

Daily: $r = 0.10/360$ and $t = 360$.

$$A = 1(1 + 0.1/360)^{360},$$
$$A = 1.1051556,$$

so $A - 1 = 0.1051556$ and the effective rate is 10.51556%.

Continuously: $r = 0.10$ and $t = 1$.

$$A = e^{0.1 \cdot 1}$$
$$A = 1.1051709,$$

so $A - 1 = 0.1051709$ and the effective rate is 10.51709%. □

Practice Problem

Compute the effective rates for an investment at 6% compounded monthly, daily, and continuously.

Answers: 6.16778%, 6.18310%, and 6.18365%

EXERCISE SET 10.2

Using formulas (10.5) and (10.6), compute the missing values.

1. $P = \$50{,}000$, $r = 0.01$, $t = 360$. $A = ?$
2. $P = \$40{,}000$, $r = 0.015$, $t = 360$. $A = ?$
3. $A = \$50{,}000$, $P = \$40{,}000$, $t = 50$. $r = ?$
4. $A = \$2000$, $P = \$1500$, $t = 80$. $r = ?$
5. $A = \$3050$, $r = 0.01$, $t = 81$. $P = ?$
6. $A = \$950$, $r = 0.02$, $t = 10$. $P = ?$
7. $A = \$750$, $P = \$600$, $r = 0.01$. $t = ?$
8. $A = \$10{,}000$, $P = \$9950$, $r = 0.005$. $t = ?$
9. How much is \$500 invested at 12% compounded annually for 10 years worth?

10. How much is \$40,500 invested at 8% compounded quarterly for 5 years worth?
11. How much is \$60,000 invested at 11% compounded monthly for 30 years worth?
12. How much is \$250 invested at 7% compounded semiannually for 8 years worth?
13. What is the present value of \$50,000 at 15% compounded monthly for 5 years?
14. What is the present value of \$10,000 at 6% compounded quarterly for 10 years?
15. What is the present value of \$25,000 at 5.2% compounded weekly for 8 years?
16. What is the effective rate of 15% compounded monthly?
17. What is the effective rate of 6% compounded daily?
18. What is the effective rate of 15% compounded continuously?
19. What is the effective rate of 6% compounded continuously?
20. How much is \$10,000 worth after continuous compounding at 10% for 10 years?
21. How much is \$2500 worth after continuous compounding at 9% for 6 months?
22. By solving formula (10.7) for P, find the present value of \$5000 compounded continuously for 5 years at 12%.
23. What is the present value of \$250 compounded continuously at 5% for 2 years?

Population growth can be modeled by the continuous interest formula (10.7) with P = original population, A = population after t periods, and r = growth rate per period.

24. In 1970 a certain town had a population of 15,000 and in 1980 it had a population of 35,000.
    - **a)** What was the rate of growth over the 10-year period?
    - **b)** Assuming the same rate of growth, what will the population be in 1990?
25. Suppose world population grows at 2% per year. How long will it take the population to double? to triple?
26. A colony of bacteria doubles in size every 4 hours.
    - **a)** What is the hourly growth rate?
    - **b)** Suppose the colony has 100 members at $t = 1$ hour. How many bacteria will be present (assuming unchecked growth) after 10 hours? after 5 days?
27. A new antibiotic will kill half the population of bacteria exposed to it every 2 hours. What is the hourly decay rate? (*Hint:* How is this different from compound growth?)

28. A certain radioactive substance has a **half-life** of 1500 years, which means that one-half of the original substance undergoes radioactive decay to a stable isotope every 1500 years. How many years does it take for a mass of 10 kilograms to decay to 1 kilogram? to 0.1 kilogram?

29. A radioactive isotope of carbon, called carbon-14, occurs naturally. Carbon-14 has a half-life of 5600 years. Since it occurs naturally, all living things that have carbon present within them also contain carbon-14. Moreover, the amount of carbon-14 is constantly renewed. However, upon death the carbon-14 begins to decay without being replenished. This fact can be used to date when a once-living thing died. Since the half-life is 5600 years, an object that contains one-half the carbon-14 naturally present must be 5600 years old. Suppose that 75% of the carbon-14 normally present is measured. How old is the object? How old is the object if it contains 25% of the carbon-14 normally present?

10.3 ORDINARY ANNUITIES

An **annuity** is a sequence of equal periodic payments. The payments can be either paid into or taken out of an account, but in both cases, the calculations utilize compound interest. We will assume the time between payments is the same as the period of compounding. (This assumption is made to simplify the calculations. It is not necessary in general.)

Almost all of us encounter annuities at one time or another; buying a home or a car, buying life insurance (whole life), or paying off an educational loan are all examples of annuities. Businesses use annuities to pay off loans or to generate funds to pay for future purchases.

We will study three different types of annuities.

1. Ordinary annuities, in which the payments are made at the end of each period.
2. Annuity due, in which the payments are made at the beginning of each period.
3. Deferred annuities, in which the stream of payments does not begin until a certain amount of time has elapsed.

Future Values and Sinking Funds

As we noted earlier, an **ordinary annuity** is an annuity in which the payments are made at the end of each period. Let's study an example.

EXAMPLE 18

Suppose you put \$1000 *each* year (at the end of the year) into a savings account that pays 10% compounded annually. How much would your account be worth at the end of 3 years?

SOLUTION

Study Fig. 10.5. Notice that

1. At the beginning, there is no money in the account.
2. After 1 year, you deposit \$1000.
3. After the second year, your original deposit has become \$1100. You add another \$1000 deposit so the account is worth \$2100.
4. At the end of the third year, the account grows from \$2100 to \$2310. You deposit another \$1000, giving you a final balance of \$3310.

Time	0	1	2	3
Old Balance	0	0	1000	2100
Interest		+ 0	+ 100	+ 210
Payment		+ 1000	+ 1000	+ 1000
New Balance		1000	2100	3310

Figure 10.5 **The behavior of an ordinary annuity.** □

The final balance of an annuity is called its **future value.** As you can see from the previous example, an annuity that extended for more than three or four periods would be very tedious to deal with. Fortunately, we can derive a formula to tell us the future value of an ordinary annuity.

Future Value (Ordinary Annuity)

$$S = R\left[\frac{(1+i)^n - 1}{i}\right] = R \cdot s_{\overline{n}|\,i} \qquad (10.8)$$

where

S = future value of the annuity,
R = periodic payment,
i = interest rate per period,
n = number of periods.

(Note the unusual notation, $R \cdot s_{\overline{n}|\,i}$. Tables of $s_{\overline{n}|\,i}$ are available in a variety of places. Here we would rather you employed your calculator because a calculator can always be used, while the tables you require may not be present.)

Before showing you how formula (10.8) can be derived, let's apply it to two examples.

EXAMPLE 19

Compute the future value of an ordinary annuity in which \$1000 is deposited yearly for 3 years into an account paying 10% compounded yearly.

SOLUTION

We have $R = \$1000$, $i = 0.10$, and $n = 3$, so

$$S = \$1000\left[\frac{(1.10)^3 - 1}{0.10}\right] \qquad (\text{or } 1000 \cdot s_{\overline{3}|\,0.10}).$$

On your calculator this becomes

1.1 [y^x] 3 [=] [−] 1 [=] [÷] 0.1 [=] [×] 1000 [=],

which gives $S = \$3310.00$ □

EXAMPLE 20

Suppose an annuity is funded by paying \$100 monthly into an account paying 1% compounded monthly. What is the future value after 10 years?

SOLUTION

We have $R = \$100$, $i = 0.01$, and $n = 10 \cdot 12 = 120$, so

$$S = \$100\left[\frac{(1.01)^{120} - 1}{0.01}\right] \qquad (\text{or } 100 \cdot s_{\overline{120}|\,0.01}).$$

On your calculator

1.01 [y^x] 120 [=] [−] 1 [=] [÷] .01 [=] [×] 100 [=],

which gives us $S = \$23{,}003.87$. □

Let's see how we can develop formula (10.8). Consider an annuity of \$100 deposited yearly for 5 years into an account paying 5% compounded yearly. Figure 10.6 shows we could find S by applying

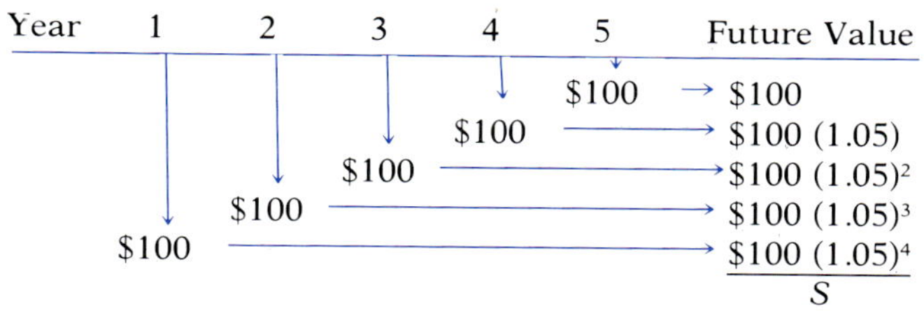

Figure 10.6 **Deriving the formula for the future value of an ordinary annuity.**

the compound interest formula (10.5) repeatedly. We have

$$S = 100(1.05)^4 + 100(1.05)^3 + 100(1.05)^2 + 100(1.05) + 100. \qquad (A)$$

The next step involves a bit of a trick, but it works nicely, so pay attention. Multiply both sides of (A) by 1.05, which gives

$$1.05 \cdot S = 100(1.05)^5 + 100(1.05)^4 + 100(1.05)^3 + 100(1.05)^2 + 100(1.05). \qquad (B)$$

Now subtract (A) from (B) to obtain

$$1.05 \cdot S - S = 100(1.05)^5 - 100,$$
$$S(1.05 - 1) = 100[(1.05)^5 - 1],$$
$$S(0.05) = 100[(1.05)^5 - 1],$$

so

$$S = 100\left[\frac{(1.05)^5 - 1}{0.05}\right],$$

as desired. This approach can be extended to any payment, R, or any interest rate, i, without any difficulty. To extend this to any number of periods, n, we must examine (A) and (B). A little thought should convince you that (A) will always start with $R(1 + i)^{n-1}$. Thus (B) will always start with $R(1 + i)^n$. Hence, when we subtract (A) from (B), we'll have $R(1 + i)^n$. Thus, by performing the subsequent steps as shown, we obtain formula (10.8).

(*Note:* If you studied the section on sequences and series in Chapter 4, you should recognize that formula (10.8) is simply the formula for the nth term of a geometric series.)

Practice Problem

Compute the future value of the following ordinary annuities.

a) A quarterly payment of $500 for 20 years, at 2% per quarter.

b) A monthly payment of $25 for 12 months, at 1% per month.

Answers: **a)** $96,885.98; **b)** $317.06

Suppose we know that after n periods we must have an amount S available. Given the interest rate per period, i, can we compute the necessary periodic payment? Clearly the answer is yes. In this case, the annuity is being used to meet a future obligation; we refer to it as a **sinking fund.** Formula (10.9) is found by solving formula (10.8) for R:

Sinking Fund (Ordinary Annuity)

$$R = S\left[\frac{i}{(1+i)^n - 1}\right] = \frac{S}{s_{\overline{n}|\,i}}, \tag{10.9}$$

where
R = sinking fund payment,
S = future value of the annuity,
n = number of periods,
i = interest rate per period.

EXAMPLE 21

Friends of yours have a baby. They wish to provide $30,000 for the baby's education 18 years from now by making monthly deposits into a savings account that pays 8% per year, compounded monthly. What should the monthly payments be?

SOLUTION

We have $S = \$30{,}000$, $i = 0.08/12$, and $n = 18 \cdot 12 = 216$, so

$$R = \$30{,}000\left[\frac{0.08/12}{\left(1+\dfrac{0.08}{12}\right)^{216} - 1}\right].$$

(Note that since 0.08/12 is a repeating decimal, we have avoided rounding it off to prevent errors.) On your calculator this is

0.08 [÷] 12 [=] [STO] [+] 1 [=] [y^x] 216 [=] [−] 1 [=] [$1/x$] [×] [RCL] [×] 30000 [=],

which yields $R = \$62.4888$. In other words, a monthly deposit of $62.49 each month for 18 years will be worth $30,000. (Notice that if you simply multiplied $62.49 by 216 months, you get only $13,497.84. The difference between this amount and $30,000 is the interest earned.) □

Practice Problem

What monthly sinking fund payment is required to provide a lump sum of $250,000 in 30 years, assuming an interest rate of 9% per year, compounded monthly?

Answer: $136.56

Present Values: Amortization

To **amortize** a loan is to pay it off. When the amount of the loan is large, the amortization is often done by using an annuity. Here, however, we want an annuity with a given present value, not future value, because the annuity must have a value equivalent to the present (current) value of the loan. The **present value** of an ordinary annuity is given by formula (10.10):

Present Value (Ordinary Annuity)

$$A = R\left[\frac{1 - (1 + i)^{-n}}{i}\right] = R \cdot a_{\overline{n}|\,i}, \qquad (10.10)$$

where

A = present value of the annuity,
R = periodic payment,
i = interest rate per period,
n = number of periods.

(Again, you can find values of $a_{n|i}$ in tables. However, the interest rate you require may not be present. Therefore, we recommend you use a calculator.)

Let's study an example before we derive the formula.

EXAMPLE 22

Jane and Jack both want to make contributions to the Rams Club. Jack has pledged $50 per year (on which the club can earn 8%) for 10 years. How much should Jane give now to match the present value of Jack's contribution?

SOLUTION

We have $R = \$50$, $i = 0.08$, and $n = 10$, so

$$A = \$50\left[\frac{1 - (1.08)^{-10}}{0.08}\right].$$

On your calculator, enter

1.08 [y^x] 10 [+/−] [+] 1 [=] [÷] 0.08 [=] [×] 50 [=],

which gives us $A = \$335.50$. In other words, Jane must give \$335.50 now to match Jack's gift. □

Practice Problem

What is the present value of an ordinary annuity that consists of quarterly payments of \$500 for 20 years, assuming an interest rate of 6% per year?

Answer: \$23,203.66

Now let's develop formula (10.10). We know that formula (10.8) gives us the **future value,** S, of an ordinary annuity. Moreover, we know the present value, say A, of a sum of money, S, can be found by formula (10.6): $A = S(1 + i)^{-n}$. Hence, to find the **present value,** A, of the annuity, we have

$$A = S(1 + i)^{-n},$$

and replacing S by its value from formula (10.8) we have

$$A = R\left[\frac{(1 + i)^n - 1}{i}\right](1 + i)^{-n},$$

so by multiplying we have

$$A = R\left[\frac{1 - (1 + i)^{-n}}{i}\right],$$

as desired.

We can also rearrange formula (10.10) to solve for R, the amount necessary to amortize a present value of A in n payments at an interest rate of i per period:

Amortization (Ordinary Annuity)

$$R = A\left[\frac{i}{1 - (1 + i)^{-n}}\right] \qquad (10.11)$$

where R = periodic payment,
A = amount to be amortized,
i = interest rate per period,
n = number of periods.

EXAMPLE 23

Joan buys a house for \$75,000. Her down payment is \$15,000, so that the balance of \$60,000 must be mortgaged. Joan's bank gives her a 30-year mortgage at an annual rate of 12%. What are the monthly payments, assuming compounding each month?

SOLUTION

We have $A = \$60{,}000$, $i = 0.12/12 = 0.01$, and $n = 12 \cdot 30 = 360$.

$$R = \$60{,}000\left[\frac{0.01}{1 - (1.01)^{-360}}\right].$$

On your calculator enter

1.01 [y^x] 360 [+/−] [=] [+/−] [+] 1 [=] [1/x] [×] 0.01 [×] 60000 [=],

which gives us $R = \$617.17$ per month. □

Practice Problem

In Example 23, find R if $i = 15\%$ per year.

Answer: \$758.67

Now we can solve the Problem from Life that began this chapter.

EXAMPLE 24

Problem from Life, part (a)

1986 OLDSMOBILE CIERA
FOR ONLY
\$9753 or
\$189.68/month

Selling price \$9753. Down payment \$2400 cash or trade-in. Amount to be financed \$7353 for 48 months. Finance charge \$1751.64, annual percentage rate 10.9%. Taxes, license, and title fee not included.

How was the payment of \$189.68 per month determined?

SOLUTION

This question is easy to answer: The monthly payment is the amount necessary to amortize \$7353. The computations can be made with formula (10.11) and $A = \$7353$, $i = 0.109/12$, and $n = 48$.

$$R = \$7353\left[\frac{0.109/12}{1-\left(1+\frac{0.109}{12}\right)^{-48}}\right].$$

On your calculator enter

0.109 [÷] 12 [=] [STO] [+] 1 [=] [y^x] 48 [+/−] [=] [+/−] [+] 1 [=] [$1/x$] [×] [RCL] [×] 7353 [=],

which gives us $R = \$189.68$ per month.

The **finance charge** of a loan is the difference between the amount borrowed and the amount repaid. In this case, the amount borrowed is \$7353, while the amount repaid is the total of the monthly payments: $48 \cdot \$189.68 = \$9104.64$. Therefore the finance charge is

$$\$9104.64 - \$7353 = \$1751.64. \quad \square$$

EXAMPLE 25

Problem from Life, part (b) What would the monthly payments be if you wanted to pay off the loan in 36 months rather than 48 months?

SOLUTION

We have the same problem except that $n = 36$. On your calculator enter

0.109 [÷] 12 [=] [STO] [+] 1 [=] [y^x] 36 [+/−] [=] [+/−] [+] 1 [=] [$1/x$] [×] [RCL] [×] 7353 [=],

which gives us

$$R = \$240.38 \text{ per month.} \quad \square$$

EXAMPLE 26

Problem from Life, part (c) What would the monthly payment be if you had \$2000 cash or trade-in instead of the \$2400?

SOLUTION

Again, we have the same problem as in Example 24 except that $A = \$7753$, not \$7353. The result will be

$$R = \$200.00 \text{ per month.} \quad \square$$

EXAMPLE 27

Problem from Life, part (d) How much cash or trade-in value is required to limit the payment to $150.00 per month?

SOLUTION

To solve this problem you must compute the present value of 48 payments of $150.00 per month. The difference between this figure and the $9753 price of the car will be the required down payment. Here we must use formula (10.10) with $R = \$150$, $n = 48$, and $i = 0.109/12$:

$$A = \$150 \cdot \left[\frac{1 - \left(1 + \frac{0.109}{12}\right)^{-48}}{0.109/12} \right].$$

On your calculator enter

0.109 [÷] 12 [=] [STO] [+] 1 [=] [y^x] 48 [+/−] [=] [+/−] [+] 1 [=] [÷] [RCL] [=] [×] 150 [=],

which gives us $A = \$5814.63$. The difference between this and the selling price of the car is

$$\$9753 - \$5814.63 = \$3938.37,$$

which is the required down payment. □

EXERCISE SET 10.3

Use formulas (10.8), (10.9), (10.10), and (10.11) to solve the following problems.

1. $R = \$100$, $i = 0.015$, $n = 20$. $S = ?$
2. $R = \$250$, $i = 0.01$, $n = 360$. $S = ?$
3. $R = \$400$, $i = 0.0125$, $n = 125$. $S = ?$
4. $R = \$50$, $i = 0.0075$, $n = 30$. $S = ?$
5. $S = \$50{,}000$, $i = 0.04/12$, $n = 125$. $R = ?$
6. $S = \$25{,}000$, $i = 0.01$, $n = 100$. $R = ?$
7. $S = \$40{,}000$, $i = 0.0125$, $n = 80$. $R = ?$
8. $S = \$250{,}000$, $i = 0.005$, $n = 480$. $R = ?$
9. $A = \$30{,}000$, $i = 0.02$, $n = 20$. $R = ?$
10. $A = \$75{,}000$, $i = 0.015$, $n = 180$. $R = ?$
11. $A = \$125{,}000$, $i = 0.0075$, $n = 360$. $R = ?$
12. $A = \$500$, $i = 0.005$, $n = 24$. $R = ?$

13. $R = \$120,\quad i = 0.04/12,\quad n = 20.\quad A = ?$
14. $R = \$250,\quad i = 0.0025,\quad n = 300.\quad A = ?$
15. $R = \$500,\quad i = 0.01,\quad n = 120.\quad A = ?$
16. $R = \$225,\quad i = 0.00125,\quad n = 360.\quad A = ?$
17. What is the value of an ordinary annuity after 20 years if \$200 is deposited at the end of each month in an account paying 12% per year, compounded monthly?
18. A couple estimates that they will need \$20,000 in cash to buy a house 5 years from today. How much should they deposit each month into an account paying 6% per year, compounded monthly?
19. Paul wants to establish a sinking fund for his retirement. He estimates that he will need \$250,000 in 20 years. How much must he pay monthly into an account paying 8% per year, compounded monthly?
20. A company estimates it will need \$500,000 in 5 years to replace some equipment. What are the required sinking fund payments assuming the account will earn 12% per year, compounded quarterly?
21. Sam and Jane were recently left a bequest of \$1,000,000 by a rich uncle. Assuming he established a sinking fund 20 years ago, how much did he deposit each week into an account paying 5.2% per year, compounded weekly?
22. Jim funded an ordinary annuity at \$100 per month for 5 years. The account payed 6% per year, compounded monthly. He then stopped making payments for 2 years, although the account continued drawing interest. He then resumed his monthly payments for another 3 years. What was the account worth after the 10 years?
23. Paul can buy a new car for \$8000. The dealer has offered Paul the car for \$225 per month for 4 years. (The loan has an interest rate of 14% per year, compounded monthly.) Should Paul pay cash or pay \$225 per month?
24. Jane is buying a house for \$60,000. Assuming she puts 20% down in cash, how much will her monthly payments be if her mortgage has a rate of 15% per year (compounded monthly) and a length of 30 years?
25. Jose bought a new color TV for \$650. He borrowed the money from his credit union at 12% per year, compounded monthly. How much will his monthly payments be if he pays off the TV in 2 years?
26. Ernie and Carmen want to buy a house. They estimate that they can afford a monthly payment of at most \$700. Assuming a 30-year mortgage at an interest rate of 13.5% per year compounded monthly, how much can they finance? Given that they have \$15,000 available for a down payment, what is the most expensive house they can afford?
27. Sue wants to establish a retirement annuity that will pay her \$2500 per month (at the end of the month) for 15 years. Sue will retire in 25 years. How much must she deposit into the account at the end of each month (for 25 years) to fund her retirement? Assume that the money will earn 10% per year, compounded monthly.

10.4

ANNUITY DUE

An **annuity due** is an annuity in which the payments are made at the **beginning** of each period rather than at the end. The authors have seen an IRA (individual retirement account) advertisement using annuity due to increase the total amount of money generated. In addition, life insurance companies use annuity due rather than an ordinary annuity.

Future Values and Sinking Funds

The following example demonstrates the difference between an ordinary annuity and annuity due.

EXAMPLE 28

Suppose you put $1000 *each* year (at the beginning of the year) into a savings account that pays 10% compounded annually. How much would your account be worth at the end of 3 years?

SOLUTION

Study Fig. 10.7. Notice that

1. At the beginning of year 1 (now), $1000 is deposited.
2. After one year, your account earns interest of $100.
3. At the beginning of the second year, you deposit another $1000.
4. At the end of the second year, your account earns interest of $210.
5. At the beginning of the third year, you deposit another $1000.
6. At the end of the third year, your account earns interest of $331, so that the final balance is $3641. (Compare this to Example 18 and Fig. 10.5, which describe an ordinary annuity.)

Year	1		2		3	
	Start	End	Start	End	Start	End
Old Balance	0	1000	1100	2100	2310	3310
Interest	0	100	0	210	0	331
Payment	1000	0	1000	0	1000	0
New Balance	1000	1100	2100	2310	3310	3641

Figure 10.7 **The behavior of an annuity due.** □

As you can see from this example, an annuity due earns an extra period of interest compared to an ordinary annuity. Therefore the future value of an annuity due is $(1 + i)$ times that of an ordinary annuity:

Future Value (Annuity Due)

$$S = R \cdot (1 + i) \cdot \left[\frac{(1 + i)^n - 1}{i}\right] = R \cdot (1 + i) \cdot s_{\overline{n}|\, i}, \qquad (10.12)$$

where
S = future value of the annuity,
R = periodic payment,
i = interest rate per period,
n = number of periods.

EXAMPLE 29

Mr. Smith will be 57 on his next birthday (April 1). He has decided to deposit \$500 each quarter until he is $59\frac{1}{2}$ into an account paying 12% per year, compounded quarterly. How much will be in the account when he is $59\frac{1}{2}$?

SOLUTION

Study the diagram in Fig. 10.8. Mr. Smith will make a total of 10 deposits. We have $n = 10$, $i = 0.12/4 = 0.03$, and $R = \$500$, so

$$S = 500 \cdot (1.03) \cdot \left[\frac{1.03^{10} - 1}{0.03}\right].$$

On your calculator this is

1.03 [y^x] 10 [=] [−] 1 [=] [÷] 0.03 [=] [×] 1.03 [×] 500 [=],

which gives us S = \$5903.90. Mr. Smith will withdraw \$5903.90 when he is $59\frac{1}{2}$ years old.

Mr. Smith's Age	57				58				59	
Month	Apr	Jul	Oct	Jan	Apr	Jul	Oct	Jan	Apr	Jul
Deposit (\$)	500	500	500	500	500	500	500	500	500	500

Figure 10.8 **Mr. Smith's annuity due account, Example 29.** □

Practice Problem

What is the future value of an annuity due if payments of \$200 are made at the beginning of each quarter for 10 years into an account earning 6% per year, compounded quarterly?

Answer: \$11,016.38

We can solve formula (10.12) for R to determine the periodic payment necessary for an annuity due to have a future value of S:

Sinking Fund (Annuity Due)

$$R = S\left\{\frac{i}{(1+i)\cdot[(1+i)^n - 1]}\right\} = \frac{S}{(1+i)\cdot s_{\overline{n}|i}}, \tag{10.13}$$

where
R = sinking fund payment,
S = future value of annuity,
i = interest rate per period,
n = number of periods.

EXAMPLE 30

A man estimates he will need \$10,000 in 5 years to start his daughter in college. How much must he invest at the beginning of each quarter if the first payment is due now and interest is 8% per year, payable quarterly?

SOLUTION

Here we have S = \$10,000, $i = 0.08/4 = 0.02$, and $n = 20$, so

$$R = \$10{,}000\left\{\frac{0.02}{(1.02)[(1.02)^{20} - 1]}\right\}.$$

On your calculator enter

1.02 [y^x] 20 [=] [−] 1 [=] [×] 1.02 [=] [1/x] [×] 0.02 [×] 10000 [=],

which gives us R = \$403.50. The man must deposit \$403.50 each quarter. □

Practice Problem

A company estimates it will need \$500,000 to upgrade its computer system in 3 years. The appropriate officers have decided to make quarterly payments, beginning now, into to a fund that pays 16% per year, compounded quarterly. What must the payments be?

Answer: $31,996.24 per quarter

Present Value

As with an ordinary annuity, the present value of an annuity due is the value of the annuity at the beginning of the term of the annuity. Formula (10.14) gives the present value of an annuity due.

Present Value (Annuity Due)

$$A = R \cdot (1 + i) \cdot \left[\frac{1 - (1 + i)^{-n}}{i}\right] = R \cdot (1 + i) \cdot a_{\overline{n}|\, i}, \qquad (10.14)$$

where
A = present value of annuity,
R = periodic payment,
i = interest rate per period,
n = number of periods.

Before we derive formula (10.14), let's demonstrate its use.

EXAMPLE 31

What is the present value of an annuity due consisting of quarterly payments of $500 for 7 years, at 12% per year, compounded quarterly?

SOLUTION

Here we have $R = \$500$, $i = 0.12/4 = 0.03$, and $n = 28$, so

$$A = \$500 \cdot (1.03) \cdot \left[\frac{1 - (1.03)^{-28}}{0.03}\right].$$

On your calculator enter

1.03 [y^x] 28 [+/−] [=] [+/−] [+] 1 [=] [÷] 0.03 [=] [×] 1.03 [×] 500 [=],

which gives us $A = \$9663.52$. □

Practice Problem

What is the present value of an annuity due consisting of monthly payments of $250, beginning now, for four years at an interest rate of 15% per year, compounded monthly?

Answer: $9095.16

We can derive formula (10.14) from formula (10.12) by remembering that the present value, A, of a sum of money, S, is defined by formula (10.6): $A = S(1+i)^{-n}$. Hence to find the present value, A, of the annuity, we have

$$A = S(1+i)^{-n}$$

and replacing S by its value from formula (10.12) we have

$$A = R \cdot (1+i) \cdot \left[\frac{(1+i)^n - 1}{i}\right] \cdot (1+i)^{-n},$$

$$A = R \cdot (1+i) \cdot \left[\frac{1-(1+i)^{-n}}{i}\right],$$

as desired.

We can also rearrange formula (10.14) to solve for R, the amount necessary to amortize a present value of A in n payments at an interest rate of i per period:

Amortization (Annuity Due)

$$R = A \cdot \left\{\frac{i}{(1+i) \cdot [1-(1+i)^{-n}]}\right\}, \qquad (10.15)$$

where
R = periodic payment,
A = amount to be amortized,
i = interest rate per period,
n = number of periods.

EXAMPLE 32

A television set may be purchased for 36 monthly payments, beginning now. The cash price is \$695.00 and the interest rate is 18% per year, compounded monthly. What is the monthly payment?

SOLUTION

We have A = \$695.00, $i = 0.18/12 = 0.015$, and $n = 36$, so

$$R = \$695 \cdot \left\{\frac{0.015}{(1.015) \cdot [1-(1.015)^{-36}]}\right\}.$$

On your calculator enter

1.015 [y^x] 36 [+/−] [=] [+/−] [+] 1 [=] [×] 1.015 [=] [1/x] [×] 0.015 [×] 695 [=],

which results in R = \$24.75 per month. □

Practice Problem

Compute the monthly payment in Example 32 for the television set to be paid off in 2 years.

Answer: \$34.18

EXERCISE SET 10.4

Use formulas (10.12), (10.13), (10.14), and (10.15) to solve the following problems.

1. $R = \$100$, $i = 0.015$, $n = 20$. $S = ?$

2. $R = \$250$, $i = 0.01$, $n = 360$. $S = ?$

3. $R = \$400$, $i = 0.0125$, $n = 125$. $S = ?$

4. $R = \$50$, $i = 0.0075$, $n = 30$. $S = ?$

5. $S = \$50{,}000$, $i = 0.04/12$, $n = 125$. $R = ?$

6. $S = \$25{,}000$, $i = 0.01$, $n = 100$. $R = ?$

7. $S = \$40{,}000$, $i = 0.0125$, $n = 80$. $R = ?$

8. $S = \$250{,}000$, $i = 0.005$, $n = 480$. $R = ?$

9. $A = \$30{,}000$, $i = 0.02$, $n = 20$. $R = ?$

10. $A = \$75{,}000$, $i = 0.015$, $n = 180$. $R = ?$

11. $A = \$125{,}000$, $i = 0.0075$, $n = 360$. $R = ?$

12. $A = \$500$, $i = 0.005$, $n = 24$. $R = ?$

13. $R = \$120$, $i = 0.08/12$, $n = 20$. $A = ?$

14. $R = \$250$, $i = 0.0025$, $n = 300$. $A = ?$

15. $R = \$500$, $i = 0.01$, $n = 120$. $A = ?$

16. $R = \$225$, $i = 0.00125$, $n = 360$. $A = ?$

17. Don Andrews just paid the last of 120 quarterly life insurance premiums of \$75 each. Had he been able to invest each payment at 8% per year, compounded quarterly, what would the total amount be?

18. What is the cash price of a washing machine that can be bought for \$35 a quarter for 3 years if the first payment is due now and interest is 24% per year, compounded quarterly?

19. The rent on a building for two years is \$4000, payable in advance. What is the equivalent monthly rental (assuming rent is due the first of each month) if interest is 13% per year, compounded monthly?

20. A \$4500 debt due in 5 years is to be discharged in five equal annual payments, the first one due immediately. Find the annual payment if interest is 10% per year.

21. Anita Jones decides to invest \$500 per quarter in an IRA beginning on her twenty-fifth birthday. She will make her last payment one quarter before her sixty-fifth birthday. Assuming the IRA will earn 9% per year, compounded quarterly, how much will she have in her account on her sixty-fifth birthday?

22. Philip Potts can buy a used car with a price of \$2400 for 20 equal monthly payments, beginning now. Assuming the interest rate is 12% per year, compounded monthly, what should each payment be?
23. Samantha Hall wants to be able to draw an annuity of \$2000 per month at the beginning of each month for 20 years after she retires. She plans to retire in 30 years. How much must she invest at the beginning of each month (starting now) to fund her retirement, assuming interest is 9% per year, compounded monthly?

10.5 DEFERRED ANNUITY

A **deferred annuity** is an ordinary annuity that begins after a number of periods have passed (been deferred). For example, a student might borrow money to go to college and begin paying the loan back after graduation. Furniture stores and automobile dealers are also frequent users of deferred annuities.

The future value, S, of a deferred annuity is the same as the future value of an ordinary annuity. Hence we can use formula (10.8) to find the future value of a deferred annuity.

EXAMPLE 33

On July 1, 1985, Mrs. Harrison decided to buy a new bedroom suite. She will begin putting \$100 each month into an account earning 15% per year, starting on December 31, 1985. She will have accumulated enough money to buy the furniture on July 31, 1986, after making the last payment. How much will she pay for the furniture?

SOLUTION

Here we have $R = \$100$, $i = 0.15/12 = 0.0125$, and $n = 8$, so

$$S = \$100 \cdot \left[\frac{(1.0125)^8 - 1}{0.0125}\right],$$

which yields $S = \$835.89$. The cash price was \$835.89.

Note: This annuity was deferred a total of 5 months. It is important for you to remember that in an ordinary annuity, the payments are made at the end of the period. □

Practice Problem

In Example 33, what would the cash price of the furniture be if the store charged interest of 18% per year, compounded monthly?

Answer: \$843.28

Determining the periodic payments for a deferred annuity with a given future value is exactly the same as for an ordinary annuity. Hence we can use formula (10.9).

EXAMPLE 34

The Anthonys want to establish an annuity for their children's college educations. They estimate they will need \$25,000 in 15 years. However, they will be unable to begin the payments for 3 years. Assuming the Anthonys expect to make quarterly payments into an account paying 9% per year, compounded quarterly, what should the payments be?

SOLUTION

The most difficult part of this problem is determining the amount of deferment. Normally, the first payment would be due at the end of the first quarter. Here, however, the Anthonys will not make a payment during the first 12 quarters. Therefore, they have deferred payment for 11 quarters. Since there are a total of 60 quarters in 15 years, by not paying for the first 11 the Anthony's must fund the \$25,000 in 49 payments. Thus we have $S = \$25{,}000$, $i = 0.09/12 = 0.0075$, and $n = 49$. Using formula (10.9) we have

$$R = \$25{,}000 \cdot \left[\frac{0.0075}{(1.0075)^{49} - 1}\right],$$

which gives us $R = \$424.07$. Each quarter, the Anthonys must deposit \$424.07 (once 3 years have elapsed). □

Practice Problem

In Example 34, suppose the Anthonys were able to begin paying after waiting only 2 years. How many payments would they make and how large would each payment be?

Answer: 53 payments of \$385.89 each

Computing the present value and the periodic payment necessary to amortize a deferred annuity is more complicated. Here the amount of time deferred will appear in the formulas. Let's take a look at an example that will help us derive the formula for the present value of a deferred annuity.

EXAMPLE 35

Jack Thompson bought a used car on July 1. The first monthly payment of \$200 is due December 31 (of the same year) and the last payment is due on July 31. Interest is 9.9% per year, compounded monthly. How much did the car cost?

SOLUTION

Let's begin by determining the length of the deferment and the length of the annuity. Study Fig. 10.9. Notice that the length of deferment is 5 months, while the length of the deferred annuity is 8 months. Therefore the total time frame is 13 months.

We want to compute the price of the car. We can do this by computing the present value of the deferred annuity. Let's think how we might do that. Suppose we could compute the value of the deferred annuity at its starting point, December 1. (Remember the first payment of an ordinary annuity occurs at the end of the first period.) Then we could use the present value formula (10.6) to compute the value on July 1. Next, how can we compute the value on December 1? We can use formula (10.10) to compute the present value of an ordinary annuity beginning December 1. Putting these steps together, we multiply the results of formula (10.10) by the results of formula (10.6) to obtain the present value.

Therefore let

$$
\begin{aligned}
A &= \text{price of car} = \text{its present value on July 1},\\
B &= \text{value of deferred annuity on December 1},\\
n &= 8 = \text{length of deferred annuity},\\
m &= 5 = \text{interval of deferment},\\
i &= 0.099/12 = 0.00825 = \text{interest rate per month},\\
R &= \text{periodic payment} = \$200 \text{ per month}.
\end{aligned}
$$

Since the present value on July 1 of a sum on December 1 is given by $A = B(1.00825)^{-5}$ and the value of the deferred annuity on

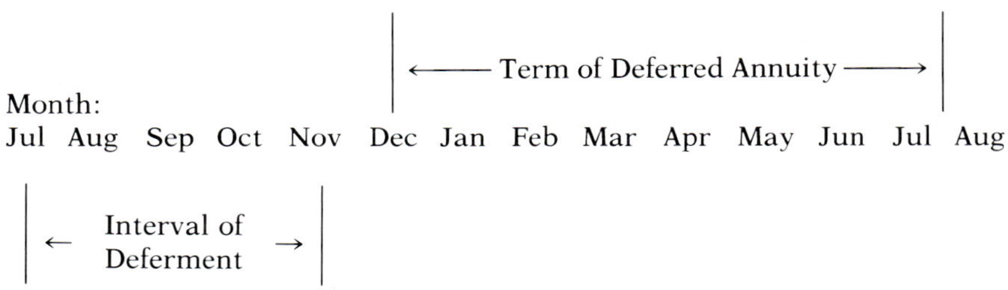

Figure 10.9 **Time schedule of deferment and payments, Example 35.**

December 1 is given by

$$B = R \cdot \left[\frac{1 - (1.00825)^{-8}}{0.00825}\right],$$

we have

$$A = \$200 \cdot (1.00825)^{-5} \cdot \left[\frac{1 - (1.00825)^{-8}}{0.00825}\right].$$

On your calculator enter

1.00825 [y^x] 8 [+/−] [=] [+/−] [+] 1 [=] [÷] 0.00825 [=] [STO] [C] 1.00825 [y^x] 5 [+/−] [=] [×] 200 [×] [RCL] [=],

which gives us $A = \$1480.13$ as the price of the car on July 1. (Note that we have to store the intermediate result. You might think this is unnecessary. Try it without storing the result and see what happens.) □

Now we can generalize this result. To find the present value of a deferred annuity, find the value of the annuity at the beginning of the annuity and then multiply it by the present value factor determined by the length of the interval of deferment. This results in formula (10.16):

Present Value (Deferred Annuity)

$$A = R \cdot (1 + i)^{-m} \cdot \left[\frac{1 - (1 + i)^{-n}}{i}\right], \tag{10.16}$$

where

A = present value of the annuity,
R = periodic payment,
i = interest rate per period,
n = number of payments,
m = interval of deferment.

Practice Problem

What would the cost of the car have been in Example 35, if the payments were \$250 each and they started on November 31?

Answer: \$2090.10

To determine the periodic payment necessary to amortize a deferred annuity, we can solve the present value formula (10.16) for

R. This gives us formula (10.17):

Amortization (Deferred Annuity)

$$R = A \cdot (1 + i)^m \cdot \left[\frac{i}{1 - (1 + i)^{-n}}\right], \tag{10.17}$$

where
R = periodic payment,
A = amount to be amortized,
i = interest rate per period,
n = number of payments,
m = interval of deferment.

EXAMPLE 36

Joanna Moody borrowed $2000 on January 1, 1986, in order to complete her degree. She has agreed to pay off the loan in 24 equal monthly payments. The first payment is due July 31, 1987. Interest is 6% per year, compounded monthly. How large is each payment?

SOLUTION

Here we have $A = \$2000$, $i = 0.06/12 = 0.005$, $n = 24$, and $m = 18$ (be careful!). Therefore

$$R = \$2000 \cdot (1.005)^{18} \cdot \left[\frac{0.005}{1 - (1.005)^{-24}}\right].$$

On your calculator enter

1.005 [y^x] 24 [+/−] [=] [+/−] [+] 1 [=] [1/x] [×] 0.005 [=] [STO] [C] 1.005 [y^x] 18 [=] [×] 2000 [×] [RCL] [=],

which gives us $R = \$96.97$. Joanna can repay the loan at $96.97 per month for 24 months, after waiting 18 months to begin payments. □

Practice Problem

In Example 36, what would the monthly payments be if

a) the payments began on January 31, 1987?

b) the interest rate were 12% per year and payments began on July 31, 1987?

Answers: **a)** $94.11; **b)** $112.61

EXERCISE SET 10.5

Use formulas (10.8), (10.9), (10.16), and (10.17) to solve the following problems.

1. $R = \$200,\quad i = 0.01,\quad n = 8,\quad m = 11.\quad S = ?\quad A = ?$
2. $R = \$300,\quad i = 0.04,\quad n = 10,\quad m = 5.\quad S = ?\quad A = ?$
3. $R = \$350,\quad i = 0.03,\quad n = 10,\quad m = 20.\quad S = ?\quad A = ?$
4. $R = \$200,\quad i = 0.06,\quad n = 8,\quad m = 20.\quad S = ?\quad A = ?$
5. $A = \$1000,\quad i = 0.03,\quad n = 20,\quad m = 6.\quad R = ?$
6. $A = \$3000,\quad i = 0.005,\quad n = 60,\quad m = 24.\quad R = ?$
7. $A = \$6500,\quad i = 0.01,\quad n = 48,\quad m = 1.\quad R = ?$
8. $A = \$1000,\quad i = 0.02,\quad n = 25,\quad m = 5.\quad R = ?$
9. Your parents invested \$1000 the day you were born in a savings account that paid 4% per year, compounded quarterly. You want to withdraw the money in 10 equal quarterly payments, beginning 3 months after your eighteenth birthday. How large are the payments?
10. What single payment now is equivalent to 40 semiannual payments of \$400 each, with the first payment due 4 years from now? Assume an interest rate of 12% per year, compounded semiannually.
11. Pete Jones borrowed \$8000 at 12% interest compounded monthly in order to buy a car. The loan is to be repaid in 48 months beginning 2 months from the day he purchased the car. What are the monthly payments? What would the payments be if the payments were due to begin 1 month after he purchased the car?
12. Tom and Elise Donovan bought \$4000 of furniture. The loan is to be repaid in 60 monthly payments at an interest rate of 13.2% per year, compounded monthly. The payments don't begin for 6 months. What are the monthly payments?
13. Sharon Sax borrowed \$1000 one year ago in order to go to school last year. Today she borrowed another \$1500. Monthly payments on both amounts will begin 12 months from today, and extend for 36 months. Interest is 7.2% per year, compounded monthly. What will her total monthly payment be?

CHAPTER REVIEW

IMPORTANT TERMS AND SYMBOLS

10.1 Simple Interest

interest A fee paid to use money.
present value The value today (represented by P) of a sum of money in the future (represented by A).

principal Amount borrowed or loaned; represented by P.
simple interest Computed assuming principal does not change. Formulas:

$$I = Prt \tag{10.1}$$
$$A = P(1 + rt) \tag{10.2}$$
$$P = \frac{A}{1 + rt} \tag{10.3}$$

where A = total amount after time t, I = interest, P = principal (present value), r = interest rate per period, and t = number of periods.

banker's discount Form of simple interest loan in which the interest is deducted from the principal *before* the money is loaned. Formula:

$$d = \frac{r}{1 - rt} \tag{10.4}$$

where d = effective (actual) rate, r = nominal (stated) rate, and t = length of loan.

time payments Can be analyzed using arithmetic sequences.

10.2 Compound Interest

compound interest Computed assuming the principal changes. Formulas:

$$A = P(1 + r)^t \tag{10.5}$$
$$P = A(1 + r)^{-t} \tag{10.6}$$

where A = total amount after time t, P = principal (present value), r = rate per period, and t = number of periods.

continuous compounding Special form of compound interest. Formula:

$$A = P \cdot e^{rt} \tag{10.7}$$

where A = total amount after time t, P = principal (present value), e = base of natural logarithms, r = rate per year, and t = number of years.

effective rate of interest Annual simple interest rate that would give the same result as the stated or nominal compound rate; used to compare different nominal rates.

10.3 Ordinary Annuities

amortization Pay off a loan with equal periodic payments.
annuity A sequence of equal periodic payments.
future value of annuity Value at end of sequence of payments.
ordinary annuity Payments are made at the end of each period.

sinking fund Series of equal periodic payments that have a known future value. Ordinary annuity formulas:

$$S = R \cdot \left[\frac{(1+i)^n - 1}{i}\right] \tag{10.8}$$

$$R = S \cdot \left[\frac{i}{(1+i)^n - 1}\right] \tag{10.9}$$

$$A = R \cdot \left[\frac{1 - (1+i)^{-n}}{i}\right] \tag{10.10}$$

$$R = A \cdot \left[\frac{i}{1 - (1+i)^{-n}}\right] \tag{10.11}$$

where S = future value of ordinary annuity, R = periodic payment, A = present value of annuity, i = interest rate per period, and n = number of periods.

10.4 Annuity Due

annuity due Annuity in which payments are made at beginning of each period. Formulas:

$$S = R \cdot (1+i) \cdot \left[\frac{(1+i)^n - 1}{i}\right] \tag{10.12}$$

$$R = S \cdot \left\{\frac{i}{(1+i) \cdot [(1+i)^n - 1]}\right\} \tag{10.13}$$

$$A = R \cdot (1+i) \cdot \left[\frac{1 - (1+i)^{-n}}{i}\right] \tag{10.14}$$

$$R = A \cdot \left\{\frac{i}{1+i) \cdot [1 - (1+i)^{-n}]}\right\} \tag{10.15}$$

where S = future value of annuity due, R = periodic payment, A = present value, i = interest rate per period, and n = number of periods.

10.5 Deferred Annuity

deferred annuity An ordinary annuity that begins after a number of periods have passed.

interval of deferment Time elapsed before deferred annuity payments begin. Deferred annuity formulas:

$$A = R \cdot (1+i)^{-m} \cdot \left[\frac{1 - (1+i)^{-n}}{i}\right] \tag{10.16}$$

$$R = A \cdot (1+i)^{m} \cdot \left[\frac{i}{1 - (1+i)^{-n}}\right] \tag{10.17}$$

where A = present value of deferred annuity, R = periodic payment, i = interest rate per period, n = number of payments, and m = interval of deferment.

REVIEW PROBLEMS

1. Tony borrowed $2500 for 9 months at 10.8% simple interest. How much interest did he pay and how much did he repay when the loan was due?
2. Suppose in Problem 1 that Tony borrowed the money with banker's discount. How much did he actually receive? How much interest did he pay? What was the effective rate?
3. Ike and Tina charged $8000. Interest is 1.75% of the unpaid balance each month. How much total interest will they pay if the charge is paid off in 4 years?
4. Tom and Jane have been bequeathed $15,000. They decide to put it into a savings account. After 5 years, how much will the account be worth assuming the interest rate is 8.4% and is compounded:

   a) quarterly **b)** monthly **c)** daily **d)** continuously?
5. In Problem 4, compute the effective rate for each rate of compounding.
6. Bob and Barbara have decided to establish a "vacation fund." They deposit $100 at the end of each month into an account earning 9.6% interest, compounded monthly. How much will they have at the end of 2 years?
7. Doug and Frank have formed a partnership. They estimate they will need more office equipment costing $25,000 in 15 months. They want to establish a sinking fund paying 9% per year, compounded monthly, to pay for the equipment. What should the monthly payments be?
8. Angie and Dick have a monthly mortgage payment of $625. The mortgage has a term of 30 years and an interest rate of 13.2% per year, compounded monthly. How much did they borrow? Assuming they made a 20% down payment, what was the total cost of the house?
9. Fred Miller wants to have $4000 cash as a down payment on a car at the end of 12 months. How much must he deposit at the beginning of each month to an account paying 6.6% per year, compounded monthly?
10. Laura and Larry have begun saving $125 out of their paychecks every month. Assuming they are paid on the first of the month, how much will they have after 3 years if interest is 6% per year, compounded monthly?
11. Arnold borrowed $5000 on June 1, 1983, to help finance his college education. He must begin paying off the loan on September 1, 1987, with quarterly payments. Interest is 5.4% per year, compounded quarterly. Assuming the loan is to be paid off in 10 years, how much should the payments be?
12. Arnold's older sister is currently paying off an educational loan with monthly payments of $55. She will make a total of 48 payments. The payments started 36 months after she borrowed the money. Assuming interest is 4.8% per year, compounded monthly, how much did she borrow originally?

11 PROBABILITY

In a certain town there are three fast food restaurants near the university: Barry's Burgers, Guido's Pizza, and Tio Pepe's tacos. During the last month, approximately 250 students per day bought dinner at each of the three restaurants. A recent survey (sponsored by the marketing club) has shown that students have a tendency to switch restaurants:

Of Barry's regular customers, 50% are loyal, 20% will switch to Guido's next month, and 30% will switch to Tio Pepe's next month.

Of Guido's regular customers, 50% are loyal, 25% will switch to Barry's next month, and 25% will switch to Tio Pepe's next month.

Of Tio Pepe's regular customers, 50% are loyal, 20% will switch to Barry's next month, and 30% will switch to Guido's next month.

a) How many customers per day can each restaurant expect next month?
b) What will the long-term market shares be? (To be solved as Example 28, Section 11.6.)

PROBLEM FROM LIFE

INTRODUCTION

Many applications of mathematics occur in situations in which there is no way of determining the exact answer to a question because of some uncertainty. To aid us in our analysis in these cases, we often use mathematical models that help us determine the probability (or likelihood) of certain events occurring.

The study of probability arose in the sixteenth and seventeenth centuries in western Europe in the analysis of certain gambling problems. In fact, it wasn't until this century that a formal (theoretical) basis for the study of probability was provided. Today, the use of probability methods is very widespread. Areas of application include quantum mechanics, reliability, statistical analysis, decision making in an uncertain environment, genetics, and (of course) gambling.

In this chapter we will study the basic concepts of probability, which will enable us to solve a variety of problems and also to prepare you for our discussions of statistics (Chapter 12) and of decision theory and game theory (Chapter 13).

Historically, there have been two basic approaches to studying probability. One is the **classical or theoretical approach,** which relies upon counting the number of ways an uncertain event might occur, assuming that all possible outcomes are equally likely, and then assigning probabilities. The second is the **empirical or frequency approach,** which uses past history to predict the likelihood of an event occurring in the future. Because of difficulties with both of these approaches, modern mathematicians have used an **axiomatic approach** based upon the use of sets.

11.1 BASICS

Experiments, Events, and Sample Spaces

We are all familiar with experiments in science. Moreover, we know that certain experiments will produce the same results when repeatedly conducted with the correct procedure. We say these experiments are **deterministic** because there is *no* uncertainty involved in them (except perhaps to the degree that the conditions for the experiment are duplicated). However, there are also experiments for which we cannot predict exactly the outcome. Some examples are flipping a coin, determining which atom will undergo radioactive decay, knowing how many textbooks will be sold each semester in a bookstore, and knowing how long a light bulb will last. An experiment in which the outcome cannot be predicted before the experiment is performed is called a **random experiment.**

We will use the concept of a random experiment to develop our approach to probability. From this point on, the term **experiment** will refer to a random experiment. Let's also agree to broaden the concept of experiment to include any activity (real or imagined) that leads to an observation (or measurement). The set of all possible outcomes of an experiment is called a **sample space** of the experiment. In many cases, we can model an experiment with more than one sample space. The choice of sample space depends upon the events we wish to analyze, as the following example demonstrates.

EXAMPLE 1 Suppose we toss two coins. What is an appropriate sample space?

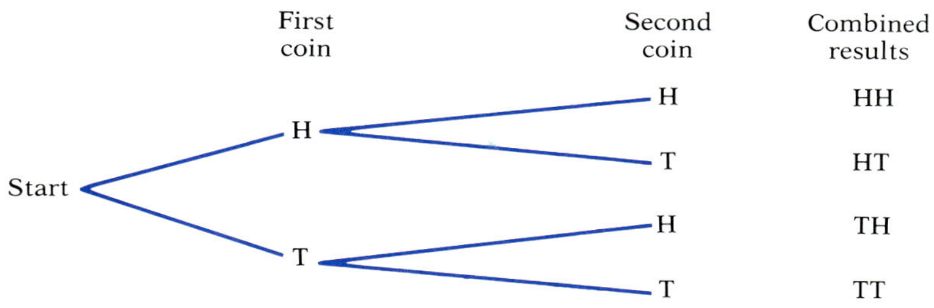

Figure 11.1 **A sample space for tossing two coins.**

SOLUTION

The correct answer depends on what we are interested in determining.

a) If we are interested in whether each coin shows a head (H) or a tail (T), we can use a **tree diagram** to determine an appropriate sample space. (See Fig. 11.1.) Thus the sample space is given by S = (HH, HT, TH, TT), so that there are four outcomes in the sample space.

b) If we are interested simply in the number of heads that appear, then an appropriate sample space is $S = \{0, 1, 2\}$, which contains only three outcomes.

c) Finally, we might be concerned only with whether the coins match (M) or not match (N), so that $S = \{M, N\}$ is an appropriate sample space. □

You should be aware that the use of certain sample spaces can lead to incorrect analysis unless you think very carefully about the situation. The distinction between the sample spaces in parts (a) and (b) of the preceding example was a crucial step in the development of probability theory. We'll discuss this in more detail later.

Practice Problem

Suppose a family has two children. What is an appropriate sample space

a) if you are interested in the sex of each child in the order of their birth?

b) if you are interested in the number of girls?

c) If you are interested in whether or not the children have the same sex?

Answers: **a)** Letting M represent a male child and F a female, we have $S = \{MM, MF, FM, FF\}$, **b)** $S = \{0, 1, 2\}$; **c)** Letting Y represent a match and N no match, we have $S = \{Y, N\}$

Consider part (a) of Example 1. Suppose we are interested in the event E = "exactly one tail occurs." A brief look at S will tell us that E occurs if either of the outcomes HT or TH occurs. We can represent this symbolically as $E = \{HT, TH\}$. Thus an **event** is a subset of the sample space S. We say that an event is **simple** when it consists of a single outcome. An event is a **compound event** when it consists of more than one outcome. We say an event **occurs** if any of the outcomes in the subset occurs.

EXAMPLE 2

Consider an ordinary deck of 52 playing cards. Such a deck contains four suits: spades, hearts, diamonds, and clubs. Suppose we draw one card and note its suit. An appropriate sample space is $S = \{Sp, He, D, C\}$. State whether each of the following events is simple or compound.

a) The suit is red.

b) The suit is black.

c) The suit is clubs.

d) The suit is not clubs.

SOLUTION

a) Since both hearts and diamonds are red, this event can be represented by $R = \{He, D\}$. This is a compound event.

b) Both spades and clubs are black, so we can represent this event by $B = \{Sp, C\}$: again, a compound event.

c) This is a simple event.

d) We can use $N = \{Sp, He, D\}$ to represent this event, so that it is compound. □

Practice Problem

A coin is flipped and the suit of a card drawn from a standard deck is noted. Which of the following are simple and which are compound events? (*Hint:* First determine an appropriate sample space.)

a) A head is observed on the coin.

b) A tail is observed and the suit is black.

c) A head is observed and the suit is hearts.

Answers: Letting H and T represent the results of the coin flip and Sp, He, D, and C represent the suit of the card, the sample space is $S = \{HSp, HHe, HD, HC, TSp, THe, TD, TC\}$, so that there are eight outcomes. Then we have **a)** compound event; **b)** compound event; **c)** simple event

Defining Probability

Let S be a sample space and let E be an arbitrary event associated with some experiment. We know that E cannot occur more than 100% of the time, nor can it occur less than 0% of the time. Thus, it should be reasonable to require that the probability of E be a real number between zero and one, inclusive. Moreover, because one of the outcomes of the experiment must occur, it should be reasonable to require that the probability of S be one. Finally, because the empty set, ϕ, can never occur, we require that the probability of ϕ be zero. (In the context of probability, we often refer to ϕ as the **impossible event.**) Thus, we have the fundamental properties of probability.

Fundamental Properties of Probability

Let S be a sample space and let E be an arbitrary event associated with some experiment. Denoting the probability of E by $P(E)$, we require that

1. $0 \leq P(E) \leq 1$.
2. $P(S) = 1$.
3. $P(\phi) = 0$.

We can state Property 2 in a different form. Suppose S is a finite sample space (it has only a finite number of outcomes). Let $e_1, e_2, \ldots, e_n$ be the outcomes of S, so that $S = \{e_1, e_2, \ldots, e_n\}$. Then $\{e_1\}$ is the simple event in which outcome e_1 occurs, $\{e_2\}$ is the simple event that e_2 occurs, etc. Then

$$P(S) = 1$$

is equivalent to

$$P(\{e_1\}) + P(\{e_2\}) + \cdots + P(\{e_n\}) = \sum_{j=1}^{n} P(\{e_j\}) = 1.$$

We have now stated three fundamental properties of probabilities, but we still don't know how to compute probabilities. (In other words, we need more information.) In many cases, we begin by assuming that all outcomes are equally likely. Thus, for example, if a sample space has three equally likely outcomes, the probability of each occurring is $\frac{1}{3}$. Be careful here! The assumption of equally likely outcomes is not always valid. For example, it would be extremely unlikely that the probability of snow in Richmond, Virginia on July 1 is the same as the probability of snow on January 1. For a finite sample space in which the outcomes are equally likely, we have the following.

Probability of Outcomes: Equal Probability Model

$$P(\{e_i\}) = \frac{1}{n}, \qquad \text{where } S = \{e_1, e_2, \ldots, e_n\}$$

Let's look at some examples of such events.

EXAMPLE 3

a) The probability of drawing the king of hearts from a 52-card deck is given by $P(\text{king of hearts}) = 1/52$, assuming that each card is as likely to be drawn as any other.

b) The probability of drawing the three of clubs from a 52-card deck is $P(\text{three of clubs}) = 1/52$ by the same reasoning as given in part (a). □

EXAMPLE 4

a) Suppose one million people each buy one raffle ticket. Then the probability of a particular person winning is given by

$$P(\text{winning}) = \frac{1}{1{,}000{,}000} = 0.000001 = 10^{-6}.$$

b) A standard die has six sides, numbered one to six. Assuming that the die is fair (not weighted or shaved), the probability of rolling a one is given by $P(\text{one}) = \frac{1}{6}$. □

Practice Problem

What are the probabilities of

a) getting a head on the toss of a fair coin?

b) drawing the ace of diamonds from a 52-card deck?

Answers: **a)** 1/2; **b)** 1/52

Of course, not all events are simple. Consider the question, What is the probability of drawing a king or queen from a standard playing deck? Here we have the event

$$E = \{\text{K-Sp, K-He, K-D, K-C, Q-Sp, Q-He, Q-D, Q-C}\},$$

where K and Q represent king and queen, respectively, and Sp, He, D, and C represent the suits spades, hearts, diamonds and clubs, respectively. We can determine the probability of E using the following rules.

> **Probability of an Arbitrary Event (E)**
>
> 1. Determine an appropriate sample space, say S, for the experiment.
> 2. Assign probabilities to the outcomes in S.
> 3. Determine the simple events whose union is E. We define the probability of E by adding the probabilities of the outcomes found in E.
> 4. For the case in which all outcomes are equally likely,
>
> $$P(E) = \frac{\text{number of outcomes in } E}{\text{number of outcomes in } S}.$$

EXAMPLE 5

Recall the situation of Example 1, in which two fair coins were flipped. Compute the probability of the event "exactly one head occurs."

SOLUTION

We can make use of the same tree diagram we used previously. We've reproduced it here as Fig. 11.2. Let E be the event that exactly one head occurs. Given that S = {HH, HT, TH, TT}, we can see that E = {HT, TH}. Thus we have

$$P(E) = \frac{\text{number of outcomes in } E}{\text{number of outcomes in } S} = \frac{2}{4} = \frac{1}{2}. \quad \square$$

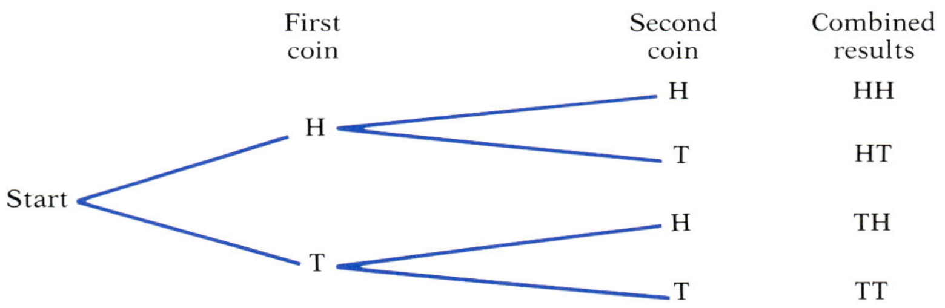

Figure 11.2 The sample space for Example 5 (flipping two coins).

EXAMPLE 6

Next, let's assume we flip three coins. Answer the following questions.

a) What is the probability of exactly three heads?
b) What is the probability of two tails and one head, in that order?
c) What is the probability of two tails and one head, in any order?

SOLUTION

As in Example 5, we can use a tree diagram to help us determine the sample space. (See Fig. 11.3.) Now we can see that S = {HHH, HHT, HTH, HTT, THH, THT, TTH, TTT}. Assuming the coins are fair, the probability of each of the outcomes in S is 1/8.

a) The event "exactly three heads" is given by E = {HHH}. Since this is a simple event, $P(E) = 1/8$.
b) The event "two tails and one head, in that order" is given by E = {TTH}, another simple event. Thus $P(E) = 1/8$.
c) The event "two tails and one head, in any order" is given by E = {HTT, THT, TTH}. Since E consists of three outcomes we have $P(E) = 3/8$. □

The analysis in these examples appears rather simple (we hope), but in fact it represents a significant step in the development of probability theory. For some period of time, people calculated the probability of observing one head when two coins are flipped as 1/3! The faulty reasoning was as follows. "When two coins are flipped, either we observe two heads, one head, or no heads. Assuming these are all equally likely simple events, the probability of one head must

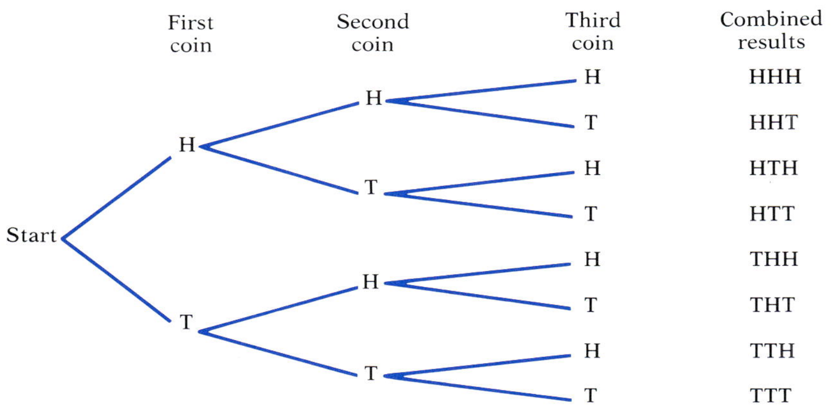

Figure 11.3 **The sample space for Example 6 (flipping three coins).**

be 1/3." We now know that these outcomes are *not* equally likely. Refer again to Example 1. The sample space $S = \{0, 1, 2\}$ is valid, but more difficult to use because the outcomes are not equally likely.

Practice Problem

Suppose a family has three children. Assuming that girls and boys are equally likely to occur at each birth, compute the probability of

a) two boys and one girl, in that order.

b) two boys and one girl, in any order.

c) at least one boy.

Answers: **a)** 1/8; **b)** 3/8; **c)** 7/8

People often refer to the **odds** of some event occurring rather than its probability. For example, we may see statements such as

The odds of a given horse winning a race are 1 to 5.
The odds of the incumbent being reelected are 2 to 1.

We can translate such statements into probability statements quite easily. Consider our first statement. It means that if the race were repeated often (under the same conditions) that the horse would win one race for every five races it didn't win. Thus the probability of the horse winning is $1 \div (1 + 5) = 1/6 \cong 0.167$. Similar reasoning will lead us to the conclusion that the probability of the incumbent winning is $2 \div (2 + 1) = 2/3 \cong 0.667$.

A general approach to converting statements about odds into statements about probability is as follows:

If the odds **in favor** of an event are "x" to "y" then the probability of the event occurring is $x/(x + y)$.

(*Note:* Be careful here. In many cases when odds are given, they are actually the odds of the event *not* happening. For example, in a horse race, the odds quoted are usually the odds of the horse not winning. Thus a horse going off at 10 to 1 odds has a probability of not winning of $10/(10 + 1) = 10/11 = 0.909$.)

Conversely, when the probability of an event occurring is given by $x/(x + y)$, then the odds in favor of the event are "x to y."

Empirical Probability

The original approach to probability was based on experience. Suppose we flipped a pair of coins 10,000 times and observed how often each event occurred (its frequency):

Event	Frequency
HH	2450
HT	2505
TH	2520
TT	2525

We could use these observed results to compute probabilities (or at least, estimates of them) as follows.

$$P(\text{HH}) = \frac{2450}{10{,}000} = 0.2450$$

$$P(\text{HT}) = \frac{2505}{10{,}000} = 0.2505$$

$$P(\text{TH}) = \frac{2520}{10{,}000} = 0.2520$$

$$P(\text{TT}) = \frac{2525}{10{,}000} = 0.2525$$

In addition, if we increased the number of flips, we would expect our (approximate) probabilities to agree more closely to predicted values.

The values we calculated above are relative frequencies. A more formal definition is as follows. Suppose we conduct an experiment n times and we note that event E occurred m times. Then the **relative frequency** of E is given by m/n. Now we can define the **empirical probability of event E:** It is the number $P(E)$ (assuming there is one) that the relative frequencies approach as n (the number of experiments) gets larger and larger. This approach is useful in many situations where calculation of probabilities by a theoretical approach is impossible. Moreover, the empirical approach is used extensively in statistics, as we'll see in Chapter 12.

EXAMPLE 7

One die is rolled 500 times with the following results.

Event	Frequency	Event	Frequency
1	90	4	82
2	101	5	88
3	79	6	60

Compute the empirical probabilities.

SOLUTION

$$P(1) = \frac{90}{500} = 0.180 \qquad P(2) = \frac{101}{500} = 0.202 \qquad P(3) = \frac{79}{500} = 0.158$$

$$P(4) = \frac{82}{500} = 0.164 \qquad P(5) = \frac{88}{500} = 0.176 \qquad P(6) = \frac{60}{500} = 0.120 \quad \square$$

EXAMPLE 8

Table 11.1 shows voter reactions to a new property tax plan according to party affiliation. Compute the empirical probabilities of a voter being (a) a Democrat and opposed to the plan, (b) an independent and in favor of the plan, (c) neutral, (d) Republican.

TABLE 11.1 Voter reactions to a new property tax plan

	Reaction			
Party affiliation	**In favor**	**Neutral**	**Opposed**	**Total**
Democratic	240	40	40	320
Republican	100	60	120	280
Independent	100	20	80	200
Total	440	120	240	800

SOLUTION

a) $P(\text{Democrat and opposed}) = 40/800 = 0.05$

b) $P(\text{independent and in favor}) = 100/800 = 0.125$

c) $P(\text{neutral}) = 120/800 = 0.15$

d) $P(\text{Republican}) = 280/800 = 0.35$ □

In situations such as shown in Example 8, there is no way to compute theoretical probabilities. Thus empirical probabilities are the only ones we can compute.

Practice Problem

Referring to Table 11.1 in Example 8, compute the empirical probabilities of a voter being

a) a Democrat and in favor of the plan.

b) an independent and opposed to the plan.

c) a Democrat.

Answers: **a)** 0.30; **b)** 0.1; **c)** 0.4

EXERCISE SET 11.1

A roulette wheel (in the United States) is marked with the numbers 1 to 36 and also 0 and 00. An experiment consists of spinning the wheel once and observing the resulting integer. Problems 1–7 refer to this situation.

1. Describe a sample space made up of equally likely simple events.
2. How many events are in the sample space?
3. What is the probability of observing a 13?
4. What is the probability of observing either 0 or 00?
5. What is the probability of observing a number (other than 0 or 00) which is even?
6. What is the probability of observing a number larger than 17?
7. Classify the events in Problems 3–6 as simple or compound.

An experiment consists of tossing four coins. Problems 8–12 refer to this experiment.

8. Find a sample space composed of equally likely simple events.
9. How many events are in the sample space?
10. What is the probability of observing exactly two heads?
11. What is the probability of observing at least one tail?
12. What is the probability of observing at least three heads?
13. Suppose two coins are tossed and that one of them is biased. Which of the following probability assignments for the simple events must be rejected? Why?

    a) $P(\text{HH}) = 0.35$, $P(\text{HT}) = 0.25$, $P(\text{TH}) = 0.30$, $P(\text{TT}) = 0.10$
    b) $P(\text{HH}) = 0.30$, $P(\text{HT}) = -0.2$, $P(\text{TH}) = 0.5$, $P(\text{TT}) = 0.4$
    c) $P(\text{HH}) = 0.35$, $P(\text{HT}) = 0.30$, $P(\text{TH}) = 0.15$, $P(\text{TT}) = 0.15$

14. In a local election there are three candidates running. The odds of some of them winning have been estimated as follows:

    Candidate A: 1 to 3
    Candidate B: 1 to 2

    a) Convert the given odds to probabilities.
    b) What is the probability of candidate B winning?

15. A basketball player has made 75 free throws in 105 attempts. What is the (approximate) empirical probability that the next free throw is successful?
16. A sample of 100 students in a university showed that 25 were taking 6 credit hours or less, 35 were taking between 7 and 12 credits, and 40 were taking 13 or more credits. Compute the (approximate) empirical probability of a student taking between 7 and 12 credit hours.

17. Four coins were flipped 500 times with the following results.

Event	Frequency
4 heads	30
3 heads	125
2 heads	175
1 head	135
0 heads	35

a) Compute the empirical probabilities of these events.
b) Compute the theoretical probabilities of these events assuming the coins are fair.

18. A sample of 200 customers in a clothing store gave the following results.

	Sex	
Age	**Male**	**Female**
Under 30	60	50
30 and over	80	10

a) What is the probability that a customer is under 30 and female?
b) What is the probability that a customer is 30 or over?

19. The results of a proposed expansion of the athletic programs at a university are shown as follows. "Lower division" includes students classified as freshmen or sophomores and "upper division" includes students classified as juniors or seniors.

	Class standing	
Reaction	**Lower division**	**Upper division**
In favor	75	80
Not sure	15	5
Opposed	30	35

a) What is the probability that a lower division student is opposed to the expansion?
b) What is the probability that an upper division student is in favor of the expansion?
c) What is the probability that a student is not sure about the expansion?

11.2

PROBABILITIES OF UNIONS, AND OF COMPLEMENTARY EVENTS

In the previous section you learned the definition of probability and saw some examples of how to compute some probabilities. There are many events, however, whose probability is awkward to compute simply by enumerating (listing) all the outcomes. The next three sections will teach you how to compute probabilities in more complex situations. Before you continue, you may wish to review the material on sets in Section 1.4.

Unions

Suppose we draw a card from a standard playing deck. Consider the following two questions:

1. What is the probability that it is either the king of hearts or a spade?
2. What is the probability that it is either a king or a spade?

Let's see how we can analyze these questions. Let A be the event that the card is the king of hearts, B be the event that the card is a spade, and C be the event that the card is a king. Because we're interested in exactly which card is drawn, there are 52 outcomes in the sample space, corresponding to the 52 cards in the deck. Then

$$A = \{\text{K-He}\},$$
$$B = \{\text{A-Sp, K-Sp, Q-Sp, J-Sp, 10-Sp, 9-Sp, 8-Sp, 7-Sp, 6-Sp, 5-Sp, 4-Sp, 3-Sp, 2-Sp}\},$$
$$C = \{\text{K-Sp, K-He, K-D, K-Ce}\}.$$

Note that the event "king of hearts or a spade" can be represented by $A \cup B$. Moreover, we can represent the event "a king or a spade" by $C \cup B$. Thus we can restate questions 1 and 2 more compactly as

1. What is $P(A \cup B)$?
2. What is $P(C \cup B)$?

To answer these questions, we need to be able to find the probability of the appropriate **union** of events.

Probability of the Union of Two Events

Let A and B be two events. Then $P(A \text{ or } B) = P(A \cup B)$ is given by one of the two formulas.

1. If $A \cap B = \phi$, then $P(A \cup B) = P(A) + P(B)$.
2. If $A \cap B \neq \phi$, then $P(A \cup B) = P(A) + P(B) - P(A \cap B)$.

Now we can answer the two questions we've been discussing. We decided that question 1 (the probability of the king of hearts or a spade) could be represented by $P(A \cup B)$, where A and B were defined appropriately. Notice that $A \cap B = \phi$ (the king of hearts is not a spade so the two sets are disjoint). Hence by formula 1 we have

$$P(A \cup B) = P(A) + P(B) = 1/52 + 13/52 = 14/52.$$

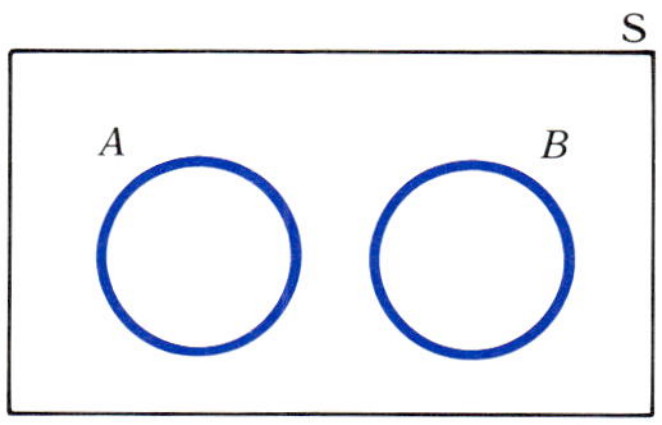

Figure 11.4 The union of disjoint events

Since the event $A \cup B$ contains precisely 14 simple events, this must be correct.

We represented question 2 (the probability of a king or a spade) by $P(C \cup B)$. Now, however, $C \cap B \neq \phi$ (because the king of spades belongs to both sets). Using formula 2,

$$P(C \cup B) = P(C) + P(B) - P(C \cap B) = 4/52 + 13/52 - 1/52 = 16/52.$$

Note that $P(C \cap B)$ is the probability that both C *and* B occur: In other words, it is the probability the card is both a king and a spade. The probability of that occurring must be 1/52.

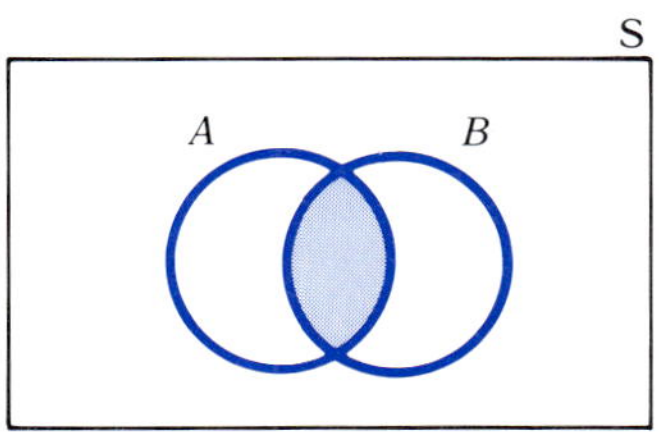

Figure 11.5 The union of nondisjoint events.

Let's see why the formulas for finding the probability of the union of two events work. First, consider the case in which the events A and B are disjoint; that is, $A \cap B = \phi$. As Fig. 11.4 shows, the probability of $A \cup B$ can be found by adding $P(A)$ and $P(B)$ because the total number of outcomes in $A \cup B$ is simply the sum of the number of outcomes in each set. Next, suppose that $A \cap B \neq \phi$. Figure 11.5 shows us that simply adding $P(A) + P(B)$ would *double count* those outcomes that occur in both sets. To correct for this, we must subtract the probability of the outcomes that occur in both sets. But this is $P(A \cap B)$, as stated in the formula.

Practice Problem

Suppose you throw one die. Compute the probability of

a) rolling a number less than 3 or more than 4.

b) rolling a 3 or a prime number.

Answer: **a)** $4/6 = 2/3$; **b)** $2/6 = 1/3$

Complementary Events

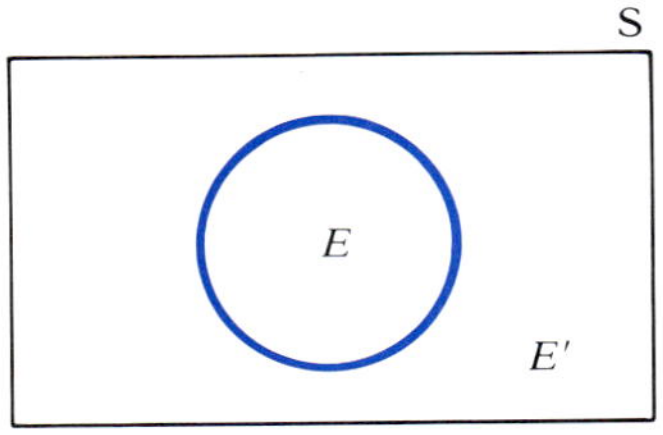

Figure 11.6 **Complementary events.**

Consider an event E in some sample space S. The event E' such that $E \cup E' = S$ and $E \cap E' = \phi$ is called the **complement of *E*.** E' is the event that E fails to occur. As demonstrated in Fig. 11.6, all the outcomes in S must be in either E or E'. (In other words, either E occurs or E' occurs.) Now, recall that $P(S) = 1$. By formula 1 for unions,

$$P(E \cup E') = P(E) + P(E')$$

but $S = E \cup E'$, so

$$P(E) + P(E') = 1.$$

Therefore it must be true that

$$P(E) = 1 - P(E') \quad \text{and} \quad P(E') = 1 - P(E).$$

Why is this of interest? In many situations, it is often easier to find the probability of the complement of an event than it is to find the probability of the event itself. By use of these results, we can then find the required probability.

EXAMPLE 9

a) Assume the probability of snow is estimated to be 0.20. What is the probability of no snow?

b) Suppose you flip five fair coins. What is the probability of obtaining at least one head?

SOLUTION

a) Probability of no snow = 1 − 0.2 = 0.8, because "no snow" is the complement of the event "snow."

b) One approach to this problem is to note that the event "at least one head" is equivalent to "one head or two heads or three heads or four heads or five heads." We can then compute the number of outcomes that correspond to each of these compound events and (eventually) obtain the desired result.

A much easier approach is to note that the complement of "at least one head" is "no heads" (in other words, all tails). This is a simple event with probability 1/32 (we'll leave it to you to prove this by drawing the appropriate tree diagram). Then

$$\begin{aligned} P(\text{at least one head}) &= 1 - P(\text{no heads}) \\ &= 1 - 1/32 \\ &= 31/32 = 0.96875. \quad \square \end{aligned}$$

Practice Problem

Compute the probabilities that

a) a family of four children has at least one girl (assuming boys and girls are equally likely).

b) it rains at least one day out of three, assuming the probability of rain each day is 0.5 (regardless of what happened the previous day).

Answers: **a)** 15/16 = 0.9375; **b)** 7/8 = 0.875

Making use of our knowledge of the probability of complementary events will enable us to gain some understanding of how redundant components increase reliability. Consider a mission in space that requires sophisticated computer backup. Suppose that a single computer has a probability of successful operation of 0.98. Now, while many people might be willing to bet some money given a probability of winning of 0.98, the situation is quite different when human lives and millions of dollars of equipment are at risk. By installing three computer systems in parallel, as shown in Fig. 11.7, we can increase greatly the probability of success. When we arrange three components (such as computers) in parallel, we assume that component two will take over when component one fails. Then if both components one and two fail, component three will take over. Thus the system will operate properly if at least one of the three components is operational.

We would like to compute the probability of successful operation. Based on the preceding discussion, this is equal to the probability that at least one component is operational. We can compute this by finding the probability of the complementary event "all components fail." Therefore (assuming that the computers act independently of one another),

$$\begin{aligned} P(\text{successful operation}) &= P(\text{at least one computer operational}) \\ &= 1 - P(\text{all three computers fail}) \\ &= 1 - 0.000008 \\ &= 0.999992. \end{aligned}$$

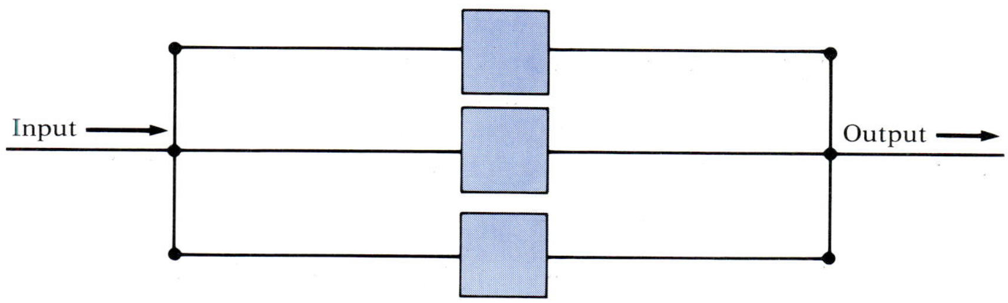

Figure 11.7 Three components in parallel increases reliability.

[We can imagine the question of whether a computer works properly or not to be equivalent to flipping a biased coin with a probability of heads of 0.98 and a probability of tails of 0.02. Then the probability of all three computers failing is equivalent to the probability of throwing three tails with our biased coin, which is (0.02)(0.02)(0.02) = 0.000008. We'll go into more detail later.]

EXERCISE SET 11.2

1. A computer system has a probability of successful operation of 0.98 during the next 100 hours. What is the probability of its failing during this time period?
2. The odds of a recession next year have been estimated to be 1 to 2. What are the odds there will be no recession?
3. A state is selected at random from the 50 states of the United States.

   a) What is the probability that it is either one of the six states in New England or a state whose name begins with a C?
   b) What is the probability that it is either one of the New England states or a state whose name begins with an M?

4. A number is to be chosen at random from the set of integers from 1 through 15. What is the probability that the number is

   a) less than 5 and even?
   b) more than 9 and odd?
   c) between 6 and 12 (inclusive) and even?

5. Four fair coins are flipped. What is the probability that there will be more heads than tails? What is the probability that the number of tails will be at least as large as the number of heads?
6. A certain part can be manufactured using either of two different machines. The probability of machine A being available is 0.9, the probability of machine B being available is 0.95 and the probability both machines are available simultaneously is 0.88. What is the probability that the part can be manufactured?

In Problems 7–13, an experiment consists of drawing one card from a 52-card deck. Compute the probability of drawing the following.

7. A king or a jack
8. A jack or a club
9. A club or a diamond
10. A heart or a four
11. A spade or an honor (an ace, king, queen, or jack)
12. A card other than a king or a jack
13. A card other than a spade or an honor

In Problems 14–20, a pair of dice are thrown. The sample space of equally likely events is as follows. Use it to compute the necessary probabilities.

First die	Second die 1	2	3	4	5	6
1	(1, 1)	(1, 2)	(1, 3)	(1, 4)	(1, 5)	(1, 6)
2	(2, 1)	(2, 2)	(2, 3)	(2, 4)	(2, 5)	(2, 6)
3	(3, 1)	(3, 2)	(3, 3)	(3, 4)	(3, 5)	(3, 6)
4	(4, 1)	(4, 2)	(4, 3)	(4, 4)	(4, 5)	(4, 6)
5	(5, 1)	(5, 2)	(5, 3)	(5, 4)	(5, 5)	(5, 6)
6	(6, 1)	(6, 2)	(6, 3)	(6, 4)	(6, 5)	(6, 6)

14. Compute the probability the sum of the dice is more than 7 or even.

15. Compute the probability that the sum of the dice is even or the dice show doubles (both dice show the same face).

16. Compute the probability that the sum of the dice is less than 7 or the dice show doubles.

17. Compute the probability the sum is 2 or 12.

18. Compute the probability the sum is neither 2 nor 12.

19. Compute the probability the sum is divisible by 3 or odd.

20. Compute the probability the sum is divisible by 5 or by 3.

11.3 CONDITIONAL PROBABILITY AND THE PROBABILITY OF INTERSECTIONS

Conditional Probability

Sometimes knowledge that an event occurred (or did not occur) may help compute the probability of another event occurring. For example, suppose a friend asks you the probability of snow today in Virginia. Your response should vary greatly, depending upon the month.

Let's look at an example that shows how knowledge of one event occurring alters the probability that another event occurs. Suppose a computer class contains 50 students. Of these, 35 are sophomores, 30 are business majors, and 10 are neither. A Venn diagram of this situation is shown in Fig. 11.8.

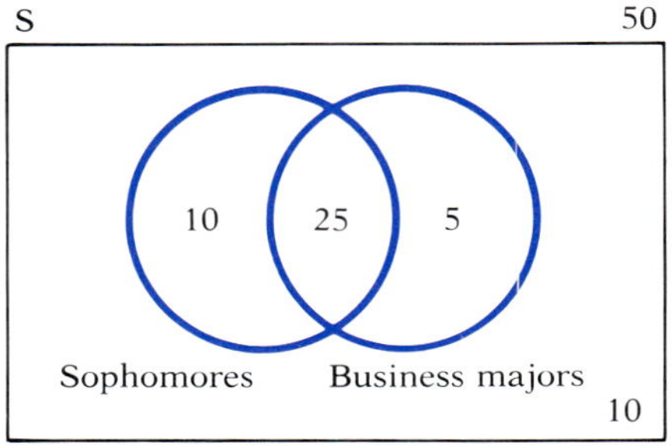

Figure 11.8 A computer class with 35 sophomores, 30 business majors, and 50 students.

To begin, let's compute the probability a student randomly selected from the class is a business major. Let B represent the set of business majors and T the set of sophomores, so the set of students who are both is the intersection of the sets B and T. Since there are 30 business majors out of the total of 50 students, we have

$$P(B) = 30/50 = 3/5 = 0.6.$$

Next, let's suppose we know the student selected is a sophomore. Now what is the probability the student is a business major? Since we are assuming the student is a sophomore, he or she must be one of the 35 students in the set T. In addition, for the student to be a business major implies the student also belongs to the set B. This means a business major would belong to both B and T, so the student would be one of the 25 students in the set $B \cap T$. Therefore the probability must be

$$P(B \text{ given } T) = 25/35 = 5/7 = 0.71429$$

because there are 25 students who are both sophomores and business majors out of the 35 sophomores. Therefore knowledge that the student is a sophomore provides us with additional information in determining if a student chosen at random is a business major.

In general, the probability of event A occurring, given that event B has occurred, is called the **conditional probability of A given B** and is denoted by $P(A \mid B)$. Conditional probabilities are computed as follows.

Conditional Probability of *A* Given *B*

$$P(A \mid B) = \frac{P(A \cap B)}{P(B)}, \qquad P(B) \neq 0$$

Think about our calculation of the probability that a student is a business major, given that he or she is a sophomore. Because we knew the student to be a sophomore, the probability of the student being a business major is different from the probability that a student chosen at random is a business major. Note in particular that knowledge of the student's class reduced the number of possibilities from 50 (the number of students in the class) to 35 (the number of sophomores in the class). When we compute the conditional probability $P(B \mid T)$ we are computing the probability that event B occurs *relative to a new sample space: in this case, the set* T.

EXAMPLE 10

In the preceding discussion we found $P(B \mid T)$ to be $5/7 = 0.71429$. Use the computational formula for conditional probability to verify this.

SOLUTION

We have $P(T) = 35/50 = 0.7$ and $P(B \cap T) = 25/50 = 0.5$. Thus

$$P(B \mid T) = \frac{P(B \cap T)}{P(T)} = \frac{0.5}{0.7} = \frac{5}{7} \cong 0.71429. \quad \square$$

EXAMPLE 11

In a given company of 1000 employees, 400 participate in the profit-sharing plan, 800 have major medical insurance, and 340 participate in both programs. What is the probability that

a) an employee chosen at random has major medical insurance?

b) an employee in the profit-sharing plan has major medical insurance?

SOLUTION

Let M represent the set of employees with major medical insurance and P represent the set of employees who participate in the profit-sharing plan. Then $M \cap P$ represents the set who participate in both programs. (See Fig. 11.9.)

a) The probability that an employee chosen at random has major medical insurance is $P(M) = 800/1000 = 0.8$.

b) The probability that an employee in the profit-sharing plan has major medical is

$$P(M \mid P) = 340/400 = 0.85$$

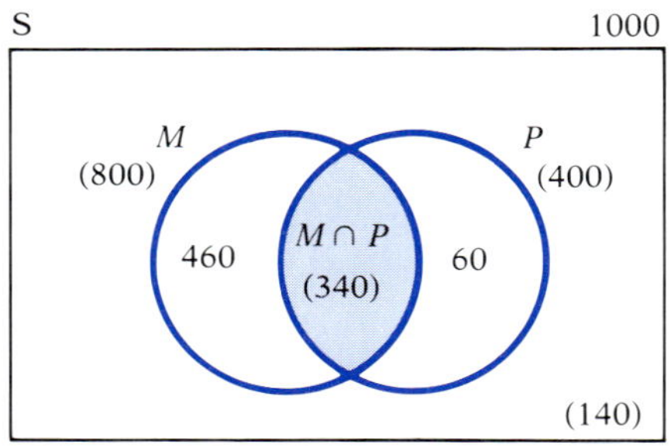

Figure 11.9 **A company of 1000 employees with profit-sharing and major medical plans.**

because there are 400 employees in the profit-sharing plan and 340 of them have major medical insurance.

We can also compute $P(M \mid P)$ using the formula for conditional probability:

$$P(M \mid P) = \frac{P(M \cap P)}{P(P)} = \frac{340/1000}{400/1000} = \frac{340}{400} = 0.85. \quad \square$$

Practice Problem

In Example 11, compute the probability that

a) an employee participates in profit-sharing given that he or she has major medical insurance.

b) an employee has major medical insurance given that he or she does not participate in the profit-sharing plan.

Answers: **a)** $P(P \mid M) = 340/800 = 0.425$;
b) $P(M \mid P') = 460/600 \cong 0.76667$

Independent Events

Knowledge that one event has occurred may not give us any information at all concerning the occurrence (or nonoccurrence) of another event. For example, suppose we flip two coins. Knowing the first coin showed heads tells us nothing about the second coin showing heads (and vice versa). A reasonable person might say the two events

$$\text{coin 1} = \text{heads} \quad \text{and} \quad \text{coin 2} = \text{heads}$$

are independent of one another. In fact, we say that events A and B are **independent** if

$$P(A \mid B) = P(A) \quad \text{or} \quad P(B \mid A) = P(B).$$

Two events that are not independent are said to be **dependent.**

EXAMPLE 12

The following data give the probability that a driver chosen at random in a city will have an accident during rush hour.

	Good weather (G)	Bad weather (G')	Total
Accident (A)	0.012	0.022	0.034
No accident (A')	0.678	0.288	0.966
Total	0.690	0.310	1.000

Are the events A (accident) and G' (bad weather) independent?

SOLUTION

The probability of an accident is given by $P(A) = 0.034$ (from the first row). The probability of an accident and bad weather is given by $P(A \cap G') = 0.022$ and the probability of bad weather is $P(G') = 0.310$. Therefore

$$\begin{aligned} P(A \mid G') &= \frac{P(A \cap G')}{P(G')} \\ &= \frac{0.022}{0.310} \\ &= 0.071 \\ &\neq P(A). \end{aligned}$$

Thus the events A (accident) and G' (bad weather) are *not* independent. □

Practice Problem

In Example 12, are the events A' (no accident) and G (good weather) independent?

Answer: no, because $P(A') = 0.966$ and $P(A' \mid G) = 0.983$

Be careful! Whether or not two events are independent does not always match your intuition. In the preceding example we would expect weather conditions to affect the accident rate, so that knowledge of the weather should be useful in predicting whether or not an accident will occur. As the next example shows, there may *not* appear to be a logical connection between the two events, at least on the surface.

EXAMPLE 13

A survey of an automobile dealer's lot showed the following results.

Number of cars with	Less than \$750 of options (L)	More than \$750 of options (L')	Total
Two doors (T)	75	125	200
Four doors (F)	100	100	200
Total	175	225	400

Are the events L and T independent?

SOLUTION

We must compute $P(L)$ and $P(L \mid T)$. From the data,

$$P(L) = 175/400 = 0.4375 \quad \text{and} \quad P(L \mid T) = 75/200 = 0.375.$$

Since $P(L) \neq P(L \mid T)$, the events L (less than \$750 of options) and T (two doors) are not independent, even though there appears to be no logical relationship between them. □

Practice Problem

In Example 13, are the events F and L independent?

Answer: no, because $P(L) = 0.4375$ and $P(L \mid F) = 0.5$

Probability of Intersections

We can make use of the concepts of conditional probability and independence to determine the probability of an intersection. Recall the formula for calculating $P(A \mid B)$:

$$P(A \mid B) = P(A \cap B)/P(B).$$

Using some simple algebra, we obtain the formula

$$P(A \cap B) = P(A \mid B) \cdot P(B).$$

However, suppose we know that A and B are independent. In such a case, $P(A \mid B) = P(A)$ and the relationship given for $P(A \cap B)$ simplifies to

$$P(A \cap B) = P(A) \cdot P(B).$$

We can summarize these results as follows.

Probability of an Intersection

1. $P(A \cap B) = P(A \mid B) \cdot P(B)$ when A and B are dependent or their relationship is unknown.
2. $P(A \cap B) = P(A) \cdot P(B)$ when A and B are independent.

One final point. By reversing our logic, we can conclude that if $P(A \cap B) = P(A) \cdot P(B)$, then A and B must be independent. Thus we have a test for independence.

Test for Independence

The events A and B are independent if $P(A \cap B) = P(A) \cdot P(B)$.

EXAMPLE 14

A company interviewed 50 candidates for computer programming positions. A total of 20 candidates had prior work experience, 15 had a professional certificate, and 10 had both. Are work experience and certification independent?

SOLUTION

Let W = the set of candidates with work experience and
C = the set of candidates with certification.

Then $P(W \cap C) = 10/50 = 0.2$, because 10 of the candidates had both. In addition, $P(W) = 20/50 = 0.4$ and $P(C) = 15/50 = 0.3$. We know that certification and work experience are independent if

$$P(W \cap C) = P(W) \cdot P(C).$$

But

$$P(W) \cdot P(C) = (0.4)(0.3) = 0.12 \neq 0.2 = P(W \cap C).$$

Hence work experience and certification are not independent. (The reason for this is that candidates with certification are more likely to possess work experience than those lacking certification.) □

Practice Problem

A company has two main divisions, Marine Products and Aeronautical Products. The probability that Marine Products will have a profit margin of at least 8% has been estimated to be 0.2 and the probability that Aeronautical Products will have a profit margin of at least 8% has been estimated to be 0.25. The probability that both achieve a profit margin of at least 8% has been estimated to be 0.05. Are the attainments of at least 8% profit in the two divisions independent?

Answer: yes

Suppose that two cards are drawn, one at a time, *without replacement* from a 52-card deck. (That is, the first card is *not* put back before the second is drawn.) Let's compute the probability both are aces.

Let A = the event "first card is an ace,"
B = the event "second card is an ace."

Events A and B are dependent because we are drawing (or sampling) without replacement. After the first card is drawn, there will be only 51 cards remaining in the deck. Moreover, if event A occurred, there will be three aces among the 51 cards. If event A did not occur, there will be four aces among the 51 cards. Thus the probability that B occurs depends very much on whether A occurred. We have $P(A) = 4/52$ and $P(B \mid A) = 3/51$ so that

$$\begin{aligned} P(A \cap B) &= P(B \mid A) \cdot P(A) \\ &= (3/51)(4/52) = 12/2652 = 0.0045. \end{aligned}$$

Now let's compute the probability that two successive aces are drawn from a 52-card deck assuming *replacement* (the first card is put back into the deck, which is then shuffled). In this case, whether or not A occurred should have no impact upon the occurrence of B, so that the events should be independent. We can check this as follows:

$$\begin{aligned} P(A \cap B) &= P(B \mid A) \cdot P(A) \\ &= (4/52)(4/52) = 16/2704 = 0.0059, \end{aligned}$$

which is indeed equal to $P(A) \cdot P(B)$. (Check it!)

EXERCISE SET 11.3

Given $P(A) = 1/2$, $P(B) = 1/5$, and $P(A \cap B) = 1/10$, find

1. $P(A')$ **2.** $P(A \cup B)$ **3.** $P(A \mid B)$ **4.** $P(B \mid A)$

5. Are A and B independent?

Given that $P(A) = 1/2$, $P(B) = 1/3$ and $P(A \cap B) = 1/4$, find

6. $P(B')$ **7.** $P(A \cup B)$ **8.** $P(B \mid A)$ **9.** $P(A \mid B)$

10. Are A and B independent?

11. At the local university, 30% of the students regularly attend the basketball games, 35% are freshmen, and 50% are upperclassmen who do not regularly attend basketball games. A student is chosen at random.

- **a)** Compute the probability the student is both a freshman and regularly attends basketball games.
- **b)** Compute the conditional probability a student regularly attends basketball games given that he is a freshman.
- **c)** Compute the conditional probability a student is an upperclassman given that he regularly attends basketball games. (*Hint:* Draw a Venn diagram.)

A new movie was shown to a test audience of 1200 people. The producers are interested in the viewers' reactions and also whether there is a relationship between a viewer's reaction and the viewer's age and sex. Each viewer recorded his or her reaction as "liked," "indifferent," or "disliked." The data are shown in Table 11.2. Problems 12–17 relate to these data.

TABLE 11.2 Reactions of a sample audience to a new movie

	Liked	Indifferent	Disliked
Males under 25	200	100	50
Males 25 and over	150	40	20
Females under 25	250	120	10
Females 25 and over	175	15	70

12. Compute the probability a viewer was male given that the viewer liked the movie.

13. Compute the probability a viewer disliked the movie given that she was female.

14. Compute the probability a viewer was female given that the viewer was indifferent to the movie.

15. Compute the probability a viewer liked the movie given that he or she was 25 or over.

16. Are the viewer's age and liking the movie independent?

17. Are the viewer's sex and disliking the movie independent?

18. Suppose two events are mutually exclusive. Are they independent or dependent?

19. In a small town, 45% of the registered voters are Republicans, 50% favor a school bond issue, and 30% are Republicans and favor the bond issue.

A registered voter is selected at random from the town.

a) What is the probability the voter is not a Republican and favors the bond issue?
b) What is the conditional probability that a voter favors the bond issue given that he or she is a Republican?
c) Are being Republican and opposing the bond issue independent?

20. A box contains 10 light bulbs, of which 8 are good and 2 are defective. A bulb is drawn from the box. If it is good, the process stops. However, if the first bulb is defective, another bulb is drawn (without replacement).

a) Compute the probability that the first bulb is defective and the second bulb is good.
b) Compute the probability that two defective bulbs are drawn.

21. Rework Problem 20 but assume the first bulb is returned to the box if it is defective.

11.4 BAYES' THEOREM

In the previous section we defined the conditional probability of an event (say, A) given that another event (B) was known to have occurred. Let's see if we can reverse the process: Can we compute the conditional probability that A occurred given that B occurred? This question has important implications, especially in the area of statistics.

Rather than just state Bayes' theorem, we'll first work an example problem without it. Next, we'll state the theorem and use it to work the same example.

EXAMPLE 15

A company has two machines that are used to manufacture a product. The first machine produces 60% of the total output and has a defective rate of 2%. The second machine produces 40% of the total output and has a defective rate of 4%. Suppose a unit of output is defective. Determine the probabilities that it came from each of the two machines.

SOLUTION

Notice that the defective rates we gave are actually conditional probabilities. We have

$$P(\text{defective} \mid \text{machine } 1) = 0.02 \quad \text{and} \quad P(\text{defective} \mid \text{machine } 2) = 0.04.$$

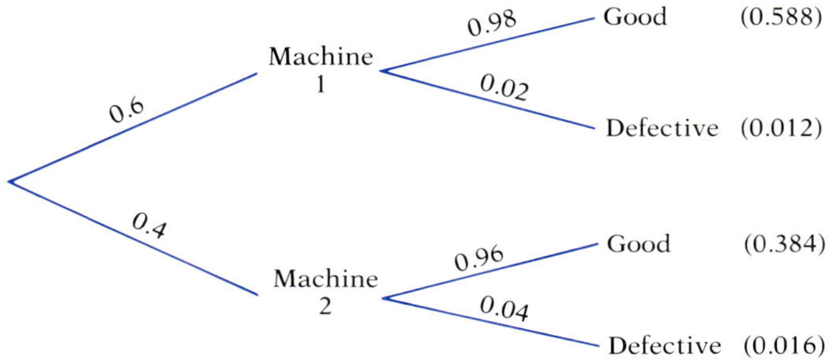

Figure 11.10 **Tree diagram for output from two machines**

We now wish to determine

$$P(\text{machine 1} \mid \text{defective}) \qquad \text{and} \qquad P(\text{machine 2} \mid \text{defective}).$$

The best approach to use is to construct a tree diagram. (See Fig. 11.10.) Notice that we drew one branch for each machine. We then marked each branch with the appropriate probability: 0.6 and 0.4, respectively. Next, at each machine we drew two branches that correspond to the product being either good or defective. Each of these branches is also marked with the appropriate probability. The last step is to multiply the probabilities along each path through the tree in order to compute the probability of every possible outcome. Thus

$$\begin{aligned}
P(\text{machine 1 and good}) &= P(\text{good} \mid \text{machine 1}) \cdot P(\text{machine 1}) \\
&= (0.98)(0.6) = 0.588, \\
P(\text{machine 1 and defective}) &= P(\text{defective} \mid \text{machine 1}) \cdot P(\text{machine 1}) \\
&= (0.02)(0.6) = 0.012, \\
P(\text{machine 2 and good}) &= P(\text{good} \mid \text{machine 2}) \cdot P(\text{machine 2}) \\
&= (0.96)(0.4) = 0.384, \\
P(\text{machine 2 and defective}) &= P(\text{defective} \mid \text{machine 2}) \cdot P(\text{machine 2}) \\
&= (0.04)(0.4) = 0.016.
\end{aligned}$$

Notice that the sum of these probabilities is 1.00. (Which should not be surprising because the four outcomes shown in the tree comprise the sample space.)

Now, by definition

$$\begin{aligned}
P(\text{machine 1} \mid \text{defective} &= \frac{P(\text{machine 1 and defective})}{P(\text{defective})} \\
&= 0.012/(0.012 + 0.016) = 0.42857
\end{aligned}$$

and

$$P(\text{machine 2} \mid \text{defective}) = \frac{P(\text{machine 2 and defective})}{P(\text{defective})}$$
$$= 0.016/(0.012 + 0.016) = 0.57143. \quad \square$$

Now, we will state Bayes' theorem (named after Thomas Bayes, an eighteenth-century mathematician).

Bayes' Theorem

Let $A_1, A_2, \ldots, A_n$ be mutually exclusive events whose union is the entire sample space. Let B be an arbitrary event in S such that $P(B) \neq 0$. Then

$$P(A_1 \mid B) = \frac{P(B \mid A_1) \cdot P(A_1)}{P(A_1) \cdot P(B \mid A_1) + P(A_2) \cdot P(B \mid A_2) + \cdots + P(A_n) \cdot P(B \mid A_n)}.$$

Similar results hold for $P(A_2 \mid B), \ldots, P(A_n \mid B)$.

Bayes theorem is useful in situations such as the one in Example 15. Its value is that it enables us to perform the calculations without having to draw probability trees (and therefore without having to calculate probabilities that aren't of interest). We'll demonstrate these advantages by first reworking Example 15 using Bayes' theorem and then analyzing another similar situation. After these examples, we'll discuss why Bayes' theorem is correct.

EXAMPLE 16

Example 15 Revisited Let

$$A_1 = \text{"machine 1,"}$$
$$A_2 = \text{"machine 2,"}$$
$$B = \text{"defective."}$$

Given that $P(A_1) = 0.6$, $P(A_2) = 0.4$, $P(B \mid A_1) = 0.02$, and $P(B \mid A_2) = 0.04$, calculate $P(A_1 \mid B)$ and $P(A_2 \mid B)$.

SOLUTION

By Bayes' theorem,

$$P(A_1 \mid B) = \frac{P(B \mid A_1) \cdot P(A_1)}{P(B \mid A_1) \cdot P(A_1) + P(B \mid A_2) \cdot P(A_2)},$$

so
$$P(A_1 \mid B) = \frac{(0.02)(0.6)}{(0.02)(0.6) + (0.04)(0.4)} = 0.42857.$$

Similarly,

$$P(A_2 \mid B) = \frac{P(B \mid A_2) \cdot P(A_2)}{P(B \mid A_1) \cdot P(A_1) + P(B \mid A_2) \cdot P(A_2)},$$

so
$$P(A_2 \mid B) = \frac{(0.04)(0.4)}{(0.02)(0.6) + (0.04)(0.4)} = 0.57143. \quad \square$$

EXAMPLE 17

Suppose a new test for the detection of pneumonia has been developed. Before being distributed, it was evaluated in a field test. The following results were obtained from a sample of 1000 people. Using other methods, it was found that 50 people actually had pneumonia. The new test indicated pneumonia in 49 of these people and it indicated pneumonia in 19 of the people who did not have it. Compute the probability that a person (chosen at random) having a positive reaction to the test actually has pneumonia.

SOLUTION

Let

A_1 = "person has pneumonia,"

A_2 = "person does not have pneumonia,"

B = "positive reaction to test."

We know $P(A_1) = 50/1000 = 0.05$ so that $P(A_2) = 0.95$. Moreover, $P(B \mid A_1) = 49/50 = 0.98$ and $P(B \mid A_2) = 19/950 = 0.02$. By Bayes' theorem,

$$P(A_1 \mid B) = \frac{(0.98)(0.05)}{(0.98)(0.05) + (0.02)(0.95)} = \frac{0.049}{0.068} = 0.72.$$

Thus even though the test is very reliable in detecting pneumonia in those who actually have it (98% reliability), the overall results are not nearly as good. This is because a large portion of the population does not have pneumonia, so that even though the test gives false positive reactions only 2% of the time, the total number of false positive reactions will be significant. $\square$

Practice Problem

Given the data in Example 17, compute the probability that a person actually has pneumonia given a negative reaction. [*Hint:* given a negative reaction, a person either has pneumonia or doesn't. Thus $P(B' \mid A) = 1 - P(B \mid A)$.]

Answer: 0.00107

Let's see why Bayes' theorem works. Recall that the events $A_1, \ldots, A_n$ are both mutually exclusive and their union is equal to the sample space S. In addition, event B is an arbitrary event in S such that $P(B) > 0$. We wish to compute the probabilities that each of the events $A_1, \ldots, A_n$ occurred given that event B occurred. In symbols, we want the values of $P(A_1 \mid B), \ldots, P(A_n \mid B)$.

To begin, we assume that event B has occurred. As Fig. 11.11 shows, this means that exactly *one* of the mutually exclusive events $A_1 \cap B$ or $A_2 \cap B$ or ... or $A_n \cap B$ occurred. (Since the sets $A_1, \ldots, A_n$ are mutually exclusive, a little thought should convince you this is reasonable.) In fact, B is equal to the union of the disjoint sets $A_1 \cap B, \ldots, A_n \cap B$. Therefore

$$\begin{aligned} P(B) &= \sum_{i=1}^{n} P(A_i \cap B) \\ &= P(A_1 \cap B) + P(A_2 \cap B) + \cdots + P(A_n \cap B). \end{aligned}$$

Now, by definition,

$$P(A_1 \mid B) = \frac{P(A_1 \cap B)}{P(B)},$$

and by our statement earlier,

$$P(A_1 \mid B) = \frac{P(A_1 \cap B)}{P(A_1 \cap B) + P(A_2 \cap B) + \cdots + P(A_n \cap B)}.$$

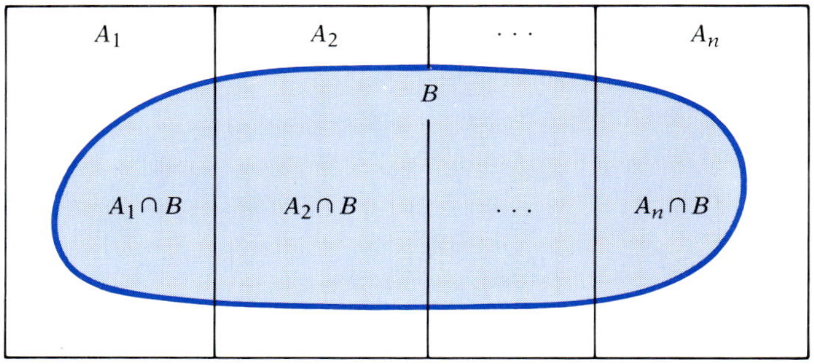

Figure 11.11 A sample space with the sets $A_1, \ldots, A_n$ and B.

In addition, we are given the conditional probabilities

$$P(B \mid A_1),\ P(B \mid A_2),\ \ldots,\ P(B \mid A_n).$$

Again, using the definition of conditional probability,

$$P(A_1 \cap B) = P(B \mid A_1) \cdot P(A_1), \quad P(A_2 \cap B) = P(B \mid A_2) \cdot P(A_2), \ \ldots, \quad P(A_n \cap B) = P(B \mid A_n) \cdot P(A_n).$$

Now all we need do is to substitute these into the expression for $P(A_1 \mid B)$ to obtain

$$P(A_1 \mid B) = \frac{P(B \mid A_1) \cdot P(A_1)}{P(B \mid A_1) \cdot P(A_1) + P(B \mid A_2) \cdot P(A_2) + \cdots + P(B \mid A_n) \cdot P(A_n)},$$

which is Bayes' theorem.

Practice Problem

A factory has two machines. The newer machine produces 75% of the output and has a defect rate of 1%, while the older machine produces 25% of the output with a defect rate of 5%. Given that a unit of output is defective, compute the probabilities that it was produced by

a) the newer machine. **b)** the older machine.

Answers: **a)** 0.375; **b)** 0.625

EXERCISE SET 11.4

For Problems 1–6, assume that $P(A_1) = 0.4$, $P(A_2) = 0.3$, and $P(A_3) = 0.3$. In addition, assume $P(B \mid A_1) = 0.1$, $P(B \mid A_2) = 0.15$, and $P(B \mid A_3) = 0.1$. Compute the following probabilities.

1. $P(A_1 \mid B)$ **2.** $P(A_2 \mid B)$

3. $P(A_3 \mid B)$ **4.** $P(A_1 \mid B')$

5. $P(A_2 \mid B')$ **6.** $P(A_3 \mid B')$

Answer Problems 7 and 8 given the following information. A university uses a combination of previous academic record, standardized test results, and recommendations to screen applicants. Past experience has shown that 50% of the total number of applicants are actually capable of performing successfully. The screening procedure approves qualified students 80% of the time and approves unqualified students 20% of the time.

7. Given that a student has been approved by the screening procedure, compute the probabilities of success and failure.

8. Given that a student was not approved by the screening procedure, compute the probabilities that the student would have been successful or would have been a failure.

Answer Problems 9 and 10 using the following information. In a certain city, 55% of the registered voters are Democrats, 35% are Republicans, and the remainder are independent. A survey has indicated that 80% of the Democrats will vote yes on a bond issue, 40% of the Republicans will vote yes, and 30% of the independents will vote yes.

9. Compute the probabilities that a voter selected at random who will vote yes is a

a) Democrat. **b)** Republican.

10. Compute the probability that a voter selected at random who will vote no is an independent.

11. A new test for a serious disease has been developed. In field trials, the test was able to detect 99.5% of those people who actually had the disease while it gave a false indication of the disease for 1% of those who did not have it. Approximately 2.5% of the population in a certain area has the disease.

a) Compute the probability that a person actually has the disease given a positive reaction.

b) Compute the probability that a person actually has the disease given a negative reaction.

12. The Second National Bank ranks applicants for automobile loans in one of three categories depending on their credit rating. The following data on the history of past loans have been gathered.

Credit rating	Percentage of applicants	Probability of default
Excellent	25%	0.05
Fair	60%	0.15
Poor	15%	0.40

Given that a loan holder defaults, compute the probabilities that his or her credit rating was excellent, fair, and poor.

13. A typing pool has four secretaries. The office manager has gathered the following data on their past performance.

Secretary	Percentage of typing	Probability of an error
1	25%	0.02
2	20%	0.04
3	30%	0.01
4	25%	0.03

Given that an error has been detected in a document, compute the probabilities that each of the four typed the document.

14. An electronic device contains parts from six different suppliers. The table following shows the number of parts manufactured by each supplier and the probability of failure during the next year. If a part fails, what is the probability that supplier B manufactured it?

Supplier	Number of parts	Probability of failure
A	30	0.002
B	25	0.004
C	20	0.005
D	10	0.008
E	5	0.010
F	10	0.010

11.5 COUNTING: PERMUTATIONS AND COMBINATIONS

We have presented the basic framework of probability and have shown you how to solve a variety of problems that require the use of the concepts of probability. However, there are many other problems that you could solve with more knowledge of how to count. For example, consider the following situation. Suppose a room contains 25 people. What is the probability that two or more people have the same birthday (without regard to year)? Most people would guess the answer to be quite small: perhaps only 0.01 or so. Actually, as we shall show later, the correct answer is 0.569.

The Multiplication Principle

To begin our discussion of counting, consider the following scenario. The local computer store advertises four different brands of computer. Each of the brands offers three models. How many types of computer (brand-model pairs) are available? As Fig. 11.12 shows, the answer is rather simple. For each brand, there are three models. Since there are four brands, there must be a total of $4 \cdot 3 = 12$ different types of computers.

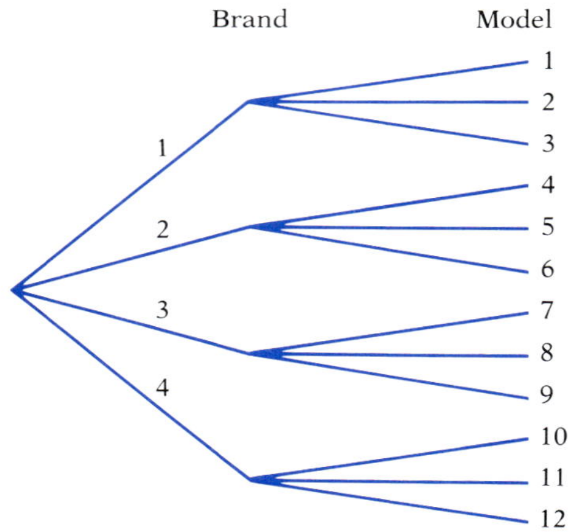

Figure 11.12 A tree diagram for choosing different computers.

This example demonstrates a fundamental principle of counting, usually called the **multiplication principle.**

Multiplication Principle

Consider two events, say A and B. If A can occur in m ways and B can occur in n ways (independent of one another), then the two events together can occur in $m \cdot n$ ways.

The multiplication principle can be extended to any (finite) number of events, assuming all the events are independent of one another.

EXAMPLE 18

A traveler must go from Richmond to Los Angeles via Chicago. There are five flights from Richmond to Chicago, each of which has eighteen connections from Chicago to Los Angeles. How many possible ways are there to make the trip?

SOLUTION

By the multiplication principle, there are $5 \cdot 18 = 90$ ways of making the trip. □

EXAMPLE 19

A basketball team has five players. How many ways could they be introduced at the start of a game?

SOLUTION

Initially, there are five choices for the first player to be introduced. For the next player there are only four choices (because one player has been introduced previously). Similarly, there are three choices for the third player, two choices for the fourth player, and one choice for the fifth player. By the multiplication principle, there are $5 \cdot 4 \cdot 3 \cdot 2 \cdot 1 = 120$ different ways the team could be introduced. □

Practice Problem

A professor must choose a main text and a workbook for a course. There are fifteen different texts that might be used and eight different workbooks. How many different ways can the professor choose the books?

Answer: 120

In many situations that involve the multiplication principle, we will encounter products such as $5 \cdot 4 \cdot 3 \cdot 2 \cdot 1$ (see Example 19). We will find it convenient to use a symbol to denote such products. For any natural number n,

$$n! = \textbf{n factorial} = n(n-1)(n-2)(n-3)\cdots(2)(1)$$

Thus $5 \cdot 4 \cdot 3 \cdot 2 \cdot 1 = 5! = 120$, while $6! = 6 \cdot 5 \cdot 4 \cdot 3 \cdot 2 \cdot 1 = 720$. In fact, a little experimentation should convince you that

$$n! = n \cdot [(n-1)!]$$

for all natural numbers $n \geq 1$. For this to make sense when $n = 1$, we define $0! = 1$.

(*Note*: Many calculators will compute factorials for you. Check your calculator to see if it does.)

Practice Problem

Compute the following: **a)** $4!$, **b)** $8!$.

Answers: **a)** 24; **b)** 40,320

Permutations and Combinations

There are two particular types of counting problem that occur very often in problems involving probability. By appropriate modification of the multiplication principle, we can analyze these situations. As the next two examples demonstrate, these two problems are related.

EXAMPLE 20

Consider the letters a, b, c, and d. How many different ways can we arrange groups of two of them without repeating the letters?

SOLUTION

There are four choices for the first letter and three for the second. Therefore the total number of arrangements is $4 \cdot 3 = 12$. A complete enumeration of these arrangements is

$$ab \quad ac \quad ad \quad ba \quad bc \quad bd \quad ca \quad cb \quad cd \quad da \quad db \quad dc \quad \square$$

EXAMPLE 21

Suppose a group of four people is playing bridge. How many different two-person pairs can be formed?

SOLUTION

The easiest approach is to simply enumerate them. Letting a, b, c, and d represent the four people, we have

$$ab \quad ac \quad ad \quad bc \quad bd \quad cd$$

so there are only six different pairs. □

In Example 20 the order of the letters was significant. We call such an arrangement a **permutation.** In Example 21 the ordering of the pairs was not important. An arrangement in which order is not significant is called a **combination.** Formulas that give the number of permutations or combinations can be derived from the multiplication principle.

Let's begin by considering permutations. Suppose we want to determine the number of different orderings (permutations) of r objects chosen from a group of n objects. By the multiplication principle, there are

n	ways to choose the first object
$n - 1$	ways to choose the second object
$n - 2$	ways to choose the third object
$\vdots$	
$n - (r - 1) = n - r + 1$	ways to choose the rth object

Therefore the total number of ways to choose all r objects must be

$$n \cdot (n - 1) \cdot (n - 2) \cdots (n - r + 1) = \frac{n!}{(n - r)!}.$$

(Check this by writing out the numerator and denominator, and then dividing out common factors.)

Permutations

The number of permutations of n objects taken r at a time $(r \leq n)$ is

$$P(n, r) = \frac{n!}{(n-r)!} = n \cdot (n-1) \cdot (n-2) \cdots (n-r+1).$$

EXAMPLE 22

Nine women have formed a softball team. How many different ways can three of them be chosen to play left fielder, center fielder, and right fielder?

SOLUTION

We can answer this by using either the multiplication principle or the formula for permutations. Using the multiplication principle, we note there are nine choices for the first outfield position, eight for the second position, and seven choices for the remaining position. Thus the number of choices is

$$9 \cdot 8 \cdot 7 = 504.$$

Using the formula for permutations, we want $P(9, 3)$ (we are choosing three outfielders from the nine women on the team) so

$$P(9, 3) = \frac{9!}{(9-3)!} = \frac{9 \cdot 8 \cdot 7 \cdot \not{6} \cdot \not{5} \cdot \not{4} \cdot \not{3} \cdot \not{2} \cdot \not{1}}{\not{6} \cdot \not{5} \cdot \not{4} \cdot \not{3} \cdot \not{2} \cdot \not{1}} = 9 \cdot 8 \cdot 7 = 504.$$

Note that we did *not* compute the values of 9! and 6! in the calculation of $P(9, 3)$. The reason is simple: It's much easier to cancel as we did than to compute 9! (= 362,880) and then divide by 6! (= 720). (Check your calculator. Many calculators will compute factorials and some will even compute the number of permutations.) □

Practice Problem

A fraternity has 15 members. In how many ways can a president, vice-president, and treasurer be chosen?

Answer: $P(15, 3) = 2730$

As we noted earlier, a combination is an arrangement in which the order of the objects does *not* matter. In other words, we are interested solely in whether an object has been included in the arrangement or not. We can count the number of combinations of r objects chosen from a group of n by counting the number of arrangements in which the order does matter (the permutations) and then dividing to eliminate multiple counting.

For example, consider the number of ways we can choose three letters from the set $\{a, b, c, d, e\}$. There are sixty different permutations (why?), as Table 11.3 shows. However, there are only ten different combinations because there are six different arrangements of each three-letter combination (since $3! = 6$, which is the number of permutations of a three-letter group). Thus, the number of combinations $= 60 \div 6 = 10$.

We can use this approach to determine the number of combinations of r objects taken from a group of n objects: Count the

TABLE 11.3 Permutations and combinations of three-letter groups

Permutation	Equivalent combination	Permutation	Equivalent combination
abc		*abd*	
acb		*adb*	
bac	*abc*	*bad*	*abd*
bca		*bda*	
cab		*dab*	
cba		*dba*	
abe		*ace*	
aeb		*aec*	
bae	*abe*	*cae*	*ace*
bea		*cea*	
eab		*eac*	
eba		*eca*	
ade		*bcd*	
aed		*bdc*	
dae	*aed*	*cbd*	*bdc*
dea		*cdb*	
ead		*dbc*	
eda		*dcb*	
bce		*cde*	
bec		*ced*	
cbe	*bce*	*dce*	*cde*
ceb		*dec*	
ebc		*ecd*	
ecb		*edc*	

number of permutations of n objects taken r at a time and then divide by the number of permutations of r objects to eliminate any multiple counting. Thus the number of combinations of n objects taken r at a time should be

$$\begin{aligned} P(n, r) \div P(r, r) &= [n \cdot (n-1) \cdot (n-2) \cdots (n-r+1)] \div r! \\ &= \frac{n \cdot (n-1) \cdot (n-2) \cdots (n-r+1)}{r \cdot (r-1) \cdot (r-2) \cdots 3 \cdot 2 \cdot 1} \\ &= \frac{n!}{(n-r)!\, r!}. \end{aligned}$$

Combinations

The number of combinations of n objects taken r at a time ($r \leq n$) is

$$C(n, r) = \frac{n!}{(n-r)!\, r!} = \frac{n \cdot (n-1) \cdot (n-2) \cdots (n-r+1)}{r \cdot (r-1) \cdot (r-2) \cdots 3 \cdot 2 \cdot 1}.$$

EXAMPLE 23

How many different thirteen-card bridge hands are there?

SOLUTION

Here we must compute $C(52, 13)$, which is given by

$$\begin{aligned} C(52, 13) &= \frac{52!}{(52-13)!\, 13!} \\ &= \frac{52 \cdot 51 \cdot 50 \cdots 3 \cdot 2 \cdot 1}{(39 \cdot 38 \cdot 37 \cdots 3 \cdot 2 \cdot 1)(13 \cdot 12 \cdot 11 \cdots 3 \cdot 2 \cdot 1)} \\ &= \frac{52 \cdot 51 \cdot 50 \cdots 40}{13 \cdot 12 \cdot 11 \cdots 1} \\ &= \text{(approximately) } 6.3501356 \times 10^{11}. \quad \square \end{aligned}$$

Practice Problem

How many ways can four people be chosen to form a special project team from a fifteen-person department?

Answer: $C(15, 4) = 455$

Practice Problem

How many different five-card poker hands are there?

Answer: $C(52, 5) = 2{,}598{,}960$

TABLE 11.4 Permutations versus combinations

The number of arrangements of n objects, taken r at a time (repetitions not allowed)	
Permutations	**Combinations**
Order is significant	Order not significant
$P(n, r) = \dfrac{n!}{(n-r)!} = n \cdot (n-1) \cdots (n-r+1)$	$C(n, r) = \dfrac{n!}{(n-r)!\,r!} = \dfrac{n \cdot (n-1) \cdots (n-r+1)}{r \cdot (r-1) \cdots 3 \cdot 2 \cdot 1}$

A comparison between permutations and combinations is shown in Table 11.4.

WARNING

BE VERY CAREFUL APPLYING COUNTING FORMULAS

Be very sure you understand the situation before using the formulas. If you aren't sure whether you're dealing with a permutation, a combination, or something else, use the multiplication principle and tree diagrams. You will be much less likely to make mistakes. (Avoid the situation of "having a formula and looking for a problem"!)

The next example demonstrates our warning.

EXAMPLE 24

In many states, license plate numbers are formed by three letters followed by three numbers. How many different license plates are possible (excluding personalized plates)?

SOLUTION

This problem requires careful analysis. It should be obvious that order matters, because ABC-123 is different from (say) CBA-123. Thus we know the answer does not involve combinations. However, it may not involve permutations either. The key is whether or not repetitions are allowed. Thus there are two potential answers to this problem.

a) Assuming repetitions are allowed: Simply use the multiplication principle. There are 26 choices for each of the first three letters and 10 choices for each of the three numbers (the digits 0–9) so we have $26 \cdot 26 \cdot 26 \cdot 10 \cdot 10 \cdot 10 = 17{,}576{,}000$.

b) Assuming repetitions are not allowed: Now we can use the multiplication principle in conjunction with the permutation formula. For the letters the number of permutations is $P(26, 3)$, which is 15,600, and for the numbers the number of permutations is $P(10, 3) = 720$. The answer is the product of these two, which is $15{,}600 \cdot 720 = 11{,}232{,}000$. □

Practice Problem

How many seven-digit telephone numbers are possible if the first digit cannot be zero and

a) only odd digits are allowed?

b) only odd digits and no repetitions are allowed?

Answers: **a)** 78,125; **b)** none

Counting and Probability

Now we'll present a series of examples showing how we can solve more complicated probability problems through the use of counting techniques.

The following example demonstrates a problem in the area of statistical quality control. Clearly, every organization that produces and markets a product must be concerned with questions of this sort.

EXAMPLE 25

A manufacturer of floppy diskettes (for microcomputers) puts a total of twenty diskettes into a package. Before shipment, two of the diskettes are randomly selected and tested. If any are found to be defective, the box is not shipped. Suppose that a box contains three defective diskettes. What is the probability that it is not shipped?

SOLUTION

It is important to consider your approach in a complicated problem before simply "diving" into it. Now, the box of diskettes won't be shipped whenever the sample contains at least one of the defective diskettes. Thus we must find the probability that the sample contains one or more of the defective diskettes. We can accomplish this by counting. More specifically, we need to count the number of samples that contain one or more defectives. We must also count the total number of samples that are possible. The required probability will be the ratio of these two numbers.

First, let's compute the total number of samples of size two. We will assume (quite reasonably) the company samples without replacement. Now, we don't care about the order in which the diskettes in the sample are chosen. Our only concern is which diskettes appear. Thus we are dealing with a combination, so that the total number of samples of size two must be

$$C(20, 2) = \frac{20!}{(20 - 2)!\,2!} = 190.$$

Next, we must compute the number of ways that a sample can contain at least one defective. This can occur in two ways: One of the diskettes in the sample can be defective or both could be defective. Suppose the sample contains one defective diskette. Then we must have chosen one of the three defective diskettes and also one of the seventeen good diskettes (because the sample contains two diskettes). There are $C(3, 1)$ ways of choosing one of the defective diskettes. For each of these ways, there are $C(17, 1)$ ways of choosing one of the good diskettes. Therefore the total number of ways to choose a sample with one defective diskette is $C(3, 1) \cdot C(17, 1) = 3 \cdot 17 = 51$ ways. (We leave the arithmetic to you.) The probability of a sample with one defective diskette is given by

$$P(1 \text{ bad}) = \frac{51}{190}.$$

Similarly, the sample could contain two defective diskettes (and no good diskettes). There are $C(3, 2) = 3$ ways this could occur. Therefore the probability of a sample with two defective diskettes is

$$P(2 \text{ bad}) = \frac{3}{190}.$$

Finally, since the events "one bad diskette" and "two bad diskettes" are disjoint, we have

$$\begin{aligned} P(\text{not shipping}) &= P(1 \text{ bad or } 2 \text{ bad}) \\ &= P(1 \text{ bad}) + P(2 \text{ bad}) \\ &= \frac{51}{190} + \frac{3}{190} = 0.28421. \quad \square \end{aligned}$$

Practice Problem

Reconsider Example 25 but now suppose the sample is of size three instead of size two. What is the probability that a box with three defective diskettes will not be shipped?

Answer: 460/1140 = 0.4035

Our next example describes a problem generally referred to as the **birthday problem.**

EXAMPLE 26

Consider a group of five people selected randomly. Let's compute the probability that two or more of them share the same birthday.

SOLUTION

Before we begin, we must make some simplifying assumptions. First, we'll rule out leap years. In addition, we'll assume that all the 365 days of the year are equally likely to be birthdays. (This isn't quite true, but it is a reasonable assumption.)

Now let's think about our approach. We have a sample space which consists of all possible five-somes of birthdays. For example, (May 10, March 21, March 31, December 6, September 4) is an outcome in the sample space. Since we must allow repetitions, the total number of outcomes is $365 \cdot 365 \cdot 365 \cdot 365 \cdot 365 = 365^5$ (a rather large number). Next, we could try to count the number of outcomes that contain two or more identical birthdays. However, we have no convenient mechanism for doing this. All is not lost, however. Maybe we can count the number of outcomes that do not contain two or more identical birthdays. In other words, rather than compute the probability of two or more people having matching birthdays, we could compute the probability of no matching birthdays. Our desired probability could then be obtained by subtracting $P(\text{no match})$ from 1.

Counting the outcomes with no matching birthdays is rather simple: We use the multiplication rule. Such outcomes have 365 choices for the first birthday, 364 choices for the next birthday (it can't match the first), 363 choices for the third birthday, 362 choices for the fourth birthday, and 361 choices for the fifth birthday. Therefore the total number of outcomes with no match is

$$365 \cdot 364 \cdot 363 \cdot 362 \cdot 361.$$

Combining this with our count of the total number of outcomes, we have

$$P(\text{no match}) = \frac{365 \cdot 364 \cdot 363 \cdot 362 \cdot 361}{365^5} = 0.97286 \qquad \text{(approximately)}$$

and therefore

$$P(\text{match}) = 1 - P(\text{no match}) = 0.02714. \quad \square$$

We can extend this analysis to groups of any size. Let's develop a general formula to help calculate $P(\text{no match})$ for any size group (of less than 366 people, since a group of this size or larger *must* contain at least one pair of matching birthdays). By the same reasoning we used above, for a group of six people

$$P(\text{no match}) = \frac{365 \cdot 364 \cdot 363 \cdot 362 \cdot 361 \cdot 360}{365^6} = 0.95954.$$

For a group of seven people,

$$P(\text{no match}) = \frac{365 \cdot 364 \cdot 363 \cdot 362 \cdot 361 \cdot 360 \cdot 359}{365^7} = 0.94376.$$

In general, then, for a group of n people, we have

$$P(\text{no match}) = \frac{365 \cdot 364 \cdot 363 \cdots (365 - n + 1)}{365^n}.$$

Although this formula does become difficult to compute for larger values of n, we now have a vehicle for computing the probability of no matching birthdays and therefore the probability of matching birthdays for groups of any desired size. In fact, for $n = 23$ (a group of size 23), the probability of no matching birthdays is less than 0.5, so that the probability of a match is (slightly) better than 0.5. For a group of size 50, the probability of matching birthdays is better than 0.95 (actually it's closer to 0.97).

We will end this section with the computation of some probabilities related to the game of poker.

EXAMPLE 27

Compute the probability of being dealt the following hands in straight draw poker (each player is dealt five cards from a standard 52-card deck): a full house (which consists of three cards of one rank and two cards of another), a flush (which consists of five cards of the same suit) and four of a kind (four of the five cards are of the same rank).

SOLUTION

First, we must determine the number of possible poker hands. Since we don't care about the order in which the cards are dealt, we are interested in the number of combinations. Thus the total number of possible five-card hands is $C(52, 5)$ which is

$$C(52, 5) = \frac{52!}{(52 - 5)!\,5!} = \frac{52 \cdot 51 \cdot 50 \cdot 49 \cdot 48}{5 \cdot 4 \cdot 3 \cdot 2 \cdot 1} = 2{,}598{,}960$$

(which you were asked to compute earlier in a Practice Problem).

Full House There are thirteen ranks of cards in poker; 2, 3, 4, 5, 6, 7, 8, 9, 10, jack, queen, king, and ace. There are also four suits: clubs, diamonds, hearts, and spades, which means that there are four cards of each rank. Thus there are $C(4, 3)$ ways to get three cards out of the four for each rank. Since there are thirteen ranks, the total number of ways to get three of one rank is $13 \cdot C(4, 3)$. In addition, to make up a full house, the remaining two cards must form a pair of another rank. Using similar reasoning, there are $12 \cdot C(4, 2)$ ways to do this (because there are twelve ranks remaining). The total number of ways to be dealt a full house is then

$$13 \cdot C(4, 3) \cdot 12 \cdot C(4, 2) = 13 \cdot 4 \cdot 12 \cdot 6 = 3744.$$

Hence,

$$P(\text{full house}) = 3744/2{,}598{,}960 = 0.00144 \qquad \text{(approximately).}$$

Flush Now we must count the number of ways to obtain five cards in the same suit. For each suit, there are $C(13, 5)$ ways to be dealt five cards from the suit. Since there are four suits, the total number of ways to be dealt a flush is then

$$4 \cdot C(13, 5) = 4 \cdot \frac{13!}{(13 - 5)!\,5!} = 4 \cdot 1287 = 5148$$

so that

$$P(\text{flush}) = 5148/2{,}598{,}960 = 0.00198 \qquad \text{(approximately).}$$

Four of a Kind Here we must count the number of ways to obtain four cards of one rank and one card from the remainder of the deck. There is only one way to be dealt all the cards of one given rank, so there are thirteen ways of being dealt four of a kind. The remaining card can be any of the other forty-eight cards in the deck, so the total number of ways to be dealt four of a kind is

$$13 \cdot 1 \cdot 48 = 624$$

which means $P(\text{four of a kind}) = 624/2{,}598{,}960 = 0.00024$ (approximately). □

Practice Problem

What is the probability of being dealt two different pairs in straight draw poker?

Answer: The number of ways of being dealt two different pairs is

$$13 \cdot C(4, 2) \cdot 12 \cdot C(4, 2) \cdot 44 = 13 \cdot 6 \cdot 12 \cdot 6 \cdot 44 = 247{,}104$$

and so $P(\text{two pair}) = 247{,}104/2{,}598{,}960 = 0.09508$.

EXERCISE SET 11.5

Evaluate the following.

1. $P(8, 3)$
2. $C(8, 3)$
3. $C(6, 2)$
4. $P(6, 2)$
5. $C(10, 4)$
6. $P(10, 4)$
7. The top management of a small firm, consisting of six individuals, must attend a banquet. Determine the number of different seating arrangements that are possible if

   a) all six people attend.
   b) only four of the six attend.

8. The championship series of the National Basketball Association, which matches the winners of the Eastern and Western Conferences, is "best of seven." This means the first team to win four games is the champion. If we let E denote the event "eastern representative wins" and W the event "western representative wins", how many different sequences of winners are possible?
9. Suppose that an elected body consists of 100 members, and 10 must be selected to form a committee. The majority party, which controls 60 of the 100 seats, will have 7 of the seats on the committee. (The other 3 seats will belong to the minority party.) How many different committees can be formed?
10. A bridge hand consists of 13 cards (from a standard 52-card deck). How many bridge hands are there that include no card higher than a nine?
11. What is the probability that a bridge hand contains

    a) no card lower than a jack?
    b) all four aces?

12. A package contains fifty light bulbs, of which five are defective. A sample of five bulbs is selected at random without replacement.

    a) How many different samples are possible?
    b) How many of the samples contain at least one defective bulb?
    c) The package will be shipped if the sample contains no more than one defective bulb. What is the probability that it will be shipped?

13. A tour of 500 people is returning from abroad. Two of the people have contracted a rare and dangerous disease. Health officials screen a sample of 20 people, selected randomly from the group. What is the probability that the disease is detected?
14. A building has ten floors. Six people get on an elevator. What is the probability that they all get off on different floors?
15. A firm employs 15 women and 10 men in one of its departments. A special committee of 5 people has been formed to study working conditions. The women have objected because the committee has no

women as members. The firm responded that the committee was chosen randomly. What is the probability that such a committee, selected randomly, would not contain any women?

11.6 MARKOV CHAINS*

Thus far we have been dealing with situations in which we were concerned with the outcome of a single experiment. However, a variety of situations occur in which we are concerned with a sequence of experiments. For example, consider a small town that has only two food stores, A and B. Currently, each store has 50% of the total volume. The management of store A has undertaken an advertising campaign to lure customers away from store B. Management expects that as a result of the advertising, 90% of store A's customers will return the next week (with the other 10% going to store B) while only 70% of store B's customers will return to store B (with the other 30% going to store A). Management would like to compute the new market shares for the next week and the two following weeks (assuming the probabilities of 90% and 70% are maintained).

Let's study this scenario. Notice that we have a system (in this case, the customers of two stores) whose behavior cannot be predicted precisely. We can, however, give probabilities that certain outcomes will occur.

To begin our analysis, we'll organize some of the problem data into a vector. We can think of the market share for each store as the probability that a customer chosen at random will shop in that store. Here each store has 50% of the total number of customers so the probability that a customer shops in a particular store is 0.50. We can think of these probabilities as describing the current condition of the system, often called its **state.** We place these probabilities into a vector called a **probability vector.** In general, each element in a probability vector gives the probability that the system will be in one of the possible conditions (states). Since the system must be in one of the states, the elements in a probability vector must sum to 1. The initial probability vector for our two stores is [0.5, 0.5], where the first entry is the probability that a random customer goes to store A and the second entry is the corresponding entry for store B. Note that the sum of these probabilities is indeed equal to 1.

Next we have a set of conditional probabilities that describes how the system can change from one period to the next, given its

* This section requires the use of vector-matrix notation. See Chapter 5.

current condition (state). These conditional probabilities are usually organized into a matrix called a **transition matrix.** The transition matrix for our example of the stores is

$$
\begin{array}{llcc}
 & & \multicolumn{2}{c}{\text{Next Week}} \\
 & & \text{Store A} & \text{Store B} \\
\text{This Week} & \begin{array}{l}\text{Store A} \\ \text{Store B}\end{array} & \multicolumn{2}{c}{\begin{bmatrix} 0.9 & 0.1 \\ 0.30 & 0.70 \end{bmatrix}}
\end{array}
$$

Here the first row of the transition matrix tells us that

$$P(\text{store A next week} \mid \text{store A this week}) = 0.9$$

and

$$P(\text{store B next week} \mid \text{store A this week}) = 0.1.$$

In general, a transition matrix must be square (because it gives the conditional probabilities of moving from any one characteristic state to any other possible state of the system). Since each entry in the matrix is a (conditional) probability, it must be between 0 and 1. Finally, each row of the matrix must sum to 1 because no matter what the system is currently doing, it must be found in one of the possible states in the next time period.

Thus we can describe our system (the customers of two stores) with a probability vector and a transition matrix.

A sequence of experiments of a system such that the system can have only a finite number of states and the probability of being in a given state after an experiment depends only upon the state of the system immediately before the experiment (given by the probability vector) and the transition matrix is called a **Markov chain.** Note especially that we didn't say anything about the transition matrix having to be the same from one experiment to the next. However, we will assume the transition matrix is constant because otherwise the analysis becomes exceedingly complicated. Thus our system is a Markov chain whose probability vector and transition matrix were given. The experiments consist of observing the fraction of customers who shop at each store during a given week.

Markov Chain

A Markov chain is a sequence of experiments of a system such that

1. the system has only a finite number of states,
2. the probability that the system is in a given state after an experiment depends only on the state of the system immediately before the experiment and the transition matrix.

Returning to our problem, we want to determine the probability vector for week 2. In other words, we want $P(\text{store A})$ and $P(\text{store B})$ for the second week. We can accomplish this by noting that

$$\begin{aligned} P(\text{store A in week 2}) = {} & P(\text{store A in week 1}) \\ & \cdot P(\text{store A in week 2} \mid \text{store A in week 1}) + P(\text{store B in week 1}) \\ & \cdot P(\text{store A in week 2} \mid \text{store B in week 1}) \end{aligned}$$

and

$$\begin{aligned} P(\text{store B in week 2}) = {} & P(\text{store A in week 1}) \\ & \cdot P(\text{store B in week 2} \mid \text{store A in week 1}) + P(\text{store B in week 1}) \\ & \cdot P(\text{store B in week 2} \mid \text{store B in week 1}). \end{aligned}$$

Moreover, we can accomplish these calculations very easily by simply multiplying the probability vector for week 1 by the transition matrix. Thus the probability vector for week 2 is

$$[0.5, 0.5]\begin{bmatrix} 0.9 & 0.1 \\ 0.30 & 0.70 \end{bmatrix} = [0.6, 0.4]$$

In other words, store A will have 60% of the customers and store B only 40% of the customers in the second week.

Given that we can find the probability vector for the second week from that of the first week by multiplying by the transition matrix, a little thought should convince us that we can find the probability for the third week from that of the second week by a similar multiplication. Hence the probability vector for week 3 is

$$[0.6, 0.4]\begin{bmatrix} 0.9 & 0.1 \\ 0.30 & 0.70 \end{bmatrix} = [0.66, 0.34].$$

Store A will have 66% of the customers and store B will have 34% of the customers in the third week.

We can use this process as often as we wish to compute the probability vector for any given week. The outcomes for the next 10 weeks are shown in Table 11.5. Notice that the probability vector seems to be approaching [0.75, 0.25]. In fact,

$$[0.75, 0.25]\begin{bmatrix} 0.9 & 0.1 \\ 0.3 & 0.7 \end{bmatrix} = [0.75, 0.25].$$

Multiplying the probability vector [0.75, 0.25] by the transition matrix results in the same probability vector. This vector is called the **steady-state probability vector** for the given transition matrix; it gives the long-run market shares.

Fortunately, we can determine the steady-state probability vector without going through the tedious calculations shown. Let

$\mathbf{v} = [v_1, v_2]$ denote the steady-state probability vector,

P denote the transition matrix.

TABLE 11.5 Probability vectors for stores A and B

Week	[Store A, Store B]	
1	[0.5, 0.5]	
2	[0.6, 0.4]	
3	[0.66, 0.34]	
4	[0.696, 0.304]	
5	[0.7176, 0.2824]	
6	[0.73056, 0.26944]	
7	[0.73834, 0.26166]	(approximately)
8	[0.743, 0.257]	(approximately)
9	[0.7458, 0.2542]	(approximately)
10	[0.7475, 0.2525]	(approximately)

Then the steady-state probability vector is the solution to the matrix equation

$$\mathbf{v}P = \mathbf{v}.$$

In the case of the two stores, $\mathbf{v} = [v_1, v_2]$ so we have

$$[v_1, v_2]\begin{bmatrix} 0.9 & 0.1 \\ 0.3 & 0.7 \end{bmatrix} = [v_1, v_2].$$

In addition, because v_1 and v_2 are probabilities, they must sum to 1. Thus the preceding system becomes

$$\begin{aligned} 0.9v_1 + 0.3v_2 &= v_1, \\ 0.1v_1 + 0.7v_2 &= v_2, \\ v_1 + \quad v_2 &= 1. \end{aligned}$$

Now, the first two equations reduce to the same equation after moving all the v's to the left-hand side. Thus we need only retain the first equation and the last. The solution is

$$\mathbf{v} = [v_1, v_2] = [0.75, 0.25].$$

Thus the long-run market shares will indeed be 75% for store A and 25% for store B.

Now we're prepared to solve the Problem from Life.

EXAMPLE 28

Problem from Life In a certain town there are three fast-food restaurants near the university: Barry's Burgers, Guido's Pizza, and

Tio Pepe's Tacos. During the last month, approximately 250 students per day bought dinner at each of the three restaurants. A recent survey (sponsored by the marketing club) has shown that students have a tendency to switch restaurants:

Of Barry's regular customers, 50% are loyal, 20% will switch to Guido's next month, and 30% will switch to Tio Pepe's next month.

Of Guido's regular customers, 50% are loyal, 25% will switch to Barry's next month, and 25% will switch to Tio Pepe's next month.

Of Tio Pepe's regular customers, 50% are loyal, 20% will switch to Barry's next month, and 30% will switch to Guido's next month.

a) How many customers per day can each restaurant expect next month?

b) What will the long-term market shares be?

SOLUTION

We can solve this problem by modeling it as a Markov chain. In this case, the states of the system correspond to the market shares of each restaurant. The transition matrix will reflect the tendencies of the students to remain "loyal" or to switch restaurants. Thus, the current probability vector is given by [0.333, 0.333, 0.334] (which does sum to 1). (In order for the probabilities to sum to 1, one of them must be 0.334. We arbitrarily chose the third probability to have this value.) In the transition matrix, we have used the order

Rows	Columns
{Barry's, Guido's, Tio Pepe's}	{Barry's, Guido's, Tio Pepe's}

The transition matrix is

$$P = \begin{bmatrix} 0.50 & 0.20 & 0.30 \\ 0.25 & 0.50 & 0.25 \\ 0.20 & 0.30 & 0.50 \end{bmatrix}.$$

Then the market shares next month are given by multiplying the probability vector [0.333, 0.333, 0.334] by the transition matrix P. The result is (approximately) [0.317, 0.333, 0.350]. In other words, next month Barry's Burgers will get 31.7% of the 750 students per day (= 238), Guido's Pizza will get 33.3% of the 750 students per day (= 250), and Tio Pepe's Tacos will get 35.0% of the 750 students per day (= 262).

To determine the long-run market shares, we must compute the steady-state probability vector. Let $\mathbf{v} = [v_1, v_2, v_3]$ denote this vector.

Then we must find the solution to the matrix equation

$$\mathbf{v}P = \mathbf{v}$$

with the side condition that

$$v_1 + v_2 + v_3 = 1.$$

Written out, this becomes

$$\begin{aligned} 0.50v_1 + 0.25v_2 + 0.20v_3 &= v_1, \\ 0.20v_1 + 0.50v_2 + 0.30v_3 &= v_2, \\ 0.30v_1 + 0.25v_2 + 0.50v_3 &= v_3, \\ v_1 + \quad v_2 + \quad v_3 &= 1. \end{aligned}$$

As in our earlier analysis, one of the first three equations is redundant. Throwing out the third one and solving for the steady-state probability vector gives us

$$\mathbf{v} = [v_1, v_2, v_3] = [0.310, 0.336, 0.354] \qquad \text{(approximately)}.$$

Thus Barry's Burgers will have 31.0% of the students or 232 per day, Guido's Pizza will have 33.6% of the students or 252 per day, and Tio Pepe's Tacos will have 35.4% of the students or 266 per day. □

Practice Problem

Given the probability vector [0.4, 0.6] and the transition matrix

$$P = \begin{bmatrix} 0.5 & 0.5 \\ 0.3 & 0.7 \end{bmatrix}.$$

a) Compute the probability vector in the next period.

b) Compute the steady-state probability vector.

Answers: **a)** [0.38, 0.62]; **b)** [0.375, 0.625]

One final point. Notice that the computation of the steady-state probability vector depends only on the transition matrix. Thus the long-run trends are totally independent of the initial probability vector. In fact, no matter what the initial probability vector is, the system will tend to the steady-state probability vector.

EXERCISE SET 11.6

Which of the following are valid probability vectors?

1. [0.5, 0.5]

2. [0.4, 0.3, 0.1]

3. [0.25, 0.25, 0.5]

4. [1, 0, 0]

5. [0.333, 0.667]

6. [0.3, −0.1, 0.8]

Which of the following are valid transition matrices?

7. $\begin{bmatrix} 0.5 & 0.5 \\ 0.2 & 0.8 \end{bmatrix}$

8. $\begin{bmatrix} 0.2 & 0.8 \\ 0.1 & 0.8 \end{bmatrix}$

9. $\begin{bmatrix} 1/3 & 1/3 & 1/3 \\ 0 & 4/5 & 1/5 \\ 1/9 & 3/9 & 5/9 \end{bmatrix}$

10. $\begin{bmatrix} 0 & 1/4 & 3/4 \\ 1/5 & 3/5 & 0 \\ 0 & 1/2 & 1/2 \end{bmatrix}$

11. $\begin{bmatrix} 1/2 & 1/2 & 1/10 \\ 1/3 & 2/3 & 0 \\ 1/8 & 3/8 & 3/8 \end{bmatrix}$

12. $\begin{bmatrix} 0.8 & 0.1 & 0.1 \\ 0.05 & 0.8 & 0.15 \\ 1/3 & 1/3 & 1/3 \end{bmatrix}$

13. In a small town there are two newspapers, the *Bugle* and the *Herald*. Currently, the *Bugle* has 60% of the readership and the *Herald* has the remainder. Readership surveys have indicated that 95% of the *Bugle's* readers will read it again the next month, while 90% of the *Herald's* readers will read it again the next month. Assuming that the total number of readers remains constant,

a) Determine the initial probability vector and transition matrix.
b) Determine the probability vector one month from now.
c) What is the steady-state probability vector?

14. In a city without cable television, three channels compete for viewers in each time-slot. The programs for the 9 P.M. slot are made-for-TV movies (channel 5), a prime-time soap opera (channel 7), and a sitcom (channel 12). A recent survey has indicated that channel 5 draws about 40% of the audience, channel 7 has 40%, and channel 12 only 20%. In addition, approximately 80% of channel 5 viewers will watch again the following week, while the remaining 20% split evenly between channels 7 and 12. Of channel 7's viewers, 70% will return the following week, while 10% switch to channel 5, and 20% to channel 12. Of channel 12's viewers, 75% return the following week, 15% go to channel 5, and 10% go to channel 7.

a) Determine the probability vector and transition matrix.
b) What will the market shares be next week?
c) What will the long-run shares be?

CHAPTER REVIEW

IMPORTANT TERMS AND SYMBOLS

11.1 Basics

event A subset of the sample space of an experiment.
Compound event: an event that consists of more than one outcome.
Simple event: an event that consists of one outcome.

(random) experiment A situation in which the outcome is uncertain.

fundamental properties of probability Let S be a sample space and let E be an arbitrary event associated with some experiment. Denote the probability of E by $P(E)$. Then

1. $0 \leq P(E) \leq 1$.
2. $P(S) = 1$.
3. $P(\phi) = 0$.

When S is finite and given by $S = \{e_1, e_2, \ldots, e_n\}$, then property 2 is equivalent to

$$P(\{e_1\}) + P(\{e_2\}) + \cdots + P(\{e_n\}) = \sum_{j=1}^{n} P(\{e_j\}) = 1.$$

Probability of each outcome $= 1/n$ for each outcome in the sample space, *assuming* equal probabilities for each outcome.
Probability of an arbitrary event, E: $P(E)$ is the sum of the probabilities of the outcomes found in E.

odds Sometimes used in place of probabilities; if the odds in favor of an event occurring are "x to y," the probability of the event is $x/(x + y)$.

probability A measure of whether or not an event will occur.
Empirical probability: based on observations of how often an event occurs, that is, based on relative frequency.

relative frequency The number of times an event occurred divided by the number of times it could have occurred; approximates the empirical probability of an event.

sample space The set of all possible outcomes of an experiment.

11.2 Probabilities of Unions and Complementary Events

complement of an event, *E* The event denoted by E' such that $E \cap E' = \phi$, $E \cup E' = S$ and $P(E) = 1 - P(E')$ (either E occurs or its complement, E', does).

probability of a union

1. $P(A \cup B) = P(A) + P(B)$ if $A \cap B = \phi$.
2. $P(A \cup B) = P(A) + P(B) - P(A \cap B)$ if $A \cap B \neq \phi$.

11.3 Conditional Probability and the Probability of Intersections

conditional probability The probability of an event given (knowing) that another event has occurred:

$$P(A \mid B) = \frac{P(A \cap B)}{P(B)}, \text{ where } A \text{ and } B \text{ are events.}$$

dependent events Events that aren't independent.
independent events Events A and B such that $P(A \mid B) = P(A)$ or $P(B \mid A) = P(B)$, that is, $P(A \cap B) = P(A) \cdot P(B)$.
probability of an intersection

1. $P(A \cap B) = P(A) \cdot P(B)$ when A and B are independent.
2. $\left.\begin{aligned} P(A \cap B) &= P(A \mid B) \cdot P(B) \\ &= P(B \mid A) \cdot P(A) \end{aligned}\right\}$ when A and B are not independent or when they are independent.

11.4 Bayes' Theorem

Bayes' theorem Let $A_1, A_2, \ldots, A_n$ be mutually exclusive events whose union is the entire sample space. Let B be an arbitrary event in S such that $P(B) > 0$. Then

$$P(A_1 \mid B) = \frac{P(B \mid A_1) \cdot P(A_1)}{P(B \mid A_1) \cdot P(A_1) + P(B \mid A_2) \cdot P(A_2) + \cdots + P(B \mid A_n) \cdot P(A_n)}.$$

Similar results hold for $P(A_2 \mid B), \ldots, P(A_n \mid B)$.

11.5 Counting: Permutations and Combinations

combination An arrangement of objects in which the order is not significant; formula for the number of combinations of n objects taken r at a time:

$$C(n, r) = \frac{n!}{(n-r)!\,r!} = \frac{n \cdot (n-1) \cdot (n-2) \cdots (n-r+1)}{r \cdot (r-1) \cdot (r-2) \cdots 3 \cdot 2 \cdot 1}.$$

multiplication principle If event A can occur in m ways and event B can occur in n ways (independent of one another), then together they can occur in $m \cdot n$ ways.
$n!$ (n factorial) $n! = n \cdot (n-1) \cdot (n-2) \cdots 3 \cdot 2 \cdot 1$, $n \geq 0$ (note that $1! = 1$ and $0! = 1$).
permutation An arrangement of objects in which order is significant; formula for the number of permutations of n objects taken r at a time:

$$\begin{aligned} P(n, r) &= \frac{n!}{(n-r)!} \\ &= n \cdot (n-1) \cdot (n-2) \cdots (n-r+1). \end{aligned}$$

11.6 Markov Chains

Markov chain A sequence of experiments of a system such that

1. the system has only a finite number of states.
2. the probability that the system is in a given state after an experiment depends only upon the state of the system immediately before the experiment and the transition matrix.

probability vector A vector of the probabilities that a Markov chain is in one of its possible states.

state One of the possible conditions that describe a system; also, the current condition of the system.

steady-state probability vector The probability vector that describes the long-run behavior of a Markov chain; it must satisfy the matrix equation $\mathbf{v} \cdot P = \mathbf{v}$ and the side condition $\sum_{i=1}^{n} v_i = 1$ where

$$\mathbf{v} = \text{steady-state probability vector}$$
$$P = \text{transition matrix of the Markov chain.}$$

transition matrix The matrix of conditional probabilities that a Markov chain in a given state will be in each of the other possible states after the next experiment.

REVIEW PROBLEMS

1. In a survey of 500 television viewers, 400 watched situation comedies, 400 watched game shows, and 450 watched shows with action. In addition, 150 watched comedies and game shows, 200 watched situation comedies and action shows, and 50 watched all three types of programs.

   a) What is the probability that a viewer watches game shows?
   b) Are the events "watches game shows" and "watches action shows" independent?
   c) What is the probability a viewer does not watch situation comedies given that he watches game shows?

2. A course in mathematics uses a self-paced approach. Data have been gathered that show the probability of passing a certain test on the first try is 0.75. If the student fails, the probability of passing on the second try is 0.9. If the student fails twice, the probability of passing on the third try is only 0.8.

   a) What is the probability a student fails the test twice?
   b) What is the probability a student fails the test three times?
   c) What is the probability a student requires no more than two tries to pass the test?

3. The personnel office of a software firm has developed a screening procedure for prospective employees. Data have been gathered that indicate the procedure correctly identifies 90% of the applicants who

become successful and 75% of those who do not become successful. Approximately 50% of the applicants have the necessary skills to become successful.

a) What is the probability an applicant predicted to be successful is actually successful?
b) What is the probability an applicant predicted to fail is successful?

4. A large department store stocks three brands of televisions. Company A produces 40% of the total, Company B produces 40% of the total, and Company C produces the remainder. Approximately 1% of the sets from Company A require service in the first year, while 1.5% of the Company B sets and 2% of the Company C sets require service. Suppose a television set fails in the first year. What is the probability that set was built by Company C?

5. A large electronics firm manufactures (among other things) electronic games. Approximately 4% of the games are defective. A random sample of size 50 is taken.

a) What is the probability that none of the items in the sample is defective?
b) What is the probability that two of the items are defective?

6. Compute the probability of being dealt a straight flush in standard draw poker. (A straight is a hand in which the five cards form a sequence in order. For example, {5-Cl, 6-H, 7-H, 8-Sp, 9-H} is a straight. A straight flush means the cards all belong to the same suit.) The ace is counted as being the highest card in a suit.

7. The cafeteria staff at a university is attempting to predict how many desserts to provide after the evening meal. Currently approximately 80% of the students eat dessert. The staff has also noticed that of those who take desserts, approximately 90% will take dessert at the next meal while of those who don't take dessert, approximately 40% will take dessert at the next meal.

a) How many students will take dessert after the next meal?
b) What are the long-run percentages of those who take dessert and those who don't?

8. A professor at a university likes to surprise his students. He may give them a pop quiz, ask the students to place the assignment on the board, or deliver a lecture. If he gives a quiz one day, the next day he always delivers a lecture. If he asks students to place the assignment on the board, there is a 25% chance of a quiz the next day, a 50% chance of a lecture, and a 25% chance of asking for students to place the assignment on the board. If he gives a lecture, there is a 40% chance of a quiz the next day, a 20% chance of a lecture, and a 40% chance of asking students to place the assignment on the board. In the long run, what percentage of the time can students expect quizzes, lectures, and being asked to place the assignment on the board?

12 STATISTICS

From past data it has been found that 70% of the people who enter a shopping mall make a purchase. On a certain day, 500 people enter the mall.

a) How many people can be expected to make a purchase?
b) What is the probability that no more than 350 people will make a purchase?
c) What is the probability that at least 360 people will make a purchase? (To be solved as Example 21, Section 12.6.)

PROBLEM FROM LIFE

INTRODUCTION

The subject of statistics is generally broken down into two broad areas. **Descriptive statistics** refers to the collection and summarization of data. **Inferential statistics** is the area that attempts to make inferences (conclusions) about a population given data that were obtained from a sample drawn from the population. Of course, when we use a sample to attempt to make inferences concerning a population, we first must describe the sample with descriptive statistics. Therefore we will begin the chapter by learning how to describe and summarize data.

12.1

PORTRAYING DATA

Suppose we have a set of data. The data might constitute a sample drawn from a population; for example, political pollsters deal with a sample drawn from the population of voters in order to attempt to predict the outcome of an election. Alternatively, the data might constitute an entire population; for example, we could find the height and weight of every member of a basketball team. When the size of the data set is small, for example, three data points (values), we can deal easily with the original data; but, when the size of the data set is larger, for example, fifty or more data points, it becomes much more difficult to get a "feel" for the data. One common approach in dealing with larger sets of data is to organize the data into a distribution. A **frequency distribution** is generated by defining categories of data and then counting the number of data points in each category.

EXAMPLE 1

A sample of 25 applicants for personal loans was taken by a consumer loan company. The amounts of money requested by the applicants were as follows. Organize these data into a frequency distribution.

\$ 900	\$1000	\$ 300	\$2000	\$ 550
500	550	1100	1000	350
450	950	300	2000	900
1900	600	1600	450	1650
1200	750	1500	750	700

SOLUTION

A frequency distribution of these data is shown in Table 12.1. An examination of Table 12.1 reveals some interesting information. More

TABLE 12.1 A frequency distribution for loan requests

Amount requested	Frequency	Cumulative frequency
\$ 300–599	8	8
600–899	4	12
900–1199	6	18
1200–1499	1	19
1500–1799	3	22
1800–2099	3	25

specifically, we can see the most commonly requested amount is less than \$600. In addition, 18 of the 25 requests were for less than \$1200 (we determine this by simply reading the last column, called the cumulative frequency). □

Table 12.1 illustrates several important points concerning frequency distributions.

1. Notice that the intervals are defined clearly. Thus there is no confusion about which interval to assign a data point.
2. We used 6 intervals. The actual number of intervals you should use is somewhat arbitrary. A good rule of thumb is to use between 5 and 15. Since we only had 25 data points, we knew that we should use no more than 5 or 6 intervals.
3. The difference between two successive lower limits of the intervals is called their **width.** We used intervals of equal width, in this case, a width of \$300. (The difference between two successive lower limits is \$300; for example, the first interval begins at \$300 and the next at \$600.)

   We calculated this width by noting that the largest loan request was \$2000 and the smallest was \$300, so that we had to cover a range of \$2000 − \$300 = \$1700. We then calculated the (approximate) width by dividing the range by the number of intervals:

   $$\frac{\$1700}{5} = \$340$$

   while

   $$\frac{\$1700}{6} = \$266.67.$$

   We then rounded this off to a width of \$300 per interval, which tells us to use 6 intervals (because with a width of \$300, 5 intervals are too few).
4. We included a column called **cumulative frequency,** which gives a cumulative total of the amount of loan requested up to and including the current interval. This column is a running total of the frequency column. For example, there were 12 loan applications for \$899 or less.

It is permissible to use intervals of unequal width. We do so when the range of the data is very large. In such cases, equal-width intervals would not work well because either too many intervals would be required or the intervals would be too wide. The next example illustrates a frequency distribution with intervals of unequal width.

EXAMPLE 2

There are 20 occupants in a certain condominium, including the owner. Their incomes last year were as follows. Construct a frequency distribution for these data.

$25,000	$30,000	$32,500	$27,500
22,500	35,000	34,000	37,500
40,000	29,000	36,000	45,000
28,000	33,000	28,500	41,000
31,500	39,500	81,000	34,500

SOLUTION

To begin, notice that all of the data are between $22,500 and $45,000 with the exception of one data point. In Table 12.2 we have shown a frequency distribution for this data using equal-width intervals, while in Table 12.3 we use unequal widths. Notice how much more information the distribution with unequal-width intervals conveys, because we are able to use intervals of smaller width. (The first distribution has intervals of width $10,000. The second distribution has intervals of width $5000 with the exception of the last interval.)

TABLE 12.2 A frequency distribution with equal-width intervals

Annual income	Frequency	Cumulative frequency
$22,500–32,499	8	8
32,500–42,499	10	18
42,500–52,499	1	19
52,500–62,499	0	19
62,500–72,499	0	19
72,500–82,499	1	20

TABLE 12.3 A frequency distribution with unequal-width intervals

Annual income	Frequency	Cumulative frequency
$22,000–26,999	2	2
27,000–31,999	6	8
32,000–36,999	6	14
37,000–41,999	4	18
42,000–46,999	1	19
47,000–81,000	1	20

□

It is also possible for a frequency distribution to consist of single elements of data, as well as intervals of data. Consider the following example.

EXAMPLE 3

A die was rolled 100 times with the following results.

3	1	4	6	1	3	1	6	1	6	4	5	3	3	2	3	1	3	5	2
2	4	4	2	4	4	2	5	5	3	3	5	5	5	5	5	4	3	2	6
5	5	1	6	5	1	6	1	6	1	1	1	5	3	3	4	2	4	4	2
6	1	5	4	6	3	4	2	5	2	4	6	2	1	5	1	5	1	6	2
5	3	5	6	1	6	1	2	2	6	4	1	2	2	2	3	2	6	4	3

Develop a frequency distribution for these data.

SOLUTION

When a die is rolled, the only possible results are the integers one through six. Thus, a frequency distribution for these data should indicate the number of times each of the integers one through six occurred. See Table 12.4. □

TABLE 12.4. A frequency distribution for single-element intervals

Result	Number of occurrences
1	18
2	18
3	15
4	15
5	19
6	15

Practice Problem

A typing pool consisting of 5 secretaries produced a total of 22 documents. The number of errors in each of the documents was as follows.

8	12	4	6	10	8	2	14	4	15	19
4	13	11	7	9	12	17	8	16	3	11

Develop a frequency distribution for these data using 6 equal-width intervals.

Answer:

Interval	Frequency	Cumulative frequency
2– 4	5	5
5– 7	2	7
8–10	5	12
11–13	5	17
14–16	3	20
17–19	2	22

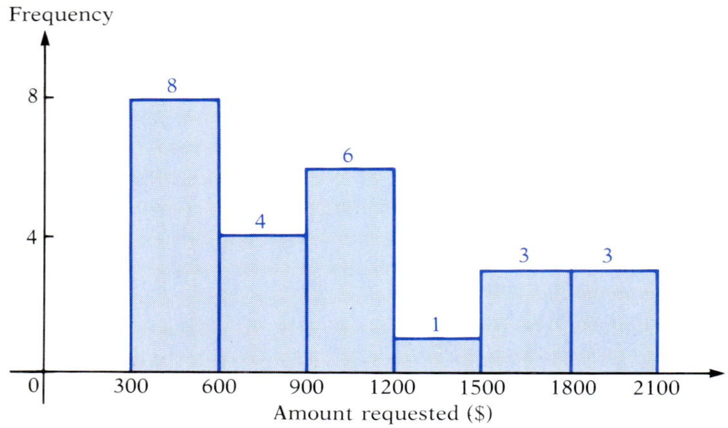

Figure 12.1 **A histogram for the data of Table 12.1.**

Once we have grouped data into a frequency distribution, we can display it using a bar graph called a **histogram.** In a histogram, the heights of the bars represent the frequencies of the intervals. The widths of the bars correspond to the widths of the intervals. Figures 12.1 and 12.2 are histograms for the distributions of Tables 12.1 and 12.3, respectively.

We can draw another graph to represent the data in a frequency distribution. A **frequency polygon** is drawn by joining the midpoints of the tops of histogram bars with straight-line segments. The midpoints of the first and last bars are joined to the horizontal axis at points that would correspond to the midpoints of extra intervals to the left of the first bar and to the right of the last bar. A frequency polygon for the data of Table 12.1 is shown in Fig. 12.3. We have

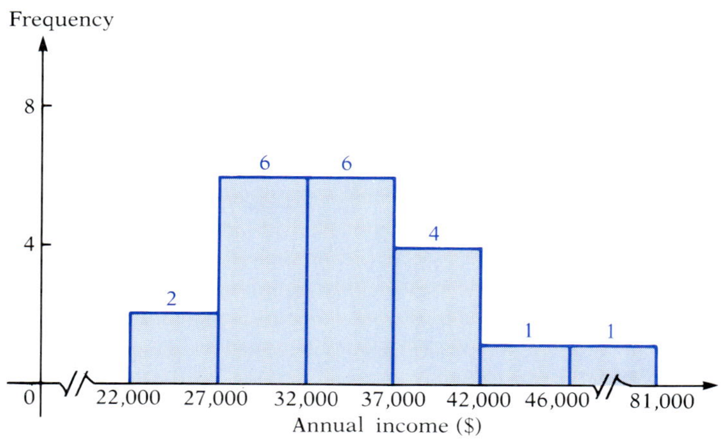

Figure 12.2 **A histogram for the data of Table 12.3.**

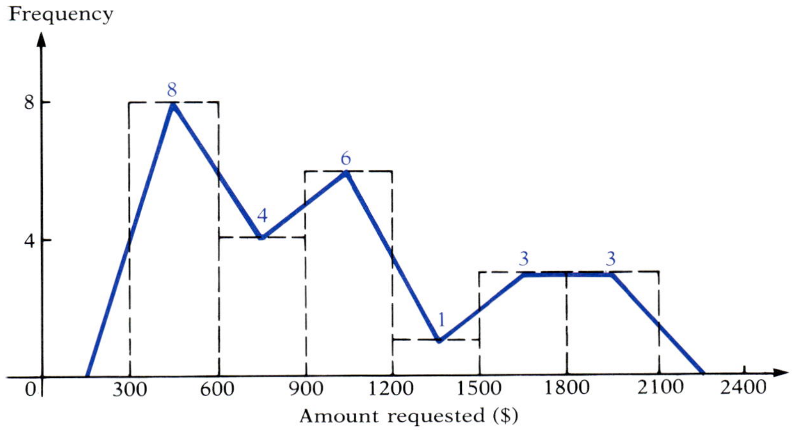

Figure 12.3 **A frequency polygon for the data of Table 12.1.**

shown the histogram bars with dotted lines in this drawing. Compare this to Fig. 12.1.

Finally, we can draw a graph to represent the cumulative frequency column of a frequency distribution. Such a graph is called a **cumulative frequency polygon** or **ogive** (pronounced: o-jive). We plot the points at the **upper limit** of each interval and not at the midpoint. A cumulative frequency polygon portrays the number of data points whose value is less than or equal to the upper limit of each interval. Figure 12.4 is a cumulative frequency polygon for the data of Table 12.1. Compare it to Fig. 12.3.

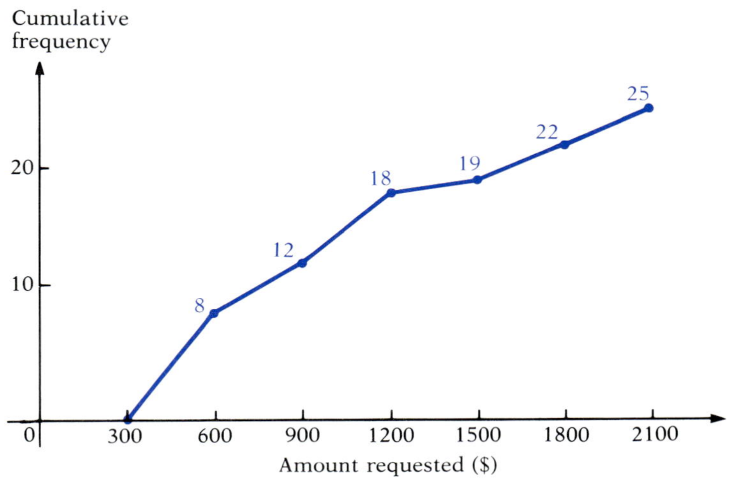

Figure 12.4 **A cumulative frequency polygon for the data of Table 12.1.**

Practice Problem

Draw a histogram, a frequency polygon, and a cumulative frequency polygon for the following frequency distribution.

Interval	Frequency	Cumulative frequency
2– 4.99	5	5
5– 7.99	2	7
8–10.99	5	12
11–13.99	5	17
14–16.99	3	20
17–19.99	2	22

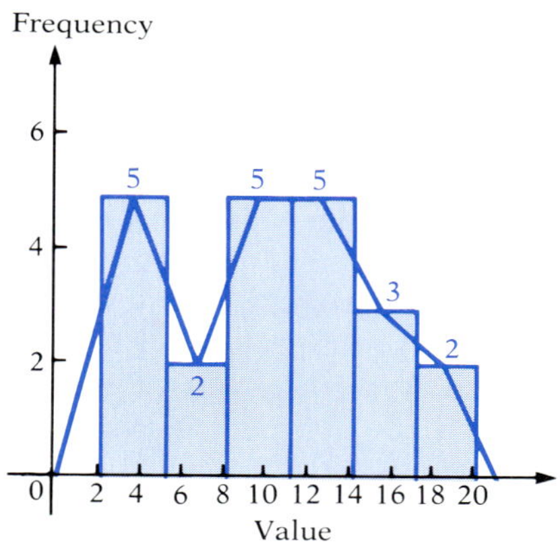

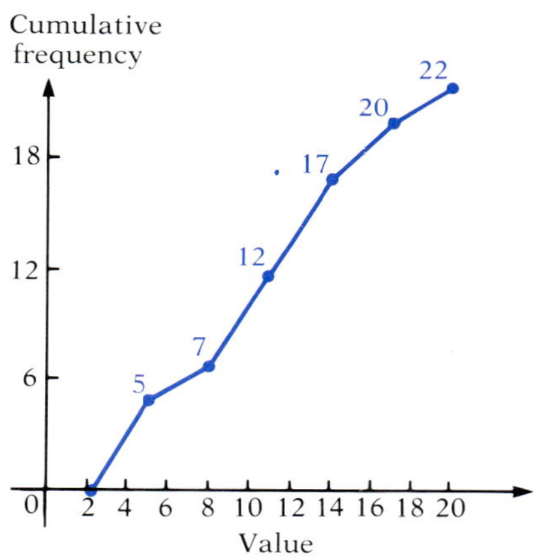

EXERCISE SET 12.1

For Problems 1–4, prepare a frequency distribution, a histogram, a frequency polygon, and a cumulative frequency polygon.

1. A sample of 24 gas stations showed the following prices for regular grade gasoline (leaded), in dollars per gallon:

1.068	1.101	1.078	1.068	1.096	1.066	1.078	0.982
1.074	0.958	0.992	1.070	1.002	1.006	1.104	1.058
1.068	1.046	1.106	1.136	1.054	1.104	1.000	1.110

2. The scores of 60 students on a college algebra test were as follows:

68	84	75	82	68	90	62	88	76	93	73	79	88	73	60
93	71	59	85	75	61	65	75	87	74	62	95	78	63	72
66	78	82	75	94	77	69	74	68	60	96	78	89	61	75
65	80	73	57	88	78	62	76	53	74	86	67	73	81	72

3. The combined Scholastic Aptitude Test (SAT) scores for a group of 80 freshmen were as follows:

1544	902	1205	880	1140	1029	734	1041	908	1306
866	1009	1040	1205	1063	1376	960	1185	1130	1334
1424	830	1190	1160	1285	1080	945	1485	1216	1003
1133	996	1271	1563	1074	1245	932	1225	1000	1145
1235	1160	1293	1005	996	1390	1014	1075	1100	785
695	1123	741	633	1561	1007	956	903	1215	979
884	1192	1473	1025	1036	1274	1198	1296	1354	1500
950	1088	934	1461	1152	1176	1410	1390	1082	833

4. The weights of 40 newborn babies at a local hospital were recorded as follows.

7.7	5.8	8.6	10.3	9.4	5.6	8.8	7.2	10.4	6.5
4.8	7.9	6.9	7.2	7.7	8.6	7.8	7.4	8.9	8.0
9.5	6.5	6.6	6.9	6.8	7.1	3.8	9.8	5.7	4.5
5.8	4.5	5.9	7.4	4.6	8.9	4.7	8.7	7.4	10.5

12.2 MEASURES OF CENTRAL TENDENCY

In the previous section we learned how to make frequency distributions and graphical presentations of data to help summarize them. In many cases, however, we want to summarize a set of data with a single number that is representative of the data set. The number typically chosen is an indication of the "location" of the data, which we call a **measure of central tendency.** There are three commonly used measures of central tendency: the mode, the median, and the (arithmetic) mean.

The Mode

The **mode** of a set of data is simply the value that occurs most often. Thus the mode of the numbers

$$5, \quad 5, \quad 7, \quad 8, \quad 6, \quad 4, \quad 2, \quad 9, \quad 5, \quad 6$$

is 5, because it occurs three times.

Practice Problem

What is the mode of the following set of data?

23, 32, 38, 27, 32, 31, 33, 26, 29

Answer: 32

The mode need not be unique. There may be two or even more modes, depending upon the data. For example, the following data

5, 6, 11, 12, 9, 7, 7, 5, 1, 12, 7, 9, 12, 10, 5

have three modes: 5, 7, and 12.

Finally, a data set will have no mode if each value occurs only once.

Practice Problem

Compute the mode for each of the following data sets.

a) 23, 32, 38, 27, 32, 31, 33, 26, 29, 23

b) 512, 545, 318, 727, 643, 589, 695

Answers: **a)** two modes: 23 and 32; **b)** no mode

The Median

The **median** is the value in the middle of a set of data that is arranged in either increasing or decreasing order. When there are an even number of values, the median is found by adding the two middle values and dividing by two. For example, given the set of data

23, 32, 38, 27, 32, 31, 33, 26, 29

we can find the median by arranging it in decreasing order and then finding the middle entry. We have

38
33
32
32
31 = median, because it is the fifth out of nine values
29
27
26
23

Similarly, given the data

$$23, \quad 32, \quad 38, \quad 27, \quad 32, \quad 31$$

we have

$$\begin{array}{l} 38 \\ 32 \\ \left.\begin{array}{l} 32 \\ 31 \end{array}\right\} \text{middle values} \quad \text{median} = \dfrac{31 + 32}{2} = 31.5 \\ 27 \\ 23 \end{array}$$

Practice Problem

Find the median of each of the following sets of data.

a) 512, 545, 318, 727, 643, 589, 695

b) 5, 5, 7, 6, 4, 3, 2, 2, 1, 7

Answers: **a)** 589; **b)** 4.5

The Mean

The **arithmetic mean** $\bar{x}$ (mean) is the most familiar measure of central tendency. It is found by simply adding all the data and dividing by the total number of data points. We can write this in compact form using summation notation as follows. Let the numbers in the data set be $x_1, x_2, \ldots, x_n$. Then

Arithmetic Mean

$$\bar{x} = \frac{x_1 + x_2 + \cdots + x_n}{n} = \frac{\sum_{i=1}^{n} x_i}{n}.$$

For example, the mean of the data set

$$23, \quad 32, \quad 38, \quad 27, \quad 32, \quad 31, \quad 33, \quad 26, \quad 29$$

is given by

$$\begin{aligned} \bar{x} &= \frac{23 + 32 + 38 + 27 + 32 + 31 + 33 + 26 + 29}{9} \\ &= 271 \div 9 \\ &= 30.11. \end{aligned}$$

EXAMPLE 4

The inhabitants of a small town reported the following incomes last year:

\$30,000	\$25,000	\$27,500	\$32,500	\$ 35,000
22,000	30,000	37,500	40,000	200,000

Compute the following measures of central tendency.

a) the mode **b)** the median **c)** the mean

SOLUTION

a) The mode is \$30,000 because it occurs twice.

b) To compute the median, we must arrange the data in either increasing or decreasing order. Using increasing order we obtain (reading across the page)

\$22,000	\$25,000	\$27,500	\$30,000	\$ 30,000
32,500	35,000	37,500	40,000	200,000

There are 10 values, so the median is the average of the fifth and sixth values. Thus the median is given by

$$\text{median} = \frac{\$30{,}000 + \$32{,}500}{2} = \$31{,}250.$$

c) To compute the mean, we simply add up all the values and divide by 10. We get

$$\bar{x} = \frac{\sum_{i=1}^{10} x_i}{n}$$

$$= \frac{\$479{,}500}{10},$$

so $$\bar{x} = \$47{,}950. \quad \square$$

Consider the results of Example 4. The mode of \$30,000 and the median of \$31,250 are in close agreement. Why is the mean so different? The reason is because the last income is \$200,000, which is considerably larger than the other incomes. *The mode and the median are not affected by extreme values, while the mean is.* Which measure of central tendency is correct? It depends upon the purpose you have in mind. In this case, if we wish to give a "typical" value for the income of the people in the town, the median is better than the mean. In general, the mean is more sensitive to extreme values in the data than either the median or the mode. Therefore, be careful when using a measure of central tendency by itself.

Practice Problem

Find the mean, median, and mode of the following data.

62 70 64 67 69 64 67 64 64 67 73 73

Answers: mode = 64, median = 67, mean = 67

One final note. All the examples we've shown so far have used numerical data. However, there are situations in which the data are **categorical,** which means that all we can do with the data is to classify them into different categories. With categorical data, the only appropriate measure is the mode. For example, suppose a class consists of 25 freshmen, 35 sophomores, 15 juniors, and 10 seniors. These data are categorical, because we can classify the students into four groups. The only valid measure of central tendency is the mode. We say the **modal group** is sophomore, because there are more sophomores in the class than there are freshmen, juniors, or seniors.

At this point, you probably think it will be obvious whether data are categorical or numerical. Unfortunately, this is not always true. Beware of the situation in which the data are **coded.** For instance, we might represent freshmen by a 1, sophomores by a 2, juniors by a 3, and seniors by a 4. Then our data set would consist of twenty-five 1's, thirty-five 2's, fifteen 3's, and ten 4's. Now you might be tempted to compute a median or a mean. *Don't!* Whether we code the data or not, they are still categorical.

The next example shows data that are categorical but can be mistakenly identified as numeric.

EXAMPLE 5

A survey of 1200 people was conducted to determine which TV channels were watched most often. The following responses were obtained.

Channel	Number watching
3	450
5	200
8	100
10	350
12	100

These data are categorical. The only valid measure of central tendency is the mode. In this case, the mode is channel 3, with 450 viewers. It would be totally inappropriate to compute the median channel or (even worse) the mean channel. □

Practice Problem

Classify the following data as categorical or numeric.

a) The final grades in a statistics course consisted of 5 A's, 7 B's, 12 C's, 10 D's, and 2 F's.

b) There were 25 men and 30 women in a mathematics class.

c) The weights of 5 people: 110 pounds, 185 pounds, 145 pounds, 200 pounds, and 160 pounds.

Answers: **a)** categorical; **b)** categorical; **c)** numeric

EXERCISE SET 12.2

Classify the data in Problems 1–7 as being either categorical or numeric.

1. Noon temperatures, in degrees Fahrenheit

2. Shirt sizes S, M, L, XL

3. Shoe sizes 6, 7, 8, 9, 10, 11

4. Sleeve lengths 33″, 34″, 35″, 36″

5. The number of people listening to each radio station

6. The number of points scored by a basketball player

7. The number of packs of cigarettes smoked coded as follows:

0 = 0 cigarettes per day
1 = between 1 and 20 cigarettes per day
2 = between 21 and 40 cigarettes per day
3 = 41 or more cigarettes per day

For the data in Problems 8–10, find the mean, median, and mode.

8. Grade point averages (GPA's):

2.50	2.25	3.50	4.00	1.50	2.10	2.50	3.30	3.80
1.75	2.60	3.20	1.95	2.40	2.90	3.15	3.35	2.75

9. Attendance at football games:

35,000	40,000	55,000	25,000	37,500	45,000
30,000	42,500	50,000	47,500	55,000	30,000
47,500	52,500	22,500	47,500	36,000	50,000

10. Money spent on groceries:

$110.00	$ 95.00	$100.00	$25.00	$75.00	$150.00
$125.00	$150.00	$ 65.00	$80.00	$35.00	$100.00
$ 95.00	$ 15.00	$140.00	$50.00	$75.00	$110.00

12.3

MEASURES OF DISPERSION

So far we've learned how to compute several measures of central tendency. Such measures give us one-number descriptions of a set of data. However, no measure of central tendency will tell us how widely varying or tightly clustered the data are. A measure of central tendency is much more representative of tightly clustered data than it is for widely varying data. We'll discuss two measures of dispersion, the range and the standard deviation. These measures will help give us a better indication of the behavior of the data in a data set.

The Range

The **range** of a set of data is simply the difference between the largest and smallest values of the data. For example, consider the data set

8 10 14 9 7 9 10 6 44

The range is $44 - 6 = 38$.

Practice Problem

Compute the range of the following data.

44 85 125 32 25 96 34 100

Answer: 100

The Standard Deviation

Another measure of dispersion is provided by the standard deviation. We define the **sample standard deviation, *s*,** as follows.

Sample Standard Deviation

$$s = \sqrt{\frac{\sum_{i=1}^{n} (x_i - \bar{x})^2}{n - 1}}$$

EXAMPLE 6

Compute the sample standard deviation for the data set

8 10 14 9 7 9 10 6 44

SOLUTION

The standard deviation is calculated as follows. First, we must compute the mean:

$$\bar{x} = \frac{8 + 10 + 14 + 9 + 7 + 9 + 10 + 6 + 44}{9} = 13.$$

Next we must subtract $\bar{x}$ from each number and square the result. The resulting numbers are called **squared deviations.** We then add up these squared deviations, divide by $(n - 1)$, and take the square root of the result.

x_i	$x_i - \bar{x}$	$(x_i - \bar{x})^2$
8	−5	25
10	−3	9
14	1	1
9	−4	16
7	−6	36
9	−4	16
10	−3	9
6	−7	49
44	31	961
	0	1122

Finally,

$$s = \sqrt{\frac{1122}{8}} = \sqrt{140.25} = 11.843 \qquad \text{(approximately).} \ \square$$

Why do we compute the standard deviation as we do? The purpose of a measure of dispersion is to give us some sense about how much the data in a data set vary. We've discussed the range. As you can see from its definition, the range depends only upon the two extreme values (the largest and the smallest) of a data set. Is there a way to consider all of the data?

Let's think about the calculations we've just completed in Example 6. We began by computing the mean. We then subtracted it from each of the data points and squared the difference. The difference between each data point and the mean tells us how far each data point is from the mean.

Why didn't we simply subtract the mean from each value and just add up the differences? The answer is because the sum of the deviations (or differences) between each point and the mean is always zero; the sum of the $(x_i - \bar{x})$ column demonstrates this fact. We can show that the sum of *any* $(x_i - \bar{x})$ column must be zero by using the properties of sums from Chapter 4. To begin, the mean of a set of data is a constant. (We just don't know its value until we compute it.)

$$\begin{aligned} \sum_{i=1}^{n} (x_i - \bar{x}) &= \sum_{i=1}^{n} x_i - \sum_{i=1}^{n} \bar{x} && \textbf{Property 3} \\ &= \sum_{i=1}^{n} x_i - n \cdot \bar{x} && \textbf{Property 1} \\ &= n \cdot \bar{x} - n \cdot \bar{x} && \textbf{Definition of } \bar{x} \\ &= 0 \end{aligned}$$

Thus we squared the differences to make them all positive. We then added up all the squared differences and divided by $(n - 1)$ to give an "average" value of the squared distance (difference) between each data point and the mean. [We divide by $(n - 1)$ rather than n because we're dealing with a sample. Using $(n - 1)$ rather than n corrects for the fact that we must "use up" some of the information in the sample to compute the sample mean, $\bar{x}$.] Finally we took the square root to convert the units from squared distance to distance. Thus the standard deviation describes the average distance of each data point from the mean.

An alternative measure of dispersion uses the absolute value of each deviation (difference between each data point and the mean) to make it positive. This leads to a measure of dispersion called the **mean absolute deviation** or MAD. The mean absolute deviation is used primarily in forecasting.

The next example demonstrates how the standard deviation measures the average distance of the data from the mean.

EXAMPLE 7

Consider the following sets of data:

Set A: 5, 5, 5, 5, 5

Set B: 3, 4, 5, 6, 7

Set C: −7, −1, 5, 11, 17

We'll leave it to you to convince yourselves that each set of data has a mean of 5. Now, let's compute the standard deviations.

Set A			Set B			Set C		
x_i	$(x_i - 5)$	$(x_i - 5)^2$	x_i	$(x_i - 5)$	$(x_i - 5)^2$	x_i	$(x_i - 5)$	$(x_i - 5)^2$
5	0	0	3	−2	4	−7	−12	144
5	0	0	4	−1	1	−1	−6	36
5	0	0	5	0	0	5	0	0
5	0	0	6	1	1	11	6	36
5	0	0	7	2	4	17	12	144
		0			10			360
$s = \sqrt{\frac{0}{5-1}}$			$s = \sqrt{\frac{10}{5-1}}$			$s = \sqrt{\frac{360}{5-1}}$		
$s = 0$			$s = 1.5811$			$s = 9.4868$		

The data sets are shown in Fig. 12.5. Note how the value of s increases as the data become more widespread.

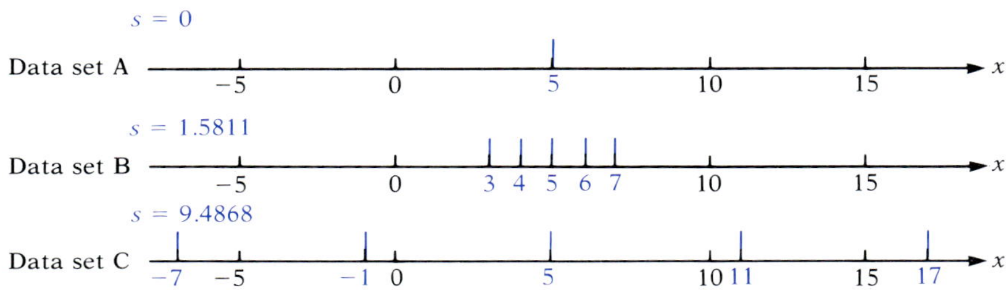

Figure 12.5 Three data sets with the same mean.

□

A more efficient formula for the standard deviation is available. We will refer to it as the **shortcut** formula:

Sample Standard Deviation-Shortcut Formula

$$s = \sqrt{\frac{\left[\sum_{i=1}^{n} (x_i^2) - \left(\sum_{i=1}^{n} x_i\right)^2 \Big/ n\right]}{n - 1}}$$

This formula is faster because we don't need to first compute the mean and then go back and subtract it from each data point. Let's see how the calculations go with the shortcut approach. We'll use the same data we used in Example 6.

EXAMPLE 8

x_i	x_i^2
8	64
10	100
14	196
9	81
7	49
9	81
10	100
6	36
44	1936
117	2643

Thus

$$\sum_{i=1}^{9} x_i = 117, \qquad \sum_{i=1}^{9} x_i^2 = 2643.$$

We then have

$$s = \sqrt{\frac{2643 - (117)^2/9}{9 - 1}} = \sqrt{140.25} = 11.843. \qquad \square$$

We've computed sample standard deviations with a denominator of $n - 1$. Almost all of the data sets that you will encounter will be samples. A different denominator is used to compute the standard deviation of a population (n, rather than $n - 1$). It is very unusual, however, to be working with an entire population. For this reason, we have emphasized the computations for samples. Unless we tell you specifically that you have a population, assume that any data set is a sample.

Practice Problem

Compute the sample standard deviation of the following set of data.

44 85 125 32 25 96 34 100

Answer: $s = 38.213$

In the formulas for the standard deviation, we computed an "average" squared distance from the mean of a data set and then took its square root. If we do not take the square root, we compute a measure called the **variance.** Thus the variance is the square of the standard deviation (or, the standard deviation is the square root of the variance). In Example 8, the sample variance is therefore 140.25.

The Coefficient of Variation

We know that a data set with a standard deviation of (say) 10 is more widely dispersed than a data set with a standard deviation of 5. However, sometimes we may wish to take into account the size (value) of the numbers in the data set. For example, suppose we had two data sets, one consisting of the heights of a group of adults and the other consisting of the heights of a group of five-year-olds. We should expect that the standard deviation of the set of adult's heights is larger than the standard deviation of the set of children's heights. (The heights of the adults could vary from under five feet to more than seven feet, while the heights of the children would probably be between three and four feet.) However, we also should expect the average height of the adults to be larger than the average height of the children. We want some method of comparing the dispersions of the data sets on a **relative** basis.

A statistic that helps us analyze situations such as this is the **coefficient of variation** or cv, defined as follows.

Coefficient of Variation

$$\text{cv} = \frac{s}{\bar{x}} \cdot 100\%$$

The coefficient of variation is a measure of **relative dispersion.** It tells us how large the standard deviation of a set of data is as a percentage of the mean of the data.

EXAMPLE 9 Compute the cv's for the data sets of Example 7.

SOLUTION

We have

Set A	Set B	Set C
$\text{cv} = \frac{0}{5} \cdot 100\%$	$\text{cv} = \frac{1.5811}{5} \cdot 100\%$	$\text{cv} = \frac{9.4868}{5} \cdot 100\%$
$\text{cv} = 0\%$	$\text{cv} = 31.622\%$	$\text{cv} = 189.736\%$

A cv of over 100% is very large. It indicates that the data are very widely dispersed compared to the mean. □

Practice Problem

Compute the coefficient of variation for each of the following data sets. Which data set has the larger relative dispersion?

a) 44 85 125 32 25 96 34 100

b) 1 5 9 4 3 1 2 9

Answers: **a)** cv = 56.507%; **b)** cv = 76.244%. The data set shown in (b) has the larger relative dispersion.

EXERCISE SET 12.3

Find the range, sample standard deviation, and coefficient of variation for the data sets in Problems 1–4.

1. 3, 3, 5, 7, 9

2. 10, 12, 20, 22, 30, 32

3. 100, 110, 120, 110, 90, 95, 100, 105

4. 25, 22, 18, 32, 27, 95, 15

5. The starting members of a basketball team had the following weights and heights:

Weight (lb)	132	161	203	245	225
Height (in)	66	72	76	79	83

a) Compute the coefficient of variation for the weights and the heights.
b) Which data set had the larger relative dispersion, weights or heights?

6. A company was worried about the absenteeism. The following data were collected over a 13-week period:

Week	1	2	3	4	5	6	7	8	9	10	11	12	13
Number absent	1	3	4	2	3	5	2	1	4	6	1	3	4

Compute the sample standard deviation and the range of the number of workers absent each week.

7. Six workers in a company were observed doing a particular operation. The times (in minutes) it took each of them to process a batch of 50 items were as follows.

12.25 10.75 11.50 12.00 11.75 10.75

The standard time for this operation is to be the mean of the observations. Acceptable times range from the mean less one standard deviation to the mean plus one standard deviation. Compute the standard time and the range of acceptable times. How many of the sample workers were within the acceptable range?

8. A bakery experienced the following daily demand for its specialty cakes.

Day	1	2	3	4	5	6	7	8	9	10
Demand	48	12	77	24	32	43	96	03	62	77

Compute the range and the sample standard deviation of the daily demand for the cakes.

12.4 THE BINOMIAL DISTRIBUTION

In Chapter 11 we discussed the concept of an empirical probability distribution: one in which the probabilities are based upon actual observations. In the remainder of this chapter, we will introduce the concept of a **theoretical probability distribution:** one in which the probabilities are based upon a theoretical (or hypothetical) model rather than on actual observations. There are many theoretical probability distributions; we will discuss four of them in this section and the next three.

When we talk about probability distributions, we are referring to a description of the possible values some variable can have and the likelihood that the variable has of assuming the different possible values. Thus probability distributions describe a variable called a random variable. A **random variable** is a variable whose value is determined by some chance (or random) process. For example, when we flip a coin, we can think of the outcome as a random variable, with possible values of heads or tails. As another example, consider a grocery store. The amount of money spent by a customer is a random variable. We might take a sample to try to estimate the distribution of this random variable. In many situations, however, there is a theoretical distribution that will provide the appropriate probability distribution for a random variable. Indeed, inferential statistics is based upon the use of many different hypothetical, theoretical distributions.

The first theoretical distribution we'll discuss is the **binomial distribution.** It is appropriate whenever a random variable is used to count the number of "successes" in a series of trials in which each trial can result in a success or a failure. An appropriate sample space in this situation is given by $S = \{0, 1, 2, \ldots, n\}$, where n is the number of trials. However, the outcomes in S are *not* equally likely.

(We've shown you one example of this in Chapter 11: flipping a coin twice in succession.)

More specifically, we describe a series of trials as a **Bernoulli process** when the following conditions hold.

Bernoulli Process

1. There are two mutually exclusive outcomes for each trial. We usually refer to these as **success** and **failure.**
2. The series of trials is a set of independent events. (The outcome from one trial has no influence on the outcome of the next trial.)
3. The probability of a success, denoted by p, is constant from trial to trial. This implies that the probability of a failure, $q = 1 - p$, is also constant.

The binomial distribution is used to compute the probability of obtaining a given number of successes in a Bernoulli process. Thus the binomial distribution is appropriate for situations in which we wish to count the number of successes. Recall from Chapter 11 that $C(n, x)$ represents the number of combinations of n things taken x at a time. The binomial distribution gives us the probability of exactly x successes in n trials with the probability of success at each trial equal to p:

Binomial Distribution

$$P(x \text{ successes in } n \text{ trials}) = C(n, x)p^x q^{n-x}, \qquad 0 \le x \le n,$$

where

$$C(n, x) = \frac{n!}{(n-x)!\,x!} = \frac{n \cdot (n-1) \cdots (n-x+1)}{x \cdot (x-1) \cdots 3 \cdot 2 \cdot 1}.$$

Let's look at some examples for which a binomial distribution is appropriate.

EXAMPLE 10

The standard example of a Bernoulli process is flipping a coin repeatedly. Suppose we flip a coin four times. Let's assume that a success = heads. Assuming the coin is fair, the probability of heads = probability of tails = 0.5. Compute the probability of observing zero, one, two, three, or four heads, where X is the number of heads.

SOLUTION

Using the binomial distribution we have

x (Number of heads)	$P(x) = C(4, x)(0.5)^x(0.5)^{n-x}$
0	$C(4, 0)(0.5^0)(0.5^4) = 0.0625$
1	$C(4, 1)(0.5^1)(0.5^3) = 0.25$
2	$C(4, 2)(0.5^2)(0.5^2) = 0.375$
3	$C(4, 3)(0.5^3)(0.5^1) = 0.25$
4	$C(4, 4)(0.5^4)(0.5^0) = 0.0625$
	1.0000

Notice that the distribution does indeed sum to 1. A table of the binomial distribution is given in Appendix C. Of course, you can use your calculator if you wish. □

If we flip a coin four times and repeat the experiment many times, we can expect the empirical distribution to be very similar to the theoretical distribution given in Example 10. However, when we perform the experiment just once, we observe exactly zero, one, two, three, or four heads. The binomial distribution tells us that we are more likely to see two heads than any other outcome, but certainly the other outcomes are possible.

A comment about notation: We represent a random variable by a capital letter. The various specific values it can have are represented by lower-case letters. Thus, in Example 10, we represented the number of heads by X. The specific number of heads in the distribution shown was represented by x.

EXAMPLE 11

Suppose we flip a biased coin so that the probability of heads = 0.6 and the probability of tails = 0.4. Compute the appropriate binomial distribution.

SOLUTION

x (Number of heads)	$P(x) = C(4, x)(0.6)^x(0.4)^{n-x}$
0	$C(4, 0)(0.6^0)(0.4^4) = 0.0256$
1	$C(4, 1)(0.6^1)(0.4^3) = 0.1536$
2	$C(4, 2)(0.6^2)(0.4^2) = 0.3456$
3	$C(4, 3)(0.6^3)(0.4^1) = 0.3456$
4	$C(4, 4)(0.6^4)(0.4^0) = 0.1296$
	1.0000

With this biased coin, we can see that outcomes with two or more heads are more likely than those with less than two heads. □

Practice Problem

Consider a gambling game in which you roll a pair of dice. If the sum of the dice is 8 or more, you win. Generate the binomial distribution for three plays of the game. (*Hint:* In Section 11.2, you were asked to compute probabilities of events similar to "the sum of the dice is 8 or more.")

Answer: We have p = probability (sum = 8 or more) = 15/36 = 0.4167, which gives us q = 0.5833. The binomial distribution is given by the following, where X is the number of wins:

x	$P(x)$
0	0.19846
1	0.42533
2	0.30385
3	0.07236

Earlier in this chapter you learned how to compute the mean and standard deviation of a sample. We can also compute the mean and standard deviation of a theoretical distribution. Because we are dealing with (theoretically) exact values and not estimates obtained from samples, we denote the mean by the Greek letter μ (mu) and the standard deviation by the Greek letter σ (sigma). (In general, we use Greek letters to denote population parameters. Lowercase Roman letters indicate sample values. Thus μ represents the population mean while $\bar{x}$ represents the mean of a sample.) For the binomial distribution it can be shown that

Binomial Distribution

Mean: $\mu = np$

Standard deviation: $\sigma = \sqrt{npq}$

where
n = number of trials.
p = probability of success.
$q = 1 - p$ = probability of failure.

EXAMPLE 12

Compute the mean and standard deviation for the binomial distribution in Example 10 (in which a fair coin was flipped four times).

SOLUTION

Here we have $n = 4$, $p = 0.5$, and $q = 0.5$, so that

$$\mu = np = 4(0.5) = 2,$$
$$\sigma = \sqrt{npq} = \sqrt{4(0.5)(0.5)} = 1. \quad \square$$

EXAMPLE 13

Compute the mean and standard deviation for the binomial distribution in Example 11 (in which a biased coin was flipped four times).

SOLUTION

With the biased coin we have $n = 4$, $p = 0.6$ and $q = 0.4$.

$$\mu = np = 4(0.6) = 2.4$$
$$\sigma = \sqrt{npq} = \sqrt{4(0.6)(0.4)} = 0.97980 \qquad \text{(approximately)}. \quad \square$$

The mean μ tells us the average number of successes we expect to see if the experiment is conducted repeatedly. Comparing Examples 12 and 13, we expect on the average to see more heads with the biased coin than with a fair coin: 2.4 as compared to 2. Please note that μ is a measure of central tendency and may not even be a possible outcome.

Practice Problem

Suppose you play a game of chance that you expect to win with probability 0.45. Assuming you play 20 times, compute the values of μ and σ.

Answer: $\mu = 20(0.45) = 9$ and $\sigma = \sqrt{20(0.45)(0.55)} = 2.22486$

EXERCISE SET 12.4

For Problems 1–6, a variety of Bernoulli processes are described. For each one, give the appropriate binomial distribution, the mean and the standard deviation.

1. A fair die is rolled five times and a success is a number divisible by 3.
2. A quiz consists of four multiple choice questions, each with five possible answers. Assume a student answers the questions randomly by guessing. A success is a correct answer.
3. A manufacturing process produces parts with a defect rate of 0.01. A sample of five items is chosen from each day's production. (We assume that the number of parts produced each day is much larger than five, so that we can assume the probabilities of a good part and of a defective part are constant.) A success is a good part.
4. Four cards are drawn from a standard 52-card deck, with replacement. A success is a spade.
5. Six babies (none of whom are brothers or sisters) are tested to determine whether they can identify unwashed articles of clothing belonging to their mothers. There are six mothers, so the probability of a success (correct identification) is 1/6 if the babies must guess.
6. A certain drug is effective 90% of the time. It is administered to a group of eight patients. A success is a cure.
7. Suppose the probability a marriage will end in divorce within five years is 0.5. What are the mean and standard deviation for the binomial distribution describing 5000 marriages?
8. An incoming freshman class at a university consists of 2000 students. The probability that an average student graduates within five years is 0.6. Compute the mean and standard deviation for the binomial distribution describing this situation.

12.5

THE POISSON DISTRIBUTION*

In this section we will describe a distribution that can be used to count the number of random events during a specified interval of time or space. For example, it can describe random variables that count the number of customers arriving at a bank in a half hour, the number of lightning flashes in ten minutes, or the number of repairs needed in five miles of highway.

* This section requires knowledge of the exponential function. See Chapter 9.

Poisson Distribution

A random variable has a **Poisson** distribution whenever the following three conditions are satisfied.

1. The probability of the occurrence of the event being counted is the same for any two intervals of equal length.
2. The occurrence or nonoccurrence of the event in any interval is independent of the occurrence or nonoccurrence of the event in any other (nonoverlapping) interval.
3. The probability of two or more events occurring in an interval becomes zero as the length of the intervals shrinks.

We can compute Poisson probabilities as follows.

Poisson Distribution

$$P(x) = \frac{\lambda^x e^{-\lambda}}{x!} \qquad \text{for } x = 0, 1, 2, \ldots,$$

where x = the specific number of events that occur.
λ = the average number of occurrences in an interval (λ is the Greek letter lambda).
e = the base of the natural logarithm system $= 2.71828\ldots.$

Let's look at some examples for which the Poisson distribution is appropriate.

EXAMPLE 14

Suppose a bank is interested in the number of customers who arrive during a 15-minute period on weekdays. Assuming the probability of a customer arriving during any 15-minute period is the same as during any other (nonoverlapping) 15-minute period and that the number of arrivals during any 15-minute period is independent of the number of arrivals during any other 15-minute period, the Poisson distribution is appropriate. Analysis of past data has shown that on the average, 7 customers arrive during every 15-minute interval. Compute the probability of X customers arriving during a 15-minute interval, for values of x from 0 to 11.

SOLUTION

The required probabilities can be found in Table 12.5. A more complete table can be found in Appendix D. Of course, you can use your calculator instead. Let's step through the calculation of $P(X = 5)$. The keystrokes are

7 [y^x] 5 [=] [×] 7 [+/−] [e^x] [=]
[÷] 5 [x!] [=] 0.12772 (approximately)

TABLE 12.5 Poisson probabilities for $\lambda = 7$

x	$P(x) = \frac{7^x e^{-7}}{x!}$	x	$P(x) = \frac{7^x e^{-7}}{x!}$
0	0.0009	6	0.1490
1	0.0064	7	0.1490
2	0.0223	8	0.1304
3	0.0521	9	0.1014
4	0.0912	10	0.0710
5	0.1277	11	0.0452 □

EXAMPLE 15

The Poisson distribution need not always be used with intervals of time; it can also refer to intervals of space (distance). Consider a company that manufactures cloth. Assuming the equipment is working normally, past data have indicated that the cloth averages 5 defects per 1000 yards, and the number of defects in any 1000-yard length is independent of the number of defects in any other 1000-yard length. Then the number of defects in a 1000-yard length, say X, is a random variable with a Poisson distribution with $\lambda = 5$. Compute the probabilities of observing x defects in a 1000-yard length of cloth for values of x from 0 to 11.

SOLUTION

The appropriate values are shown in Table 12.6. Again, more data can be found in Appendix D or by use of your calculator.

Notice that the probability of at least one defect is given by

$$\begin{aligned} P(\text{at least one defect}) &= 1 - P(\text{no defects}) \\ &= 1 - P(0) \\ &= 1 - 0.0067 = 0.9933. \end{aligned}$$

Thus a piece of cloth with no defects is quite rare.

TABLE 12.6 Poisson probabilities for $\lambda = 5$

x	$P(x) = \dfrac{5^x e^{-5}}{x!}$	x	$P(x) = \dfrac{5^x e^{-5}}{x!}$
0	0.0067	6	0.1462
1	0.0337	7	0.1044
2	0.0842	8	0.0653
3	0.1404	9	0.0363
4	0.1755	10	0.0181
5	0.1755	11	0.0082 □

Practice Problem

Suppose the number of arrivals to a restaurant averages ten customers per hour, independent of the number of arrivals during any other one-hour period. What are the probabilities of

a) no arrivals in one hour?

b) exactly ten arrivals in one hour?

Answers: **a)** 0.000045; **b)** 0.1251

Sometimes we know the average rate of a Poisson process, but we want to compute probabilities for a distribution with a different interval. For instance, in Example 15 we knew the average number of defects per 1000-yard length of cloth was 5. If we were discussing a 500-yard length of cloth, the appropriate defect rate would be 2.5 defects per 500-yard length. If we were discussing a 2000-yard length of cloth, the appropriate defect rate would be 10 defects per 2000 yard length.

Practice Problem

Suppose the number of arrivals to a restaurant averages ten customers per hour, independent of the number of arrivals during any other one-hour period. Compute the probabilities of

a) no arrivals in one-half hour.

b) exactly five arrivals in one-half hour.

Answers: with $\lambda = 5$ per one-half hour, we have **a)** 0.0067; **b)** 0.1755

The mean and standard deviation for the Poisson distribution are

Poisson Distribution

Mean: $\mu = \lambda$
Standard deviation: $\sigma = \sqrt{\lambda}$

In other words, λ is indeed the average value of a Poisson distribution.

A very useful property of the Poisson distribution is that it can be used to approximate the binomial distribution. As you can see from Appendix C, we have included values of the binomial distribution only for $n = 25$ or less. However, when n becomes large and p becomes small, the Poisson distribution can be used to approximate binomial probabilities by setting $\lambda = np$. The general rule of thumb is that the approximation will be useful for $n \geq 20$ and $p \leq 0.05$. The following example illustrates this approximation.

EXAMPLE 16

Let's compare a situation in which the random variable has a binomial distribution with the Poisson approximation. Suppose a grocery store manager has determined that approximately 5% of the store's customers purchase more than \$100 of groceries. The manager wants to compute the probabilities that

a) at least one person in a sample of size 20 purchases more than \$100 of groceries,

b) at least one person in a sample of size 100 purchases more than \$100 of groceries.

SOLUTION

Assuming that there are a large number of regular customers, the answer to the manager's questions should be computed using the binomial distribution. Let X be the number of customers who purchase more than \$100 worth of groceries.

a) The probability of no purchases of more than \$100 in a sample of size 20 is $P(X = 0) = (0.95)^{20} = 0.3585$. Hence the probability of at least one purchase of more than \$100 is given by $P(X > 0) = 1 - 0.3585 = 0.6415$. To approximate this with a Poisson distribution, set $\lambda = np = 20 \cdot 0.05 = 1$. The Poisson distribution

TABLE 12.7 Comparison of binomial probabilities with Poisson approximation for $n = 20$ and $p = 0.05$

x	Binomial probability	Poisson approximation
0	0.3585	0.3679
1	0.3774	0.3679
2	0.1887	0.1839
3	0.0596	0.0613
4	0.0133	0.0153
5	0.0022	0.0031
6	0.0003	0.0005

tells us $P(X = 0) = 0.3679$, which is quite close to the true value. Table 12.7 shows a more complete comparison between the binomial (true) probabilities and the Poisson approximation.

b) The table in Appendix C does not contain probability values for sample sizes of 100 (nor do most tables of the binomial distribution). To use the Poisson approximation, set $\lambda = 100 \cdot 0.05 = 5$. The Poisson value $P(X = 0)$ can be found in Appendix D; it is 0.0067. Hence the probability of at least one customer in 100 purchasing more than \$100 of groceries is approximately $1 - 0.0067 = 0.9933$. □

Practice Problem

Referring to Example 16, use the Poisson approximation to compute the probability that no more than 2 customers out of 100 buy more than \$100 of groceries.

Answer: 0.1246

EXERCISE SET 12.5

On the average, six people per hour use the self-service banking facility at a department store during prime shopping hours. Assuming the number of customers per hour has a Poisson distribution, what is the probability of each of the following statements?

1. Exactly six people will use the facility during a randomly selected hour.
2. Fewer than five people will use the facility during a randomly selected hour.

3. No one will use the facility during a randomly selected hour.
4. No one will use the facility during a randomly selected 10-minute interval. (*Hint*: If the facility averages six people per hour, how many does it average in 10 minutes?)
5. A new automated production process has been averaging one breakdown per day. Management is concerned about the possibility of three or more breakdowns in one day. Assuming the number of breakdowns has a Poisson distribution, what is the probability of three or more breakdowns during a randomly selected day?
6. Jobs arrive at a computer center at an average of 8 per 5-minute interval. Assuming the number of jobs has a Poisson distribution, what is the probability that 7 jobs arrive during a randomly selected interval? What is the probability that more than 10 jobs arrive during a randomly selected interval?
7. An accounting firm has determined that approximately 1% of a company's accounts receivable balances are incorrect. A sample of 150 accounts has been selected for audit. What is the probability that none of the accounts selected will show errors? What is the probability that no more than two of the accounts selected will show errors?
8. An insurance company is considering writing coverage for a relatively rare condition for a certain group. The probability that a randomly selected individual will have the condition is 0.001. The group contains 5000 people. What is the probability that no more than two people in the group have the condition?

12.6 THE NORMAL DISTRIBUTION

Continuous Random Variables

The previous two sections have described distributions for discrete random variables. A **discrete random variable** is a random variable that can have *only finitely or countably many distinct values*. For example, we might count the number of occurrences of some event. In this section and the next we will discuss continuous random variables. A **continuous random variable** can take on any value in some interval (or set of intervals). For example, the number of ounces in a can of soup labeled "12-ounces," the lifetime of a light bulb, and the weight of an individual are all continuous random variables. *The key idea here is that continuous random variables can be any one of the real numbers within some interval because we can always refine the measurement.*

We must treat continuous random variables differently than discrete random variables. Because continuous random variables can take an infinite number of values, we can no longer list each value of the random variable and then its associated probability (or give a formula to compute it). In fact, the question, What is the probability that a continuous random variable has the specific value x is meaningless. Why? Suppose we measure the weight of an object. Let's assume you have a scale accurate to 0.01 pounds, over the range 0 to 300 pounds. Then there are 30,000 possible results for the object's weight. Assuming all outcomes are equally probable, the probability of observing any one weight is $1 \div 30{,}000 = 0.0000333$. Now, suppose you were to use the scale at a scientific laboratory, which is accurate to 0.001 pounds. In this case, there would be 300,000 possible outcomes for the object's weight. The probability of observing any one weight would be $1 \div 300{,}000 = 0.00000333$. Moreover, we can continue this process, because it is always possible to conceive of a more accurate scale. Thus the number of possible weights is limitless, and the probability of observing any one weight is zero.

For continuous random variables we must use an approach different from that for discrete random variables. We compute the probability that a continuous random variable has a value within a **specified interval.** This is done by defining a function called a **probability density function** and then finding the area under the function over the specified interval. For example, suppose that a random variable has the probability density function shown in Fig. 12.6. The probability that the random variable is between 5 and 10 is equal to the shaded area. Now, to do this generally requires the use of integral calculus. However, for several continuous distributions, we can either use tables to look up the appropriate areas or the areas can be computed in simple fashion.

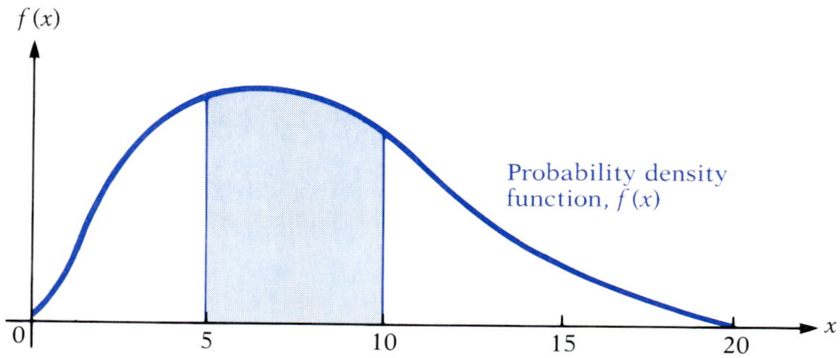

Figure 12.6 **The probability a continuous random variable occurs in an interval is the area under the density function over that interval.**

Normal Distributions and the Standard Normal Distribution

The **normal distribution** is the most important distribution in statistics. Not only is it important in its own right, but it can be used to approximate many other distributions. It is a continuous distribution with a bell-shaped density function. There are two parameters that completely describe a normal distribution: its mean μ and its standard deviation σ. Figure 12.7 shows the probability density function for three normal distributions with different values of μ and σ. The height of the density function for the normal distribution is given by the following equation.

Density Function, Normal Distribution

Let X be a normally distributed random variable. Then the density function of X is given by

$$f(x) = \frac{1}{\sigma\sqrt{2\pi}} e^{-[(x-\mu)/\sigma]^2}$$

where $\pi = 3.14159$ (approximately),
$e = 2.71828$ (approximately).

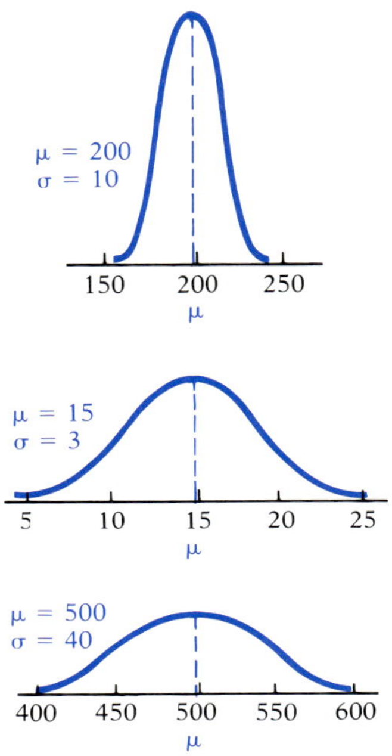

Figure 12.7 Three different normal distributions.

The normal distribution has several important properties, which we summarize as follows.

Properties of the Normal Distribution

1. Density functions of normal distributions are bell-shaped and symmetric with respect to the vertical line $x = \mu$.
2. The shape of a normal distribution is completely determined by its mean (μ) and standard deviation (σ). The value of μ determines the location of the distribution and the value of σ determines how widely the distribution is spread.
3. The total area between the x-axis and the density function curve is 1.
4. Regardless of the values of μ and σ, approximately 68.3% of the area between the x-axis and the density function lies within an interval of 1 standard deviation above and below the mean, approximately 95.5% of the area lies within an interval of 2 standard deviations above and below the mean, and approximately 99.7% of the area lies within an interval of 3 standard deviations above and below the mean. (See Fig. 12.8.)

In order to work with a normal distribution, we must be able to find areas under various parts of the distribution. This can be done rather easily by using the table given in Appendix E and making some simple calculations. The table in Appendix E is based on the density function of a normal distribution with a mean of 0 and a standard deviation of 1. The procedure for using Appendix E is illustrated by the following examples.

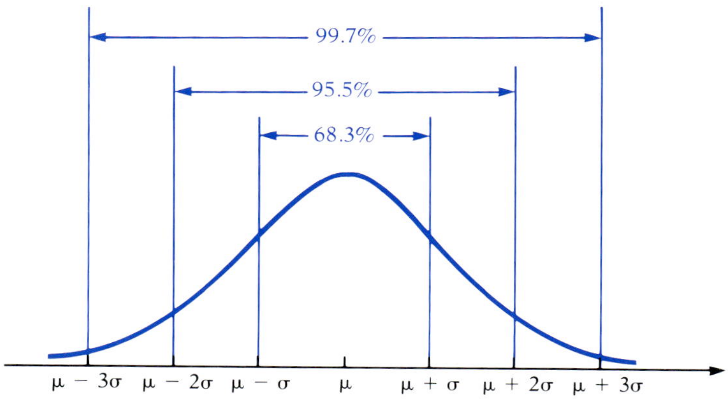

Figure 12.8 **Areas under a normal distribution.**

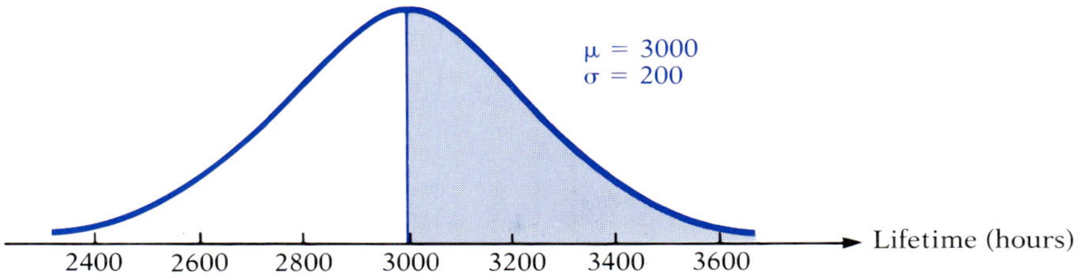

Figure 12.9 Finding the probability of a lifetime of more than 3000 hours.

EXAMPLE 17

The lifetime of an electrical component, X, has been found to be normally distributed with mean $\mu = 3000$ hours and standard deviation $\sigma = 200$ hours. Find

a) the probability the component lasts at least 3000 hours,

b) the probability the component lasts 3500 hours or less.

SOLUTION

Whenever you are working with a normal distribution, we highly recommend you draw a quick sketch of the distribution. We will do so for each of the parts of this example.

a) We wish to find $P(X \geq 3000)$. As shown in Fig. 12.9, this is equivalent to finding the shaded area, which is the area to the right of the mean of 3000 hours. We know the total area under the normal distribution must be equal to 1. Moreover, the distribution is symmetric about the mean, so the area to the left of the mean must be equal to the area to the right of the mean, and therefore both areas must be exactly equal to 0.5. Thus the probability the component lasts at least 3000 hours must be 0.5, because this is the same as finding the area to the right of the mean.

b) To find the probability the component lasts 3500 hours or less, once again we begin by drawing a graph. (See Fig. 12.10.) Here,

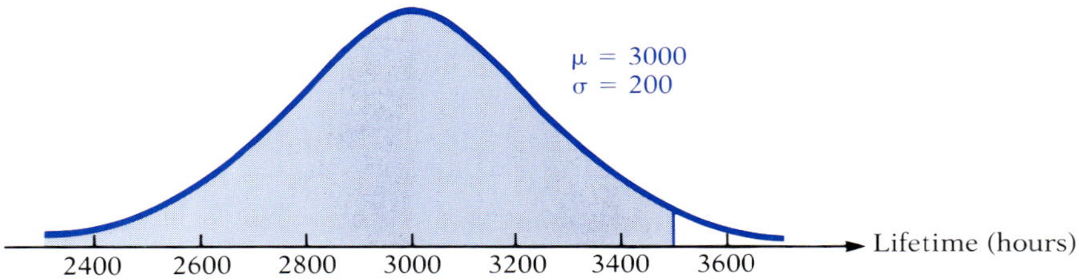

Figure 12.10 Finding the probability of a lifetime of 3500 hours or less.

we must use the table in Appendix E to find the required probability. This table is based upon the **standard normal distribution** which has mean 0 and standard deviation 1. Thus we must convert from our given normal distribution to the standard normal distribution. We do this by subtracting 3000 from 3500 and dividing the result by the standard deviation of 200, giving us a ***z*-score:**

$$z = \frac{3500 - 3000}{200} = \frac{500}{200} = 2.5$$

Thus 3500 corresponds to a z-score of 2.5 (given the mean of 3000 hours and standard deviation of 200 hours). Referring to Appendix E, we see the corresponding area is 0.9938. What does this tell us? The table in Appendix E gives us the area to the left of a given z-score (which is the probability of a z-score having the given value or less). Thus the probability of a z-score being 2.5 or less is 0.9938. But 3500 hours corresponds to a z-score of 2.5. Therefore, the probability that the component lasts 3500 hours or less is 0.9938. □

As Example 17 demonstrates, we must convert values from the given normal distribution to z-scores in order to use Appendix E. This is equivalent to asking the question, How many standard deviations above (or below) the mean of the given normal distribution is the value x? To see why, consider the manner in which the z-score is computed.

z-Score

$$z = \frac{x - \mu}{\sigma}$$

When we subtract μ from the given value x, we find out how far x is above or below the mean. Then we divide by the standard deviation. This converts the distance from the mean to new scale in which the unit of measure is the standard deviation. Thus in Example 17 we converted 3500 into a z-score of 2.5, which means that 3500 is 2.5 standard deviations above the mean of its distribution. Since the mean of that distribution is 3000 and the standard deviation is 200, we should have

$$3000 + 2.5(200) \overset{\checkmark}{=} 3500.$$

Our use of the normal distribution in Appendix E is based upon the following result. Suppose X is a normally distributed random variable with mean μ and standard deviation σ. Then the random variable Z defined by $Z = (X - \mu)/\sigma$ is normally distributed with a mean of zero and a standard deviation of one. Thus, for example, we can find probabilities such as $P(X > x)$ by transforming from X to Z: in other words, $P(X > x) = P(Z > z)$ where $z = (x - \mu)/\sigma$.

Practice Problem

Given a normally distributed random variable with mean 10 and standard deviation 2. Convert the following to z-scores.

a) 10 **b)** 13 **c)** 9

Answers: **a)** $z = 0$; **b)** $z = 1.5$; **c)** $z = -0.5$

EXAMPLE 18

As in Example 17, the lifetime of an electrical component, X, is normally distributed with mean 3000 hours and standard deviation 200 hours. Find the probability that

a) the lifetime exceeds 2800 hours,

b) the lifetime is between 3000 hours and 3500 hours.

SOLUTION

a) We must determine the probability the lifetime exceeds 2800 hours. (See Fig. 12.11.) Now we are dealing with a point to the left of the mean, so our z-score should be negative. We begin by converting 2800 hours into a z-score. We have

$$z = \frac{2800 - 3000}{200} = \frac{-200}{200} = -1.$$

The negative value for the z-score tells us that we have a point located to the left of the mean. Now, Appendix E tells us the area to the left of a z-score. For a z-score of -1, we find 0.1587.

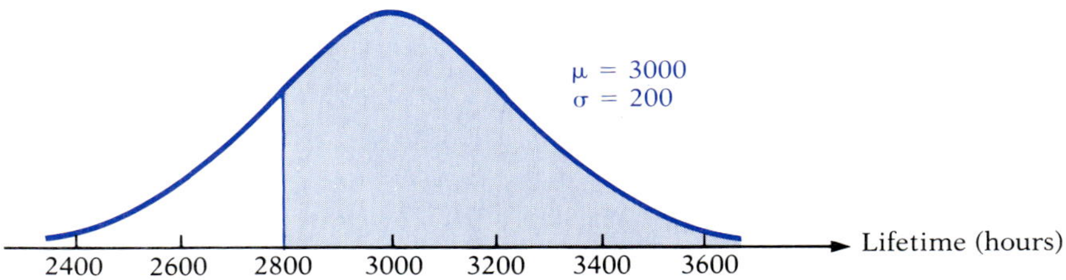

Figure 12.11 Finding the probability of a lifetime of more than 2800 hours.

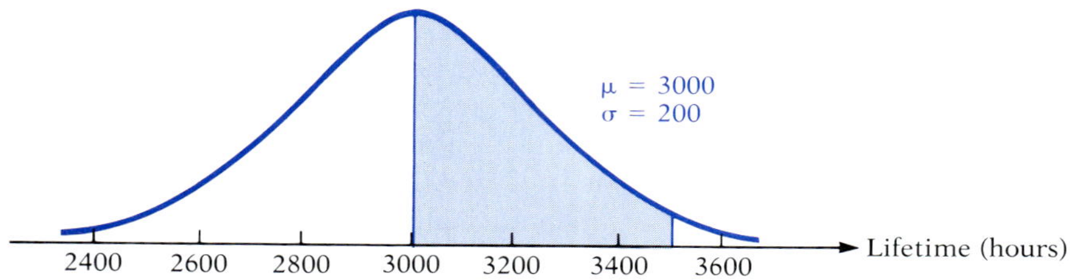

Figure 12.12 Finding the probability of a lifetime between 3000 and 3500 hours.

However, we want the area to the right of this z-score. Since the total area under a normal density function is 1, we can obtain the area we want by subtracting 0.1587 from 1. We have

$$\begin{aligned} P(\text{lifetime} > 2800 \text{ hours}) &= 1 - 0.1587 \\ &= 0.8413. \end{aligned}$$

b) To determine the probability that the lifetime is between 3000 and 3500 hours, we must find the area under the density function between 3000 and 3500 hours. (See Fig. 12.12.) As before, we begin by converting 3500 hours to a z-score:

$$z = \frac{3500 - 3000}{200} = 2.5$$

(which we obtained in Example 17). Now, we do *not* want the entire area to the left of a z-score of 2.5. We want the area between a z-score of 2.5 and the mean. We can find this area by subtracting the area to the left of the mean from the area to the left of a z-score of 2.5. According to Appendix E, the area to the left of a z-score of 2.5 is 0.9938. Since the area to the left of the mean is 0.5, we have

$$P(3000 \le \text{lifetime} \le 3500) = 0.9938 - 0.5 = 0.4938. \quad \square$$

Practice Problem

Referring to Examples 17 and 18, what are the probabilities that

a) the lifetime of the component is no more than 3000 hours?

b) the lifetime is between 3000 and 3200 hours?

c) the lifetime is less than 2800 hours?

Answers: **a)** 0.5; **b)** 0.3413; **c)** 0.1587

We can summarize these results as follows. To work with a given normal distribution, we must convert it to the standard normal distribution, which has mean 0 and a standard deviation 1. This is done by computing the difference between a value x and μ and then dividing by σ to obtain a z-score:

$$z = \frac{\text{difference between } x \text{ and } \mu}{\text{standard deviation}} = \frac{x - \mu}{\sigma}.$$

Remember that Appendix E gives the area to the left of each z-score.

EXAMPLE 19

On standard IQ (intelligence quotient) tests, the mean is 100 and the standard deviation is 15. The distribution of scores is (almost) normal. How high would a person's IQ have to be to place the person in the upper 1% of the population?

SOLUTION

To answer this question, we must reverse our logic a bit. Here we are given the area and we must work backwards to the z-score and then to the value of x. To be in the upper 1% means the score must be in the extreme right-hand tail of the distribution, as Fig. 12.13 shows us. This means that there must be an area of 0.01 to the right of our desired IQ score. Since the total area under the density function is 1, the area to the left of the desired IQ must be $1 - 0.01 = 0.99$. Thus we must look in Appendix E to see what z-score gives us an area of 0.99. The closest value of z is 2.33. In other words, the z-score of the desired IQ is 2.33. But the way we would compute this z-score is

$$z = \frac{\text{IQ} - 100}{15},$$

so

$$\text{IQ} = 15z + 100, \quad \textbf{Solve for IQ.}$$

$$\text{IQ} = 15 \cdot 2.33 + 100, \quad \textbf{Substitute for } \boldsymbol{z}.$$

$$\text{IQ} = 134.95.$$

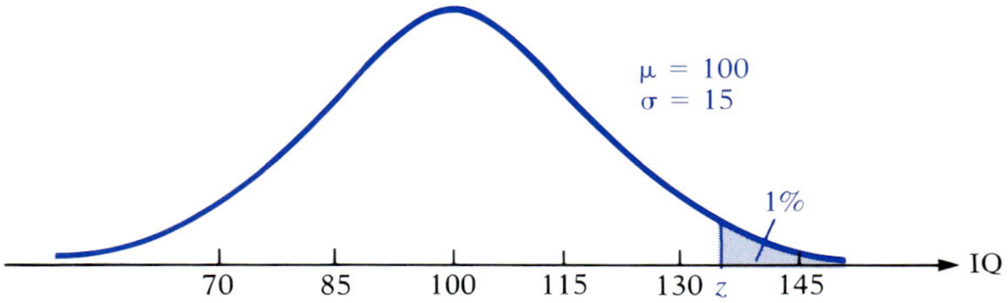

Figure 12.13 **Determining the IQ for the upper 1% of the population.**

In other words, a person with an IQ of 134.95 or larger will be in the upper 1% of the population. (We can check this by converting 134.95 to a z-score and showing that only 1% of the area of the standard normal lies to the right of this point.) □

EXAMPLE 20

A certain organization requires an IQ score at least 4 standard deviations above average to be eligible for membership. What is the minimum score required for membership?

SOLUTION

This is actually quite simple to compute. A score of 4 standard deviations above average means we want to find the IQ with a z-score of 4. Again,

$$z = \frac{\text{IQ} - 100}{15},$$

so

$$\text{IQ} = 15z + 100,$$

$$\text{IQ} = 15 \cdot 4 + 100, \qquad \textbf{Substitute for } z.$$

$$\text{IQ} = 160.$$

Hence, the desired IQ score is 160. (This is a rather exclusive club!) □

Practice Problem

Referring to Examples 19 and 20, what IQ score is

a) the cutoff for the upper 10% of the population?

b) three standard deviations above average?

Answers: **a)** the z-score is 1.28, so IQ = 119.2; **b)** 145

Approximating the Binomial Distribution

We know that computing binomial distributions for large values of n is quite difficult. However, when both p and q are not too different from 0.5 and n is large, we can use the normal distribution to approximate the binomial. The rule of thumb is the approximation is useful when both np and nq are at least 10 and n is at least 30. The approximation is performed by setting

$$\mu = np \quad \text{and} \quad \sigma = \sqrt{npq}.$$

Then simple convert the values of x (the number of events) to z-scores and proceed.

We can use this approximation to solve the Problem from Life.

EXAMPLE 21

Problem from Life From past data it has been found that 70% of the people who enter a shopping mall make a purchase. On a certain day, 500 people enter the mall.

a) How many people can be expected to make a purchase?

b) What is the probability that no more than 350 people will make a purchase?

c) What is the probability that at least 360 people will make a purchase?

SOLUTION

The scenario described here is a Bernoulli process. Thus, the number of people who make a purchase, X, has a binomial distribution with $n = 500$, $p = 0.7$, and $q = 0.3$. The table in Appendix C does not include values for an n of 500. Since we have $np = 500 \cdot 0.7 = 350$ and $nq = 500 \cdot 0.3 = 150$, we can use the normal distribution to approximate the binomial distribution. To find the approximating normal distribution, set

$$\mu = np = 500 \cdot 0.7 = 350,$$
$$\sigma = \sqrt{npq} = \sqrt{500 \cdot 0.7 \cdot 0.3} = \sqrt{105} = 10.247.$$

a) The number of people that can be expected to make a purchase is simply the mean of the binomial distribution. Thus we can expect 350 people to make a purchase.

b) The probability that no more than 350 people make a purchase is approximately 0.5, since 350 is the mean of the approximating normal distribution.

c) The probability that at least 360 people make purchases is approximately $P(X \geq 360)$, where X is normally distributed with mean $\mu = 350$ and standard deviation $\sigma = 10.247$. As Fig. 12.14 indicates, we want to find the shaded area to the right of 360.

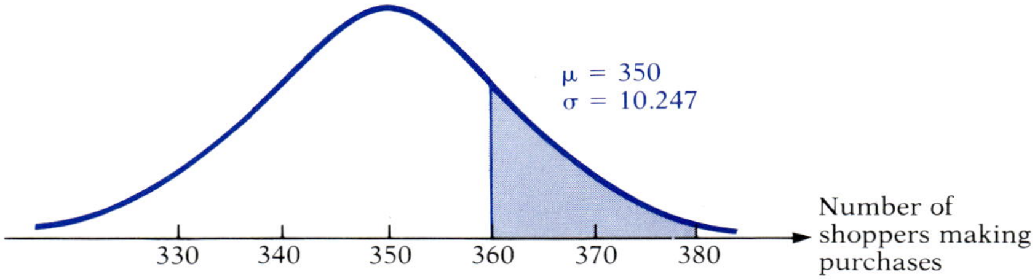

Figure 12.14 **Using the normal distribution to approximate the binomial distribution.**

This is done by computing

$$z = \frac{x - \mu}{\sigma} = \frac{360 - 350}{10.247} = 0.98 \qquad \text{(to two decimal places).}$$

Using Appendix E, we see the area to the left of $z = 0.98$ is 0.8365. Hence the area to the right of $z = 0.98$ must be given by $1 - 0.8365 = 0.1635$. Therefore the probability of at least 360 people making purchases is approximately 0.1635. □

Practice Problem

Referring to Example 20, what is the probability that at least 345 people make purchases?

Answer: 0.6879

EXERCISE SET 12.6

A particular type of automobile is listed as weighing 2800 pounds. Its weight is actually normally distributed with a mean of 2800 pounds and a standard deviation of 25 pounds. What is the probability that the car weighs as described in problems 1–3?

1. The car weighs at least 2800 pounds.
2. The car weighs no more than 2850 pounds.
3. The car weighs between 2775 and 2825 pounds.

The number of ounces of soup in a can is normally distributed with a mean of 8 ounces and a standard deviation of 0.01 ounces. What are the probabilities listed in Problems 4–6?

4. The can contains more than 8.01 ounces of soup.
5. The can contains less than 7.995 ounces of soup.
6. The can contains between 7.995 and 8.005 ounces of soup.

The sales of a product are approximately normally distributed with a mean of 10,000 units per month and a standard deviation of 1500 units per month.

7. What is the probability the company will sell more than 12,000 units in any given month?
8. What is the probability the company will sell between 9000 and 11,000 units in any given month?
9. The company wants to produce enough units to have a probability of 0.9 that it can meet monthly demand. How many units should it produce?

The scores on a standardized college admissions test are approximately normally distributed, with a mean of 450 and a standard deviation of 100.

10. What percentage of the people taking the test score between 400 and 500?

11. Suppose someone receives a score of 600. What percentage of those taking the test score below this? above this?

12. Suppose a particular university refuses to admit anyone with a score below 425. What percentage of the people taking the test are not acceptable to the university?

A university has discovered that approximately 15% of the students registered for courses will receive at least one grade of W ("Withdraw"). Suppose in a given semester there are 15,000 students registered.

13. What is the probability that at least 1800 students will receive W's?

14. What is the probability that between 2000 and 2400 students will receive W's?

15. What is the probability that no more than 2300 students will receive W's?

A company is using a controversial advertising campaign to market a new product. It has been found that approximately 60% of potential buyers like the ads, while 40% of the potential buyers are strongly opposed to them. A sample of 100 people has been polled about their reactions to the ads.

16. What is the probability that between 50 and 70 people will like the ads?

17. What is the probability that no more than 55 people will like the ads?

18. What is the probability that at least 50 people will like the ads?

12.7 THE EXPONENTIAL DISTRIBUTION*

When the number of events or successes has a Poisson distribution (see Section 12.5), then the length of time (or space) between events has an **exponential distribution.** Since time (or space) is continuous, the exponential distribution is continuous. Therefore, as with the normal (or any other continuous) distribution, it is not meaningful to compute the probability of the next event occurring at some exact amount of time in the future (for example, in exactly 1 minute). Rather, we must ask questions related to intervals; for example, we

* This section requires knowledge of the exponential function. See Chapter 9.

might try to compute the probability that the next event occurs within 1 minute.

Suppose the number of events in a given interval of length t has a Poisson distribution with a mean of λ. Then it can be shown that the time between events, T, has an exponential distribution with mean $1/\lambda$ and standard deviation $\sqrt{1/\lambda}$. The density function of the exponential distribution with mean $1/\lambda$ is given by

$$f(t) = \begin{cases} \lambda e^{-\lambda t} & \text{for } t > 0 \\ 0 & \text{for } t \leq 0 \end{cases}$$

Density function, exponential distribution

The exponential distribution satisfies the following (which can be shown using calculus).

Exponential Distribution

Let T be the time at which the next event occurs. Then

1. P(the next event *occurs within* an interval of length t) $= P(T \leq t) = 1 - e^{-\lambda t}$. (See Fig. 12.15.)
2. P(the next event *does not occur within* an interval of length t) $= P(T > t) = e^{-\lambda t}$. (See Fig. 12.16.)

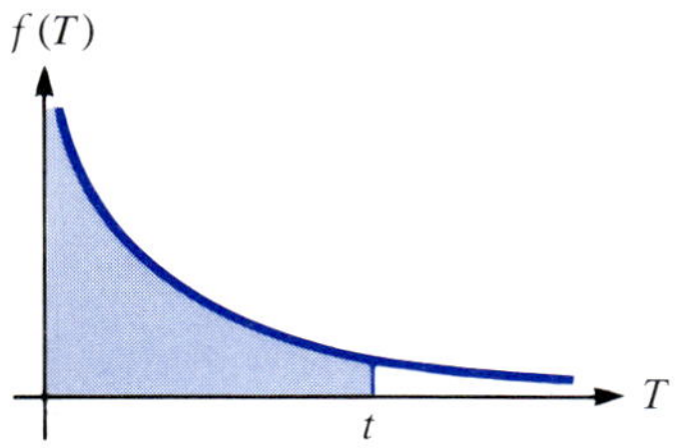

Figure 12.15 Exponential distribution: $P(T \leq t)$.

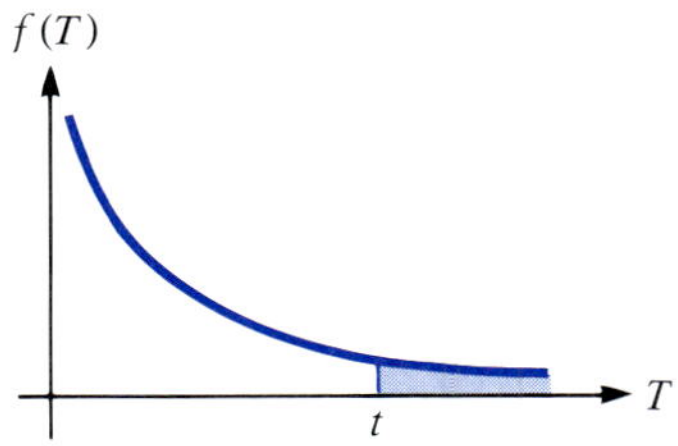

Figure 12.16 Exponential distribution: $P(T > t)$.

EXAMPLE 22

Customers call a machine repair service at a rate of five calls per hour. Beginning at a random point in time, compute the probability that the next call occurs within one-half hour.

SOLUTION

We have an average of five calls per hour, so $\lambda = 5$. Then

$$\lambda t = \text{average calls per half-hour} = 5 \cdot 0.5 = 2.5.$$

Therefore

$$\begin{aligned} P(T \le \text{one-half hour}) &= 1 - e^{-\lambda t} \\ &= 1 - e^{-2.5} \\ &= 1 - 0.08208 = 0.91792. \end{aligned}$$

This can be computed using a calculator quite easily. Enter

[2.5] [+/−] [e^x] [+/−] [+] [1] [=]

If your calculator has parentheses, you can enter the operations in natural order:

[1] [−] [(] [2.5] [+/−] [e^x] [)] [=] □

EXAMPLE 23

Compute the probability the next call does not occur for at least 12 minutes. (See Example 22.)

SOLUTION

Again, with an average of five calls per hour, $\lambda = 5$. Then

$$\begin{aligned} \lambda t &= \text{average calls per 12 minutes} \\ &= 5 \cdot 0.2 \qquad \textbf{(12 minutes = 0.2 hours)} \\ &= 1. \end{aligned}$$

Then $P(T > 12 \text{ minutes}) = e^{-\lambda t} = e^{-1} = 0.37688$ □

Practice Problem

Trucks arrive at a loading dock with an average of four per hour. Compute the probability that beginning at a random point in time

a) the next truck arrives within one-half hour,

b) the next truck does not arrive within the next hour.

Answers: **a)** 0.86466; **b)** 0.01832

EXERCISE SET 12.7

On the average, a ship arrives at a loading dock every second day.

1. What is the probability that after the departure of a ship four days will pass before the next one arrives?
2. What is the probability that after the departure of a ship no more than two days will pass before the next one arrives?
3. What is the probability that after the departure of a ship no more than one day will pass before the next one arrives?

A 1000-yard roll of cloth contains 10 flaws on the average. What is the probability the first flaw occurs as stated in Problems 4–6?

4. Occurs within the first 50 yards.

5. Does not occur for 150 yards.

6. Occurs within 200 yards.

Jobs arrive at a computer system at an average of 120 per hour. Assuming a job has just been completed, what are the following probabilities?

7. The next job does not arrive for ten minutes.

8. The next job arrives within one minute.

9. The next job arrives between two minutes and six minutes. (*Hint:* Find the probability the job arrives within six minutes and subtract the probability that it arrives within two minutes.)

CHAPTER REVIEW

IMPORTANT TERMS AND SYMBOLS

Introduction

descriptive statistics The collection and summarization of data.
inferential statistics Using statistical methodology to make conclusions about a population (usually based on sample results).

12.1 Portraying Data

frequency How often a given value or range of values occurred.
Cumulative frequencies: A running total of the frequency column in a frequency distribution.
Cumulative frequency polygon (ogive): A graph of the cumulative frequencies; the values are graphed against the upper limit of the categories.
frequency distribution A summary table in which categories of data, frequencies of each category, and cumulative frequencies are listed.
frequency polygon A graph drawn by joining the midpoints of a histogram with straight lines.
histogram A bar graph whose heights represent frequencies.

12.2 Measures of Central Tendency

mean The arithmetic average of the data. Formula:

$$\overline{x} = \frac{\sum_{i=1}^{n} x_i}{n}.$$

median The value in the middle of a data set arranged in ascending or descending order.

mode The value that occurs most often.

12.3 Measures of Dispersion

coefficient of variation The ratio of the standard deviation to the mean (expressed in percent). Formula:

$$\text{cv} = \frac{s}{\overline{x}} \cdot 100\%.$$

range The difference between the largest and smallest data points.

standard deviation Measures how far (on the average) the data vary from the mean. Formulas (for sample standard deviation):

$$s = \sqrt{\frac{\sum_{i=1}^{n} (x_i - \overline{x})^2}{n - 1}} \quad \text{(definition, sample standard deviation)}$$

$$s = \sqrt{\frac{\left[\sum_{i=1}^{n} (x_i^2) - \left(\sum_{i=1}^{n} x_i\right)^2 \Big/ n\right]}{n - 1}} \quad \text{(shortcut formula, sample standard deviation)}$$

variance The square of the standard deviation.

12.4 The Binomial Distribution

Bernoulli process A set of repeated experiments with either "success" or "failure" as outcome of each experiment; each experiment is independent of all others. The probability of success (p) and the probability of failure (q) are constant for each experiment.

binomial distribution Computes the probability of obtaining a given number of successes in a Bernoulli process;

$$P(x \text{ successes in } n \text{ trials}) = C(n, x)p^x q^{n-x}, \qquad x = 0, 1, \ldots, n,$$

where p = probability of a success, $q = 1 - p$ = probability of a failure (values given in Appendix C).

Mean of a binomial distribution: $\mu = np$.

Standard deviation of a binomial distribution: $\sigma = \sqrt{npq}$.

random variable A variable whose value is determined by some chance (or random) process.

theoretical probability distribution A distribution in which the probabilities are based upon a theoretical (or hypothetical) model.

12.5 The Poisson Distribution

approximating binomial distribution with Poisson Set $\lambda = np$; valid for $n \geq 20$ and $p \leq 0.05$.

Poisson distribution A model in which the probability of occurrence of an event is the same for any two intervals of equal length; the probability of occurrence or nonoccurrence of an event is independent of its occurrence in other nonoverlapping intervals; the probability of two or more events occurring in an interval becomes 0 as the length of the interval shrinks;

$$P(x \text{ occurrences}) = \frac{\lambda^x e^{-\lambda}}{x!}, \qquad x = 0, 1, 2, \ldots,$$

where λ = average rate (values given in Appendix D).
Mean of a Poisson distribution: $\mu = \lambda$.
Standard deviation of a Poisson distribution: $\sigma = \sqrt{\lambda}$.

12.6 The Normal Distribution:

approximating binomial distribution with normal distribution Set $\mu = np$ and $\sigma = \sqrt{npq}$; valid for $n \geq 30$ and $np, nq \geq 10$.

continuous random variable A random variable that can take on any value in an interval or set of intervals.

discrete random variable A random variable that can take on only finitely or countably many values.

normal distribution A distribution with the bell-shaped density function given by

$$f(x) = \frac{1}{\sigma\sqrt{2\pi}} e^{-[(x-\mu)/\sigma]^2},$$

where μ = mean, σ = standard deviation.
Standard normal distribution: Has mean 0 and standard deviation 1; Appendix E gives areas to the left of z, where

$$z = \frac{x - \mu}{\sigma}.$$

probability density function Used to compute the probability that a continuous random variable is within a specified interval; the probability is given by the area under the density function over the interval.

12.7 The Exponential Distribution

exponential distribution A continuous distribution that gives the probability that the time (space) between events is within some interval of time (space) when the count of occurrences of the event is Poisson; has density function

$$f(t) = \begin{cases} \lambda e^{-\lambda t} & \text{for } t > 0 \\ 0 & \text{for } t \leq 0 \end{cases}$$

where mean $= 1/\lambda$.

Properties: Let T be the time between successive occurrences of an event. Then

$$P(T \leq t) = 1 - e^{-\lambda t},$$
$$P(T > t) = e^{-\lambda t}.$$

REVIEW PROBLEMS

Make a frequency distribution and draw a histogram, frequency polygon, and a cumulative frequency polygon for the data sets in Problems 1 and 2.

1. Number of refrigerators sold per week:

0	5	1	3	4	5	2	4	5	2
3	2	0	3	4	2	4	0	1	2
2	4	3	1	3	2	5	0	5	2
1	3	1	1	0	2	1	3	2	3
2	3	1	3	2	2	3	4	1	3

2. Monthly living expenses ($):

170	180	205	230	240	315	360	290	240	240	230	195
180	205	215	245	265	330	400	335	260	270	255	220
250	280	295	290	330	420	390	290	280	230	210	195

For the data sets in Problems 3 and 4, compute the mean, median, mode, range, sample standard deviation, and coefficient of variation.

3. Battery life (in hours):

200 225 190 215 230 195 210 190

4. Weight of coffee jars (ounces):

8.20 7.90 7.85 8.10 8.25 8.00 7.95 8.15 8.10 7.90

When a particular machine is functioning properly, only 2% of the items produced are defective. Assuming the machine is functioning properly, solve Problems 5–8.

5. Suppose a sample of two items is taken. What is the probability none are defective?
6. Suppose a sample of five items is taken. What is the probability that more than one is defective?
7. What is the expected number of defective items in a sample of 200?
8. What is the standard deviation of the number of defective items in a sample of 200?
9. Suppose a salesperson makes a sale on 35% of the calls.

   a) If the salesperson makes three calls per day, what is the probability of more than four sales in a five-day week?
   b) Assume the salesperson makes a commission of $100 per sale. Assuming the salesperson works 50 weeks per year, how many sales are expected annually? What is the expected annual income?

During rush hour in a city, accidents occur at a rate of 3 per hour. Given that the morning rush hour is $1\frac{1}{2}$ hours and the evening rush hour is 2 hours, solve Problems 10–13.

10. On a randomly chosen day, what is the probability there will be no accidents during the morning rush hour?
11. What is the probability of three accidents during the evening rush hour?
12. What is the probability of four or fewer accidents during the morning rush hour?
13. What is the probability there will be no accidents at all?

A given process has a defect rate (when properly adjusted) of 0.5%. Assume a random sample of size 500 is taken when the process is adjusted properly.

14. What is the probability the sample contains no defects?
15. What is the probability the sample contains at least four defects?
16. What is the probability the sample contains three defects?

A company produces parts whose diameters are normally distributed with an average of 20 millimeters and a standard deviation of 0.02 millimeters.

17. What proportion of parts will have diameters between 19.97 and 20.03 millimeters?
18. A part is unusable if its diameter is less than 19.95 or greater than 20.05 millimeters. What proportion of parts will be unusable?
19. Suppose 10,000 parts are produced per day. How many of these would you expect to have diameters between 19.98 and 20.02 millimeters?

Suppose you toss a fair coin 1000 times. What is the probability of obtaining the following results?

20. At least 490 heads.
21. At most 525 tails.
22. Between 475 and 525 tails.

Phone calls come into a taxicab dispatcher's office at a rate of 5 calls per 15-minute interval. What is the probability that after receiving a call, the next call

23. Will not arrive for 4 minutes?

24. Will arrive within 2 minutes?

25. Will arrive within 5 minutes?

13

DECISION THEORY AND GAME THEORY

Brazell Development Company has recently purchased some land on which to build condominiums. The company can build one of three developments: small, medium, or large. The decision is complicated by an uncertain economy. Brazell could lose a significant amount of money if it were to build a large development followed by low demand. Conversely, if the company were to be conservative and build a small development followed by large demand, the subsequent profits would be much smaller than they could have been. After some analysis and thought, management has prepared the following table, which shows the anticipated profit for every possible combination of building option and future demand.

Anticipated profits

	Demand		
Options	Low	Medium	High
Small	\$400,000	\$400,000	\$400,000
Medium	\$100,000	\$600,000	\$600,000
Large	−\$300,000	\$300,000	\$900,000

The best current estimates for demand are as follows: $P(\text{low}) = 0.2$, $P(\text{medium}) = 0.35$ and $P(\text{high}) = 0.45$. What is the recommended decision? (To be solved as Example 7, Section 13.2.)

PROBLEM FROM LIFE

INTRODUCTION

In this chapter we will discuss two techniques that are widely used in decision making. Both techniques deal with situations in which your knowledge is imperfect. The first technique, called decision theory, is appropriate when you can describe the situation with a probability distribution. We will use decision theory to solve the Problem from Life (see Section 13.2). The second technique is called game theory; it is appropriate when you must make a decision in competition with (thinking) adversaries.

13.1 EXPECTED VALUE

Consider a basketball player who makes an average of 75% of her free throws (foul shots). Suppose she were to attempt 100 free throws. How many do you think she would make? A reasonable estimate would be that we expect her to make 75% of the 100 attempts or 75 free throws. This is an example of an expected value. In Chapter 12 we introduced the concept of the mean of a probability distribution and gave you formulas to compute the mean for several different distributions. Another term for mean is expected value. (We'll show you why they are equivalent after we present the formula.) In this chapter, we will find it necessary to compute the expected value (mean) of an arbitrary probability distribution. Of course, most of these situations will be more complicated than the case of our basketball player, so we need a more formal approach.

We define the **expected value** of a probability distribution for a discrete random variable as follows.

Suppose the random variable, X, has n possible numeric values, $x_1, x_2, \ldots, x_n$. In addition, the probabilities of these outcomes are, respectively, $p_1, p_2, \ldots, p_n$. Then the expected value of the distribution is given by

Expected Value

$$\begin{aligned} \text{EV} &= x_1p_1 + x_2p_2 + \cdots + x_np_n \\ &= \sum_{i=1}^{n} x_ip_i \end{aligned}$$

Now let's see why the EV is equal to the mean of a probability distribution. First, we need to present a variation of an arithmetic average called a **weighted average.** Suppose that $x_1, x_2, \ldots, x_n$ are the possible values of a random variable and that $w_1, w_2, \ldots, w_n$ are the *weights* of these values. Then the weighted average of the random variable is given by

$$\bar{x}_w = \frac{x_1w_1 + x_2w_2 + \cdots + x_nw_n}{w_1 + w_2 + \cdots + w_n} = \frac{\sum_{i=1}^{n} x_iw_i}{\sum_{i=1}^{n} w_i}.$$

(Grades in college courses are usually computed by weighted averages, in which tests have a certain weight, homework has a certain weight, and the final exam has a certain weight.) Weighted averages are often computed with weights that sum to 1, because then no division is required.

Now look at the formula for the EV. Notice that each possible value of the random variable (the x_i's) is multiplied by the probability that it occurs. But this is simply a weighted average in which the probabilities are the weights. Thus the more probable a value x_i is, the more weight it carries in computing the EV.

Now let's look at an example of computing an EV.

EXAMPLE 1

Ann is considering buying a raffle ticket for \$1. The grand prize is a color television worth \$500. There will be 5000 tickets sold, each with an equal chance of winning. Compute the expected value of one ticket.

SOLUTION

We have an experiment with two outcomes: winning or losing. We can think of the random variable X as being the value of each outcome. If Ann's ticket is drawn, she wins a net of \$500 − \$1 = \$499; if her ticket is not drawn, she wins a net of \$0 − \$1 = −\$1. Since there are 5000 tickets, the probability of Ann's ticket being drawn is 1/5000 = 0.0002. Therefore the probability of her ticket not being drawn is 1 − 0.0002 = 0.9998. We have the distribution as follows.

Outcome	Probability	Value of outcome (\$)
Wins	0.0002	499
Loses	0.9998	−1

The expected value for this distribution is

$$\text{EV} = \$499(0.0002) + (-\$1)(0.9998) = \$0.0998 - \$0.9998 = -\$0.90.$$

In other words, Ann can expect to lose \$0.90, or 90 cents. □

Note in Example 1 that it is impossible to actually lose 90 cents in this raffle. Clearly Ann will either lose \$1 or win \$499. However, if she were to buy raffle tickets of this type over a very long period of time, on the average she would lose 90 cents for each ticket she purchased. (Her total net winnings divided by the total number of tickets bought would be −90 cents.) Thus the expected value of a distribution does *not* have to be one of the possible outcomes. For example, when someone says the typical family has 2.1 children, the person is using an expected value, but it is obviously impossible for a family to have 2.1 children.

Practice Problem

Suppose you buy a raffle ticket for 50 cents. The prize for the winning ticket is worth \$100; there will be 1000 tickets sold. What is the expected value of the ticket?

Answer: −\$0.40

The next example demonstrates how we can use expected values to analyze decisions whose outcomes are uncertain.

EXAMPLE 2

A construction company can bid on a building project with potential profits of \$500,000 or on a road surfacing contract with potential profits of \$750,000. It is also possible to bid on both projects. If both bids are successful, the company will have to use overtime and subcontract part of the work so that the total profit would be only \$650,000. Based on past experience, the company estimates a probability of 0.6 of getting the building project and 0.3 of getting the road surfacing contract. Assuming the outcomes of the bids are independent, on which job or jobs should the company bid?

SOLUTION

The company can bid on the building project only, bid on the road project only, or bid on both. Here the random variable is the value of each possible outcome to making a bid. We will make a decision by computing the expected value for each alternative and choosing the one with the largest expected value.

If the company bids on the building project, the outcomes are a profit of \$500,000 with a probability of 0.6 or a profit of \$0 with a probability of $1 - 0.6 = 0.4$. Hence the expected value is

$$\begin{aligned} \text{EV(building only)} &= \$500{,}000(0.6) + \$0(0.4) \\ &= \$300{,}000. \end{aligned}$$

Similarly, the company can bid on the road project with outcomes of a profit of \$750,000 with a probability of 0.3 or a profit of \$0 with a probability of $1 - 0.3 = 0.7$. The expected value is

$$\begin{aligned} \text{EV(road only)} &= \$750{,}000(0.3) + \$0(0.7) \\ &= \$225{,}000 \end{aligned}$$

Finally, the company can bid on both projects. In this case there are four possible outcomes. The probability distribution is given as follows.

Outcome	Probability	Value ($)
Get both jobs	0.18	650,000
Building only	0.42	500,000
Road only	0.12	750,000
Neither job	0.28	0

To obtain these probabilities, assume the outcomes of the bids are independent. Also,

$$P(\text{no building}) = 1 - 0.6 = 0.4,$$
$$P(\text{no road}) = 1 - 0.3 = 0.7.$$

Hence

$$P(\text{both}) = P(\text{building}) \cdot P(\text{road}) = (0.6)(0.3) = 0.18,$$
$$P(\text{building only}) = P(\text{building}) \cdot P(\text{no road}) = (0.6)(0.7) = 0.42,$$
$$P(\text{road only}) = P(\text{no building}) \cdot P(\text{road}) = (0.4)(0.3) = 0.12,$$
$$P(\text{neither}) = P(\text{no building}) \cdot P(\text{no road}) = (0.4)(0.7) = 0.28.$$

The expected value of a bid on both jobs is thus

$$\begin{aligned} \text{EV(both)} &= \$650{,}000(0.18) + \$500{,}000(0.42) \\ &\quad + \$750{,}000(0.12) + \$0(0.28) \\ &= \$417{,}000. \end{aligned}$$

We can see that bidding on both projects has the largest expected value. Therefore the company should bid on both projects. □

Practice Problem

Suppose in Example 2 that the probability of winning the building project is 0.5 and the probability of winning the road project is 0.2. What are the expected values and what should the company do?

Answers: EV(building only) = $250,000, EV(road only) = $150,000, and EV(both) = $340,000. Bid on both projects.

Now let's look at some more examples of computing expected values.

EXAMPLE 3

Compute the expected number of boys in a family with three children of different ages (to rule out twins or triplets).

SOLUTION

The sample space S of all possible outcomes is given by

$$S = \{BBB, BBG, BGB, GBB, BGG, GBG, GGB, GGG\}.$$

Here the random variable is the number of boys. Assuming the probability of a boy is equal to the probability of a girl at any single birth, all of the events in the sample space are equally likely. Hence, the following chart gives the probabilities of 0, 1, 2, or 3 boys.

Outcome (number of boys)	0	1	2	3
Probability	0.125	0.375	0.375	0.125

Now we can compute the expected number of boys:

$$\begin{aligned} EV &= 0(0.125) + 1(0.375) + 2(0.375) + 3(0.125) \\ &= 0.375 + 0.750 + 0.375 \\ &= 1.5. \end{aligned}$$

In other words, a three-child family can expect to have an average of 1.5 boys. □

Practice Problem

What is the expected number of girls in a four-child family (assuming no multiple births)?

Answer: 2

EXAMPLE 4

A roulette wheel (in the United States) is numbered from 1 through 36 and also includes 0 and 00. In one form of roulette, you bet \$1 on "even." You win your dollar plus an additional dollar if one of the 18 even numbers occurs. You lose your dollar if one of the 18 odd numbers, 0, or 00 occurs. Compute the expected value of this game.

SOLUTION

Notice that there are 38 possible outcomes to a spin of the roulette wheel: the numbers 1–36, 0, and 00. Assuming these are all equally likely, $P(\text{winning}) = 18/38 = 0.47368$ (approximately) and $P(\text{losing}) = 20/38 = 0.52632$ (approximately). The random variable is the net return: For winning it's \$1 and for losing it's $-\$1$. Hence the expected value of the game is

$$\begin{aligned} EV(\text{``evens''}) &= \$1(0.47368) + (-\$1)(0.52632) \\ &= -\$0.05264. \end{aligned}$$

In other words, you lose on the average 5.264 cents every time you play the game. □

Practice Problem

In Europe, roulette wheels are numbered from 1 through 36 and also include 0, but do not include 00. What is the expected value of betting "evens"?

Answer: $-\$0.02702$ or -2.702 cents

A game that has an expected value of zero is called a **fair** game. Games in gambling casinos in which you bet against the casino are not fair because the casino must make a profit. Losing 5.264 cents per dollar bet (on the average) may not seem like a significant amount, but when large sums are wagered, the return is quite large. The casino can expect to win 5.264 cents per dollar or 5.264% of the amount wagered. Thus, if \$1 million were bet, the casino could expect to be \$52,640 ahead. In general, assuming you wish to play at all, this means you should quit when you win in a game with a negative expected value.

EXAMPLE 5

Bob has just purchased a new stereo system for his car for \$1000. An insurance company charges a premium of \$60 for one year's coverage, claiming a probability of 0.05 the system will be stolen sometime during the year. What is the expected value of the insurance policy?

SOLUTION

There are only two outcomes here: Either the system is stolen or it isn't. The probabilities for these events are 0.05 and 0.95, respectively. The random variable is Bob's return. Now, suppose the system is stolen. Then Bob's return is \$1000 (for a new system) − \$60 = \$940. If the system is not stolen, then Bob's return is −\$60 (the cost of the insurance). Thus the expected value of the policy is

$$\begin{aligned} \text{EV(insurance)} &= \$940(0.05) + (-\$60)(0.95) \\ &= \$47 - \$57 \\ &= -\$10. \end{aligned}$$

Thus, on the average, Bob loses an average of \$10 per year for every year the insurance is in force. □

Practice Problem

Suppose in Example 5 the probability of the stereo system being stolen during the year is 0.06. Compute the expected value of the policy.

Answer: \$0

In general, we can anticipate the expected value of an insurance policy being negative, because the insurance company must cover expenses and make a profit. Why do people buy insurance? Because most people are willing to pay a (relatively) small amount of money to avoid the catastrophic loss of a large sum. Thus in Example 5 Bob may be willing to pay $60 per year (and therefore lose approximately $10 per year) to avoid having to pay $1000 for a new stereo system in the event of a theft.

EXERCISE SET 13.1

In Problems 1–5, find the expected value of each of the following distributions.

1.

Outcome	1	2	3
Probability	0.3	0.4	0.3

2.

Outcome	1	3	5	7
Probability	0.3	0.25	0.25	0.2

3.

Outcome	40	50	60	70
Probability	0.05	0.35	0.45	0.15

4.

Outcome	−10	20	30	100	120
Probability	0.2	0.25	0.15	0.2	0.2

5.

Outcome	−30	−10	10	30	50
Probability	0.3	0.3	0.05	0.05	0.3

6. Suppose you play a game that involves flipping a coin. You win $5 on heads and lose $5 on tails. Compute the expected value of the game if

a) the coin is fair,
b) the coin is biased with a probability of 0.6 of heads.

7. Suppose you draw 3 cards at random (without replacement) from a 52-card deck. What is the expected number of

a) red cards? **b)** spades? **c)** kings?

8. Consider a game in which you roll a single die and receive the dollar amount corresponding to the number of dots on the face that is turned up. What fee should you be charged so that the game is fair?

9. In roulette, it is possible to bet $1 on a single number from 1 to 36. If the number turns up, you win back your dollar plus an additional $35. Otherwise, you lose your dollar. Compute the expected value of the game

a) for the American roulette wheel with 1–36, 0, and 00,
b) for the European roulette wheel with 1–36 and 0.

10. An insurance company has written 10 policies of $50,000, 100 of $10,000, and 1000 of $5000 for people of age 25. Their past data have shown the probability of dying during the twenty-sixth year of life is 0.005. How much money can the company expect to pay out during the coming year on these policies?

11. A baker bakes bread at the beginning of each day. He realizes a profit of $0.25 for each loaf he sells and a loss of $0.10 for each loaf not sold (because the bread ages and may not be sold at full price after the first day). From past records, the baker estimates the following probability distribution for the daily demand for bread.

Quantity demanded	Probability
100	0.1
200	0.3
300	0.4
400	0.15
500	0.05

a) What is the expected daily demand for bread?
b) What is the expected daily profit if the baker makes 300 loaves?

12. Find the expected number of boys in a family of five children with no multiple births.

13. A raffle offers a first prize of $1000, two second prizes of $100, and five third prizes of $25. If 20,000 tickets are sold at 25 cents each, what are the expected winnings for a single ticket?

14. Suppose you receive notification in the mail of a contest from a magazine company. The grand prize is worth $50,000, there are two second prizes of $25,000, five third prizes of $5,000, and 20 fourth prizes of $500. A total of 1,000,000 entries have been received. What is the expected value of a single entry? Is it worth the 22 cents it would cost you for a stamp to enter the contest?

15. A large consulting firm has just been awarded a contract for a fixed price of $250,000. Assuming that all goes well, the company expects to make $75,000 profit. If minor difficulties are encountered, the company will make $25,000 profit and if major difficulties are encountered, the company will lose $50,000. Compute the expected profits if the probability of all going well is 0.5, the probability of minor difficulties is 0.25, and the probability of major difficulties is 0.25.

16. The championship series in the National Basketball Association is "best of seven." In other words, the first team to win four games wins the championship. Suppose team A has a probability of 0.6 of winning each individual game and team B has a probability of 0.4 of winning each game. What is the expected number of games in the championship series? (*Hint:* Either A or B must win. Moreover, the series must take between four games and seven games. Find the probability that each team wins the championship series in a series of 4, of 5, of 6, or of 7 games).

13.2 DECISION MAKING

In the previous section you learned how to compute expected values. Next, we'll show you how to use expected values to make decisions when you are faced with uncertainty. Let's begin with a simple example.

EXAMPLE 6

Ann Jones has some money to invest and can choose either a mutual fund or a speculative stock. Her broker has advised her that the mutual fund will earn \$1000 profit during a time of economic growth and will earn \$250 during a recession. The stock will earn \$2000 during a time of economic growth and will lose \$1500 during a recession. Her broker estimates that the probability of economic growth is 0.5 and the probability of recession is 0.5. How should Ann Jones invest her money?

SOLUTION

Notice that Ann Jones would have no problem if she were to know the state of the economy. Thus the difficulty arises because of the uncertainty of the economy. In this case, she is uncertain about the exact **state of nature.** In other words, Ann does not know which of the possible alternatives (states of nature) will occur. The first step in decision making is to determine the possible states of nature. After determining the possible states of nature, you should list the actions you can control. These various actions are called **strategies.** Here Ann Jones has two strategies: invest in the mutual fund or invest in the speculative stock. The results of each strategy under each state of nature are called **payoffs** and are usually arranged in a **payoff table.** The payoff table for Ann Jones is shown in Table 13.1.

Once the payoff table has been constructed, the decision maker can use a variety of approaches. For instance, suppose Ann Jones is an optimist; she assumes the state of nature is favorable. In this case, she should invest in the speculative stock because this strategy results in the largest profit.

Of course, Ann Jones might also be a pessimist and assume the least favorable state of nature will occur. In that case, she should invest in the mutual fund.

As you might guess, Ann could also use the probability distribution given her by her broker to calculate expected values for the

TABLE 13.1 Payoff table for Ann Jones, Example 6

	States of nature	
Strategies	**Growth**	**Recession**
Invest in mutual fund	\$1000	\$250
Invest in speculative stock	\$2000	−\$1500

two strategies. She could then choose the strategy that will result in the largest expected value. Here we have

$$\begin{aligned} \text{EV(mutual fund)} &= (\$1000)0.5 + (\$250)0.5 \\ &= \$625, \\ \text{EV(speculative stock)} &= (\$2000)0.5 + (-\$1500)0.5 \\ &= \$250. \end{aligned}$$

The maximum expected profit occurs if she invests in the mutual fund. □

Practice Problem

Suppose in Example 6 that the probability of economic growth is 0.8 and the probability of recession is 0.2. What strategy should Ann Jones select to maximize her expected profit? What is the expected profit?

Answers: invest in the speculative stock; EV = \$1300

The process for making decisions by maximizing expected values is summarized as follows.

Making Decisions by Maximizing Expected Values

1. Determine the possible states of nature.
2. List all relevant strategies.
3. For each strategy, determine the payoff for each of the possible states of nature. Organize the payoffs in a payoff table.
4. Determine the probability distribution for the possible states of nature. Use the distribution to compute the expected value for each strategy.
5. Choose the strategy with the maximum expected value.

Note: We wish to emphasize that maximizing expected monetary payoff is a commonly used approach to decision making, but certainly not the only one. If a person is *risk averse,* he or she may wish to avoid a strategy that has the potential of a large loss. Thus, a risk averse person may choose a strategy in order to preclude a significant loss from occurring, even though another strategy with a larger expected payoff is available. Similarly, a *risk taker* may prefer a strategy with the possibility of making a large profit, even though another strategy with a larger expected payoff is available. For more details, consult a more advanced discussion of decision making.

Now let's look at some examples of applying this process. The first example is the Problem from Life.

EXAMPLE 7

Problem from Life Brazell Development Company has recently purchased some land on which to build condominiums. The company can build one of three developments: small, medium, or large. The decision is complicated by an uncertain economy. Brazell could lose a significant amount of money if it were to build a large development followed by low demand. Conversely, if the company were to be conservative and build a small development followed by large demand, the subsequent profits would be much smaller than they could have been. After some analysis and thought, management has prepared the following table, which shows the anticipated profit for every possible combination of building option and future demand.

Anticipated profits

	Demand		
Options	**Low**	**Medium**	**High**
Small	\$400,000	\$400,000	\$400,000
Medium	\$100,000	\$600,000	\$600,000
Large	−\$300,000	\$300,000	\$900,000

The best current estimates for demand are as follows: $P(\text{low}) = 0.2$, $P(\text{medium}) = 0.35$, and $P(\text{high}) = 0.45$. What is the recommended decision?

SOLUTION

In this case, the states of nature, the strategies, and the payoff table have been determined. The states of nature are the three possible demands that can occur. The strategies are the three building options. The payoff table consists of the anticipated profits. Thus we need only compute the expected values of the three strategies.

$$\begin{aligned}
\text{EV(build small)} &= (400{,}000)0.2 + (400{,}000)0.35 + (400{,}000)0.45 \\
&= \$400{,}000, \\
\text{EV(build medium)} &= (100{,}000)0.2 + (600{,}000)0.35 + (600{,}000)0.45 \\
&= \$500{,}000, \\
\text{EV(build large)} &= (-300{,}000)0.2 + (300{,}000)0.35 + (900{,}000)0.45 \\
&= \$450{,}000.
\end{aligned}$$

The largest expected value is \$500,000 for the strategy "build a medium development." Thus the Brazell Company should choose this strategy. □

Practice Problem

Suppose the probability estimates in Example 7 are $P(\text{low}) = 0.4$, $P(\text{medium}) = 0.25$, and $P(\text{high}) = 0.35$. What is the recommended decision?

Answer: build either small or medium with EV = \$400,000

EXAMPLE 8

A baker bakes bread at the beginning of each day. He realizes a profit of \$0.25 for each loaf he sells and a loss of \$0.10 for each loaf not sold (because the bread ages and may not be sold at full price after the first day). From records of past demand, he estimates the following probability distribution for the daily demand for bread (in loaves).

Quantity demanded	Probability
About 100	0.1
About 200	0.3
About 300	0.4
About 400	0.2

(He knows that the quantity demanded on any given day will probably not be an even multiple of 100. However, because of the manner in which the bread is made and baked, he must make an amount that is an even multiple of 100.)

Compute the best strategy for the baker: How many loaves should he bake?

SOLUTION

First of all, we must identify the states of nature. Here there are four, which correspond to the four possible levels of demand for bread.

Next we must determine all relevant strategies. The baker has four strategies: He can bake 100, 200, 300, or 400 loaves of bread. (As we mentioned, he must bake bread in even multiples of 100 loaves.)

Our next task is to determine the payoffs. Since there are four strategies and four states of nature, the payoff table will have sixteen entries. There must be an entry for every possible combination of state of nature and strategy. We'll take the payoffs in order by strategy.

Suppose the baker bakes 100 loaves. Then he will always sell all 100 loaves, because demand is never less than 100. However, he can never sell any more than the 100 loaves. Thus his profits will be (100)($0.25) = $25 regardless of the level of demand.

Suppose the baker bakes 200 loaves. Then there are two possible results: He may sell either 100 loaves (when demand is exactly 100) or 200 loaves (when demand is 200 or more). When the demand is 100 loaves, his profits are

$$(100)(\$0.25) + (100)(-\$0.10) = \$15,$$

and when the demand is 200 or more loaves, his profits are

$$(200)(\$0.25) = \$50.$$

Suppose he bakes 300 loaves. Then there are three possible results. When the demand is 100 loaves, he will sell 100 loaves. When the demand is 200 loaves, he will sell 200 loaves. When demand is 300 or more, he will sell 300 loaves. The profits are as follows.

Demand 100 loaves: $(100)(\$0.25) + (200)(-\$0.10) = \$5$,

Demand 200 loaves: $(200)(\$0.25) + (100)(-\$0.10) = \$40$,

Demand 300 or more loaves: $(300)(\$0.25) = \$75$.

Finally, suppose he bakes 400 loaves. Then there are four possible results: He will sell 100, 200, 300, or all 400 loaves, depending upon the exact amount of demand. The profits are computed as follows.

Demand 100 loaves: $(100)(\$0.25) + (300)(-\$0.10) = -\$5$,

Demand 200 loaves: $(200)(\$0.25) + (200)(-\$0.10) = \$30$,

Demand 300 loaves: $(300)(\$0.25) + (100)(-\$0.10) = \$65$,

Demand 400 loaves: $(400)(\$0.25) = \$100$.

At this point we can construct the payoff table.

	States of nature: demand			
Strategies	**100**	**200**	**300**	**400**
Bake 100	\$25	\$25	\$25	\$25
Bake 200	\$15	\$50	\$50	\$50
Bake 300	\$ 5	\$40	\$75	\$75
Bake 400	−\$ 5	\$30	\$65	\$100

Since we were given the probability distribution, the last step in the decision making procedure is to compute the expected values of the strategies.

$$\begin{aligned}
\text{EV(bake 100)} &= \$25, \\
\text{EV(bake 200)} &= (\$15)(0.1) + (\$50)(0.9) = \$46.50, \\
\text{EV(bake 300)} &= (\$5)(0.1) + (\$40)(0.3) + (\$75)(0.6) \\
&= \$57.50, \\
\text{EV(bake 400)} &= (-\$5)(0.1) + (\$30)(0.3) + (\$65)(0.4) + (\$100)(0.2) \\
&= \$54.50.
\end{aligned}$$

The largest expected value is \$57.50 for the strategy "bake 300 loaves," so the baker should adopt this strategy. His profits will average \$57.50 per day. □

Practice Problem

What should the baker in Example 8 do if the probability distribution for demand is given by $P(100) = 0.25$, $P(200) = 0.25$, $P(300) = 0.25$, and $P(400) = 0.25$? What is the expected value of this strategy?

Answers: bake 300 loaves; EV = \$48.75

EXERCISE SET 13.2

In Problems 1–3, use the payoff table that follows. Determine the strategy for each problem with the largest expected value using the probabilities shown for the states of nature.

	States of nature			
Strategies	**State 1**	**State 2**	**State 3**	**State 4**
1	−\$300	\$300	\$ 600	\$ 900
2	−\$100	\$200	\$ 300	\$1500
3	−\$500	\$500	\$1000	\$1000

1. Use probability of each state = 0.25.
2. Use P(state 1) = 0.4, P(state 2) = 0.2, P(state 3) = 0.2, P(state 4) = 0.2.
3. Use P(state 1) = 0.2, P(state 2) = 0.3, P(state 3) = 0.3, P(state 4) = 0.2.
4. Sue Martinez owns a small business. She must decide on the level of insurance coverage for her computer system. She can either purchase no insurance, purchase a low-cost policy, or purchase a comprehensive policy. There are two states of nature: There will be no severe electrical storms, in which case the system will operate flawlessly; or the system will be heavily damaged by severe electrical storms. The payoff table is as follows.

	States of nature	
Strategies	**No storms**	**Severe storms**
No insurance	\$ 0	−\$10,000
Low-cost policy	−\$250	−\$ 2500
Comprehensive policy	−\$500	−\$ 600

What strategy should she pursue if the probability of severe electrical storms is

a) 0.01? **b)** 0.05?

5. An investment analyst estimates there is approximately a 50% chance of an upturn in the chemical industry during the third quarter of the year, with the probabilities of no change and downturn being about equal. A client is considering either investing \$20,000 in a mutual fund specializing in chemical industry stocks or investing in corporate AAA-rated bonds yielding 10% per year. If the chemical industry experiences an upturn during the third quarter, the value of the mutual fund shares (including dividends) will increase by 18% during the next year. If there is no change during the third quarter, the value of the fund will increase by 4% during the next year. If there is a downturn, the value will decrease by 10%.

a) Construct a payoff table.
b) What is the best investment decision?

6. A charter boat owner must decide how many spare propellers to buy at \$200 each for the upcoming season. The owner has contracted to replace the engine next year, so that unused propellers will have no value. The cost of not having a spare propeller available when needed is \$400, which includes the cost of special handling and lost cruise time. Based on previous experience, the owner estimates the probability distribution for the number of spare propellers needed during a cruise season to be as follows.

Number of spare propellers needed	Probability
0	0.25
1	0.35
2	0.30
3	0.10

How many spare propellers should the owner order?

7. Ace Manufacturing produces (among other things) small marine engines. Management must decide among three strategies. The first is to go to a new design. If the new engines work as promised, sales will increase by \$100,000. However, if the new engines work poorly, sales will decrease by \$50,000. The second strategy is to modify the current design. If successful, sales will increase by \$50,000. If unsuccessful, sales will decrease by \$10,000. The third strategy is to continue with the current design. Management believes the probability of the new engines and the modification being successful is somewhere between 0.4 and 0.6. What is the best strategy assuming the probability of the new design and the modified design being successful is

   a) 0.4? **b)** 0.6?

8. A promoter is planning a big concert for the middle of the summer. He can either plan for an indoor concert or an outdoor concert. If the weather is good, the outdoor concert will generate \$75,000 profit and the indoor concert will generate \$25,000 profit. If the weather is poor, the outdoor concert will lose \$25,000 and the indoor concert will generate \$30,000 profit. Past weather service records indicate that the probability of the weather being good during the middle of the summer is 0.7. What should the promoter do?

9. A college fraternity is planning to sell hot dogs at the annual Spring Fling. The hot dogs cost \$1.50 per package and will be sold for \$2.50 per package. At past affairs, demand ranged between 100 and 130 packages. The best estimates for the probabilities are as follows.

Demand	Probability
100	0.30
110	0.20
120	0.30
130	0.20

All hot dogs that aren't used will be donated to a charity. How many packages should the fraternity order?

13.3

DECISION TREES

Many decision-making problems are complicated by the fact that the eventual economic consequences are related to a sequence of decisions and random events. We can use an approach called **decision tree** analysis to help us in problems that require a sequence of decisions to be made. Before we deal with such problems, however, let's analyze a simpler situation that requires only one decision to be made. This will demonstrate how to draw a decision tree and how to make use of it.

Let's reconsider the situation described in Example 6 in the previous section. Ann Jones, an investor, can choose between a mutual fund or a speculative stock. Subsequently, the economy will either grow or go into recession with equal probabilities. The payoff table is

	States of nature	
Strategies	**Growth**	**Recession**
Invest in mutual fund	\$1000	\$250
Invest in speculative stock	\$2000	−\$1500

Ann's problem can be represented by a decision tree, as shown in Fig. 13.1. Note that the tree shows the natural progression that would occur in the decision-making process. First, Ann must choose her strategy (mutual fund or stock). Then one of the states of nature (growth or recession) will occur. Notice that every possible sequence of events is represented in the tree.

The lines in the decision tree are called **branches.** Points at which lines intersect are called **nodes.** More specifically, there are **decision nodes** (points where decisions must be made) and **state-of-nature nodes** (points where a random process to determine a state of nature will occur). Decision nodes are represented by squares and state-of-nature nodes are represented by circles.

We analyze a decision tree by working backwards. At each state-of-nature node we use the probability distribution to compute an expected value. Thus at node 2 in Fig. 13.1 the expected value is

$$\text{EV(node 2)} = (\$1000)(0.5) + (\$250)(0.5) = \$625,$$

while at node 3 the expected value is

$$\text{EV(node 3)} = (\$2000)(0.5) + (-\$1500)(0.5) = \$250.$$

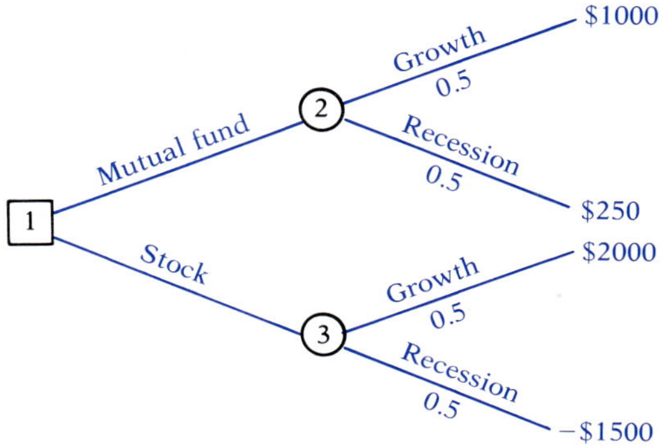

Figure 13.1 **Decision tree for Ann Jones' investment problem.**

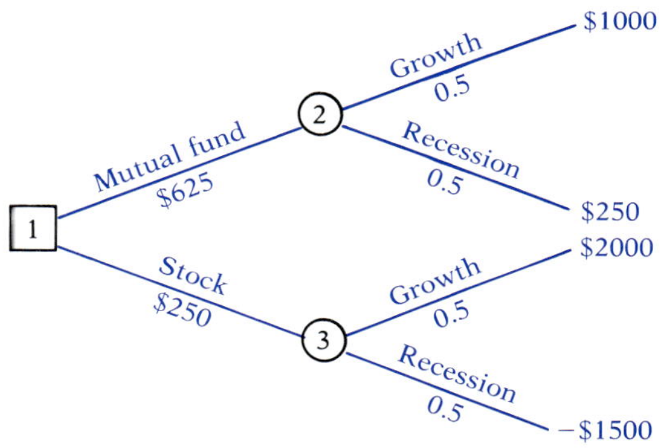

Figure 13.2 **Decision tree for Ann Jones' problem with expected values shown.**

We can continue backwards through the tree by viewing the decision node (node 1) as shown in Fig. 13.2. Here we have replaced all the branches at nodes 2 and 3 by their expected values. We want to maximize expected values, so we should select the strategy (branch) from node 1 with the largest value. This is, of course, "invest in mutual fund." □

Now let's consider a more complicated problem in which a sequence of decisions must be made. We'll use the same approach as before to analyze it: We start at the right of the tree and work our way backwards. At state-of-nature nodes we compute expected values and at decision nodes we select the branch with the largest value.

EXAMPLE 9

A computer manufacturer has been presented with a proposal for a new computer and must decide whether or not to develop it. The project will cost $200,000, and it has a probability of success of 0.7.

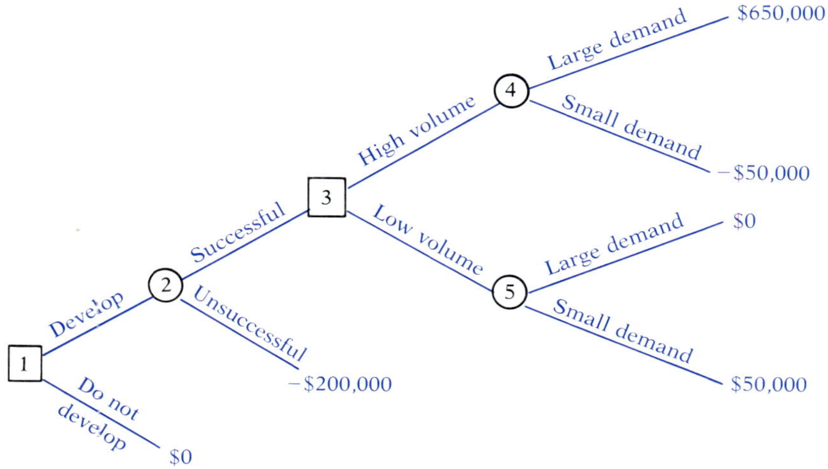

Figure 13.3 **Decision tree for computer manufacturer's problem.**

If the project is unsuccessful, it will be terminated. If the project is successful, the manufacturer must then decide to manufacture at low volume or high volume. The profits will depend on demand. If demand is large, then low volume will give gross profits of \$200,000 and high volume will give gross profits of \$850,000. If demand is small, then low volume will give gross profits of \$250,000 and high volume will give gross profits of \$150,000. The probability of large demand is estimated to be 0.4 and the probability of small demand is estimated to be 0.6. What should the manufacturer do?

SOLUTION

Before we can analyze the decision tree we must construct it. (See Fig. 13.3.) Note that the profit figures associated with large and small demands are net figures, which can be obtained simply by subtracting the \$200,000 project cost from the gross figures given in the problem description.

Now let's do the analysis. We begin with the state-of-nature nodes 4 and 5. At node 4 we have

$$\text{EV(node 4)} = (\$650{,}000)(0.4) + (-\$50{,}000)(0.6) = \$230{,}000,$$

while at node 5 we have

$$\text{EV(node 5)} = (\$0)(0.4) + (\$50{,}000)(0.6) = \$30{,}000.$$

These results are shown in Fig. 13.4, in which nodes 4 and 5 and their associated state-of-nature branches have been eliminated ("pruned"). Thus at node 3 we have a choice of the strategy "high

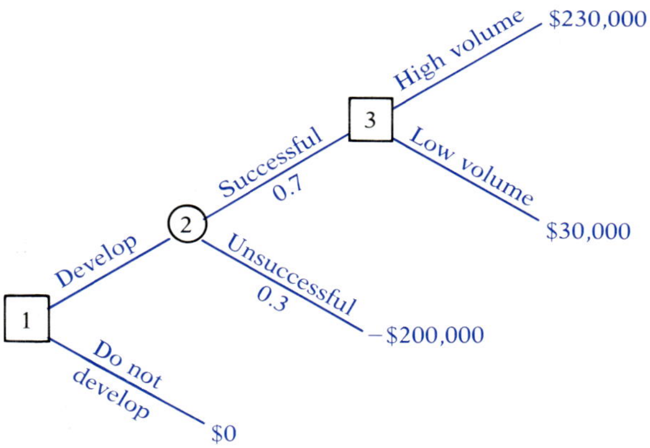

Figure 13.4 **Second decision tree, computer manufacturer's problem.**

volume" with an expected value of \$230,000 and the strategy "low volume" with an expected value of \$30,000. Our choice must be "high volume." This is reflected in Fig. 13.5. Notice that we have now eliminated decision node 3 and passed on the expected value of the appropriate strategy to the state-of-nature branch labeled "successful."

At node 2 we must compute an expected value. We have

$$\text{EV(node 2)} = (\$230{,}000)(0.7) + (-\$200{,}000)(0.3) = \$101{,}000.$$

Finally, at node 1, the computer manufacturer has a choice between the strategy "develop" with an expected value of \$101,000 (which we just computed at node 2) and the strategy "don't develop" with a value of \$0. The correct choice is "develop."

Let's review the results. First, the computer manufacturer should attempt to develop the new product. If successful, it should then use a strategy of "high volume." If the project is unsuccessful, then it should be terminated.

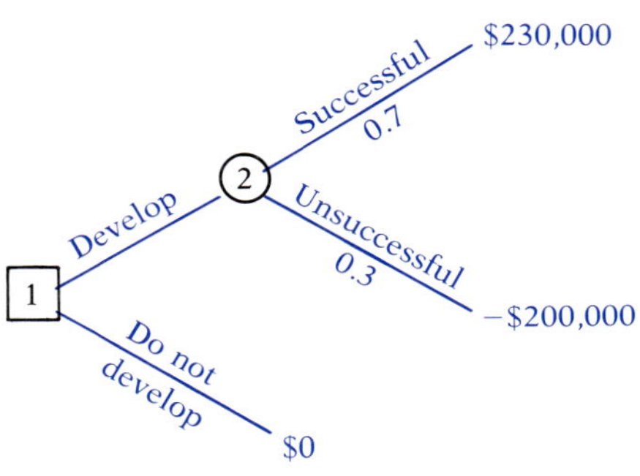

Figure 13.5 **Third decision tree, computer manufacturer's problem.**

□

Practice Problem

Suppose the probability of large demand in Example 9 is only 0.1 while the probability of small demand is 0.9. What is the best strategy for the manufacturer?

Answer: do not develop the project

As a final example, let's consider a more complicated situation, in which two decisions must be made.

EXAMPLE 10

The manager of research and development for Protean Products must decide whether or not to fund a project to develop a new synthetic oil. The project will be considered a major technical success, a minor technical success, or a failure. The company will realize profits of \$175,000 if the project is a major technical success. A minor technical success will be worth \$30,000 profit; a failure will cost \$100,000.

Preliminary development involving a prototype oil will be conducted to help the manager of research and development make her decision. Two different approaches are possible: an extensive one and a cursory one. The extensive approach will cost \$25,000, and the cursory approach will cost \$10,000. However, the extensive approach yields a better prediction of the eventual outcome of the project.

The extensive approach has three possible outcomes:

1. The prototype works well at all temperatures.
2. The prototype works well only at temperatures above 10° F.
3. The prototype works poorly at all temperatures.

Based on past experience, the technical staff has estimated the following probabilities concerning the quality of predictions and the likelihood of the possible outcomes for the extensive approach.

Quality of predictions: Extensive approach	Likelihood of outcomes: Outcome	Likelihood of outcomes: Probability
$P(\text{major success} \mid \text{outcome } 1) = 0.70$ $P(\text{minor success} \mid \text{outcome } 1) = 0.25$ $P(\text{failure} \mid \text{outcome } 1) = 0.05$	outcome 1	0.20
$P(\text{major success} \mid \text{outcome } 2) = 0.25$ $P(\text{minor success} \mid \text{outcome } 2) = 0.55$ $P(\text{failure} \mid \text{outcome } 2) = 0.20$	outcome 2	0.50
$P(\text{major success} \mid \text{outcome } 3) = 0.05$ $P(\text{minor success} \mid \text{outcome } 3 = 0.30$ $P(\text{failure} \mid \text{outcome } 3) = 0.65$	outcome 3	0.30

The cursory approach has only two possible outcomes:

4. The prototype works reasonably well.
5. The prototype works poorly.

The estimates of the technical staff concerning the quality of the predictions of the cursory approach and the likelihood of the outcomes 4 and 5 are as follows.

Quality of predictions: Cursory approach	Likelihood of outcomes	
	Outcome	**Probability**
P(major success | outcome 4) = 0.4 P(minor success | outcome 4) = 0.3 P(failure | outcome 4) = 0.3	outcome 4	0.6
P(major success | outcome 5 = 0.25 P(minor success | outcome 5) = 0.30 P(failure | outcome 5) = 0.45	outcome 5	0.4

What strategy should the manager of research and development follow?

SOLUTION

As before, we will analyze the situation with a decision tree. Here, however, we have additional information provided by the development of the prototype. Notice that we are given the probabilities of observing different outcomes of the prototype development and the probabilities of the eventual outcome of the project based on the results of the prototype approach chosen.

The decision tree is shown in Fig. 13.6. The two branches emanating from node 1 correspond to the two approaches for the prototype: the extensive and the cursory. Thus the first decision involves the choice of approach for the prototype. The three branches following node 2 correspond to the three possible outcomes of the extensive approach; the two branches following node 3 correspond to the two possible outcomes of the cursory approach. Note that we've represented nodes 2 and 3 as state-of-nature nodes because the decision maker cannot control the outcomes. Once the outcome of the prototype approach is known, a decision must be made whether to fund the project. The branches following nodes 9, 10, 11, 12, and 13 represent the three possible outcomes of the project (major success, minor success, or failure). Thus this project requires two decisions to be made. First, the manager of research and development must choose the prototype approach to use. Second, she must make a decision whether to continue with the project based on the results of the prototype experiment.

We analyze this problem starting at the end (right) of the decision tree. Nodes 9, 10, 11, 12, and 13 are state-of-nature nodes corresponding to the eventual outcome of the project. The expected values for these nodes are as follows.

$$
\begin{aligned}
\text{EV(node 9)} &= 0.7(175{,}000) + 0.25(30{,}000) + 0.05(-100{,}000) \\
&= \$125{,}000, \\
\text{EV(node 10)} &= 0.25(175{,}000) + 0.55(30{,}000) + 0.20(-100{,}000) \\
&= \$40{,}250 \\
\text{EV(node 11)} &= 0.05(175{,}000) + 0.30(30{,}000) + 0.65(-100{,}000) \\
&= -\$47{,}250, \\
\text{EV(node 12)} &= 0.4(175{,}000) + 0.3(30{,}000) + 0.3(-100{,}000) \\
&= \$49{,}000, \\
\text{EV(node 13)} &= 0.25(175{,}000) + 0.30(30{,}000) + 0.45(-100{,}000) \\
&= \$7750.
\end{aligned}
$$

We can use these figures to make the necessary decisions at nodes 4, 5, 6, 7, and 8. (See Fig. 13.7.) At every node the choice is between funding the project or not funding. At nodes 4, 5, 7, and 8, funding the project is the better decision. At node 6, not funding the project is the better decision. Once decisions are made, we can "pass" the value of the decision through the appropriate decision node to the branch to the left. Thus we pass \$125,000 through node 4, \$40,250 through node 5, \$0 through node 6, \$49,000 through node 7, and \$7750 through node 8. The results of this are shown in Fig. 13.8.

As Fig. 13.8 indicates, nodes 2 and 3 are state-of-nature nodes. Now that we've pruned the tree, we can compute the expected value at each of these nodes. Notice that we included the cost of the prototype development approach.

$$
\begin{aligned}
\text{EV(node 2)} &= 0.2(125{,}000) + 0.5(40{,}250) + 0.3(0) - 25{,}000 \\
&= \$45{,}125 - \$25{,}000 \\
&= \$20{,}125, \\
\text{EV(node 3)} &= 0.6(49{,}000) + 0.4(7750) - 10{,}000 \\
&= \$32{,}500 - \$10{,}000 \\
&= \$22{,}500.
\end{aligned}
$$

Thus at node 1, the decision must be made between the extensive approach with an expected value of \$20,125 and the cursory approach with an expected value of \$22,500. The choice that maximizes expected value is the cursory approach. We can summarize our overall strategy.

1. Choose the cursory prototype development.
2. Make the choice to fund the project or not based on the results of the prototype development:

Result of prototype development	Best strategy	Expected value
Prototype works reasonably well	Fund project	$49,000
Prototype works poorly	Fund project	$ 7,750

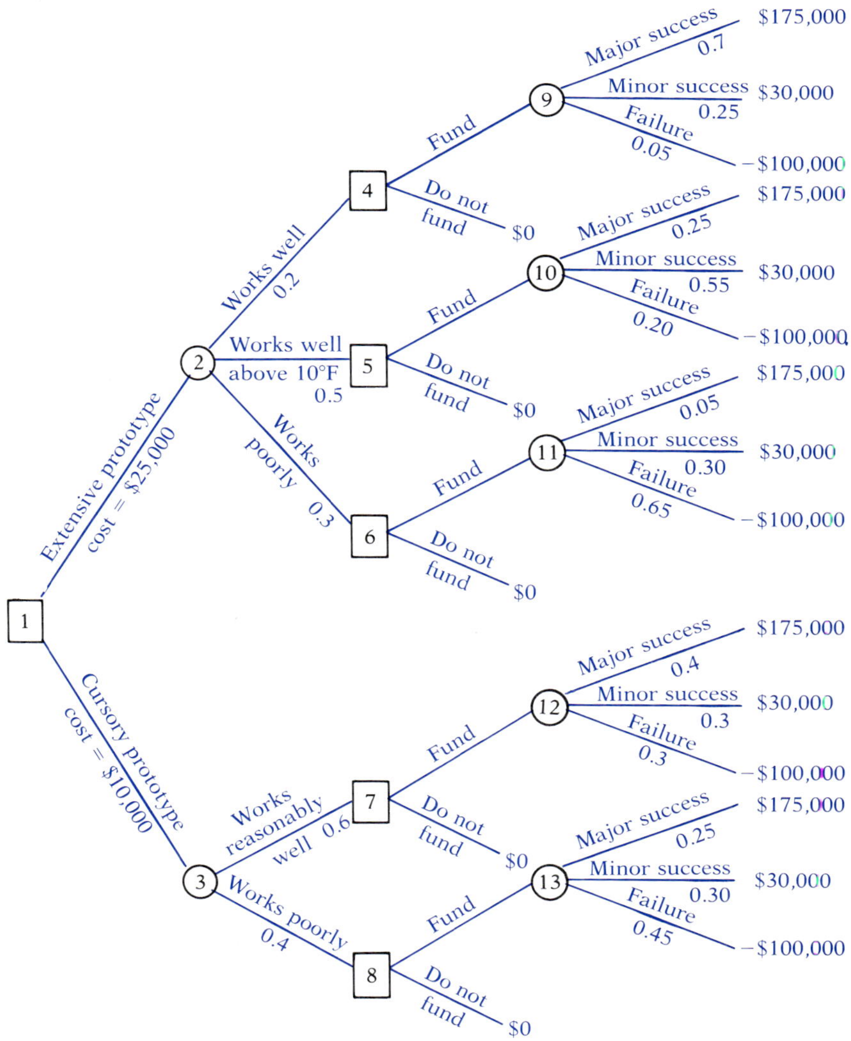

Figure 13.6 **Decision tree for Protean Products, Example 10.**

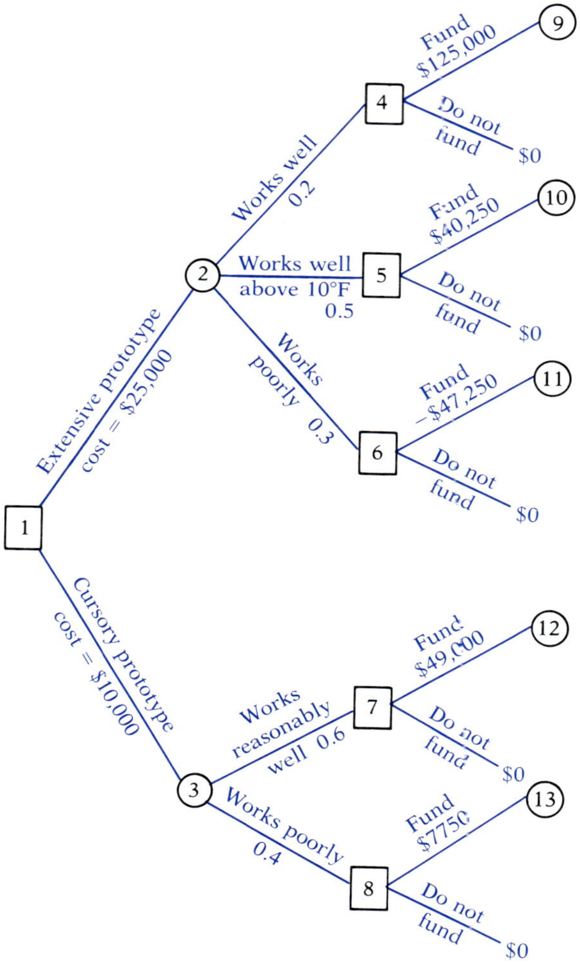

Figure 13.7 **Decision tree for Protean Products with expected values for nodes 9, 10, 11, 12, and 13.**

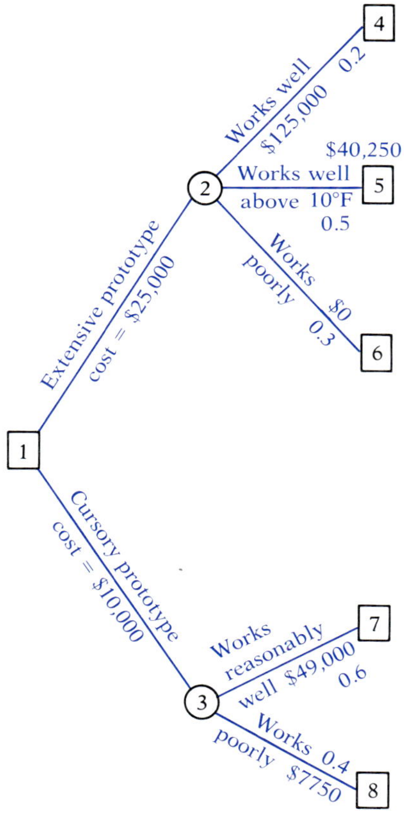

Figure 13.8 **Decision tree, Protean Products, after pruning to the right of the first decision.** □

Practice Problem

Suppose the probability estimates for the outcome of the prototype development were

$$P(\text{outcome } 1) = 0.15,$$
$$P(\text{outcome } 2) = 0.5,$$
$$P(\text{outcome } 3) = 0.35,$$
$$P(\text{outcome } 4) = 0.5,$$
and
$$P(\text{outcome } 5) = 0.5.$$

What are the best strategies to follow in Example 10? What are the expected values?

Answers: fund cursory prototype and fund project in each case; EV(extensive prototype) = \$13,875 and EV(cursory prototype) = \$18,375

EXERCISE SET 13.3

1. What should Protean Products do (see Example 10) if the extensive prototype development costs \$30,000 and the cursory prototype development costs \$11,000?
2. What should Protean Products do (see Example 10) if the probability of a major success is 0.5, the probability of a minor success is 0.35, and the probability of a failure is 0.15 given outcome 1 (the extensive prototype works well)?
3. Sam Spade, the owner of Sam's Sporting Goods, is considering adding a line of skis and accessories. He estimates his profits will be \$7000 if the snowfall is heavy and \$2000 if the snowfall is moderate. He will lose \$9000 if the snowfall is light. Long-term trends indicate that $P(\text{heavy snowfall}) = 0.4$, $P(\text{moderate snowfall}) = 0.3$, $P(\text{light snowfall}) = 0.3$.

   a) What is Sam's best decision?

   b) Suppose Sam waits to check the temperature pattern in September before making a decision. Estimates of the conditional probabilities associated with an unseasonably cold September are $P(\text{heavy snowfall} \mid \text{cold}) = 0.50$, $P(\text{moderate} \mid \text{cold}) = 0.25$, and $P(\text{light snowfall} \mid \text{cold}) = 0.25$. What should Sam do if September is unseasonably cold? What if September is not unseasonably cold?

4. Mary Caldwell, a buyer for a department store, has been offered a shipment of 50 medium-quality stereo systems manufactured abroad for \$5000. However, past experience has indicated that shipments from this manufacturer have a distinct possibility of having been damaged in transit; this occurs about 25% of the time. If the shipment is damaged, Mary can sell the systems for a total of \$4000. If no damage has occurred, she can resell the shipment for a total of \$8000.

   a) What should she do?

   b) Suppose that Mary can have one system uncrated and tested for \$25. Past experience has shown the probability of the shipment being good given that a randomly selected system is good is 0.75, while the probability of the shipment being good given that a randomly selected system is damaged is 0.25. What should Mary do if the uncrated system is damaged? What should Mary do if the uncrated system is good? Is the payment worthwhile? (*Hint:* Consider the expected value of her best decision both with and without the test).

13.4 GAME THEORY

Strictly Determined Games

In the previous sections of this chapter you learned how decision theory can help make decisions in a variety of situations in which you are faced with uncertainty. In this section we'll introduce you to **game theory,** which can be used to help make decisions in situations in which you are faced with a thinking adversary.

Game theory was developed in the 1940s to help analyze competitive situations involving conflicting interests. We could also use it to analyze recreational games. In game theory we assume there are two (or more) persons with different objectives. We also assume each person knows the other person's objectives. We'll limit our discussion to **two-person games:** those with just two persons (or sides).

A game is often represented by a **game** or **payoff matrix.** For example, the payoff matrix shown in Table 13.2 represents a two-person game involving players A and B. Each player has two choices or strategies. Player A can choose either row 1 or row 2, while player B can choose column 1 or column 2. Thus Table 13.2 is an example of a 2 × 2 game. In general, if A has m strategies and B has n strategies, the game is said to be $m \times n$.

Each number in the matrix is the payoff that results when each player chooses a strategy. A positive number indicates player B pays money to player A; a negative indicates player A pays money to player B. For example, suppose A chooses row 2 and B chooses column 1. Then B must pay A $2.

Practice Problem

What is the result if A chooses row 1 and B chooses column 2?

Answer: A must pay B $3

TABLE 13.2 An example of a payoff matrix for a two-person game

$$
\begin{array}{ccc}
 & & \text{Player B} \\
 & & \begin{array}{cc} 1 & \;\;2 \end{array} \\
\text{Player A} & \begin{array}{c} 1 \\ 2 \end{array} & \begin{bmatrix} 5 & -3 \\ 2 & 1 \end{bmatrix}
\end{array}
$$

Notice in our example that A's winnings are equal to B's losses and vice versa. Such a game is called a **zero-sum game.** There are also **nonzero-sum games** in which money can enter the game from outside. In nonzero-sum games it is possible for all players to win (or lose) simultaneously. We will limit our discussion to zero-sum games.

Practice Problem

Classify the following as zero-sum or nonzero-sum games:

a) You flip a coin with a friend. You win \$1 on heads and lose \$1 on tails.

b) You play poker with four friends.

c) You invest money in the stock market.

Answers: **a)** and **b)** are zero-sum games; **c)** is a nonzero-sum game

Now let's think about determining strategies. Consider the game in Table 13.2. Player A can choose either row 1 or row 2. The largest payoff for A occurs if he chooses row 1 and B chooses column 1, namely, \$5. However, if A continues to choose row 1, then B will catch on (sooner or later) and then play column 2, which will result in A losing \$3. Thus a strategy of attempting to always maximize his gain won't work very well for A, because B is simultaneously trying to maximize his gain. A more conservative strategy will work much better. Player A can play row 2 and never make less than \$1! If player B plays column 1, A will win \$2 and if B plays column 2 then A will win \$1. Thus A should choose row 2 to ensure that he wins at least \$1. We say that Player A is maximizing his minimum gain by choosing row 2. This is called a **maximin** strategy.

Now let's look at it from B's point of view. If B plays column 1, the best he can do is lose \$2 (to A). However, if B plays column 2, he will either win \$3 (if A chooses row 1) or he will lose \$1 (if A chooses row 2). Thus B should always play column 2. Player B is minimizing his maximum loss. This is called a **minimax** strategy.

This pair of strategies, (2, 2), is called a **saddle point** of the game. A saddle point is the smallest number in its row and the largest number in its column. A game with a saddle point is said to be **strictly determined,** because the optimal (best) strategy for both players is to choose the saddle point. (The saddle point is the optimal strategy because we assume both players realize that their adversary will act rationally. Thus player A will utilize the maximin strategy and player B will utilize the minimax strategy. These are conservative strategies that are adopted because of the presence of the thinking adversary.) The **value** of the game is the value of the saddle point; in the case of the game in Table 13.2, the value is $1.

Let's look at some examples of games and try to determine which ones have saddle points.

EXAMPLE 11

Does the following game have a saddle point? If so, what are the optimal strategies and the value of the game?

$$\begin{array}{cc} & \text{Player B} \\ & \begin{array}{cc} 1 & 2 \end{array} \\ \text{Player A} \begin{array}{c} 1 \\ 2 \end{array} & \begin{bmatrix} 0 & 1 \\ -3 & 10 \end{bmatrix} \end{array}$$

SOLUTION

This game has a saddle point. A's optimal strategy is row 1; B's optimal strategy is column 1. The value of the game is $0, so that the game is fair. □

EXAMPLE 12

Does the following game has a saddle point? If so, what are the optimal strategies and the value of the game?

$$\begin{array}{cc} & \text{Player B} \\ & \begin{array}{ccc} 1 & 2 & 3 \end{array} \\ \text{Player A} \begin{array}{c} 1 \\ 2 \\ 3 \end{array} & \begin{bmatrix} 5 & 8 & 7 \\ -1 & -3 & 10 \\ 2 & 12 & -6 \end{bmatrix} \end{array}$$

SOLUTION

This game has a saddle point. A's optimal strategy is row 1; B's optimal strategy is column 1. The value of the game is $5, so it is not fair. Note that our criterion for finding the saddle point, a row minimum that is also a column maximum, is not limited to 2×2 games. It works for this game, which is 3×3. □

EXAMPLE 13

Does the following game have a saddle point? If so, what are the optimal strategies and the value of the game?

$$\begin{array}{cc} & \text{Player B} \\ & \begin{array}{cc} 1 & 2 \end{array} \\ \text{Player A} \begin{array}{c} 1 \\ 2 \end{array} & \begin{bmatrix} 0 & 1 \\ 2 & 0 \end{bmatrix} \end{array}$$

SOLUTION

This game does not have a saddle point, so it is not strictly determined. (There is no payoff in the matrix that is simultaneously a row minimum and a column maximum.) □

Practice Problem

Which of the following games are strictly determined? What is the value of each strictly determined game?

a) Player B (columns 1, 2), Player A (rows 1, 2)

$$\begin{bmatrix} 3 & 2 \\ 2 & -4 \end{bmatrix}$$

b) Player B (columns 1, 2, 3), Player A (rows 1, 2, 3)

$$\begin{bmatrix} -2 & -4 & -6 \\ 1 & 3 & -8 \\ -9 & 10 & -9 \end{bmatrix}$$

c) Player B (columns 1, 2), Player A (rows 1, 2)

$$\begin{bmatrix} 2 & 5 \\ 9 & -3 \end{bmatrix}$$

Answers: **a)** and **b)** are strictly determined; the value of **a)** is \$2 and the value of **b)** is −\$6

Sometimes a game isn't as complicated as it looks. Consider the following 5 × 2 game:

Player B (columns 1, 2), Player A (rows 1–5)

$$\begin{bmatrix} 4 & 10 \\ 8 & 6 \\ 6 & 12 \\ 10 & 14 \\ 8 & 8 \end{bmatrix}$$

Player A has five strategies. However, not all of them are worthwhile. In fact, row 4 has a larger payoff in each column than any other row. Thus row 4 **dominates** all the other rows. This means they can be eliminated, because player A will choose only row 4. Moreover, since player A will choose row 4 exclusively, player B will choose column 1. The optimal strategy is (4, 1); the value of the game is \$10.

If you study this game, you will notice that there is a saddle point in row 4 and column 1. However, it is far easier to see this after we've eliminated the dominated rows.

Practice Problem

Reduce the following game by eliminating the dominated rows.

$$\text{Player A}\quad \begin{array}{c} \\ 1\\2\\3\\4 \end{array} \overset{\text{Player B}}{\begin{array}{cc} 1 & 2 \\ \left[\begin{array}{cc} 3 & 2 \\ 5 & 3 \\ 1 & 4 \\ 3 & 5 \end{array}\right] \end{array}}$$

Answers: row 2 dominates row 1 and row 4 dominates row 3. The resulting game is

$$\text{Player A}\quad \begin{array}{c} 2\\4 \end{array} \overset{\text{Player B}\quad 1\quad 2}{\begin{bmatrix} 5 & 3 \\ 3 & 5 \end{bmatrix}}$$

We can also find dominated columns by looking for a column whose entries are always *less* than those of another column. The following example demonstrates column dominance.

EXAMPLE 14

Check the following game for columns that are dominated.

$$\text{Player A}\quad \begin{array}{c} 1\\2\\3 \end{array} \overset{\text{Player B}\quad 1\quad 2\quad 3\quad 4\quad 5}{\begin{bmatrix} 8 & 10 & 13 & 16 & 9 \\ 10 & 12 & 6 & 15 & 10 \\ 4 & 9 & 18 & 20 & 6 \end{bmatrix}}$$

SOLUTION

Every number in column 1 is less than the corresponding entry in column 2, so column 1 dominates column 2. Similarly, column 3 dominates column 4. Finally, column 1 dominates column 5. Thus we can eliminate columns 2, 4, and 5. The resulting game has only two columns and three rows:

$$\text{Player A}\quad \begin{array}{c} 1\\2\\3 \end{array} \overset{\text{Player B}\quad 1\quad 3}{\begin{bmatrix} 8 & 13 \\ 10 & 6 \\ 4 & 18 \end{bmatrix}} \quad \square$$

Practice Problem

Reduce the following game matrix by eliminating dominated columns. Does this game have a saddle point? If so, what is the value of the game?

$$\begin{array}{cc} & \text{Player B} \\ & \begin{array}{ccccc} 1 & 2 & 3 & 4 & 5 \end{array} \\ \text{Player A} \begin{array}{c} 1 \\ 2 \end{array} & \begin{bmatrix} 3 & 5 & 1 & -6 & -2 \\ 5 & 7 & 2 & 9 & -1 \end{bmatrix} \end{array}$$

Answers: column 5 dominates columns 1, 2, and 3; saddle point in row 2 and column 5; value of game is $-\$1$

As you check for dominated rows and columns, you should be aware that when you eliminate a dominated row (or column), you may then discover that additional rows or columns have become dominated and can be eliminated. The following example demonstrates this process on a 5 × 4 game.

EXAMPLE 15

Eliminate all dominated rows and columns in the following 5 × 4 game:

$$\begin{array}{cc} & \text{Player B} \\ & \begin{array}{cccc} 1 & 2 & 3 & 4 \end{array} \\ \text{Player A} \begin{array}{c} 1 \\ 2 \\ 3 \\ 4 \\ 5 \end{array} & \begin{bmatrix} 100 & 80 & 150 & 50 \\ 150 & 55 & 100 & 80 \\ 150 & 55 & 100 & 25 \\ 125 & 60 & 100 & 80 \\ 90 & 70 & 125 & 75 \end{bmatrix} \end{array}$$

SOLUTION

First, note that column 4 dominates both columns 1 and 3. This reduces the game to 5 × 2, as follows.

$$\begin{array}{cc} & \text{Player B} \\ & \begin{array}{cc} 2 & 4 \end{array} \\ \text{Player A} \begin{array}{c} 1 \\ 2 \\ 3 \\ 4 \\ 5 \end{array} & \begin{bmatrix} 80 & 50 \\ 55 & 80 \\ 55 & 25 \\ 60 & 80 \\ 70 & 75 \end{bmatrix} \end{array}$$

Now row 1 dominates row 3 and row 4 dominates row 2 (but they do not in the original matrix). This reduces the game to 3 × 2, as shown below, which is much easier to analyze. (No, it does not have a saddle point.)

$$\begin{array}{cc} & \text{Player B} \\ & \begin{array}{cc} 2 & 4 \end{array} \\ \text{Player A} \begin{array}{c} 1 \\ 4 \\ 5 \end{array} & \begin{bmatrix} 80 & 50 \\ 60 & 80 \\ 70 & 75 \end{bmatrix} \end{array} \quad \square$$

Practice Problem

Reduce the following game by using dominance to eliminate rows and columns.

$$\begin{array}{cc} & \text{Player B} \\ & \begin{array}{cccc} 1 & 2 & 3 & 4 \end{array} \\ \text{Player A} \begin{array}{c} 1 \\ 2 \\ 3 \\ 4 \end{array} & \begin{bmatrix} 10 & 13 & 11 & -2 \\ 14 & 6 & -5 & 5 \\ 9 & 10 & 5 & -4 \\ 15 & 4 & -3 & 3 \end{bmatrix} \end{array}$$

Answer: First, row 1 dominates row 3. Then column 4 dominates both columns 1 and 2, which reduces to the 3×2 game

$$\begin{array}{cc} & \text{Player B} \\ & \begin{array}{cc} 3 & 4 \end{array} \\ \text{Player A} \begin{array}{c} 1 \\ 2 \\ 4 \end{array} & \begin{bmatrix} 11 & -2 \\ -5 & 5 \\ -3 & 3 \end{bmatrix} \end{array}$$

Games Without Saddle Points: Mixed Strategies

We know how to solve games with a saddle point. For such games, both players have a single optimal strategy. We often refer to a single strategy as a **pure** strategy. Unfortunately, not every game has a saddle point. For games that do not possess a saddle point, the players will not have pure strategies. Thus each would have a **mixed** strategy which dictates that certain rows and columns should be played a given fraction of the time.

In a mixed strategy, it is important for the players not to follow a specific pattern (for example, row 1, then row 2, then row 3, then row 1, etc.) because a thinking opponent would be quick to take advantage of the pattern. Thus a mixed strategy consists of finding the appropriate probability to tell how often each row (column) should be played. It is then necessary to use a randomizing device (for example, a spinner) to determine which row (column) would be played next. Because the calculations can become messy, we will limit ourselves to finding the solutions to 2 × 2 games. (Of course, we may start with a larger game that is reduced to a 2 × 2 game after eliminating dominated rows and columns.)

Consider the game

$$\begin{array}{cc} & \text{Player B} \\ & \begin{array}{cc} 1 & 2 \end{array} \\ \text{Player A} \begin{array}{c} 1 \\ 2 \end{array} & \begin{bmatrix} 3 & 6 \\ 5 & 4 \end{bmatrix} \end{array}$$

There is no saddle point, so this game is not strictly determined. In a mixed strategy, player A should determine the probabilities that describe how often each row should be played. Let

$$\begin{aligned} a &= \text{probability with which A chooses row 1,} \\ 1 - a &= \text{probability with which A chooses row 2.} \end{aligned}$$

Now, suppose B plays column 1. Then A's expected return is

$$\text{EV(col. 1)} = 3a + 5(1 - a),$$

and when B plays column 2, A's expected return is

$$\text{EV(col. 2)} = 6a + 4(1 - a).$$

For A to maximize his minimum expected profits, the two expected values given must be equal. If they aren't, B can lower A's return by choosing the column with the lower expected value. Hence, we have

$$\begin{aligned} \text{EV(col. 1)} &= \text{EV(col. 2)}, \\ 3a + 5(1 - a) &= 6a + 4(1 - a), \\ 5 - 2a &= 4 + 2a, \\ 1 &= 4a, \\ a &= 0.25. \end{aligned}$$

In other words, A should play row one 25% of the time and row two 75% (1 − 0.25 = 0.75) of the time. The expected value of the game can be found by substituting the values of a and $1 - a$ into either EV(col. 1) or EV(col. 2). Using the former gives us

$$\begin{aligned} \text{EV(game)} &= \text{EV(col. 1)}, \\ &= 3(0.25) + 5(0.75), \\ &= \$4.50. \end{aligned}$$

Now let's look at how we compute the probabilities player B should use.

EXAMPLE 16

In the game we just analyzed, what probabilities should B use?

SOLUTION

Let b = probability with which B chooses column 1,
$1 - b$ = probability with which B chooses column 2.

Suppose A plays row 1. The B's expected return is

$$\text{EV(row 1)} = -3b - 6(1 - b).$$

(The values for B are negative because they represent gains for A and therefore losses for B.) Similarly, B's expected return when A plays row 2 is

$$\text{EV(row 2)} = -5b - 4(1 - b).$$

For B to minimize his maximum loss, these expected values must be equal. Setting the expected values equal to one another yields

$$\begin{aligned} \text{EV(row 1)} &= \text{EV(row 2)}, \\ -3b - 6(1 - b) &= -5b - 4(1 - b), \\ 3b - 6 &= -b - 4, \\ 4b &= 2, \\ b &= 0.5. \end{aligned}$$

B should play each column 50% of the time. □

Now let's look at another game with a mixed strategy.

EXAMPLE 17

A friend of yours offers you the following bet. Each of you must call either heads or tails. When you call heads and she calls tails, she'll give you \$3; when you call tails and she calls heads she'll give you \$1. However, when you both match, then you give her \$2. What is your optimal strategy? Is this game fair?

SOLUTION

We can answer the second question by answering the first. We can model this situation as a game. You each have two strategies: call heads (H) or tails (T). Thus the game will be 2×2. The entries in the payoff table are your winnings.

		Friend H	Friend T
You	H	−2	3
	T	1	−2

Let

$$h = \text{probability for you to play heads},$$
$$1 - h = \text{probability for you to play tails}.$$

Then

$$\text{EV(col. 1)} = -2h + 1(1 - h),$$
$$\text{EV(col. 2)} = 3h - 2(1 - h).$$

Setting these two expected values equal gives

$$-2h + 1(1 - h) = 3h - 2(1 - h),$$

so

$$h = 0.375,$$
$$1 - h = 0.625.$$

You should play heads 37.5% of the time and tails 62.5% of the time.

We can compute the expected value of the game using the expected value of the first column:

$$\text{EV(game)} = \text{EV(col. 1)} = -\$2(0.375) + \$1(0.625),$$
$$\text{EV(game)} = -\$0.125.$$

You can expect to lose \$0.125 or 12.5 cents (on the average) each time you play the game (with your optimal strategy). You might question your friendship, because the game is not fair. □

Practice Problem

In Example 17, what are the probabilities for your "friend"?

> Answer: She should play heads 62.5% of the time and tails 37.5% of the time.

We can also use game theory to help us analyze situations that do not actually involve a thinking adversary. The following example discusses a situation in which decision theory would appear to be the appropriate approach. However, decision theory requires that we have probability distributions available to describe the states of nature. When these probabilities are not available, we can use game theory (at least some of the time) to help us make the decision.

EXAMPLE 18

Sam has a concession at Carter-Finley Stadium for the sale of sunglasses and umbrellas. When it rains, he can sell 1000 umbrellas; when it is sunny, he can sell 200 umbrellas and 2000 sunglasses. He pays \$0.50 for the umbrellas and \$0.20 for the sunglasses. He sells the umbrellas for \$1 and the sunglasses for \$0.50. He is willing to invest \$500. Assuming everything that isn't sold is a total loss, what should he purchase to maximize his minimum expected profit?

SOLUTION

If Sam had reliable probability estimates for the weather, he could use decision theory (and maximize his expected profits). However, we can assume that reliable probabilities aren't available. We can model Sam's problem as a game as follows. He has two strategies: He can buy for rain (buy 1000 umbrellas) or for sun (200 umbrellas and 2000 sunglasses). His "opponent" (the weather) can either be rainy or sunny.

When Sam buys for rain, he buys 1000 umbrellas, at \$0.50 each, for a total cost of \$500. When the weather is rainy, he sells all 1000 umbrellas at \$1 each, so his net profit is \$1000 − \$500 or \$500. When the weather is sunny, he can sell only 200 umbrellas at \$1 each, so his net profit is \$200 − \$500 or −\$300.

When Sam buys for sun, he buys 200 umbrellas at \$0.50 each and 2000 sunglasses at \$0.20 each for a total cost of \$500. When the weather is rainy, he sells 200 umbrellas at \$1 each for a net profit of \$200 − \$500 or −\$300. When the weather is sunny, he sells 200 umbrellas at \$1 each and 2000 sunglasses at \$0.50 each for a net profit of \$1200 − \$500 or \$700.

The game matrix is therefore

$$\begin{array}{ccc} & & \text{Weather} \\ & & \begin{array}{cc} \text{Rainy} & \text{Sunny} \end{array} \\ \text{Buy for} & \begin{array}{c} \text{Rain} \\ \text{Sun} \end{array} & \begin{bmatrix} 500 & -300 \\ -300 & 700 \end{bmatrix} \end{array}$$

Since the game does not have a saddle point, Sam should use a mixed strategy. Let

$$r = \text{probability he buys for rain}$$
$$1 - r = \text{probability he buys for sun.}$$

When the weather is rainy, his expected profits are

$$\text{EV(rainy)} = 500r - 300(1 - r).$$

Similarly, his expected profits for sunny weather are

$$\text{EV(sunny)} = -300r + 700(1 - r).$$

Equating the two expected values and solving for r gives us

$$500r - 300(1 - r) = -300r + 700(1 - r),$$

so

$$r = 0.556$$

and

$$1 - r = 0.444.$$

Sam should buy for rain 55.6% of the time and buy for sun 44.4% of the time. As an alternative, he can put 55.6% of his money into rainy day gear and 44.4% of his money into sunny day gear. He will then earn the value of the game on a very steady basis.

The value of the game is (using the expected value for rainy weather)

$$\text{EV(game)} = \text{EV(rainy)} = \$500(0.556) - \$300(0.444),$$
$$\text{EV(game)} = \$144.80. \quad \square$$

Practice Problem

In Example 18, assuming Sam buys 55.6% rainy weather gear and 44.4% sunny weather gear, what does he buy?

Answer: 645 umbrellas and 888 sunglasses

EXERCISE SET 13.4

Solve the following games by finding the optimal strategy for each player. In addition, find the value of the game.

1. Player B (columns 1, 2); Player A (rows 1, 2)
$$\begin{bmatrix} 0 & 1 \\ -3 & 10 \end{bmatrix}$$

2. Player B (columns 1, 2); Player A (rows 1, 2)
$$\begin{bmatrix} 5 & 2 \\ -3 & -1 \end{bmatrix}$$

3. Player B (columns 1, 2); Player A (rows 1, 2)
$$\begin{bmatrix} 0 & 1 \\ 2 & 0 \end{bmatrix}$$

4. Player B (columns 1, 2); Player A (rows 1, 2)
$$\begin{bmatrix} 2 & 0 \\ 0 & 2 \end{bmatrix}$$

5. Player B (columns 1, 2); Player A (rows 1, 2)
$$\begin{bmatrix} -1 & 0 \\ 0 & -2 \end{bmatrix}$$

6. Player B (columns 1, 2); Player A (rows 1, 2)
$$\begin{bmatrix} 8 & 1 \\ 4 & 6 \end{bmatrix}$$

7. Player B (columns 1, 2); Player A (rows 1, 2)
$$\begin{bmatrix} 7/8 & -3/4 \\ -1/2 & 2/3 \end{bmatrix}$$

8. Player B (columns 1, 2); Player A (rows 1, 2)
$$\begin{bmatrix} 1/2 & 1 \\ 5/6 & 2/3 \end{bmatrix}$$

9. Player B (columns 1, 2); Player A (rows 1, 2)
$$\begin{bmatrix} -2 & 1/2 \\ 3 & -1 \end{bmatrix}$$

10. Player B (columns 1, 2); Player A (rows 1, 2)
$$\begin{bmatrix} 2 & -3/2 \\ 4 & -1 \end{bmatrix}$$

11.

		Player B 1	Player B 2
Player A	1	2	5
	2	3	6
	3	5	4

12.

		Player B 1	Player B 2
Player A	1	6	1
	2	−1	−2
	3	4	3

13. Two students decide to play the following game, rather than study. Each one shows either one or two fingers. When there is a match, B wins an amount equal to the total number of fingers shown. When there is no match, A wins an amount equal to the total number of fingers shown.

a) Determine the payoff matrix.
b) Find the optimal strategy for each player and the value of the game.

14. Repeat Problem 13 assuming each player shows one or three fingers.

15. Ace Electronics has decided to promote its new personal computers with an advertising campaign featuring both TV ads and ads in certain computer journals. Unfortunately, its leading competitor, Consolidated Electronics, has decided to announce its new computers at the same time. The data in the payoff matrix show the increased (decreased) sales in millions for the two companies.

		Consolidated TV	Consolidated Journals
Ace	TV	2.0	−1.4
	Journals	−1.0	1.0

Find the appropriate advertising strategy for Ace and the value of the game.

16. In a certain city there are only two pizza parlors, Tony's and Sal's. Tony must decide what the price of his large pizza should be. The numbers in the following payoff matrix show profits for the two establishments (positive numbers are in Tony's favor). What should Tony charge?

		Sal's Price 7.00	Sal's Price 8.00
Tony's Price	7.00	300	750
	8.00	−1000	500

17. Precision Manufacturing is considering the problem of buying raw materials for the next quarter. About 150 tons are required during a normal quarter, but extremes of as little as 100 tons or as much as 200 tons have occurred. In addition, the price per ton tends to fluctuate; possible prices are expected to be \$20, \$30, or \$40 a ton. The raw materials can be purchased now, however, at \$20 per ton. To complicate matters, Precision is moving its operations after the coming quarter and any excess material would have no value. What should Precision do? (*Hint:* Solve as a game. All entries in the payoff table are costs, so they should carry a negative sign.)

CHAPTER REVIEW

IMPORTANT TERMS AND SYMBOLS

13.1 Expected Values

expected value (EV) The (weighted) average value of a random variable with a given probability distribution. Formula:

$$\text{EV} = \sum_{i=1}^{n} x_i \cdot p_i = x_1 \cdot p_1 + x_2 \cdot p_2 + \cdots + x_n \cdot p_n,$$

where x_i = possible values of the random variable,
p_i = probability that x_i is the value.

fair game Has expected value of zero.

13.2 Decision Theory

decision making by maximizing expected values

1. Determine the possible states of nature.
2. List all relevant strategies.
3. For each strategy, determine the payoff for each of the possible states of nature. Organize the payoffs in a payoff table.
4. Determine the probability distribution for the possible states of nature. Use the distribution to compute the expected value for each strategy.
5. Choose the strategy with maximum expected value.

payoff The result of choosing a particular strategy when one of the states of nature holds (occurs).

payoff table Table that contains all possible payoffs.

states of nature Possible alternative values or conditions of "nature."
strategy One of the alternatives that the decision maker can choose.

13.3 Decision Trees

backwards analysis Method of analyzing decision trees; compute expected value at state-of-nature nodes and choose best decision at decision nodes while going through tree from right to left.
branch Line in a decision tree that connects nodes.
decision tree A diagram that represents all possible combinations of the states of nature and the strategies in the sequence in which they will occur.
node Point of intersection of two or more branches.
Decision node: Node at which a decision must be made.
State-of-nature node: Node at which a random process determines which state-of-nature will occur.

13.4 Game Theory

domination Situation in which the size of a game can be reduced because one strategy is always better than (dominates) one or more other strategies.
Column domination: All numbers in one column are equal to or less than all numbers in dominated column; the dominated column can be removed from further consideration.
Row domination: All numbers in one row are equal to or larger than all numbers in dominated row; the dominated row can be removed from further consideration.
game A particular situation involving two or more players (or adversaries).
Two-person game: Game with only two persons (adversaries).
game matrix Common method of representing games in which the payoffs are given in a matrix whose rows and columns correspond to the various strategies of the players.
game theory An approach to making decisions in situations involving thinking adversaries.
nonzero-sum game A game in which the total winnings of all players are not equal to the total losses of all players.
player One of the participants (adversaries, sides) in a game.
saddle point A value in a game matrix which is the smallest in its row and the largest in its column; provides optimal strategy for each player.
strategies The various alternatives that a player can choose.
Mixed strategies: No single strategy is optimal; player should determine the probability (fraction of time) for using each strategy.
Optimal strategy: The best strategy for one player.
Pure strategy: A single strategy which is optimal.
strictly determined game A game that has a saddle point (and therefore a pure strategy for each player).
value of game In a two-person game, value of player A's winnings.
Fair game: has value of zero.
zero-sum game Game in which the total of all players' winnings are equal to the total of all players' losses.

REVIEW PROBLEMS

Compute the expected value of the following distributions.

1.

x	$p(x)$
2	0.15
4	0.25
6	0.25
8	0.15
20	0.20

2.

x	$p(x)$
−90	0.05
15	0.20
30	0.25
100	0.40
200	0.10

3. You draw two cards (without replacement) from a standard 52-card playing deck. Determine

a) the expected number of hearts,
b) the expected number of "face" cards (jacks, queens or kings).

4. In Keno (game of chance), the house has a pot containing 80 balls (numbered from 1 to 80). You buy a chance for $1 and note the number of one of the balls. Twenty numbers (from 1 to 80) are then selected randomly. If your number is among the ones selected, you receive $3.20 back. If not, you lose your $1 bet. Compute the expected value of the game.

5. A manufacturing company is facing the possibility of a strike against one of its key suppliers. The company can either buy one month's or three month's supply. If there is no strike, buying three month's supply will entail an extra $500 in holding costs. If there is a strike, buying one month's supply will cost the company approximately $3000 in penalty costs. Assume the "correct" decisions (buying one month's supply and no strike or three month's supply with strike) cost the company nothing. What should the company do if the probability of a strike is

a) 10%? **b)** 50%?

6. Acme Manufacturing is considering purchasing an insurance policy for its new office building. The policy has an annual cost of $10,000. If minor fire damage occurs to the office building, a cost of $100,000 is anticipated, while a major fire would cost $200,000. The probability of no fire within the next year is estimated to be 0.96, the probability of a minor fire is estimated to be 0.03, and the probability of a major fire is estimated to be 0.01. Should Acme buy the insurance?

7. A student must take two final examinations on the first day of finals. Because of a previous commitment, her study time is limited to 12 hours. She must decide whether to study equal amounts of time for each exam or split her time 8 hours–4 hours. Depending upon the difficulty of the exams, she expects the following total changes in her class averages.

		States of Nature		
		Exam 1 hard	Exam 2 hard	Both hard
Strategies	8 hr on first 4 hr on second	+20	−5	+5
	4 hr on first 8 hr on second	−10	+25	+5
	6 hr on each	+15	+15	+10

What should she do if the probability of only exam 1 hard is 0.40, the probability of only exam 2 hard is 0.35 and the probability of both hard is 0.25?

8. A contract research firm has received two RFP's (requests for proposals). The company must decide whether to bid on only the first, on only the second, on both or on neither. The payoff matrix is shown below.

		States of Nature: Get Contract			
		One only	Two only	Both	Neither
Strategies	Bid one only	$50,000	—	—	−$5,000
	Bid two only	—	$60,000	—	−$7,500
	Bid both	$40,000	$50,000	$100,000	−$10,000
	Bid neither	$0	$0	$0	$0

Management's best estimates of the probabilities of winning the contracts are as follows.

Bid one only: $P(\text{contract}) = 0.3, P(\text{no contract}) = 0.7$
Bid two only: $P(\text{contract}) = 0.2, P(\text{no contract}) = 0.8$
Bid both: $P(\text{contract 1 only}) = 0.25, P(\text{contract 2 only}) = 0.2$
$P(\text{both contracts}) = 0.05, P(\text{neither}) = 0.50$

What is management's best strategy?

Solve the following games by finding each player's optimal strategy and the value of the game.

9.

Player B columns 1, 2; Player A rows 1, 2:

$$\begin{array}{cc} & \text{Player B} \\ & \begin{array}{cc} 1 & 2 \end{array} \\ \text{Player A} \begin{array}{c} 1 \\ 2 \end{array} & \begin{bmatrix} 3 & 7 \\ 5 & 2 \end{bmatrix} \end{array}$$

10.

$$\begin{array}{cc} & \text{Player B} \\ & \begin{array}{cc} 1 & 2 \end{array} \\ \text{Player A} \begin{array}{c} 1 \\ 2 \end{array} & \begin{bmatrix} 1/3 & 2/3 \\ 2/3 & -1 \end{bmatrix} \end{array}$$

11.

$$\begin{array}{cc} & \text{Player B} \\ & \begin{array}{ccc} 1 & 2 & 3 \end{array} \\ \text{Player A} \begin{array}{c} 1 \\ 2 \\ 3 \end{array} & \begin{bmatrix} 6 & -16 & -18 \\ 0 & 12 & -24 \\ -8 & 4 & -10 \end{bmatrix} \end{array}$$

12. Sam and Dave have been watching wrestling on TV and are thoroughly bored. Sam suggests a game in which they try to form words. Dave will choose either the letter a or the letter i and Sam will choose either f, t, or x. If the two letters form a word, Dave will win \$1 or \$4 (if the word is a noun or pronoun). If the two letters don't form a word, Sam will win \$2. Determine the optimal strategy for each player. Is the game fair?

APPENDIXES

APPENDIX A	Common logarithms
APPENDIX B	Natural logarithms
APPENDIX C	Binomial distribution
APPENDIX D	Poisson distribution
APPENDIX E	Normal distribution

APPENDIX A Common logarithms

N	0	1	2	3	4	5	6	7	8	9
0		0000	3010	4771	6021	6990	7782	8451	9031	9542
1	0000	0414	0792	1139	1461	1761	2041	2304	2553	2788
2	3010	3222	3424	3617	3802	3979	4150	4314	4472	4624
3	4771	4914	5051	5185	5315	5441	5563	5682	5798	5911
4	6021	6128	6232	6335	6435	6532	6628	6721	6812	6902
5	6990	7076	7160	7243	7324	7404	7482	7559	7634	7709
6	7782	7853	7924	7993	8062	8129	8195	8261	8325	8388
7	8451	8513	8573	8633	8692	8751	8808	8865	8921	8976
8	9031	9085	9138	9191	9243	9294	9345	9395	9445	9494
9	9542	9590	9638	9685	9731	9777	9823	9868	9912	9956
10	0000	0043	0086	0128	0170	0212	0253	0294	0334	0374
11	0414	0453	0492	0531	0569	0607	0645	0682	0719	0755
12	0792	0828	0864	0899	0934	0969	1004	1038	1072	1106
13	1139	1173	1206	1239	1271	1303	1335	1367	1399	1430
14	1461	1492	1523	1553	1584	1614	1644	1673	1703	1732
15	1761	1790	1818	1847	1875	1903	1931	1959	1987	2014
16	2041	2068	2095	2122	2148	2175	2201	2227	2253	2279
17	2304	2330	2355	2380	2405	2430	2455	2480	2504	2529
18	2553	2577	2601	2625	2648	2672	2695	2718	2742	2765
19	2788	2810	2833	2856	2878	2900	2923	2945	2967	2989
20	3010	3032	3054	3075	3096	3118	3139	3160	3181	3201
21	3222	3243	3263	3284	3304	3324	3345	3365	3385	3404
22	3424	3444	3464	3483	3502	3522	3541	3560	3579	3598
23	3617	3636	3655	3674	3692	3711	3729	3747	3766	3784
24	3802	3820	3838	3856	3874	3892	3909	3927	3945	3962
25	3979	3997	4014	4031	4048	4065	4082	4099	4116	4133
26	4150	4166	4183	4200	4216	4232	4249	4265	4281	4298
27	4314	4330	4346	4362	4378	4393	4409	4425	4440	4456
28	4472	4487	4502	4518	4533	4548	4564	4579	4594	4609
29	4624	4639	4654	4669	4683	4698	4713	4728	4742	4757
30	4771	4786	4800	4814	4829	4843	4857	4871	4886	4900
31	4914	4928	4942	4955	4969	4983	4997	5011	5024	5038
32	5051	5065	5079	5092	5105	5119	5132	5145	5159	5172
33	5185	5198	5211	5224	5237	5250	5263	5276	5289	5302
34	5315	5328	5340	5353	5366	5378	5391	5403	5416	5428
35	5441	5453	5465	5478	5490	5502	5514	5527	5539	5551
36	5563	5575	5587	5599	5611	5623	5635	5647	5658	5670
37	5682	5694	5705	5717	5729	5740	5752	5763	5775	5786
38	5798	5809	5821	5832	5843	5855	5866	5877	5888	5899
39	5911	5922	5933	5944	5955	5966	5977	5988	5999	6010
40	6021	6031	6042	6053	6064	6075	6085	6096	6107	6117
41	6128	6138	6149	6160	6170	6180	6191	6201	6212	6222
42	6232	6243	6253	6263	6274	6284	6294	6304	6314	6325
43	6335	6345	6355	6365	6375	6385	6395	6405	6415	6425
44	6435	6444	6454	6464	6474	6484	6493	6503	6513	6522
45	6532	6542	6551	6561	6571	6580	6590	6599	6609	6618
46	6628	6637	6646	6656	6665	6675	6684	6693	6702	6712
47	6721	6730	6739	6749	6758	6767	6776	6785	6794	6803
48	6812	6821	6830	6839	6848	6857	6866	6875	6884	6893
49	6902	6911	6920	6928	6937	6946	6955	6964	6972	6981
50	6990	6998	7007	7016	7024	7033	7042	7050	7059	7067
N	0	1	2	3	4	5	6	7	8	9

APPENDIX A Common logarithms (cont.)

N	0	1	2	3	4	5	6	7	8	9
50	6990	6998	7007	7016	7024	7033	7042	7050	7059	7067
51	7076	7084	7093	7101	7110	7118	7126	7135	7143	7152
52	7160	7168	7177	7185	7193	7202	7210	7218	7226	7235
53	7243	7251	7259	7267	7275	7284	7292	7300	7308	7316
54	7324	7332	7340	7348	7356	7364	7372	7380	7388	7396
55	7404	7412	7419	7427	7435	7443	7451	7459	7466	7474
56	7482	7490	7497	7505	7513	7520	7528	7536	7543	7551
57	7559	7566	7574	7582	7589	7597	7604	7612	7619	7627
58	7634	7642	7649	7657	7664	7672	7679	7686	7694	7701
59	7709	7716	7723	7731	7738	7745	7752	7760	7767	7774
60	7782	7789	7796	7803	7810	7818	7825	7832	7839	7846
61	7853	7860	7868	7875	7882	7889	7896	7903	7910	7917
62	7924	7931	7938	7945	7952	7959	7966	7973	7980	7987
63	7993	8000	8007	8014	8021	8028	8035	8041	8048	8055
64	8062	8069	8075	8082	8089	8096	8102	8109	8116	8122
65	8129	8136	8142	8149	8156	8162	8169	8176	8182	8189
66	8195	8202	8209	8215	8222	8228	8235	8241	8248	8254
67	8261	8267	8274	8280	8287	8293	8299	8306	8312	8319
68	8325	8331	8338	8344	8351	8357	8363	8370	8376	8382
69	8388	8395	8401	8407	8414	8420	8426	8432	8439	8445
70	8451	8457	8463	8470	8476	8482	8488	8494	8500	8506
71	8513	8519	8525	8531	8537	8543	8549	8555	8561	8567
72	8573	8579	8585	8591	8597	8603	8609	8615	8621	8627
73	8633	8639	8645	8651	8657	8663	8669	8675	8681	8686
74	8692	8698	8704	8710	8716	8722	8727	8733	8739	8745
75	8751	8756	8762	8768	8774	8779	8785	8791	8797	8802
76	8808	8814	8820	8825	8831	8837	8842	8848	8854	8859
77	8865	8871	8876	8882	8887	8893	8899	8904	8910	8915
78	8921	8927	8932	8938	8943	8949	8954	8960	8965	8971
79	8976	8982	8987	8993	8998	9004	9009	9015	9020	9025
80	9031	9036	9042	9047	9053	9058	9063	9069	9074	9079
81	9085	9090	9096	9101	9106	9112	9117	9122	9128	9133
82	9138	9143	9149	9154	9159	9165	9170	9175	9180	9186
83	9191	9196	9201	9206	9212	9217	9222	9227	9232	9238
84	9243	9248	9253	9258	9263	9269	9274	9279	9284	9289
85	9294	9299	9304	9309	9315	9320	9325	9330	9335	9340
86	9345	9350	9355	9360	9365	9370	9375	9380	9385	9390
87	9395	9400	9405	9410	9415	9420	9425	9430	9435	9440
88	9445	9450	9455	9460	9465	9469	9474	9479	9484	9489
89	9494	9499	9504	9509	9513	9518	9523	9528	9533	9538
90	9542	9547	9552	9557	9562	9566	9571	9576	9581	9586
91	9590	9585	9600	9605	9609	9614	9619	9624	9628	9633
92	9638	9643	9647	9652	9657	9661	9666	9671	9675	9680
93	9685	9689	9694	9699	9703	9708	9713	9717	9722	9727
94	9731	9736	9741	9745	9750	9754	9759	9763	9768	9773
95	9777	9782	9786	9791	9795	9800	9805	9809	9814	9818
96	9823	9827	9832	9836	9841	9845	9850	9854	9859	9863
97	9868	9872	9877	9881	9886	9890	9894	9899	9903	9908
98	9912	9917	9921	9926	9930	9934	9939	9943	9948	9952
99	9956	9961	9965	9969	9974	9978	9983	9987	9991	9996
100	0000	0004	0009	0013	0017	0022	0026	0030	0035	0039
N	0	1	2	3	4	5	6	7	8	9

APPENDIX B Natural logarithms

n	$\log_e n$	n	$\log_e n$	n	$\log_e n$
0.0		4.5	1.5041	9.0	2.1972
0.1	7.6974	4.6	1.5261	9.1	2.2083
0.2	8.3906	4.7	1.5476	9.2	2.2192
0.3	8.7960	4.8	1.5686	9.3	2.2300
0.4	9.0837	4.9	1.5892	9.4	2.2407
0.5	9.3069	5.0	1.6094	9.5	2.2513
0.6	9.4892	5.1	1.6292	9.6	2.2618
0.7	9.6433	5.2	1.6487	9.7	2.2721
0.8	9.7769	5.3	1.6677	9.8	2.2824
0.9	9.8946	5.4	1.6864	9.9	2.2925
1.0	0.0000	5.5	1.7047	10	2.3026
1.1	0.0953	5.6	1.7228	11	2.3979
1.2	0.1823	5.7	1.7405	12	2.4849
1.3	0.2624	5.8	1.7579	13	2.5649
1.4	0.3365	5.9	1.7750	14	2.6391
1.5	0.4055	6.0	1.7918	15	2.7081
1.6	0.4700	6.1	1.8083	16	2.7726
1.7	0.5306	6.2	1.8245	17	2.8332
1.8	0.5878	6.3	1.8405	18	2.8904
1.9	0.6419	6.4	1.8563	19	2.9444
2.0	0.6931	6.5	1.8718	20	2.9957
2.1	0.7419	6.6	1.8871	25	3.2189
2.2	0.7885	6.7	1.9021	30	3.4012
2.3	0.8329	6.8	1.9169	35	3.5553
2.4	0.8755	6.9	1.9315	40	3.6889
2.5	0.9163	7.0	1.9459	45	3.8067
2.6	0.9555	7.1	1.9601	50	3.9120
2.7	0.9933	7.2	1.9741	55	4.0073
2.8	1.0296	7.3	1.9879	60	4.0943
2.9	1.0647	7.4	2.0015	65	4.1744
3.0	1.0986	7.5	2.0149	70	4.2485
3.1	1.1314	7.6	2.0281	75	4.3175
3.2	1.1632	7.7	2.0412	80	4.3820
3.3	1.1939	7.8	2.0541	85	4.4427
3.4	1.2238	7.9	2.0669	90	4.4998
3.5	1.2528	8.0	2.0794	95	4.5539
3.6	1.2809	8.1	2.0919	100	4.6052
3.7	1.3083	8.2	2.1041		
3.8	1.3350	8.3	2.1163		
3.9	1.3610	8.4	2.1282		
4.0	1.3863	8.5	2.1401		
4.1	1.4110	8.6	2.1518		
4.2	1.4351	8.7	2.1633		
4.3	1.4586	8.8	2.1748		
4.4	1.4816	8.9	2.1861		

APPENDIX C Binomial distribution

		p										
n	*X*	.1	.2	.25	.3	.4	.5	.6	.7	.75	.8	.9
1	0	.9000	.8000	.7500	.7000	.6000	.5000	.4000	.3000	.2500	.2000	.1000
	1	.1000	.2000	.2500	.3000	.4000	.5000	.6000	.7000	.7500	.8000	.9000
2	0	.8100	.6400	.5625	.4900	.3600	.2500	.1600	.0900	.0625	.0400	.0100
	1	.1800	.3200	.3750	.4200	.4800	.5000	.4800	.4200	.3750	.3200	.1800
	2	.0100	.0400	.0625	.0900	.1600	.2500	.3600	.4900	.5625	.6400	.8100
3	0	.7290	.5120	.4219	.3430	.2160	.1250	.0640	.0270	.0156	.0080	.0010
	1	.2430	.3840	.4219	.4410	.4320	.3750	.2880	.1890	.1406	.0960	.0270
	2	.0270	.0960	.1406	.1890	.2880	.3750	.4320	.4410	.4219	.3840	.2430
	3	.0010	.0080	.0156	.0270	.0640	.1250	.2160	.3430	.4219	.5120	.7290
4	0	.6561	.4096	.3164	.2401	.1296	.0625	.0256	.0081	.0039	.0016	.0001
	1	.2916	.4096	.4219	.4116	.3456	.2500	.1536	.0756	.0469	.0256	.0036
	2	.0486	.1536	.2109	.2646	.3456	.3750	.3456	.2646	.2109	.1536	.0486
	3	.0036	.0256	.0469	.0756	.1536	.2500	.3456	.4116	.4219	.4096	.2916
	4	.0001	.0016	.0039	.0081	.0256	.0625	.1296	.2401	.3164	.4096	.6561
5	0	.5905	.3277	.2373	.1681	.0778	.0313	.0102	.0024	.0010	.0003	.0000
	1	.3281	.4096	.3955	.3602	.2592	.1563	.0768	.0284	.0146	.0064	.0005
	2	.0729	.2048	.2637	.3037	.3456	.3125	.2304	.1323	.0879	.0512	.0081
	3	.0081	.0512	.0879	.1323	.2304	.3125	.3456	.3087	.2637	.2048	.0729
	4	.0004	.0064	.0146	.0283	.0768	.1563	.2592	.3602	.3955	.4096	.3281
	5	.0000	.0003	.0010	.0024	.0102	.0313	.0778	.1681	.2373	.3277	.5905
6	0	.5314	.2621	.1780	.1176	.0467	.0156	.0041	.0007	.0002	.0001	.0000
	1	.3543	.3932	.3560	.3025	.1866	.0938	.0369	.0102	.0044	.0015	.0001
	2	.0984	.2458	.2966	.3241	.3110	.2344	.1382	.0595	.0330	.0154	.0012
	3	.0146	.0819	.1318	.1852	.2765	.3125	.2765	.1852	.1318	.0819	.0146
	4	.0012	.0154	.0330	.0595	.1382	.2344	.3110	.3241	.2966	.2458	.0984
	5	.0001	.0015	.0044	.0102	.0369	.0938	.1866	.3025	.3560	.3932	.3543
	6	.0000	.0001	.0002	.0007	.0041	.0156	.0467	.1176	.1780	.2621	.5314
7	0	.4783	.2097	.1335	.0824	.0280	.0078	.0016	.0002	.0001	.0000	.0000
	1	.3720	.3670	.3115	.2471	.1306	.0547	.0172	.0036	.0013	.0004	.0000
	2	.1240	.2753	.3115	.3177	.2613	.1641	.0774	.0250	.0115	.0043	.0002
	3	.0230	.1147	.1730	.2269	.2903	.2734	.1935	.0972	.0577	.0287	.0026
	4	.0026	.0287	.0577	.0972	.1935	.2734	.2903	.2269	.1730	.1147	.0230
	5	.0002	.0043	.0115	.0250	.0774	.1641	.2613	.3177	.3115	.2753	.1240
	6	.0000	.0004	.0013	.0036	.0172	.0547	.1306	.2471	.3115	.3670	.3720
	7	.0000	.0000	.0001	.0002	.0016	.0078	.0280	.0824	.1335	.2097	.4783

(Cont.)

Adapted from Donald B. Owen, *Handbook of Statistical Tables*, Copyright © 1962, U.S. Department of Energy; published by Addison-Wesley, Reading, Mass. Adapted from Neil Weiss and Matthew Hassett, *Introductory Statistics*. Reading, Mass.: Addison-Wesley, 1982. Reprinted with permission.

APPENDIX C Binomial distribution (cont.)

		p										
n	X	.1	.2	.25	.3	.4	.5	.6	.7	.75	.8	.9
8	0	.4305	.1678	.1001	.0576	.0168	.0039	.0007	.0001	.0000	.0000	.0000
	1	.3826	.3355	.2670	.1977	.0896	.0312	.0079	.0012	.0004	.0001	.0000
	2	.1488	.2936	.3115	.2965	.2090	.1094	.0413	.0100	.0038	.0011	.0000
	3	.0331	.1468	.2076	.2541	.2787	.2188	.1239	.0467	.0231	.0092	.0004
	4	.0046	.0459	.0865	.1361	.2322	.2734	.2322	.1361	.0865	.0459	.0046
	5	.0004	.0092	.0231	.0467	.1239	.2188	.2787	.2541	.2076	.1468	.0331
	6	.0000	.0011	.0038	.0100	.0413	.1094	.2090	.2965	.3115	.2936	.1488
	7	.0000	.0001	.0004	.0012	.0079	.0312	.0896	.1977	.2670	.3355	.3826
	8	.0000	.0000	.0000	.0001	.0007	.0039	.0168	.0576	.1001	.1678	.4305
9	0	.3874	.1342	.0751	.0404	.0101	.0020	.0003	.0000	.0000	.0000	.0000
	1	.3874	.3020	.2253	.1556	.0605	.0176	.0035	.0004	.0001	.0000	.0000
	2	.1722	.3020	.3003	.2668	.1612	.0703	.0212	.0039	.0012	.0003	.0000
	3	.0446	.1762	.2336	.2668	.2508	.1641	.0743	.0210	.0087	.0028	.0001
	4	.0074	.0661	.1168	.1715	.2508	.2461	.1672	.0735	.0389	.0165	.0008
	5	.0008	.0165	.0389	.0735	.1672	.2461	.2508	.1715	.1168	.0661	.0074
	6	.0001	.0028	.0087	.0210	.0743	.1641	.2508	.2668	.2336	.1762	.0446
	7	.0000	.0003	.0012	.0039	.0212	.0703	.1612	.2668	.3003	.3020	.1722
	8	.0000	.0000	.0001	.0004	.0035	.0176	.0605	.1556	.2253	.3020	.3874
	9	.0000	.0000	.0000	.0000	.0003	.0020	.0101	.0404	.0751	.1342	.3874
10	0	.3487	.1074	.0563	.0282	.0060	.0010	.0001	.0000	.0000	.0000	.0000
	1	.3874	.2684	.1877	.1211	.0403	.0098	.0016	.0001	.0000	.0000	.0000
	2	.1937	.3020	.2816	.2335	.1209	.0439	.0106	.0014	.0004	.0001	.0000
	3	.0574	.2013	.2503	.2668	.2150	.1172	.0425	.0090	.0031	.0008	.0000
	4	.0112	.0881	.1460	.2001	.2508	.2051	.1115	.0368	.0162	.0055	.0001
	5	.0015	.0264	.0584	.1029	.2007	.2461	.2007	.1029	.0584	.0264	.0015
	6	.0001	.0055	.0162	.0368	.1115	.2051	.2508	.2001	.1460	.0881	.0112
	7	.0000	.0008	.0031	.0090	.0425	.1172	.2150	.2668	.2503	.2013	.0574
	8	.0000	.0001	.0004	.0014	.0106	.0439	.1209	.2335	.2816	.3020	.1937
	9	.0000	.0000	.0000	.0001	.0016	.0098	.0403	.1211	.1877	.2684	.3874
	10	.0000	.0000	.0000	.0000	.0001	.0010	.0060	.0282	.0563	.1074	.3487
11	0	.3138	.0859	.0422	.0198	.0036	.0005	.0000	.0000	.0000	.0000	.0000
	1	.3835	.2362	.1549	.0932	.0266	.0054	.0007	.0000	.0000	.0000	.0000
	2	.2131	.2953	.2581	.1998	.0887	.0269	.0052	.0005	.0001	.0000	.0000
	3	.0710	.2215	.2581	.2568	.1774	.0806	.0234	.0037	.0011	.0002	.0000
	4	.0158	.1107	.1721	.2201	.2365	.1611	.0701	.0173	.0064	.0017	.0000
	5	.0025	.0388	.0803	.1321	.2207	.2256	.1471	.0566	.0268	.0097	.0003
	6	.0003	.0097	.0268	.0566	.1471	.2256	.2207	.1321	.0803	.0388	.0025
	7	.0000	.0017	.0064	.0173	.0701	.1611	.2365	.2201	.1721	.1107	.0158
	8	.0000	.0002	.0011	.0037	.0234	.0806	.1774	.2568	.2581	.2215	.0710
	9	.0000	.0000	.0001	.0005	.0052	.0269	.0887	.1998	.2581	.2953	.2131
	10	.0000	.0000	.0000	.0000	.0007	.0054	.0266	.0932	.1549	.2362	.3835
	11	.0000	.0000	.0000	.0000	.0000	.0005	.0036	.0198	.0422	.0859	.3138

APPENDIX C Binomial distribution (cont.)

		p										
n	*X*	.1	.2	.25	.3	.4	.5	.6	.7	.75	.8	.9
12	0	.2824	.0687	.0317	.0138	.0022	.0002	.0000	.0000	.0000	.0000	.0000
	1	.3766	.2062	.1267	.0712	.0174	.0029	.0003	.0000	.0000	.0000	.0000
	2	.2301	.2835	.2323	.1678	.0639	.0161	.0025	.0002	.0000	.0000	.0000
	3	.0852	.2362	.2581	.2397	.1419	.0537	.0125	.0015	.0004	.0001	.0000
	4	.0213	.1329	.1936	.2311	.2128	.1208	.0420	.0078	.0024	.0005	.0000
	5	.0038	.0532	.1032	.1585	.2270	.1934	.1009	.0291	.0115	.0033	.0000
	6	.0005	.0155	.0401	.0792	.1766	.2256	.1766	.0792	.0401	.0155	.0005
	7	.0000	.0033	.0115	.0291	.1009	.1934	.2270	.1585	.1032	.0532	.0038
	8	.0000	.0005	.0024	.0078	.0420	.1208	.2128	.2311	.1936	.1329	.0213
	9	.0000	.0001	.0004	.0015	.0125	.0537	.1419	.2397	.2581	.2362	.0852
	10	.0000	.0000	.0000	.0002	.0025	.0161	.0639	.1678	.2323	.2835	.2301
	11	.0000	.0000	.0000	.0000	.0003	.0029	.0174	.0712	.1267	.2062	.3766
	12	.0000	.0000	.0000	.0000	.0000	.0002	.0022	.0138	.0317	.0687	.2824
13	0	.2542	.0550	.0238	.0097	.0013	.0001	.0000	.0000	.0000	.0000	.0000
	1	.3672	.1787	.1029	.0540	.0113	.0016	.0001	.0000	.0000	.0000	.0000
	2	.2448	.2680	.2059	.1388	.0453	.0095	.0012	.0001	.0000	.0000	.0000
	3	.0997	.2457	.2517	.2181	.1107	.0349	.0065	.0006	.0001	.0000	.0000
	4	.0277	.1535	.2097	.2337	.1845	.0873	.0243	.0034	.0009	.0001	.0000
	5	.0055	.0691	.1258	.1803	.2214	.1571	.0656	.0142	.0047	.0011	.0000
	6	.0008	.0230	.0559	.1030	.1968	.2095	.1312	.0442	.0186	.0058	.0001
	7	.0001	.0058	.0186	.0442	.1312	.2095	.1968	.1030	.0559	.0230	.0008
	8	.0000	.0011	.0047	.0142	.0656	.1571	.2214	.1803	.1258	.0691	.0055
	9	.0000	.0001	.0009	.0034	.0243	.0873	.1845	.2337	.2097	.1535	.0277
	10	.0000	.0000	.0001	.0006	.0065	.0349	.1107	.2181	.2517	.2457	.0997
	11	.0000	.0000	.0000	.0001	.0012	.0095	.0453	.1388	.2059	.2680	.2448
	12	.0000	.0000	.0000	.0000	.0001	.0016	.0113	.0540	.1029	.1787	.3672
	13	.0000	.0000	.0000	.0000	.0000	.0001	.0013	.0097	.0238	.0550	.2542
14	0	.2288	.0440	.0178	.0068	.0008	.0001	.0000	.0000	.0000	.0000	.0000
	1	.3559	.1539	.0832	.0407	.0073	.0009	.0001	.0000	.0000	.0000	.0000
	2	.2570	.2501	.1802	.1134	.0317	.0056	.0005	.0000	.0000	.0000	.0000
	3	.1142	.2501	.2402	.1943	.0845	.0222	.0033	.0002	.0000	.0000	.0000
	4	.0349	.1720	.2202	.2290	.1549	.0611	.0136	.0014	.0003	.0000	.0000
	5	.0078	.0860	.1468	.1963	.2066	.1222	.0408	.0066	.0018	.0003	.0000
	6	.0013	.0322	.0734	.1262	.2066	.1833	.0918	.0232	.0082	.0020	.0000
	7	.0002	.0092	.0280	.0618	.1574	.2095	.1574	.0618	.0280	.0092	.0002
	8	.0000	.0020	.0082	.0232	.0918	.1833	.2066	.1262	.0734	.0322	.0013
	9	.0000	.0003	.0018	.0066	.0408	.1222	.2066	.1963	.1468	.0860	.0078
	10	.0000	.0000	.0003	.0014	.0136	.0611	.1549	.2290	.2202	.1720	.0349
	11	.0000	.0000	.0000	.0002	.0033	.0222	.0845	.1943	.2402	.2501	.1142
	12	.0000	.0000	.0000	.0000	.0005	.0056	.0317	.1134	.1802	.2501	.2570
	13	.0000	.0000	.0000	.0000	.0001	.0009	.0073	.0407	.0832	.1539	.3559
	14	.0000	.0000	.0000	.0000	.0000	.0001	.0008	.0068	.0178	.0440	.2288

(Cont.)

APPENDIX C Binomial distribution (cont.)

		p										
n	*X*	.1	.2	.25	.3	.4	.5	.6	.7	.75	.8	.9
15	0	.2059	.0352	.0134	.0047	.0005	.0000	.0000	.0000	.0000	.0000	.0000
	1	.3432	.1319	.0668	.0305	.0047	.0005	.0000	.0000	.0000	.0000	.0000
	2	.2669	.2309	.1559	.0916	.0219	.0032	.0003	.0000	.0000	.0000	.0000
	3	.1285	.2501	.2252	.1700	.0634	.0139	.0016	.0001	.0000	.0000	.0000
	4	.0428	.1876	.2252	.2186	.1268	.0417	.0074	.0006	.0001	.0000	.0000
	5	.0105	.1032	.1651	.2061	.1859	.0916	.0245	.0030	.0007	.0001	.0000
	6	.0019	.0430	.0917	.1472	.2066	.1527	.0612	.0116	.0034	.0007	.0000
	7	.0003	.0138	.0393	.0811	.1771	.1964	.1181	.0348	.0131	.0035	.0000
	8	.0000	.0035	.0131	.0348	.1181	.1964	.1771	.0811	.0393	.0138	.0003
	9	.0000	.0007	.0034	.0116	.0612	.1527	.2066	.1472	.0917	.0430	.0019
	10	.0000	.0001	.0007	.0030	.0245	.0916	.1859	.2061	.1651	.1032	.0105
	11	.0000	.0000	.0001	.0006	.0074	.0417	.1268	.2186	.2252	.1876	.0428
	12	.0000	.0000	.0000	.0001	.0016	.0139	.0634	.1700	.2252	.2501	.1285
	13	.0000	.0000	.0000	.0000	.0003	.0032	.0219	.0916	.1559	.2309	.2669
	14	.0000	.0000	.0000	.0000	.0000	.0005	.0047	.0305	.0668	.1319	.3432
	15	.0000	.0000	.0000	.0000	.0000	.0000	.0005	.0047	.0134	.0352	.2059
20	0	.1216	.0115	.0032	.0008	.0000	.0000	.0000	.0000	.0000	.0000	.0000
	1	.2702	.0576	.0211	.0068	.0005	.0000	.0000	.0000	.0000	.0000	.0000
	2	.2852	.1369	.0669	.0278	.0031	.0002	.0000	.0000	.0000	.0000	.0000
	3	.1901	.2054	.1339	.0716	.0123	.0011	.0000	.0000	.0000	.0000	.0000
	4	.0898	.2182	.1897	.1304	.0350	.0046	.0003	.0000	.0000	.0000	.0000
	5	.0319	.1746	.2023	.1789	.0746	.0148	.0013	.0000	.0000	.0000	.0000
	6	.0089	.1091	.1686	.1916	.1244	.0370	.0049	.0002	.0000	.0000	.0000
	7	.0020	.0545	.1124	.1643	.1659	.0739	.0146	.0010	.0002	.0000	.0000
	8	.0004	.0222	.0609	.1144	.1797	.1201	.0355	.0039	.0008	.0001	.0000
	9	.0001	.0074	.0271	.0654	.1597	.1602	.0710	.0120	.0030	.0005	.0000
	10	.0000	.0020	.0099	.0308	.1171	.1762	.1171	.0308	.0099	.0020	.0000
	11	.0000	.0005	.0030	.0120	.0710	.1602	.1597	.0654	.0271	.0074	.0001
	12	.0000	.0001	.0008	.0039	.0355	.1201	.1797	.1144	.0609	.0222	.0004
	13	.0000	.0000	.0002	.0010	.0146	.0739	.1659	.1643	.1124	.0545	.0020
	14	.0000	.0000	.0000	.0002	.0049	.0370	.1244	.1916	.1686	.1091	.0089
	15	.0000	.0000	.0000	.0000	.0013	.0148	.0746	.1789	.2023	.1746	.0319
	16	.0000	.0000	.0000	.0000	.0003	.0046	.0350	.1304	.1897	.2182	.0898
	17	.0000	.0000	.0000	.0000	.0000	.0011	.0123	.0716	.1339	.2054	.1901
	18	.0000	.0000	.0000	.0000	.0000	.0002	.0031	.0278	.0669	.1369	.2852
	19	.0000	.0000	.0000	.0000	.0000	.0000	.0005	.0068	.0211	.0576	.2702
	20	.0000	.0000	.0000	.0000	.0000	.0000	.0000	.0008	.0032	.0115	.1216

APPENDIX C Binomial distribution (cont.)

		p										
n	*X*	.1	.2	.25	.3	.4	.5	.6	.7	.75	.8	.9
25	0	.0718	.0038	.0008	.0001	.0000	.0000	.0000	.0000	.0000	.0000	.0000
	1	.1994	.0236	.0063	.0014	.0000	.0000	.0000	.0000	.0000	.0000	.0000
	2	.2659	.0708	.0251	.0074	.0004	.0000	.0000	.0000	.0000	.0000	.0000
	3	.2265	.1358	.0641	.0243	.0019	.0001	.0000	.0000	.0000	.0000	.0000
	4	.1384	.1867	.1175	.0572	.0071	.0004	.0000	.0000	.0000	.0000	.0000
	5	.0646	.1960	.1645	.1030	.0199	.0016	.0000	.0000	.0000	.0000	.0000
	6	.0239	.1633	.1828	.1472	.0442	.0053	.0002	.0000	.0000	.0000	.0000
	7	.0072	.1108	.1654	.1712	.0800	.0143	.0009	.0000	.0000	.0000	.0000
	8	.0018	.0623	.1241	.1651	.1200	.0322	.0031	.0001	.0000	.0000	.0000
	9	.0004	.0294	.0781	.1336	.1511	.0609	.0088	.0004	.0000	.0000	.0000
	10	.0001	.0118	.0417	.0916	.1612	.0974	.0212	.0013	.0002	.0000	.0000
	11	.0000	.0040	.0189	.0536	.1465	.1328	.0434	.0042	.0007	.0001	.0000
	12	.0000	.0012	.0074	.0268	.1140	.1550	.0760	.0115	.0025	.0003	.0000
	13	.0000	.0003	.0025	.0115	.0760	.1550	.1140	.0268	.0074	.0012	.0000
	14	.0000	.0001	.0007	.0042	.0434	.1328	.1465	.0536	.0189	.0040	.0000
	15	.0000	.0000	.0002	.0013	.0212	.0974	.1612	.0916	.0417	.0118	.0001
	16	.0000	.0000	.0000	.0004	.0088	.0609	.1511	.1336	.0781	.0294	.0004
	17	.0000	.0000	.0000	.0001	.0031	.0322	.1200	.1651	.1241	.0623	.0018
	18	.0000	.0000	.0000	.0000	.0009	.0143	.0800	.1712	.1654	.1108	.0072
	19	.0000	.0000	.0000	.0000	.0002	.0053	.0442	.1472	.1828	.1633	.0239
	20	.0000	.0000	.0000	.0000	.0000	.0016	.0199	.1030	.1645	.1960	.0646
	21	.0000	.0000	.0000	.0000	.0000	.0004	.0071	.0572	.1175	.1867	.1384
	22	.0000	.0000	.0000	.0000	.0000	.0001	.0019	.0243	.0641	.1358	.2265
	23	.0000	.0000	.0000	.0000	.0000	.0000	.0004	.0074	.0251	.0708	.2659
	24	.0000	.0000	.0000	.0000	.0000	.0000	.0000	.0014	.0063	.0236	.1994
	25	.0000	.0000	.0000	.0000	.0000	.0000	.0000	.0001	.0008	.0038	.0718

APPENDIX D Poisson distribution

$$\Pr(X = k) = \frac{e^{-\lambda}\lambda^k}{k!}$$

k	λ = .1	.2	.3	.4	.5	.6	.7	.8	.9	1.0
0	.9048	.8187	.7408	.6703	.6065	.5488	.4966	.4493	.4066	.3679
1	.0905	.1637	.2222	.2681	.3033	.3293	.3476	.3595	.3659	.3679
2	.0045	.0164	.0333	.0536	.0758	.0988	.1217	.1438	.1647	.1839
3	.0002	.0011	.0033	.0072	.0126	.0198	.0284	.0383	.0494	.0613
4	.0000	.0001	.0003	.0007	.0016	.0030	.0050	.0077	.0111	.0153
5	.0000	.0000	.0000	.0001	.0002	.0004	.0007	.0012	.0020	.0031
6	.0000	.0000	.0000	.0000	.0000	.0000	.0001	.0002	.0003	.0005
7	.0000	.0000	.0000	.0000	.0000	.0000	.0000	.0000	.0000	.0001
8	.0000	.0000	.0000	.0000	.0000	.0000	.0000	.0000	.0000	.0000

k	λ = 1.5	2	3	4	5	6	7	8	9	10
0	.2231	.1353	.0498	.0183	.0067	.0025	.0009	.0003	.0001	.0000
1	.3347	.2707	.1494	.0733	.0337	.0149	.0064	.0027	.0011	.0005
2	.2510	.2707	.2240	.1465	.0842	.0446	.0223	.0107	.0050	.0023
3	.1255	.1804	.2240	.1954	.1404	.0892	.0521	.0286	.0150	.0076
4	.0471	.0902	.1680	.1954	.1755	.1339	.0912	.0573	.0337	.0189
5	.0141	.0361	.1008	.1563	.1755	.1606	.1277	.0916	.0607	.0378
6	.0035	.0120	.0504	.1042	.1462	.1606	.1490	.1221	.0911	.0631
7	.0008	.0034	.0216	.0595	.1044	.1377	.1490	.1396	.1171	.0901
8	.0001	.0009	.0081	.0298	.0653	.1033	.1304	.1396	.1318	.1126
9	.0000	.0002	.0027	.0132	.0363	.0688	.1014	.1241	.1318	.1251
10	.0000	.0000	.0008	.0053	.0181	.0413	.0710	.0993	.1186	.1251
11	.0000	.0000	.0002	.0019	.0082	.0225	.0452	.0722	.0970	.1137
12	.0000	.0000	.0001	.0006	.0034	.0113	.0264	.0481	.0728	.0948
13	.0000	.0000	.0000	.0002	.0013	.0052	.0142	.0296	.0504	.0729
14	.0000	.0000	.0000	.0001	.0005	.0022	.0071	.0169	.0324	.0521
15	.0000	.0000	.0000	.0000	.0002	.0009	.0033	.0090	.0194	.0347
16	.0000	.0000	.0000	.0000	.0000	.0003	.0014	.0045	.0109	.0217
17	.0000	.0000	.0000	.0000	.0000	.0001	.0006	.0021	.0058	.0128
18	.0000	.0000	.0000	.0000	.0000	.0000	.0002	.0009	.0029	.0071
19	.0000	.0000	.0000	.0000	.0000	.0000	.0001	.0004	.0014	.0037
20	.0000	.0000	.0000	.0000	.0000	.0000	.0000	.0002	.0006	.0019
21	.0000	.0000	.0000	.0000	.0000	.0000	.0000	.0001	.0003	.0009
22	.0000	.0000	.0000	.0000	.0000	.0000	.0000	.0000	.0001	.0004
23	.0000	.0000	.0000	.0000	.0000	.0000	.0000	.0000	.0000	.0002
24	.0000	.0000	.0000	.0000	.0000	.0000	.0000	.0000	.0000	.0001
25	.0000	.0000	.0000	.0000	.0000	.0000	.0000	.0000	.0000	.0000

APPENDIX E Normal distribution. Areas under the standard normal curve (areas to the left).

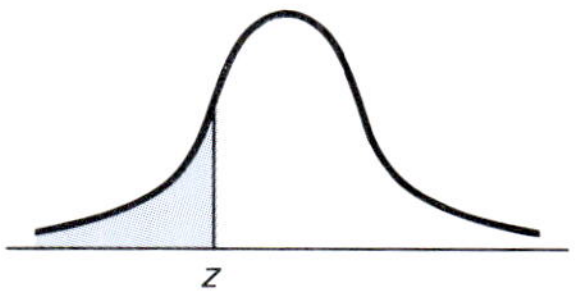

z	0	1	2	3	4	5	6	7	8	9
−3.0*	.0013	.0013	.0013	.0012	.0012	.0011	.0011	.0011	.0010	.0010
−2.9	.0019	.0018	.0017	.0017	.0016	.0016	.0015	.0015	.0014	.0014
−2.8	.0026	.0025	.0024	.0023	.0023	.0022	.0021	.0020	.0020	.0019
−2.7	.0035	.0034	.0033	.0032	.0031	.0030	.0029	.0028	.0027	.0026
−2.6	.0047	.0045	.0044	.0043	.0041	.0040	.0039	.0038	.0037	.0036
−2.5	.0062	.0060	.0059	.0057	.0055	.0054	.0052	.0051	.0049	.0048
−2.4	.0082	.0080	.0078	.0075	.0073	.0071	.0069	.0068	.0066	.0064
−2.3	.0107	.0104	.0102	.0099	.0096	.0094	.0091	.0089	.0087	.0084
−2.2	.0139	.0136	.0132	.0129	.0125	.0122	.0119	.0116	.0113	.0110
−2.1	.0179	.0174	.0170	.0166	.0162	.0158	.0154	.0150	.0146	.0143
−2.0	.0228	.0222	.0217	.0212	.0207	.0202	.0197	.0192	.0188	.0183
−1.9	.0287	.0281	.0274	.0268	.0262	.0256	.0250	.0244	.0239	.0233
−1.8	.0359	.0351	.0344	.0336	.0329	.0322	.0314	.0307	.0301	.0294
−1.7	.0446	.0436	.0427	.0418	.0409	.0401	.0392	.0384	.0375	.0367
−1.6	.0548	.0537	.0526	.0516	.0505	.0495	.0485	.0475	.0465	.0455
−1.5	.0668	.0655	.0643	.0630	.0618	.0606	.0594	.0582	.0571	.0559
−1.4	.0808	.0793	.0778	.0764	.0749	.0735	.0721	.0708	.0694	.0681
−1.3	.0968	.0951	.0934	.0918	.0901	.0885	.0869	.0853	.0838	.0823
−1.2	.1151	.1131	.1112	.1093	.1075	.1056	.1038	.1020	.1003	.0985
−1.1	.1357	.1335	.1314	.1292	.1271	.1251	.1230	.1210	.1190	.1170
−1.0	.1587	.1562	.1539	.1515	.1492	.1469	.1446	.1423	.1401	.1379
− .9	.1841	.1814	.1788	.1762	.1736	.1711	.1685	.1660	.1635	.1611
− .8	.2119	.2090	.2061	.2033	.2005	.1977	.1949	.1922	.1894	.1867
− .7	.2420	.2389	.2358	.2327	.2296	.2266	.2236	.2206	.2177	.2148
− .6	.2743	.2709	.2676	.2643	.2611	.2578	.2546	.2514	.2483	.2451
− .5	.3085	.3050	.3015	.2981	.2946	.2912	.2877	.2843	.2810	.2776
− .4	.3446	.3409	.3372	.3336	.3300	.3264	.3228	.3192	.3516	.3121
− .3	.3821	.3783	.3745	.3707	.3669	.3632	.3594	.3557	.3520	.3483
− .2	.4207	.4168	.4129	.4090	.4052	.4013	.3974	.3936	.3897	.3859
− .1	.4602	.4562	.4522	.4483	.4443	.4404	.4364	.4325	.4286	.4247
− .0	.5000	.4960	.4920	.4880	.4840	.4801	.4761	.4721	.4681	.4641

(Cont.)

Adapted from *Probability with Statistical Applications*, second edition, by F. Mosteller, R. E. K. Rourke, and G. B. Thomas, Jr. Reading, Mass.: Addison-Wesley, 1970, p. 473.

* For $z \leq -4$ the areas are 0 to four decimal places.

APPENDIX E Normal distribution (cont.)

z	0	1	2	3	4	5	6	7	8	9
.0	.5000	.5040	.5080	.5120	.5160	.5199	.5239	.5279	.5319	.5359
.1	.5398	.5438	.5478	.5517	.5557	.5596	.5636	.5675	.5714	.5753
.2	.5793	.5832	.5871	.5910	.5948	.5987	.6026	.6064	.6103	.6141
.3	.6179	.6217	.6255	.6293	.6331	.6368	.6406	.6443	.6480	.6517
.4	.6554	.6591	.6628	.6664	.6700	.6736	.6772	.6808	.6844	.6879
.5	.6915	.6950	.6985	.7019	.7054	.7088	.7123	.7157	.7190	.7224
.6	.7257	.7291	.7324	.7357	.7389	.7422	.7454	.7486	.7517	.7549
.7	.7580	.7611	.7642	.7673	.7704	.7734	.7764	.7794	.7823	.7852
.8	.7881	.7910	.7939	.7967	.7995	.8023	.8051	.8078	.8106	.8133
.9	.8159	.8186	.8212	.8238	.8264	.8289	.8315	.8340	.8365	.8389
1.0	.8413	.8438	.8461	.8485	.8508	.8531	.8554	.8577	.8599	.8621
1.1	.8643	.8665	.8686	.8708	.8729	.8749	.8770	.8790	.8810	.8830
1.2	.8849	.8869	.8888	.8907	.8925	.8944	.8962	.8980	.8997	.9015
1.3	.9032	.9049	.9066	.9082	.9099	.9115	.9131	.9147	.9162	.9177
1.4	.9192	.9207	.9222	.9236	.9251	.9265	.9279	.9292	.9306	.9319
1.5	.9332	.9345	.9357	.9370	.9382	.9394	.9406	.9418	.9429	.9441
1.6	.9452	.9463	.9474	.9484	.9495	.9505	.9515	.9525	.9535	.9545
1.7	.9554	.9564	.9573	.9582	.9591	.9599	.9608	.9616	.9625	.9633
1.8	.9641	.9649	.9656	.9664	.9671	.9678	.9686	.9693	.9699	.9706
1.9	.9713	.9719	.9726	.9732	.9738	.9744	.9750	.9756	.9761	.9767
2.0	.9772	.9778	.9783	.9788	.9793	.9798	.9803	.9808	.9812	.9817
2.1	.9821	.9826	.9830	.9834	.9838	.9842	.9846	.9850	.9854	.9857
2.2	.9861	.9864	.9868	.9871	.9875	.9878	.9881	.9884	.9887	.9890
2.3	.9893	.9896	.9898	.9901	.9904	.9906	.9909	.9911	.9913	.9916
2.4	.9918	.9920	.9922	.9925	.9927	.9929	.9931	.9932	.9934	.9936
2.5	.9938	.9940	.9941	.9943	.9945	.9946	.9948	.9949	.9951	.9952
2.6	.9953	.9955	.9956	.9957	.9959	.9960	.9961	.9962	.9963	.9964
2.7	.9965	.9966	.9967	.9968	.9969	.9970	.9971	.9972	.9973	.9974
2.8	.9974	.9975	.9976	.9977	.9977	.9978	.9979	.9979	.9980	.9981
2.9	.9981	.9982	.9982	.9983	.9984	.9984	.9985	.9985	.9986	.9986
3.0†	.9987	.9987	.9987	.9988	.9988	.9989	.9989	.9989	.9990	.9990

† For $z \geq 4$ the areas are 1 to four decimal places.

ANSWERS TO ODD-NUMBERED PROBLEMS

CHAPTER 1

Exercise Set 1.2, page 12

1. 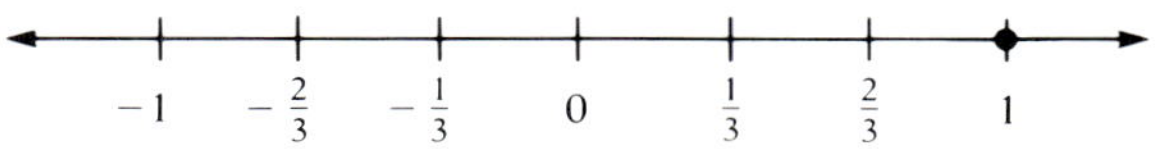

3. 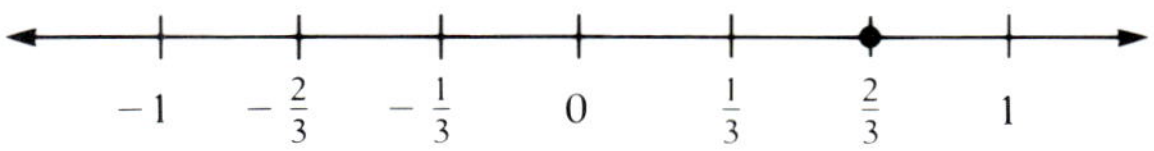

5. -8 **7.** 1 **9.** -14 **11.** 5 **13.** -3 **15.** -6 **17.** -24 **19.** 0 **21.** $\frac{9}{5}$
23. $\frac{8}{45}$ **25.** $\frac{12}{5}$ **27.** $\frac{27}{8}$ **29.** $\frac{20}{27}$ **31.** $\frac{12}{7}$

Exercise Set 1.3, page 19

1. 4.34 **3.** 6.904 **5.** 3.08 **7.** -6361.566 **9.** 2510.548 **11.** 1.976 **13.** -73.332
15. 103.000 **17.** -125.000 **19.** 61.88

Exercise Set 1.4, page 30

1. False **3.** True **5.** True **7.** False **9.** True **11.** True **13.** False **15.** False
17. 21 **19.** 5 **21.** 20 **23.** 53 **25.** 55 **27.** 60 **29.** 51 **31.** 25 **33.** 1
35. 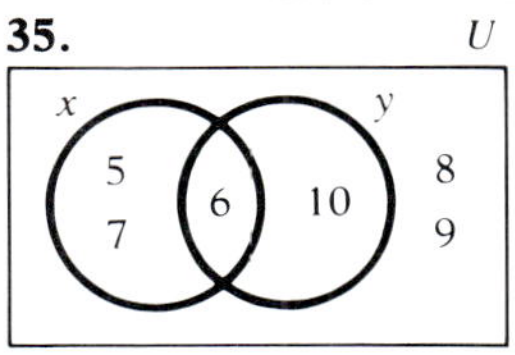

37. $\{5, 6, 7, 10\}$ $x \cup y$ **39.** $\{5, 7\}$ $x \cap y'$

Chapter 1 Review Problems, page 34

1. 18.330 **3.** −1.327 **5.** 0.133 **7.** −0.071 **9.** True **11.** False **13.** True **15.** True **17.** False **19.** False **21.** True **23.** False **25.** False

27.

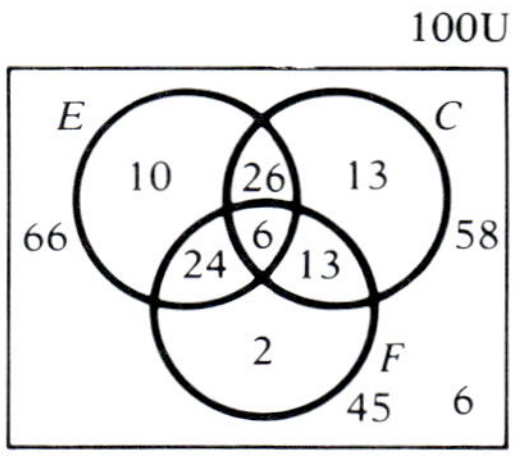

27. a) 10

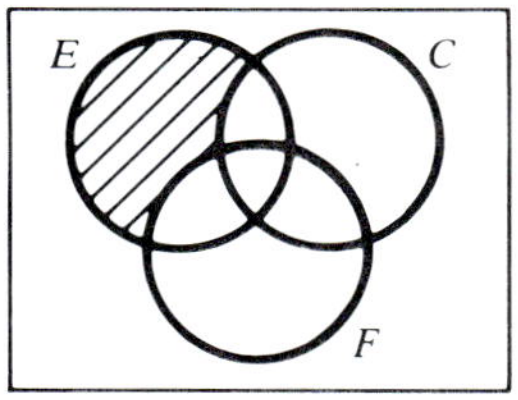

b) 13

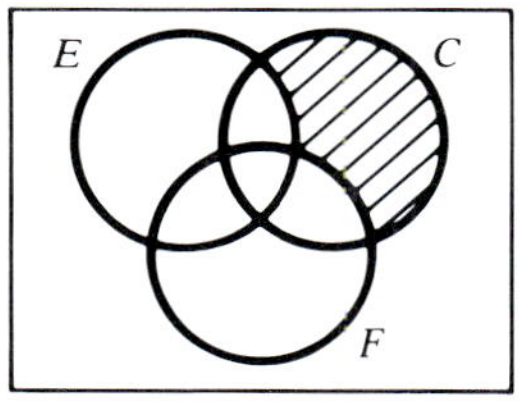

c) 2

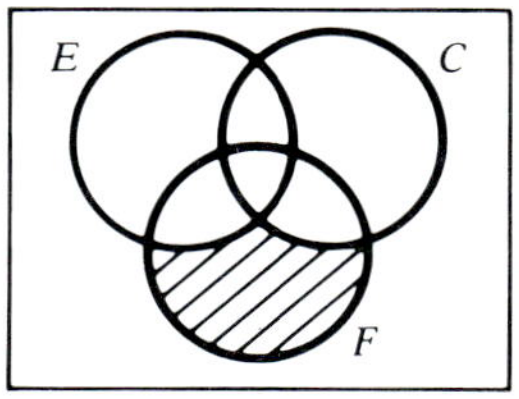

d) 26

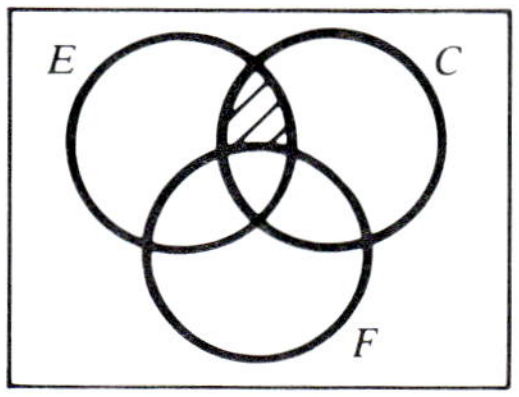

e) 13

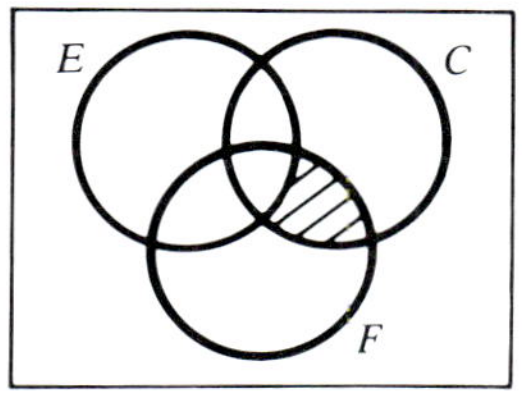

f) 42

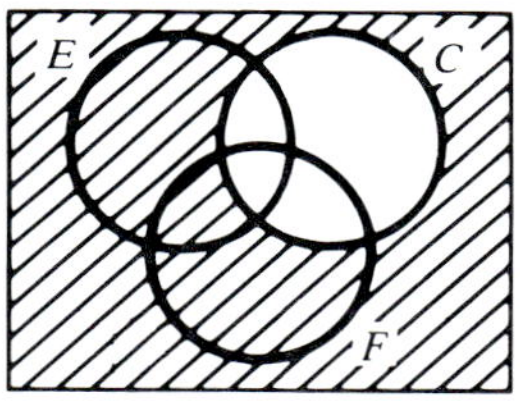

g) 6

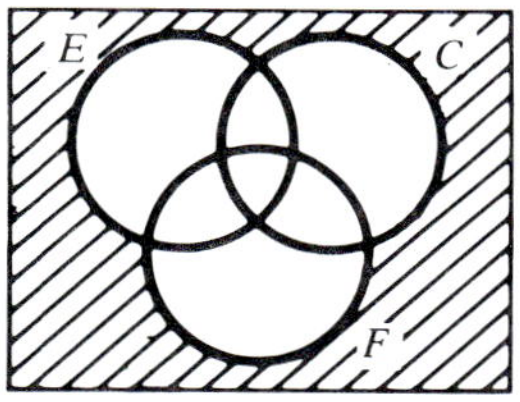

CHAPTER 2

Exercise Set 2.1, page 47

1. 3.71293 **3.** 778.688 **5.** 2.137812 **7.** 1.2286597 **9.** 1.5524633 **11.** 0.6034647 **13.** −0.0012842 **15.** −3.6059572 **17.** x **19.** $1/x^4$ **21.** $9x^6$ **23.** $257.5x^8$ **25.** $11x^{1/3}$ **27.** $4x^{5/12}$ **29.** x^{18} **31.** x^4 **33.** $0.522758x^{1/12}$ **35.** $25/x^{4/3}$ **37.** x^2y^4 **39.** $y^{3/8}/x^{1/2}$ **41.** $531441a^4b^{12/5}$ **43.** $1/u^5v^4$ **45.** x^3/y^6 **47.** y^8/x^6 **49.** $1/v^2$ **51.** $1.2247449x^{1/3}y^{1/48}$ **53.** y^{10-m} **55.** $0.5x^{K-3/2}$

Exercise Set 2.2, page 61

1. $2a + 2b$ **3.** $2a^2x^2 + 2b^2x^2$ **5.** $x^2 + 2xy + y^2$ **7.** $x^3 + x^2y + xy + y^2$ **9.** $x^2 + 2xy + y^2$ **11.** $c^2 + cd - 12d^2$ **13.** $28m^4 + 23m^2n^2 - 15n^4$ **15.** $x^2y + x^2y^2 + xy^2 - y^4$ **17.** $(4x - 3y)(2x + y)$ **19.** $(2 + 9x^4)(1 + 2x^2)(1 - 2x^2)$ **21.** $(4x + 3y)(4x - 3y)$ **23.** $(4x + 5y)(4x + 5y)$ **25.** $xy(3x - 2y)(3x + 2y)$ **27.** $(m + 9)(m - 5)$ **29.** $(3cu - 2z^3)(2cu + 3z^3)$ **31.** $3(c - 5)(c + 8)$ **33.** $u(3c + 1)(c + 7)$ **35.** $x^3(7 + 6z)(7 - 6z)$ **37.** $(2x - y)(2x + y) - z(2y + z)$ **39.** $(x^3 - y^3)(x^3 + y^3)$ **41.** $2(3u - 7v)(u + 2v)$ **43.** $4x^2(5x^2 + 4y^2)(5x^2 - 4y^2)$

Exercise Set 2.3, page 69

1. False **3.** False **5.** True **7.** ab^2 **9.** $\frac{3}{4}$ **11.** $\frac{(a+3)(a-5)}{2}$ **13.** $\frac{x+y}{x-2y}$

15. $\frac{5a+y}{4a+3y}$ **17.** $\frac{x+1}{x+5}$ **19.** $\frac{7x-1}{10}$ **21.** $\frac{22x-97}{30}$ **23.** $\frac{bx}{(x+b)(x-b)}$ **25.** $\frac{2y^2}{(y-3)(y+2)}$

27. $\frac{x^2-2x-5}{(x-2)(x+1)(x-3)}$ **29.** $\frac{4a-2}{(a-4)(a-3)}$ **31.** $\frac{(x+y-a)(x+y+a)}{xy^3a^3}$ **33.** $1/x$

Chapter 2 Review Problems, page 72

1. 74.088 **3.** -1.7099759 **5.** $1/b^3$ **7.** x^{12} **9.** $x^{2/3m+7/3}$ **11.** $3x^3+3x^2+3x+7$

13. x^3-x+6 **15.** $x^6-2x^5-x^4+9x^3-10x^2+x+2$ **17.** $9ab(c-3ab)$ **19.** $(x+9)(x-9)$

21. $(x+3)(x+3)$ **23.** $(2x+y)(x-5y)$ **25.** $3x(2x-1)(x+3)$ **27.** $20.25y^2$ **29.** $\frac{2(x+2)}{x-2}$

31. $\frac{2x^2+3x^2y-4x+1}{3y^2}$ **33.** $\frac{3m^2+8m+12}{3(m-3)(m+2)}$ **35.** $\frac{(x+4y)(8x)(x+1)(x+1)+4(x+1)(4y-x)}{(4y-x)(x+4y)(x+4y)}$

CHAPTER 3

Exercise Set 3.1, page 80

1. $c(x) = 1.50x + 2.00$, $c(6) = \$11.00$ **3.** $200.00 **5.** Rich = 37 years, Janet = 5 years

Exercise Set 3.2, page 91

1. $x = 3$

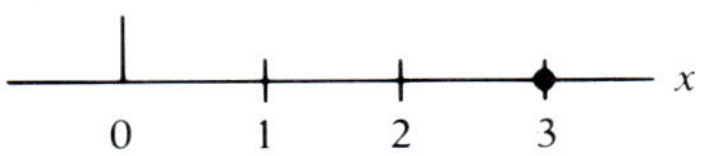

3. $x = 1\frac{3}{4}$

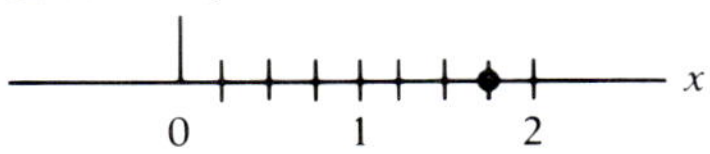

5. $y \le 1$

7. $x = 0$

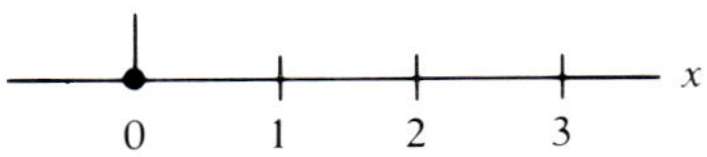

9. $T = -4.5$

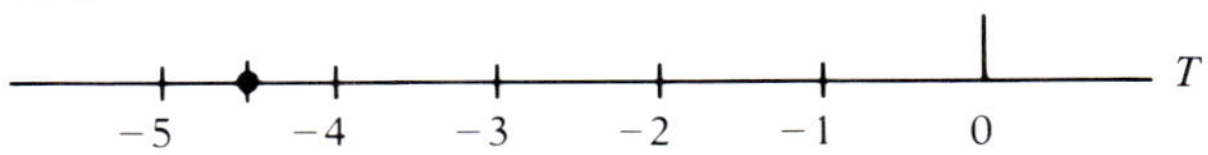

11. $n = 1$

13. $w \ge \frac{3}{4}$

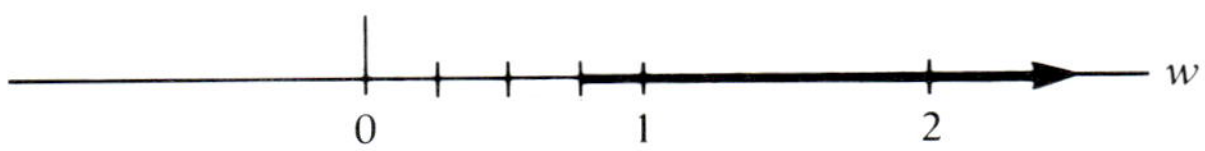

15. $-\frac{1}{7} < x < \frac{16}{7}$

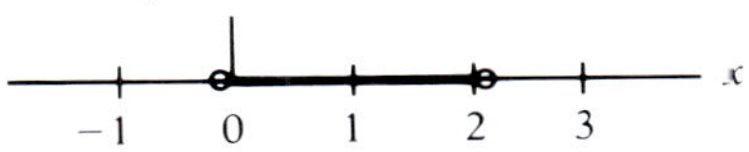

17. $\frac{2}{15} > y \ge -\frac{1}{3}$

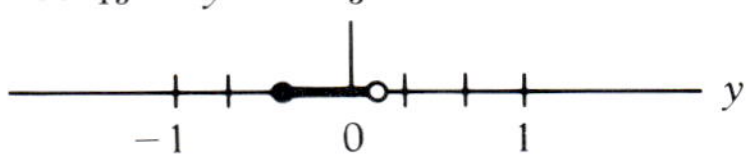

19. $y = \frac{15-3x}{2}$

21. $F = \frac{9}{5}C + 32$ **23.** $c = \frac{a}{2} + \frac{d}{2} - \frac{b}{4}$ **25.** $-11.992 \le F \le 107.006$ **27.** $x = 2308$
29. 707.5 additional miles

Exercise Set 3.3, page 104

1. $x = \pm 2$ **3.** $x = \pm 1.3229$ **5.** $x = 0.828, x = -4.828$ **7.** $x = -4, x = 3$ **9.** $x = 1.863, x = -2.863$ **11.** $x = \frac{1}{5}, x = -1$ **13.** $x = \pm\frac{7}{3}$ **15.** $x = 1.257, x = -1.591$

17. $3 < x < 4$

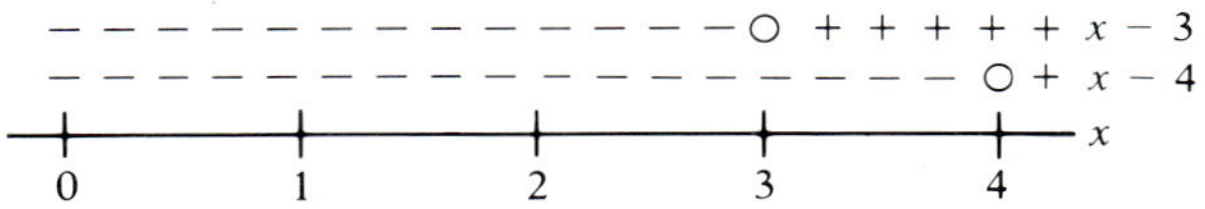

19. $x < 2$ or $a > 3$

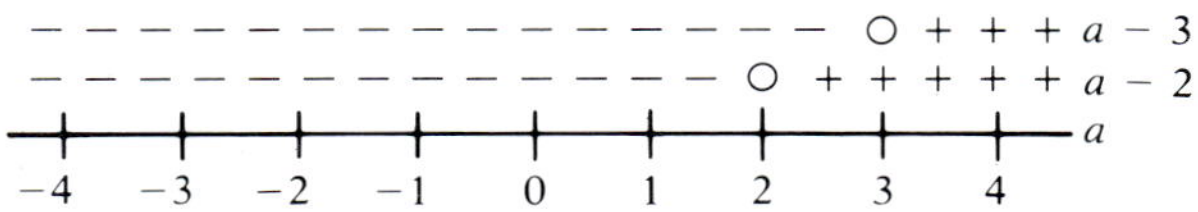

21. $-3 < c < -\frac{1}{2}$ True for all real numbers.

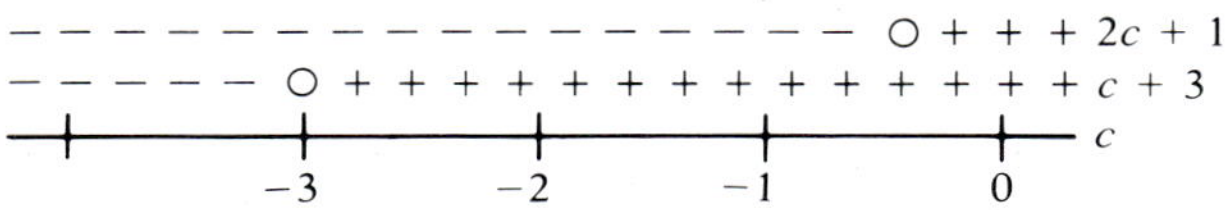

23. No solution.

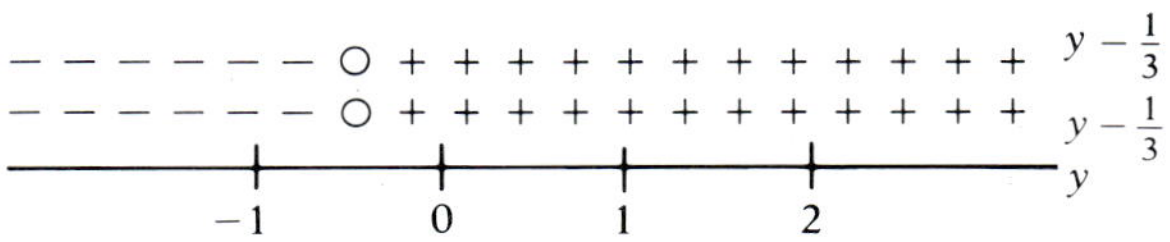

25. Price ≈ \$38.11

Exercise Set 3.4, page 120

1. Slope -2, y intercept $(0, 3)$, down to the right

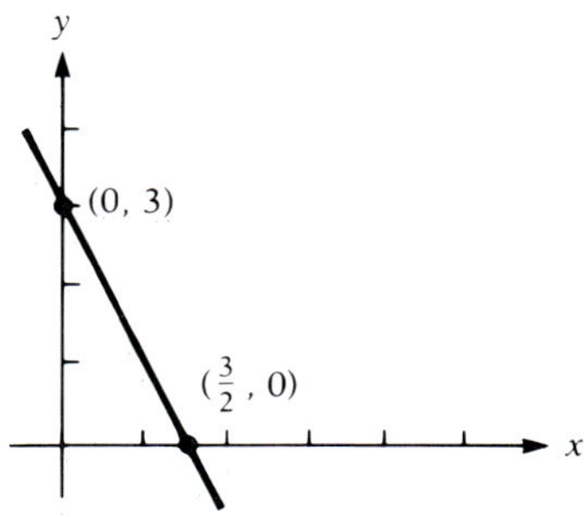

3. Slope 4, y intercept $(0, 1)$, up to the right

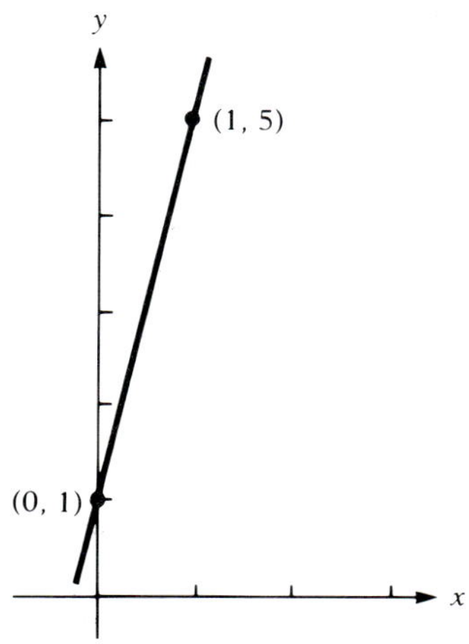

5. Slope 7, y intercept (0, 3), up to the right

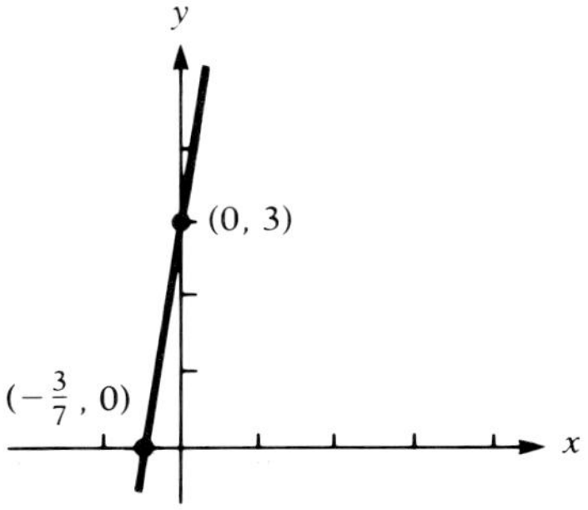

7. Slope 1, y intercept (0, −15), up to the right

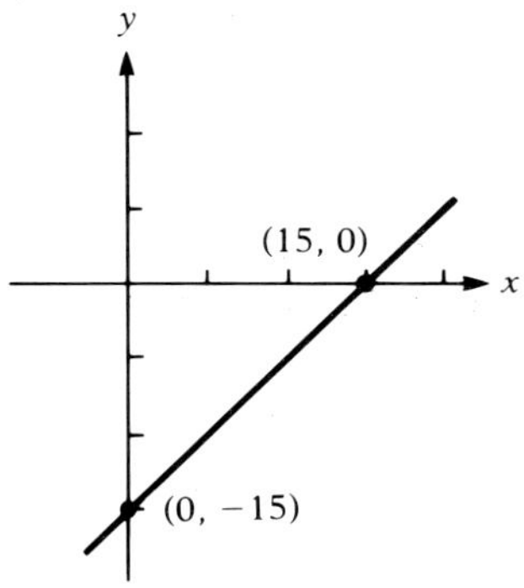

9. Slope $\frac{1}{2}$, y intercept $(0, \frac{7}{4})$, up to the right

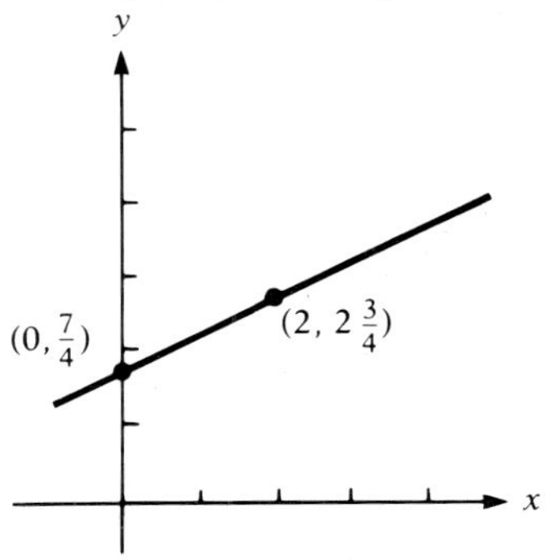

11. Undefined, no y intercept, vertical line

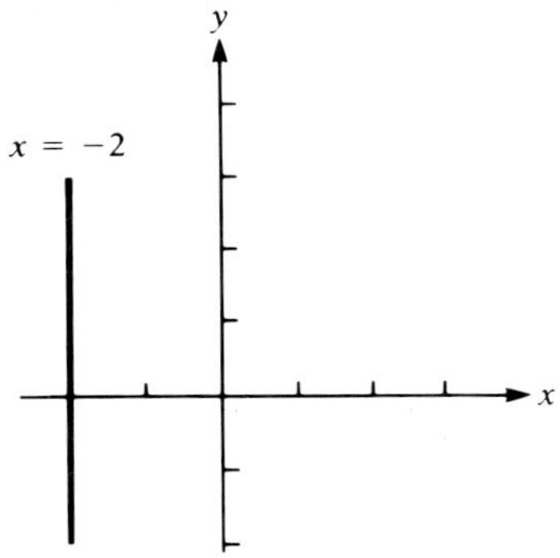

13. Slope 0, y intercept (0, 2), horizontal line

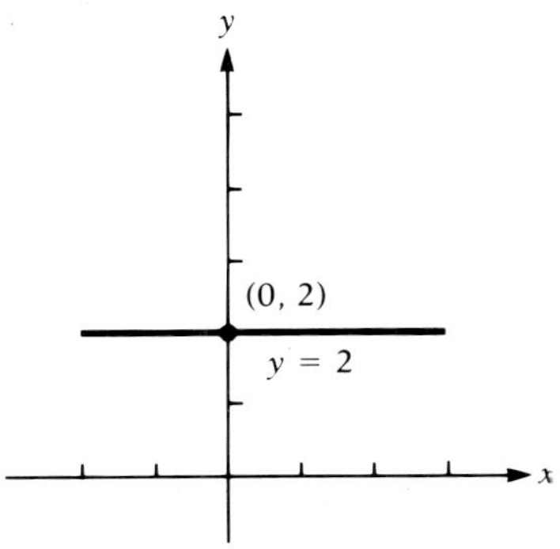

15. Slope $-\frac{3}{4}$, y intercept $(0, \frac{25}{4})$, down to the right

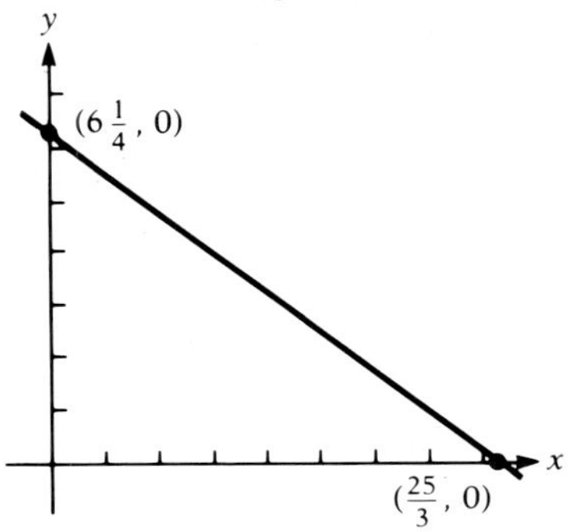

17. Slope −4, y intercept (0, 8), down to the right

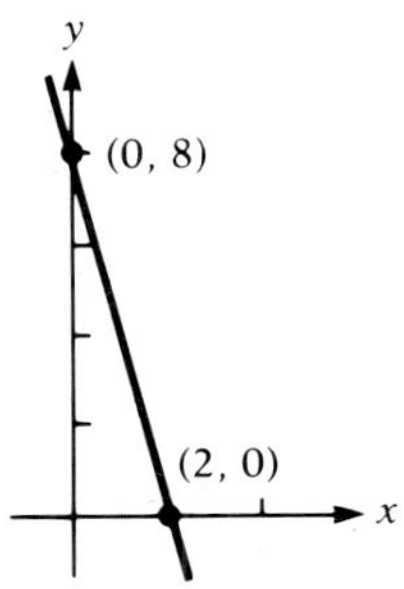

19. $y = 2x - 1$, $2x - y = 1$ **21.** $y = -x + 7$, $x + y = 7$ **23.** $y = -\frac{1}{5}x - \frac{3}{5}$, $x + 5y = -3$
25. $y = x + 3$, $x - y = -3$ **27.** $y = 3$ **29.** $y = x + 1$, $x - y = -1$
31. $y = -\frac{1}{3}x + \frac{7}{3}$, $x + 3y = 7$ **33.** $y = -\frac{1}{7}x - \frac{26}{7}$, $x + 7y = -26$ **35.** $y = \frac{5}{7}x$, $5x - 7y = 0$
37. $y = 1$ **39.** $y = -3$ **41.** $x = 4$, vertical line

43.

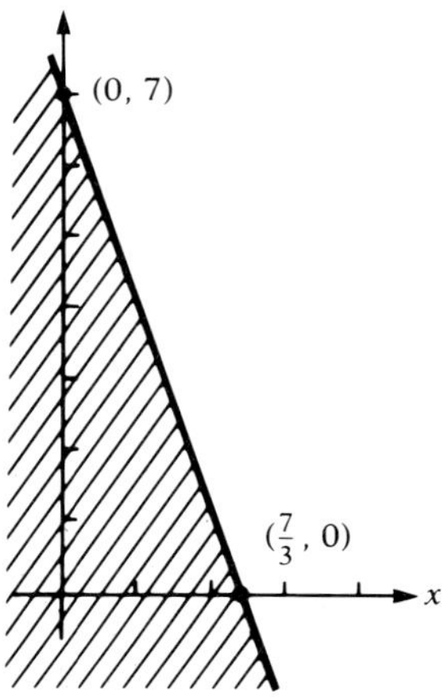

45.

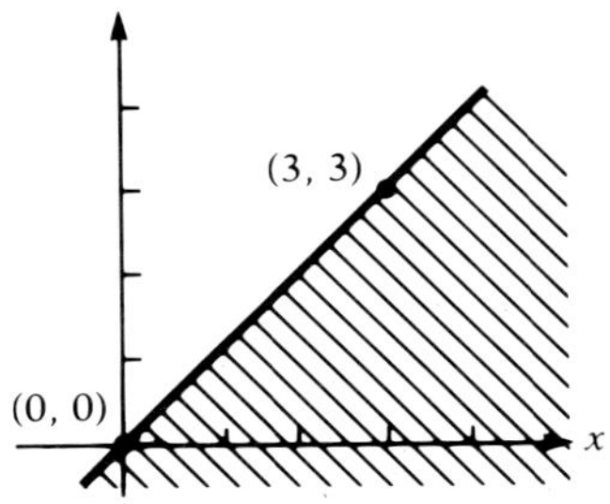

47.

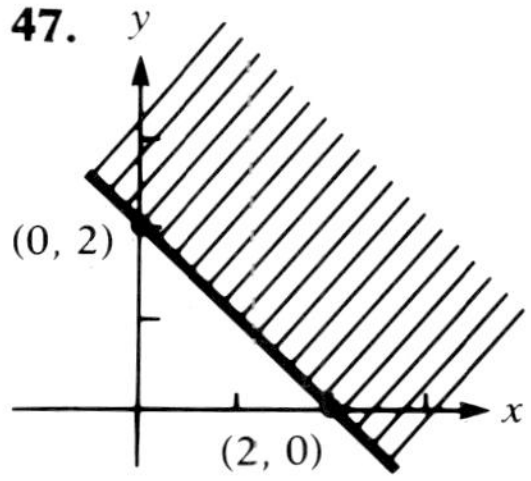

49.

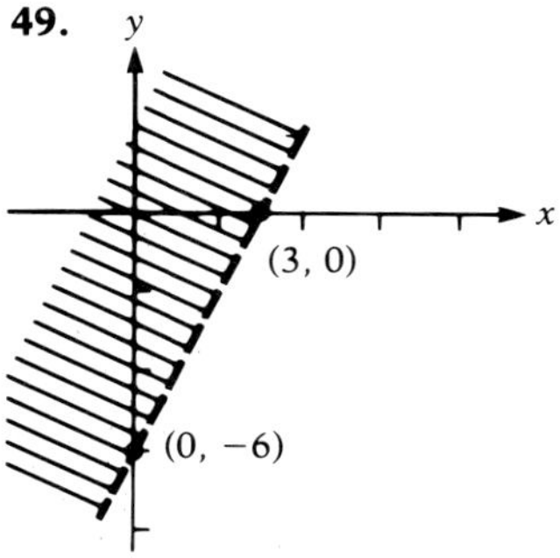

51. a) $y = -1100x + 5500$
b) $220
c) Value

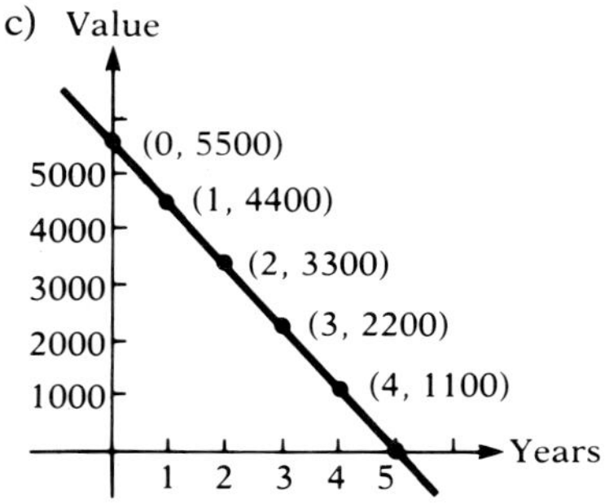

53. a) $y = -0.00125x + 3.25$
b) 480, 880
c) $3.25

Chapter 3 Review Problems, page 125

1. $-\frac{5}{2}$ **3.** $-\frac{6}{73}$ **5.** $y > \frac{1}{6}$ **7.** 65/13.9 **9.** $x = -2, x = 7$ **11.** $x = 2.618, x = 0.382$
13. $x = 1.744, x = -0.344$ **15.** $-0.37 < x < 0.27$
17. Slope 4, y intercept (0, 2), up to the right

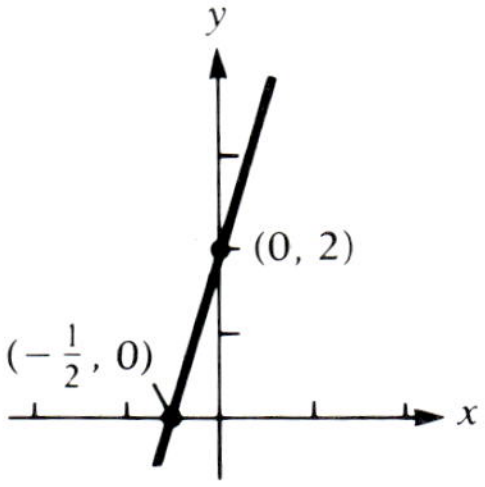

19. Slope −2, y intercept (0, 5), down to the right

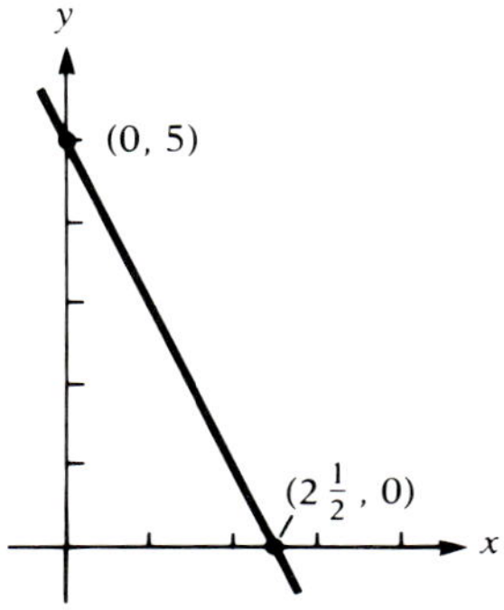

21. $y = 3x - 2, 3x - y = 2$ **23.** $y = 0x + 5, y = 5$ **25.** $y = -x - 5, x + y = -5$

27.

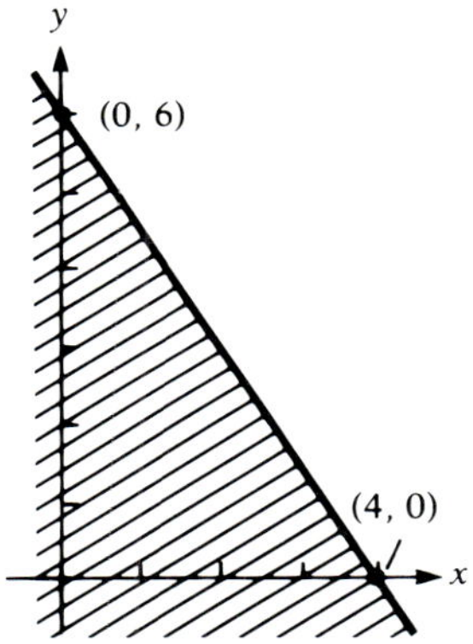

29.

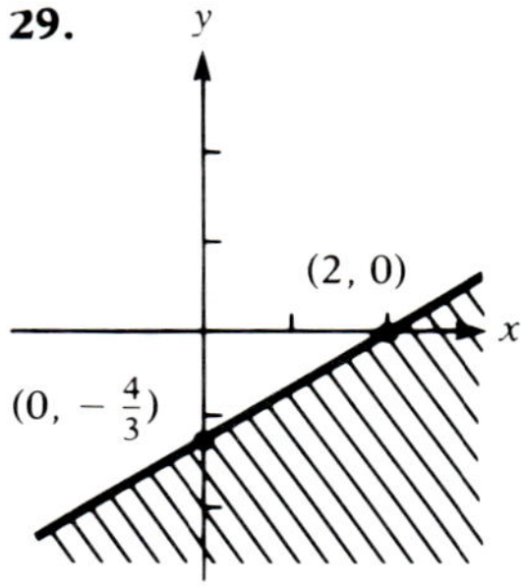

CHAPTER 4

Exercise Set 4.1, page 137

1. f, x, y **3.** q, y, z **5.** q, h, L **7.** 16 **9.** 36 **11.** 3 **13.** 0 **15.** 48 **17.** $a^2 - 1$
19. $a + 7$ **21.** 8 **23.** 3.125 **25.** 112 **27.** 2 **29.** 20 **31.** $4(x + 2)$ **33.** 1
35. $2(a + 2)$ **37.** $x^2 + 10x + 24$ **39.** $x + 10$ **41.** $f(x = x^3, g(x) = x^2 + 2x + 1$
43. $f(x) = x^3 + 3x, g(x) = 2x^4 - x^2$

Exercise Set 4.2, page 160

1. $x = 7, x = -1$ **3.** $x = -14, x = 10$ **5.** $x = -\frac{1}{2}, x = -\frac{7}{4}$

7. $y = |2x|$

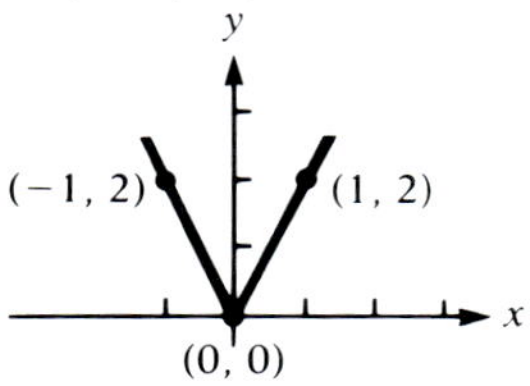

9. $y = |-2 - 3x|$

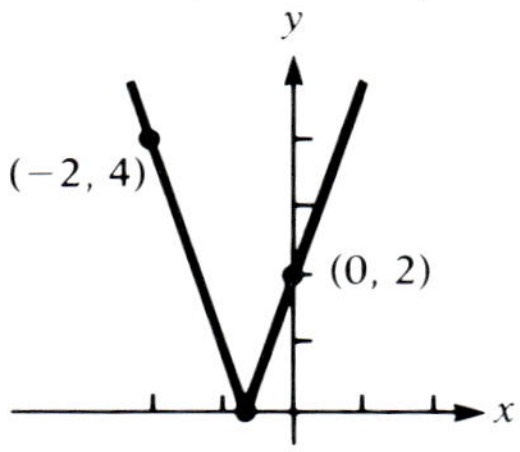

11. $y = 3x - |4 - x|$

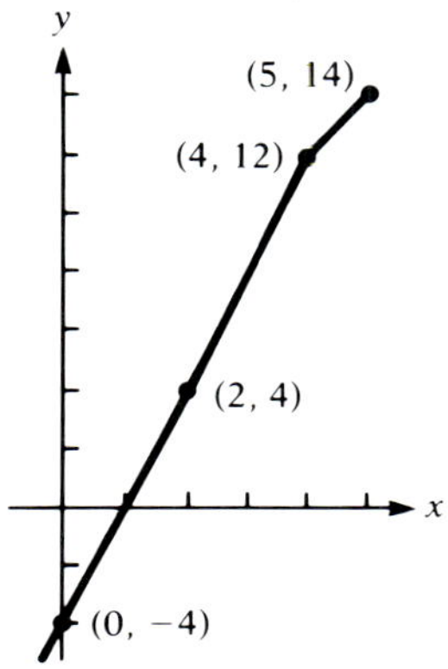

13. $y = x^2 - 4$

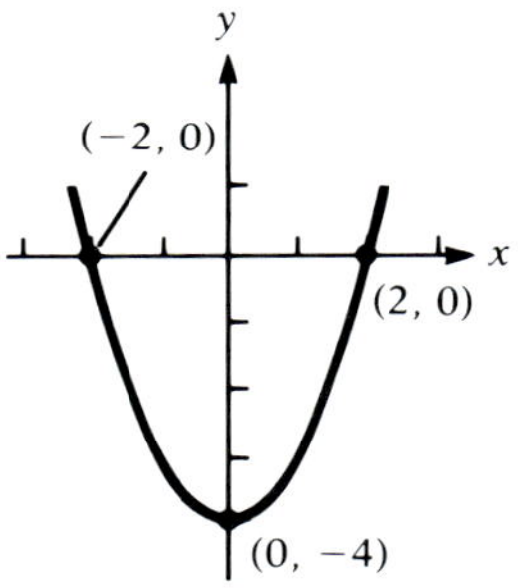

15. $y = 4x^2 - 7$

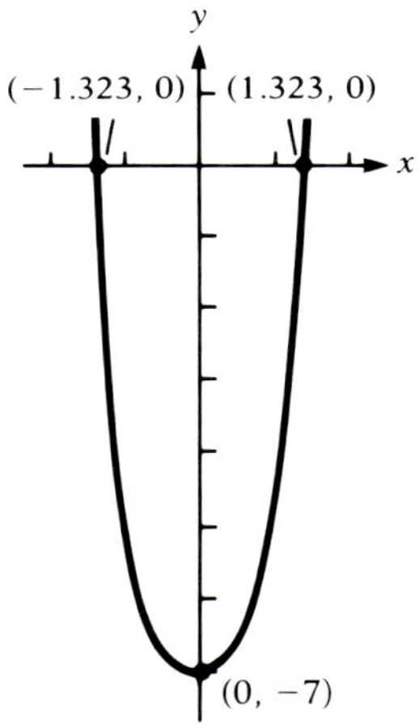

17. $y = (x - 3)^2$

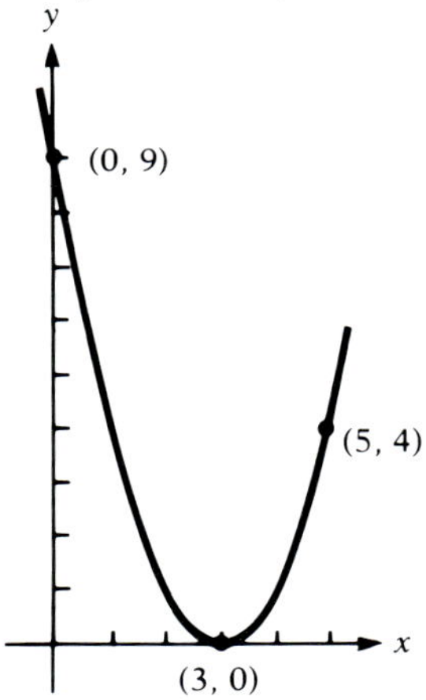

19. $y = -2(x + 3)^2$

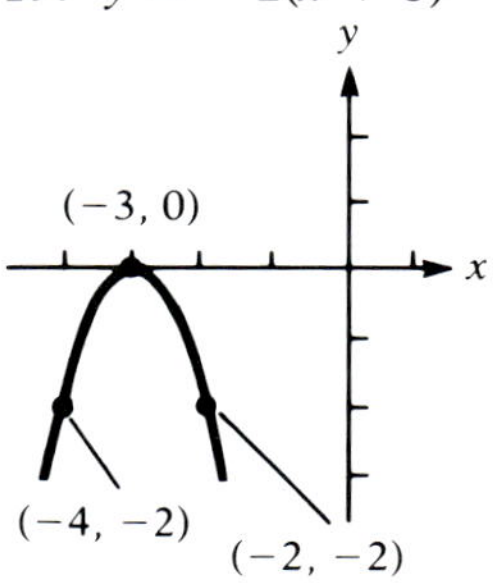

21. $y = x^2 + 5x + 6$

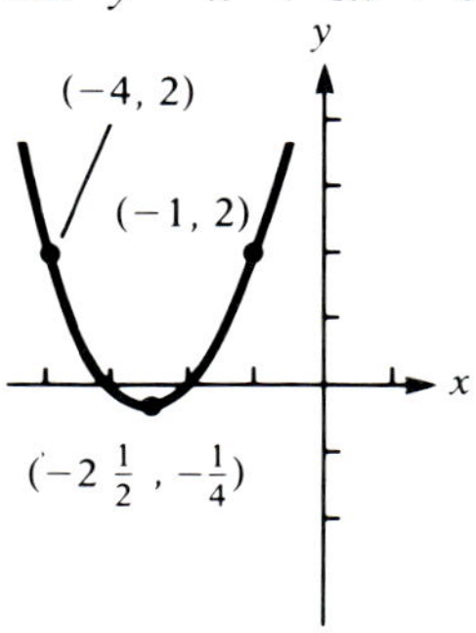

23. $y = 15x^2 + 13x - 6$

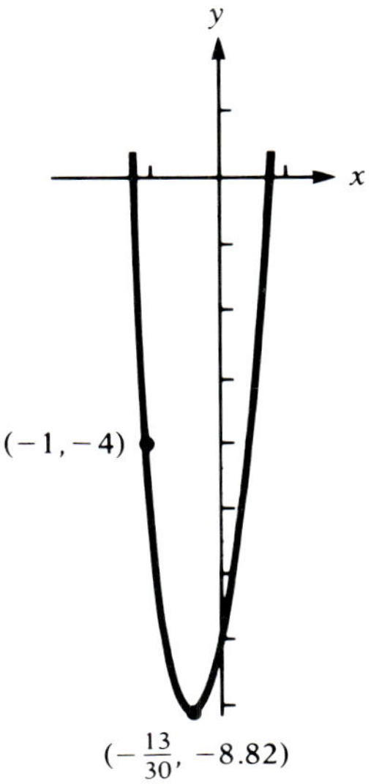

25. $y = \dfrac{2}{x}$

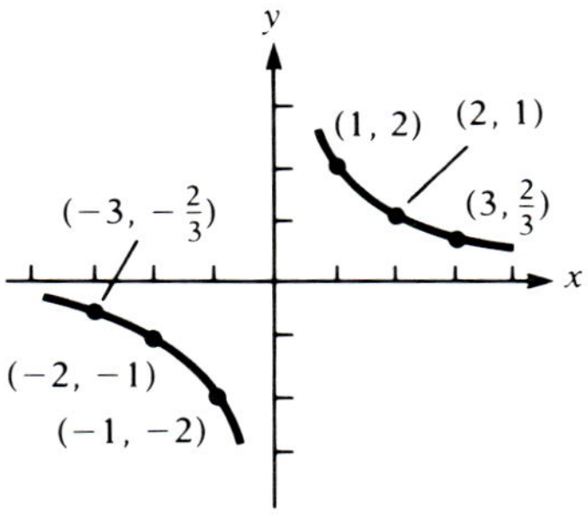

27. $y = \dfrac{4}{1 - x}$

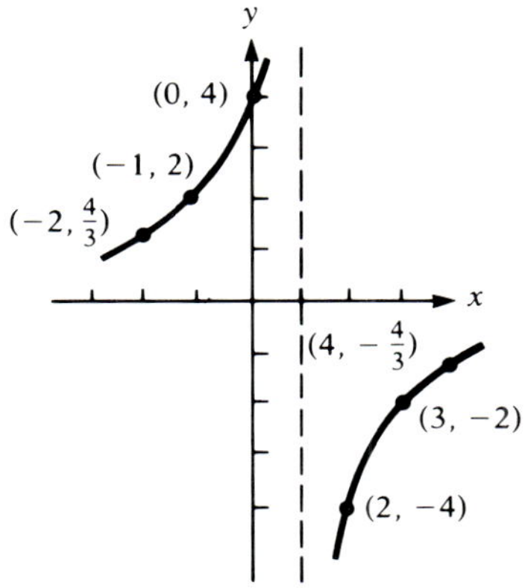

29. $y = \dfrac{x}{x - 2}$

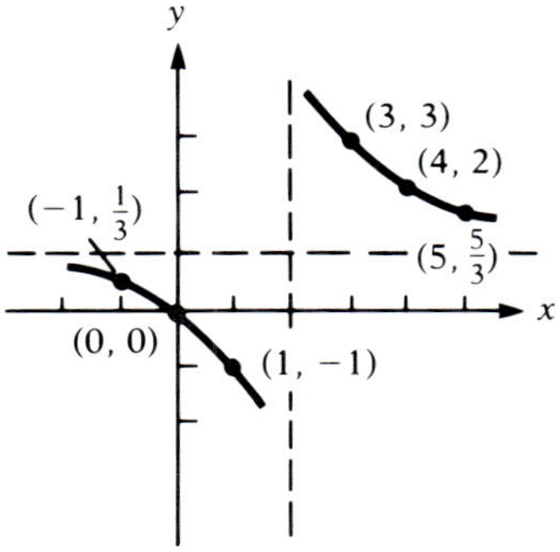

31. $y = \dfrac{x + 1}{x - 1}$

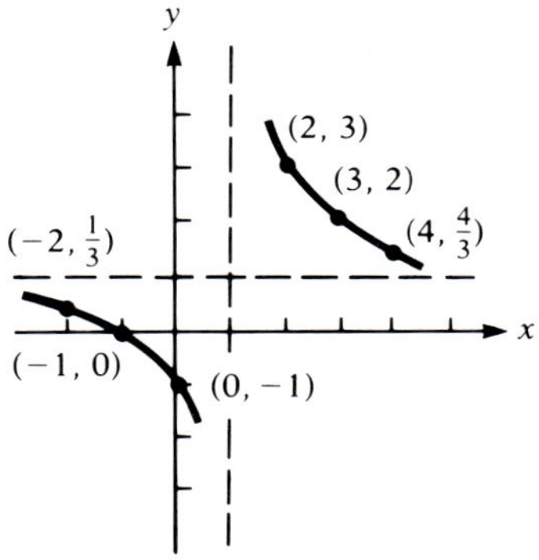

33. $y = \dfrac{4x - 2}{2x - 2}$

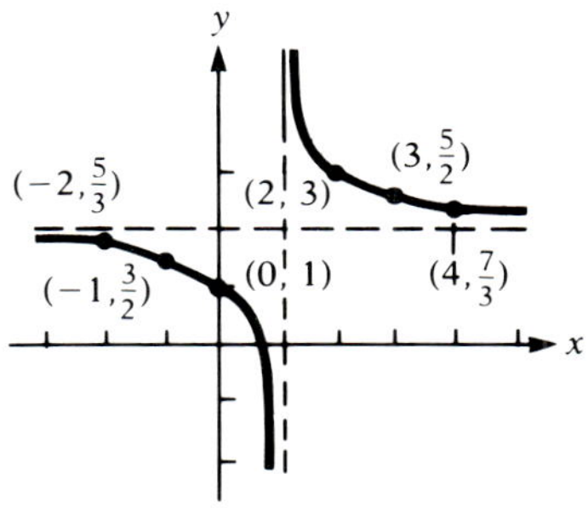

35. $y = f(x) = \begin{cases} 3x + 2, & 0 \le x < 2 \\ 2x^2, & 2 \le x \le 4 \end{cases}$

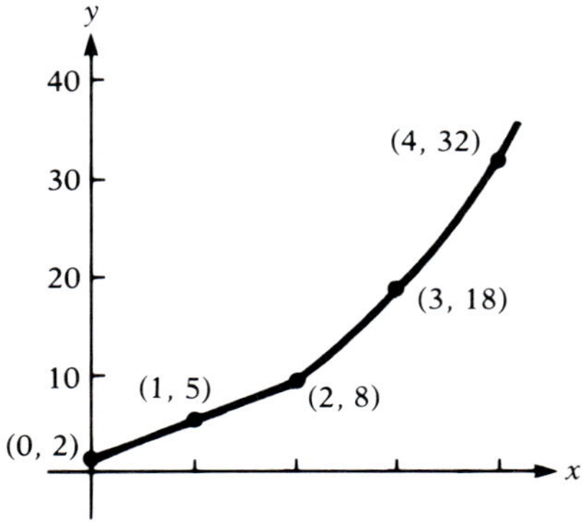

Exercise Set 4.3, page 166

1. 160 volts **3.** 360 errors **5.** 36.818 min **7.** 54 ft **9.** 31.25 miles **11.** 4 sq. in. **13.** $x = 225$; yes

Exercise Set 4.4, page 177

1. Arithmetic; 6, 10, 14, 18 **3.** Neither; $\frac{3}{2}, 2, \frac{9}{4}, \frac{12}{5}$ **5.** Geometric; 4, 16, 64, 256 **7.** Geometric; $-5, 25, -125, 625$ **9.** Geometric; $-150, 900, -5400, 32400$ **11.** Geometric; $9, -27, 81, -243$ **13.** Arithmetic; 4 **15.** Geometric; 2 **17.** Neither **19.** Geometric; -3 **21.** Arithmetic; 2 **23.** Geometric; $2x$ **25.** 58 **27.** -44 **29.** 5.5 **31.** 20 **33.** 114 **35.** 180 **37.** -250 **39.** 64 **41.** 243 **43.** 121 **45.** 328 **47.** 6.7222

49. $\sum_{i=1}^{7} i^{1/2} = 1^{1/2} + 2^{1/2} + 3^{1/2} + 4^{1/2} + 5^{1/2} + 6^{1/2} + 7^{1/2}$

51. $\sum_{i=5}^{10} f(i + 3) = f(5 + 3) + f(6 + 3) + f(7 + 3) + f(8 + 3) + f(9 + 3) + f(10 + 3)$

53. $\sum_{i=2}^{5} \left(4 - \frac{3}{i}\right) i^3 = \left(4 - \frac{3}{2}\right)2^3 + \left(4 - \frac{3}{3}\right)3^3 + \left(4 - \frac{3}{4}\right)4^3 + \left(4 - \frac{3}{5}\right)5^3$ **55.** 22.5 **57.** 112 **59.** 0. **61.** 120 **63.** 21,779 **65.** a) 19,750 b) 39,500 c) 750 every 6 months **67.** $a_5 = \$8,192.00$, $a_{10} = \$2,684.35$ **69.** 80,000 **71.** 1988: \$881,170.84, 1992: 1,386,539.40, $S_5 = \$4,057,594.52$, $S_9 = \$8,774,367.53$

Chapter 4 Review Problems, page 183

1. 2 **3.** 20 **5.** $2(x^2 + 2x - 1)$ **7.** $2x(x - 1)$ **9.** $\dfrac{2x - 1}{2x^2 - 1}$ **11.** 1 **13.** $2(2x - 2a - 1)^2 - 1$ **15.** $x = 6, x = 2$

17. $y = -\frac{2}{3}x + 1$

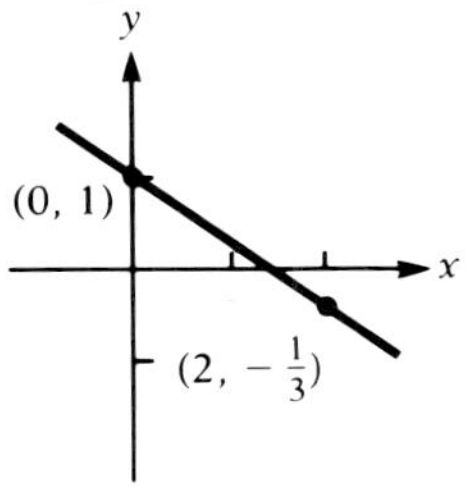

19. $3x - 3y = 15$

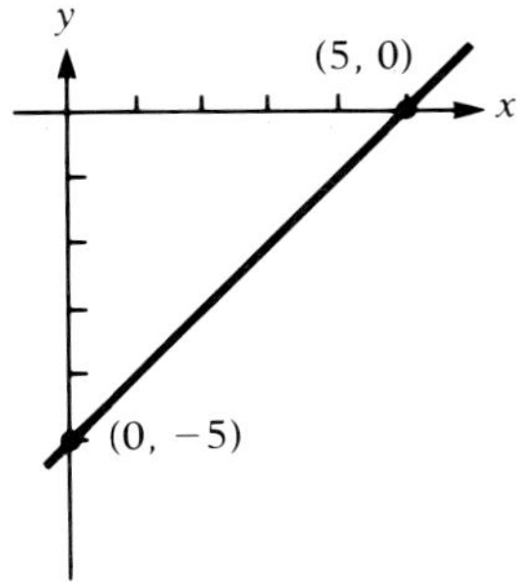

21. $y = 2x - |3 - x|$

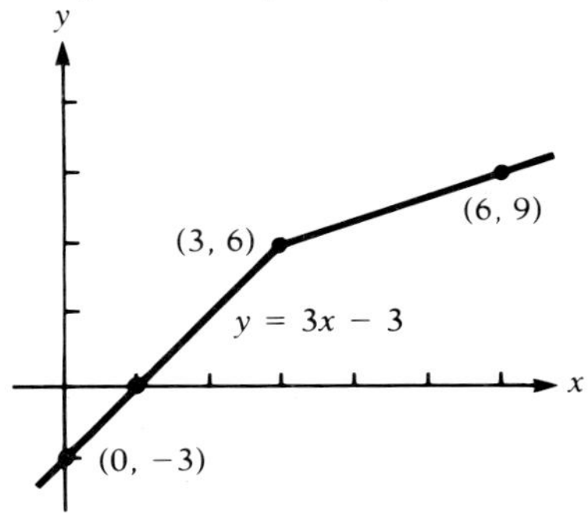

23. $y = -(x - 2)^2$

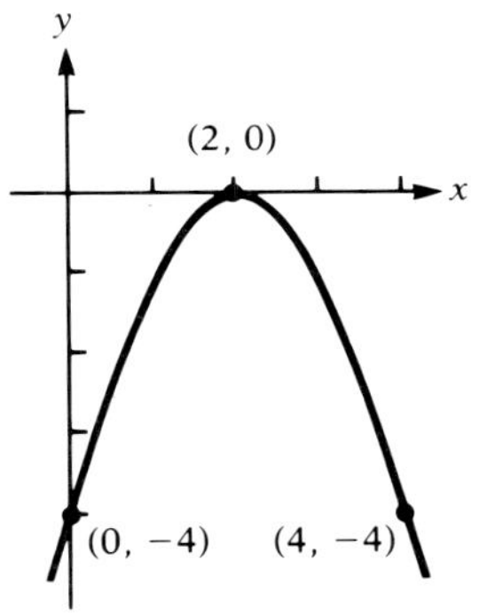

25. $y = -(4 - 3x)^2$

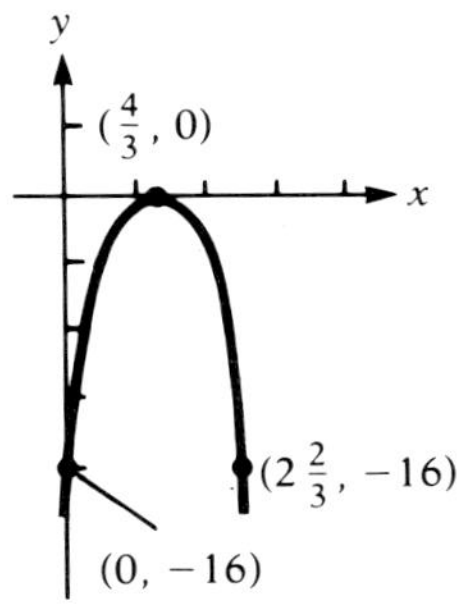

27. $y = \dfrac{3x}{x - 1}$

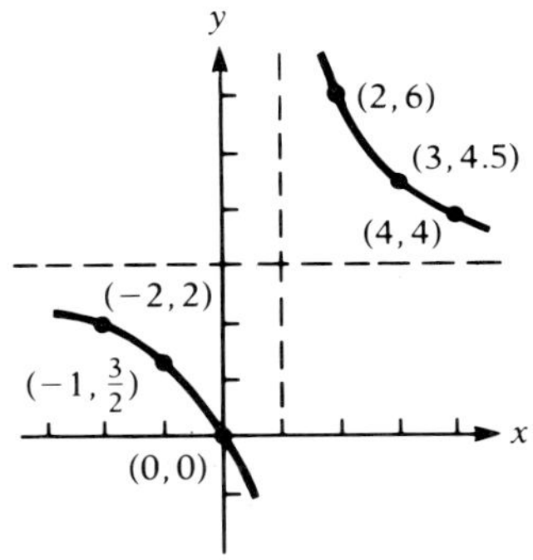

29. $y = \dfrac{x}{|x|}$

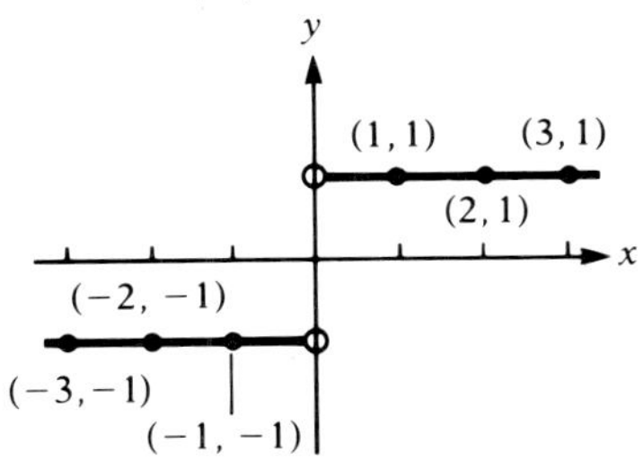

31. $y = x^4 - x^3 + x^2 - x + 2$

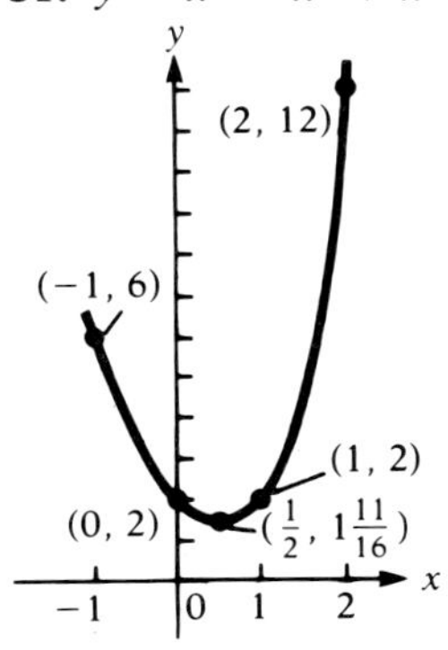

33. 120 volts **35.** 1012.4 lb **37.** Arithmetic; 4, 2, 0, −2 **39.** Neither; −1, 6, 19, 38
41. Neither; 5, 125/8, 3125/27, 78125/64 **43.** Geometric; $-2/x$ **45.** Arithmetic; −150
47. 12.7628 **49.** 115 **51.** 68.019 **53.** 8025.2
55. $[2(1) - 3]^2 + [2(2) - 3]^2 + [2(3) - 3]^2 + [2(4) - 3]^2 + [2(5) - 3]^2$
57. $g[f(2 - 1)] + g[f(3 - 1)] + g[f(4 - 1)] + g[f(5 - 1)]$ **59.** −96 **61.** 640

CHAPTER 5

Exercise Set 5.1, page 195

1. 2 **3.** 2 **5.** 3 **7.** (3, 6, 9) **9.** (5, 9, 4) **11.** (1.3, 5) **13.** False **15.** True
17. (2, 4) **19.** (3, 4.6, −6) **21.** (8.8, −22) **23.** (−4, −8, 6, −4) **25.** (5, −1)
27. (3, 3.8, 3) **29.** (11, 5) **31.** (17, 22) **33.** (8, −7, 2.4) **35.** (−8.8, 0.7, 0.8, 35)
37. (8, −7) **39.** (0.1, −1.4, 1.62) **41.** 16 **43.** 10.2 **45.** 1.6 **47.** −0.42
49. (2, 3) = (2, 3) **51.** (7, 5) = (7, 5) **53.** (16, 8) = (16, 8) **55.** 16 = 16 **57.** 31 = 31 = 31
59. a) John = (10, 3, 1), Paul = (20, 5, 2.5) b) (30, 8, 3.5) c) (10, 2, 1.5)
d) John = \$47.60, Paul = \$86.50, both = \$134.10

Exercise Set 5.2, page 214

1. 2×2 **3.** 3×2 **5.** 4×3 **7.** $\begin{bmatrix} 3 & 3 \\ 7 & 10 \end{bmatrix}$ **9.** $\begin{bmatrix} 4 & 9 & 9 \\ 9 & 1 & 1 \end{bmatrix}$ **11.** $\begin{bmatrix} 11 & 5 & 4 \\ 9 & 1 & 10 \\ 9 & 11 & 15 \end{bmatrix}$

13. True **15.** True **17.** $\begin{bmatrix} 2 & 4 \\ 0.6 & -8 \end{bmatrix}$ **19.** $\begin{bmatrix} 3 & 6 & 12 \\ 0.9 & 3 & 1.5 \end{bmatrix}$ **21.** $\begin{bmatrix} -1 & -3 & -4 \\ -2 & 0 & -7 \\ -5 & -6 & -9 \end{bmatrix}$

23. $\begin{bmatrix} -1 & 1 \\ -1 & -2 \end{bmatrix}$ **25.** $\begin{bmatrix} -2 & -5 & -1 \\ -3 & 1 & -1 \end{bmatrix}$ **27.** $\begin{bmatrix} 9 & -1 & -4 \\ 5 & 1 & -4 \\ -1 & -1 & -3 \end{bmatrix}$ **29.** $\begin{bmatrix} -1 & 4 \\ -2 & 0 \end{bmatrix}$

31. $\begin{bmatrix} 0 & 0.8 & 1.4 \\ 1.1 & 0.1 & -0.3 \end{bmatrix}$ **33.** $AB = 15$, $BA = \begin{bmatrix} 7 & 14 & 28 \\ 0 & 0 & 0 \\ 2 & 4 & 8 \end{bmatrix}$

35. $AB = \begin{bmatrix} 47 & 25 & 33 \\ 48 & 39 & 42 \\ 128 & 61 & 72 \end{bmatrix}$, $BA = \begin{bmatrix} 14 & 30 & 54 \\ 24 & 39 & 62 \\ 44 & 48 & 105 \end{bmatrix}$

37. $AB = \begin{bmatrix} -6 & 7 & 4 \\ -5 & 7 & 7 \\ -10 & 15 & 9 \end{bmatrix}$, $BA = \begin{bmatrix} 1 & 3 & 4 \\ 1 & -3 & 3 \\ 6 & 3 & 12 \end{bmatrix}$

39. $\begin{bmatrix} 1 & 2 \\ 3 & 4 \end{bmatrix} = \begin{bmatrix} 1 & 2 \\ 3 & 4 \end{bmatrix}$ **41.** $\begin{bmatrix} 3 & 1 \\ 7 & 7 \end{bmatrix} = \begin{bmatrix} 3 & 1 \\ 7 & 7 \end{bmatrix}$ **43.** $\begin{bmatrix} 8 & 6 \\ 20 & 18 \end{bmatrix} = \begin{bmatrix} 8 & 6 \\ 20 & 18 \end{bmatrix}$

45. $\begin{bmatrix} 65 & 35 \\ 129 & 67 \end{bmatrix} = \begin{bmatrix} 65 & 35 \\ 129 & 67 \end{bmatrix}$ **47.** $\begin{bmatrix} 16 & 10 \\ 56 & 42 \end{bmatrix} = \begin{bmatrix} 16 & 10 \\ 56 & 42 \end{bmatrix}$

49. $\begin{bmatrix} 20 & 10 \\ 44 & 18 \end{bmatrix} = \begin{bmatrix} 20 & 10 \\ 44 & 18 \end{bmatrix} = \begin{bmatrix} 20 & 10 \\ 44 & 18 \end{bmatrix}$ **51.** Rows = income
Cols = car pooled or not car pooled

$$\begin{bmatrix} 150 & 450 \\ 250 & 400 \\ 100 & 650 \end{bmatrix}$$

Chapter 5 Review Problems, page 218

1. 2 **3.** 3 **5.** (4, 1.1) **7.** (0.2, 12, 7) **9.** (0.5, 5, −0.27, 8) **11.** False **13.** False **15.** (4, −0.6) **17.** (4, −4, −0.8, 8) **19.** (0.7, −6) **21.** (2, 7, 7) **23.** 2 × 3

25. $\begin{bmatrix} -1 & 5 \\ 8 & 7 \end{bmatrix}$ **27.** $\begin{bmatrix} 0.4 & 5 \\ 7 & 0.2 \end{bmatrix}$ **29.** $\begin{bmatrix} 4 & 1 & 0 \\ 3 & 5 & 5 \end{bmatrix}$ **31.** True **33.** True **35.** False

37. a) Tom = (2, 1, 1, 1)
Ann = (1, 1, 3, 2)
b) \$9.89, \$16.00, Ann, \$25.89

CHAPTER 6

Exercise Set 6.2, page 235

1. $x = 2, y = 1$, unique **3.** $x = 2, y = -\frac{2}{3}$, unique **5.** $x = 3, y = -2$, unique **7.** $x = 0, y = 3$, unique **9.** $x = 8 + 3y, y = -\frac{1}{3}(8 - x)$, dependent **11.** Inconsistent, no solution **13.** $x = 1.7 + y, z = 2.3 - y$, dependent **15.** $x = 3, y = 1$, unique **17.** $x = 3, y = 1$, unique **19.** Inconsistent, no solution **21.** a) $p = \$1.50$ b) price = \$1.324, $d(p) = 6690$

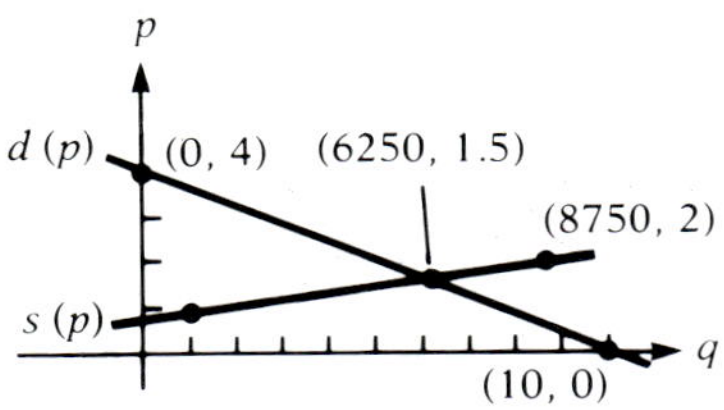

23. $x = 50$ grams, $y = 50$ grams **25.** $x = 8$ purses, $y = 12$ briefcases

Exercise Set 6.3, page 253

1. $\left[\begin{array}{cc|c} 3 & -5 & 10 \\ 4 & -2 & 17 \end{array}\right]$ **3.** $\left[\begin{array}{cc|c} 6 & -5 & 2 \\ 2 & -1 & 4 \end{array}\right]$ **5.** $\left[\begin{array}{ccc|c} 2 & 9 & -2 & 27 \\ 3 & -1 & 10 & 32 \end{array}\right]$

7. $\left[\begin{array}{cccc|c} 1 & -3 & 2 & 1 & 15 \\ 3 & -8 & 0 & 5 & 29 \\ 6 & -16 & 2 & -4 & 36 \\ 0 & 1 & -3 & -18 & 44 \end{array}\right]$ **9.** $x = \frac{34}{31}, y = -\frac{20}{31}$ **11.** $x = \frac{13}{2}, y = -5$

13. $x = \frac{14}{9}, y = \frac{46}{27}$ **15.** $x = \frac{39}{7} + \frac{3}{7}z, y = -\frac{22}{7} + \frac{15}{7}z$ **17.** $x = -\frac{7}{13}, y = \frac{54}{13}, z = -\frac{9}{13}$ **19.** $x = 1, y = 2, z = 3$ **21.** $w = -\frac{42}{11} + \frac{6}{11}z, x = \frac{50}{11} - \frac{4}{11}z, y = \frac{124}{11} - \frac{13}{11}z$ **23.** $x = \$5000, y = \$5000$ **25.** $x = 97.5, y = 32.5, z = 78$

Exercise Set 6.4, page 264

1. $\begin{bmatrix} -2 & \frac{3}{2} \\ 1 & -\frac{1}{2} \end{bmatrix}$ **3.** $\begin{bmatrix} -1 & 2 \\ 2 & -3 \end{bmatrix}$ **5.** $\begin{bmatrix} 4 & -1 & -2 \\ -2 & 1 & \frac{1}{2} \\ -1 & 0 & 1 \end{bmatrix}$ **7.** $\begin{bmatrix} -1 & 2 & 3 & -2 \\ 1 & -3 & -2 & 5 \\ -1 & 3 & 3 & -5 \\ 1 & -2 & -3 & 3 \end{bmatrix}$

9. $\left[\begin{array}{cc|cc} 1 & 2 & 1 & 0 \\ 0 & 0 & -1 & 1 \end{array}\right]$, singular **11.** $\left[\begin{array}{cc|cc} 1 & \frac{2}{3} & \frac{1}{3} & 0 \\ 0 & 0 & -2 & 1 \end{array}\right]$, singular

13. $\left[\begin{array}{ccc|ccc} 1 & 0 & 2 & 2 & -1 & 0 \\ 0 & 1 & -0.5 & -\frac{3}{2} & 1 & 0 \\ 0 & 0 & 0 & -2 & 0 & 1 \end{array}\right]$, singular

15. $\left[\begin{array}{cccc|cccc} 1 & 0 & 0 & \frac{1}{3} & 0 & \frac{1}{3} & -\frac{2}{3} & 0 \\ 0 & 1 & 0 & -\frac{1}{12} & -\frac{1}{4} & \frac{1}{6} & \frac{1}{6} & 0 \\ 0 & 0 & 1 & \frac{1}{12} & \frac{1}{4} & -\frac{1}{6} & \frac{1}{3} & 0 \\ 0 & 0 & 0 & 0 & -1 & -1 & -1 & 0 \end{array}\right]$, singular

17. $\begin{bmatrix} \frac{1}{6} & -\frac{1}{6} \\ \frac{1}{8} & \frac{1}{8} \end{bmatrix}$ **19.** $\begin{bmatrix} 1 & 2 & 2 \\ 0 & 0 & 0 \\ 1 & 1 & 1 \end{bmatrix}$, singular

21. $\begin{bmatrix} 1 & -1 & 1 & -1 \\ 0 & 1 & -1 & 1 \\ 0 & 0 & 1 & -1 \\ 0 & 0 & 0 & 1 \end{bmatrix}$

23. $\begin{bmatrix} \frac{1}{3} & \frac{1}{3} \\ -\frac{1}{3} & \frac{1}{6} \end{bmatrix}\begin{bmatrix} 6 \\ 6 \end{bmatrix} = \begin{bmatrix} 4 \\ -1 \end{bmatrix}$ **25.** $\begin{bmatrix} \frac{2}{5} & \frac{1}{5} \\ -\frac{3}{25} & \frac{1}{25} \end{bmatrix}\begin{bmatrix} 5 \\ 10 \end{bmatrix} = \begin{bmatrix} 4 \\ -\frac{1}{5} \end{bmatrix}$

27. $\begin{bmatrix} \frac{1}{3} & \frac{1}{3} \\ -\frac{1}{12} & \frac{1}{6} \end{bmatrix}\begin{bmatrix} 8 \\ 13 \end{bmatrix} = \begin{bmatrix} 7 \\ \frac{3}{2} \end{bmatrix}$

29. a) $\begin{bmatrix} \frac{8}{66} & \frac{2}{22} \\ -\frac{7}{66} & \frac{1}{22} \end{bmatrix}\begin{bmatrix} 6 \\ 15 \end{bmatrix} = \begin{bmatrix} \frac{138}{66} \\ \frac{3}{66} \end{bmatrix}$ b) $\begin{bmatrix} \frac{8}{66} & \frac{2}{22} \\ -\frac{7}{66} & \frac{1}{22} \end{bmatrix}\begin{bmatrix} -6 \\ 24 \end{bmatrix} = \begin{bmatrix} \frac{96}{66} \\ \frac{114}{66} \end{bmatrix}$

31. $\begin{bmatrix} 1 & -\frac{5}{2} \\ 0 & 0 \end{bmatrix}\begin{bmatrix} \frac{1}{2} & 0 \\ 7 & 1 \end{bmatrix}$, no solution

33. $\begin{bmatrix} \frac{5}{9} & \frac{197}{27} & -\frac{193}{54} & \frac{77}{27} \\ \frac{1}{3} & \frac{75}{27} & -\frac{78}{54} & \frac{30}{27} \\ \frac{7}{9} & \frac{13}{27} & -\frac{20}{54} & \frac{7}{27} \\ -\frac{1}{9} & \frac{2}{27} & -\frac{1}{54} & -\frac{1}{27} \end{bmatrix}\begin{bmatrix} 15 \\ 29 \\ 36 \\ -44 \end{bmatrix} = \begin{bmatrix} -34.22 \\ 15.33 \\ 0.89 \\ 1.44 \end{bmatrix}$

$\begin{bmatrix} \frac{5}{9} & \frac{197}{27} & -\frac{193}{54} & \frac{77}{27} \\ \frac{1}{3} & \frac{75}{27} & -\frac{78}{54} & \frac{30}{27} \\ \frac{7}{9} & \frac{13}{27} & -\frac{20}{54} & \frac{7}{27} \\ -\frac{1}{9} & \frac{2}{27} & -\frac{1}{54} & -\frac{1}{27} \end{bmatrix}\begin{bmatrix} 0 \\ -7 \\ -41 \\ -41 \end{bmatrix} = \begin{bmatrix} -21.46 \\ -5.78 \\ 1.19 \\ 1.76 \end{bmatrix}$

35. a) $\begin{bmatrix} 1 & -\frac{2}{3} \\ -\frac{1}{2} & \frac{1}{2} \end{bmatrix}\begin{bmatrix} 72 \\ 96 \end{bmatrix} = \begin{bmatrix} 8 \\ 12 \end{bmatrix}$ b) $\begin{bmatrix} 1 & -\frac{2}{3} \\ -\frac{1}{2} & \frac{1}{2} \end{bmatrix}\begin{bmatrix} 72 \\ 84 \end{bmatrix} = \begin{bmatrix} 16 \\ 6 \end{bmatrix}$

Exercise Set 6.5, page 270

1. a) $A = \begin{bmatrix} 0.2 & 0.7 \\ 0.5 & 0.1 \end{bmatrix}$ b) $\begin{bmatrix} 428.34 \\ 332.38 \end{bmatrix}$ **3.** $\begin{bmatrix} 50.9 \\ 85.2 \\ 70.19 \end{bmatrix}$

Chapter 6 Review Problems, page 273

1. $x = \frac{20}{7}, y = \frac{18}{7}$, unique

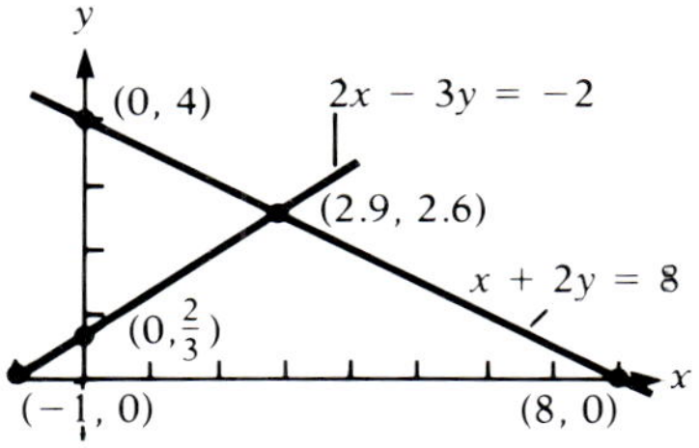

3. $x = \frac{36}{13}, y = \frac{19}{13}$, unique

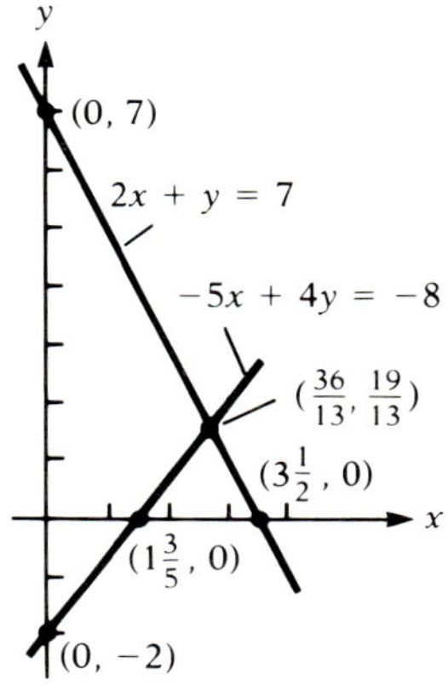

5. $x = \frac{23}{5}, y = \frac{16}{5}$, unique

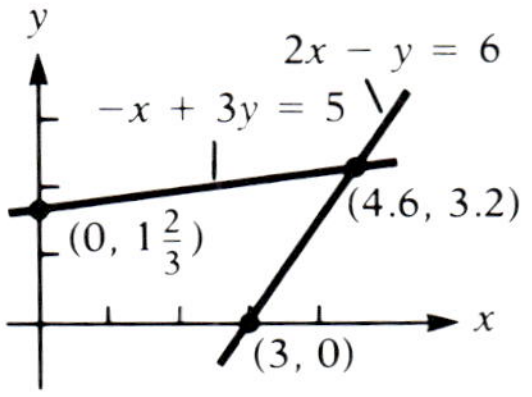

7. $x = 2.2, y = -1.4$, inverse square matrix only, unique

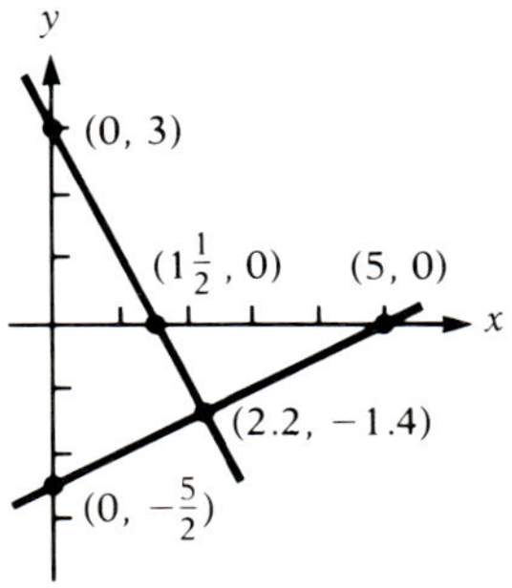

9. $x = \dfrac{27 - 11z}{7}, y = \dfrac{5z - 4}{7}$, inverse square matrix only, dependent

11. $x = 5.2, y = 3.2, z = 1.6$, unique **13.** $y = 10 - 6x, z = 18 - 10x$, dependent

15. $x = -1.5, y = 14, z = 24$, unique

17. a) $A = \begin{bmatrix} 0.25 & 0.5 \\ 0.48 & 0.33 \end{bmatrix}$ b) $x \approx \begin{pmatrix} 106.41 \\ 115.62 \end{pmatrix}$

$$x \approx \begin{pmatrix} 121.79 \\ 132.70 \end{pmatrix}$$

CHAPTER 7

Exercise Set 7.1, page 282

1.

$$\begin{array}{lrl} \text{Maximize} & P = 65D + 55C & \quad P = \text{profit} \\ \text{Subject to} & 6D + 5C \le 600 & \quad D = \text{downhill} \\ & 1D + 2C \le 400 & \quad C = \text{cross-country} \\ & D \ge 0 & \\ & C \ge 0 & \end{array}$$

3.

$$\begin{array}{lrl} \text{Maximize} & R = 0.15A + 0.10B & \quad R = \text{return} \\ \text{Subject to} & A + B \le 50{,}000 & \quad A = \text{option } A \\ & A \ge 0 & \quad B = \text{option } B \\ & B \ge 0 & \end{array}$$

$$\begin{array}{lrl} \text{Minimize} & R = 25A + 15B & \quad R = \text{risk} \\ \text{Subject to} & 0.15A + 0.10B \ge 0.13\,(50{,}000) & \quad A = \text{option } A \\ & A + B \le 50{,}000 & \quad B = \text{option } B \\ & A \ge 0 & \\ & B \ge 0 & \end{array}$$

5.

$$\begin{array}{lrl} \text{Maximize} & L = 0.50(5000)R + 0.40(8000)T & \quad L = \text{listeners 15 or below} \\ \text{Subject to} & 250R + 525T \le 1500 & \quad R = \text{radio ads} \\ & 5000R + 8000T \ge 25{,}000 & \quad T = \text{TV ads} \\ & R \ge 0 & \\ & T \ge 0 & \end{array}$$

7.

$$\begin{array}{lrl} \text{Maximize} & P = 40C + 50S & \quad P = \text{profit} \\ \text{Subject to} & C + S \le 250 & \quad C = \text{corn} \\ & 4C + 5S \le 600 & \quad S = \text{soybeans} \\ & 2C + 9S \le 1500 & \\ & C \ge 0 & \\ & S \le 0 & \end{array}$$

9.

$$\begin{array}{lrl} \text{Minimize} & C = 100S + 175D + 850F & \quad C = \text{cost} \\ \text{Subject to} & S + D + F \ge 100 & \quad S = \text{single} \\ & F \ge 0.15(S + D + F) & \quad D = \text{double} \\ & S \ge 0 & \quad F = \text{fixed drives} \\ & D \ge 0 & \\ & F \ge 0 & \end{array}$$

Exercise Set 7.2, page 296

1. Max, $x = 4$, $y = 2$, $P = 24$

3. Multiple solutions min, $x = 2.4$, $y = 2.4$, $P = 48$; min, $x = 6$, $y = 0$, $P = 48$

5. Max, $x = \frac{45}{17}$, $y = \frac{72}{7}$, $P = 124.41$ **7.** Min, $x = 2$, $y = 0$, $C = 10$

9. Max, $x = \frac{40}{7}$, $y = \frac{33}{7}$, $P = 279.29$ **11.** Min, $x = \frac{6}{5}$, $y = \frac{6}{5}$, $C = 48$

13. Max, $c = \frac{18}{7}$, $d = \frac{30}{7}$, $P = \$1414\frac{2}{7}$, yes, max, $c = \frac{36}{7}$, $d = \frac{18}{7}$, $P = \$1478\frac{4}{7}$

15. Max, f = disks = $857\frac{1}{7}$, d = diskettes = $2571\frac{3}{7}$, $P = \$11{,}785.71$

17. Min, s = senior = 0, j = junior = 25, $C = \$750{,}000$

19. Max, $A = 93.75$, $B = 156.25$, $GR = 2656.25$

Chapter 7 Review Problems, page 299

1. Minimize $C = 20{,}000A + 13{,}500B$ $\quad A$ = refinery A
Subject to $200A + 300B \geq 2000$ $\quad B$ = refinery B
$400A + 150B \geq 1750$ $\quad C$ = cost
$600A + 300B \geq 3000$
$A \geq 0, B \geq 0$

3. Maximize $R = 1.75x + 1.50y$ $\quad x$ = mixture 1
Subject to $12x + 18y \leq 6000$ $\quad y$ = mixture 2
$2x + 3y \leq 1500$ $\quad R$ = Revenue
$8x + 2y \leq 3000$
$x \geq 0, y \geq 0$

5. Maximize $P = 65W + 70C$ $\quad W$ = wheat
Subject to $W + C \leq 500$ $\quad C$ = corn
$100W + 125C \leq 60{,}000$ $\quad P$ = profit
$8W + 6C \leq 3{,}500$
$W \geq 0, C \geq 0$

7. Max, $x = 2\frac{1}{2}$, $y = 0$, $P = 7\frac{1}{2}$, unique **9.** Min, $x = 1.8$, $y = 8.4$, $P = 18.6$, unique
11. Infeasible, can't be greater than e_1 and less than e_2 and e_3 **13.** Max, $x = 9$, $y = 0$, $p = 36$, unique
15. Min, $A = 2.5$, $B = 5$, $C = 117{,}500$, unique **17.** Max, $x = 333\frac{1}{3}$, $y = 166\frac{2}{3}$, $R = 833\frac{1}{3}$, unique
19. Max, $W = 100$, $C = 400$, $P = \$34{,}500$, unique

CHAPTER 8

Exercise Set 8.1, page 318

1. $x = 4$, $y = 2$, $P = 24$ **3.** $x = 4$, $y = 0$, $P = 8$
5. $x = 2.6471$, $y = 4.2353$, $P = 124.4118$, optimal **7.** $x = 5\frac{1}{3}$, $y = 2\frac{1}{3}$, $P = 18\frac{1}{3}$, optimal

9. x = downhill, y = cross-country
Maximize $P = 65x + 55y$
Subject to $6x + 5y \leq 900$
$2x + 2y \leq 600$
$x \geq 0, y \geq 0$
Optimal, $x = 0$, $y = 180$, $P = 9900$

11. x = deluxe, y = ne plus ultra
Maximize $P = 0.95x + y$
Subject to $0.3x + 0.25y \leq 275$
$0.2x + 0.25y \leq 200$
$x \geq 0, y \geq 0,$
Optimal, $x = 750$, $y = 200$, $P = 912.50$

Exercise Set 8.2, page 328

1. Optimal, $x = 10$, $y = 0$, $P = 10$
3. Optimal, $x = 5$, $y = 0$, $P = 10$
5. Optimal, $x = 9$, $y = 5$, $P = 27.5$
7. Optimal, $x = 1.7143$, $y = 4.7143$, $P = 4.2857$
9. Optimal, $x = 450$, $y = 0$, $P = 3600$

11. x = fabric from S, y = fabric from B
Maximize $P = 6x + 8y$
Subject to $0.40x + 0.3y \leq 6000$
$0.15x + 0.25y \leq 5000$
$0.25x + 0.25y \leq 5000$
$x \geq 1500, y \geq 0$
Optimal, $x = 0$, $y = 20{,}000$, $P = 160{,}000$

Exercise Set 8.3, page 334

1. Optimal, $x = 2.4$, $y = 2.4$, $C = 48$

3. Optimal, $x = 2.5$, $y = 1.5$, $C = 38$

5. Optimal, $x = 1.8$, $y = 0.8$, $C = 47$

7. Optimal, $x = 0$, $y = 0$, $z = 20$, $C = 40$

9. x = shares A, y = shares B, z = shares C

Minimize $R = 2.5x + 2.8y + 2z$

Subject to
$$\begin{aligned} 0.10x + 0.08y + 0.08z &\geq 850 \\ 0.07x + 0.12y + 0.09z &\geq 820 \\ 100x + 50y + 50z &\leq 10{,}000 \\ x \geq 0, y \geq 0, z &\geq 0 \end{aligned}$$

Infeasible.

11. x = gal A, y = gal B, z = gal C

Minimize $C = 23.75x + 21.5y + 20z$

Subject to
$$\begin{aligned} x + y + z &\geq 10 \\ 4x + 2y + 2z &\leq 30 \\ 2x + 3y + 2z &\leq 25 \\ 3x + y + 2z &\leq 18 \\ 3x + 4y + 4z &\leq 38 \\ x \geq 0, y \geq 0, z &\geq 0 \end{aligned}$$

Optimal, $x = 2$, $y = 4$, $z = 4$, $C = \$213.50$

Exercise Set 8.4, page 344

1. Minimize $C = 10a + 8b$

Subject to
$$\begin{aligned} 2a + b &\geq 4 \\ a + 2b &\geq 4 \\ a \geq 0, b &\geq 0 \end{aligned}$$

3. Minimize $C = 32a + 40b$

Subject to
$$\begin{aligned} 8a + 5b &\geq 2 \\ 4a + 8b &\geq 1 \\ a \geq 0, b &\geq 0 \end{aligned}$$

5. Minimize $C = 24a + 2b + 4c$

Subject to
$$\begin{aligned} 6a + b &\geq 3 \\ 3a - b + c &\geq -2 \\ a \geq 0, b \geq 0, c &\geq 0 \end{aligned}$$

7. Minimize $C = -a + 50b + 4c$

Subject to
$$\begin{aligned} -a + 5b &\geq 1 \\ -a + 10b + c &\geq 1 \\ a \geq 0, b \geq 0, c &\geq 0 \end{aligned}$$

9. Minimize $C = 8a - 2b - 8c$

Subject to
$$\begin{aligned} a - 4b - 4c &\geq 2 \\ a + b - 6c &\geq -1 \\ -2a - b + 2c &\geq 1 \\ a \geq 0, b \geq 0, c &\geq 0 \end{aligned}$$

11. \$11, \$0

13. \$0, \$0, \$1.50, \$5.25

Chapter 8 Review Problems, page 346

1. Optimal, $x = 1\frac{1}{3}$, $y = 3\frac{1}{3}$, $P = 22$

3. Optimal, $x = 14$, $y = 6$, $P = 416$

5. x = type O in thousands, y = type N in thousands

Maximize $P = 80x + 60y$

Subject to
$$\begin{aligned} x + y &\leq 75 \\ 4x + 3.5y &\leq 210 \\ x &\leq 50 \\ y &\leq 50 \\ x \geq 0, y &\geq 0 \end{aligned}$$

$x = 50$

$y = 2.857$

$P = \$4171.42$

7. Minimize $C = -20a + 40c$

Subject to
$$\begin{aligned} 4a - 11b + 5c &\geq 4 \\ -5a + 4b + 4c &\geq 6 \\ a \geq 0, b \geq 0, c &\geq 0 \end{aligned}$$

9. Minimize $C = 16x + 14y + 8z$

Subject to
$$\begin{aligned} 2x + 8y - 6z &\geq 2 \\ 4x - 2y + 4z &\geq 4 \\ -4x + 6y + 8z &\geq -6 \\ x \geq 0, y \geq 0, z &\geq 0 \end{aligned}$$

CHAPTER 9

Exercise Set 9.1, page 356

1.

x	-3	0	3
$f(x)$	$\frac{1}{27}$	1	27

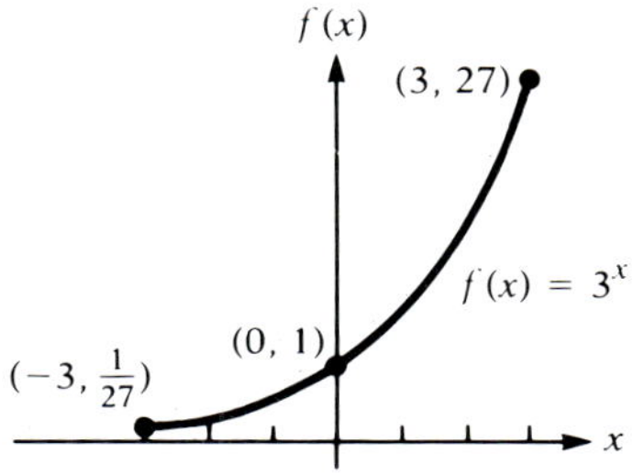

3.

x	-3	0	3
$f(x)$	0.19	1	5.2

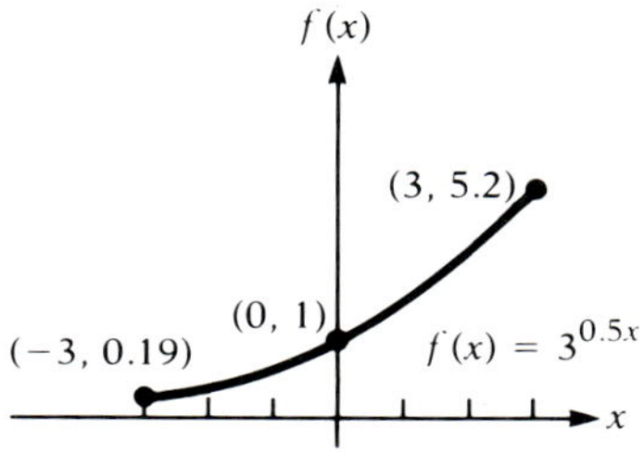

5.

x	-3	0	3
$g(x)$	64	1	$\frac{1}{64}$

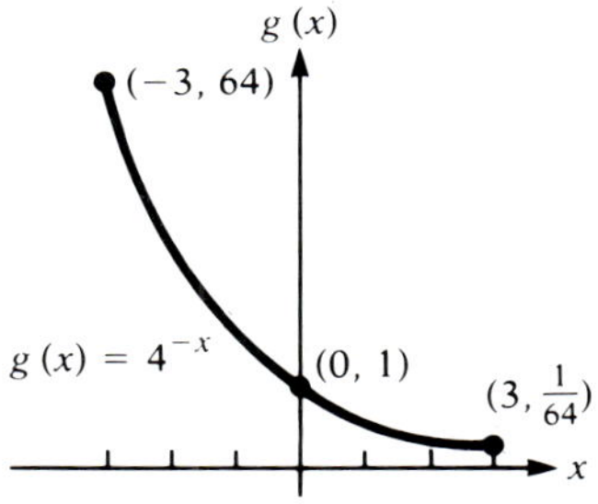

7.

x	-3	0	3
$f(x)$	9	1	0.11

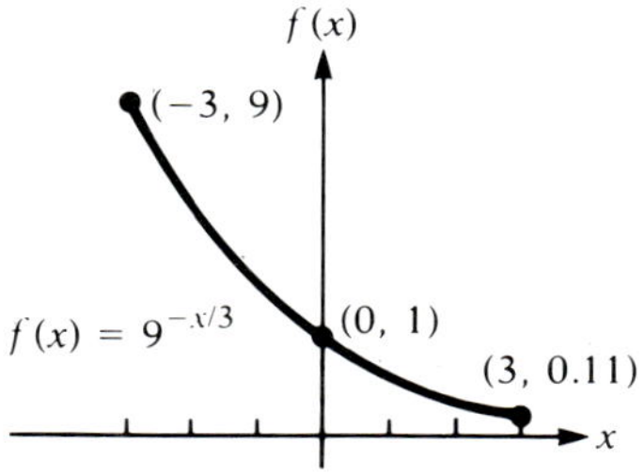

9.

x	-3	0	3
$g(x)$	0.223	1	4.48

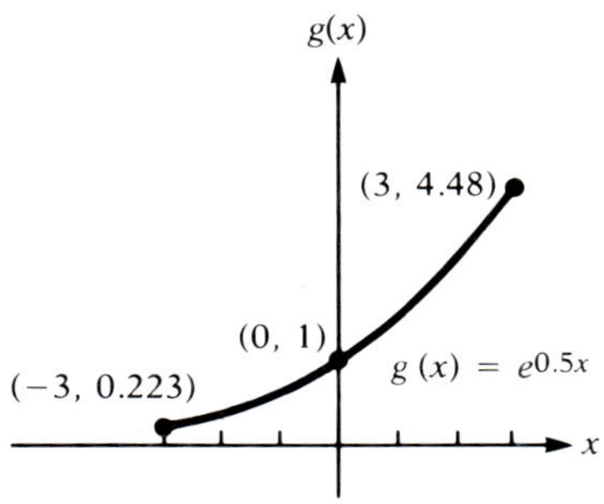

11.

x	-3	0	3
$h(x)$	0.1	2	40.2

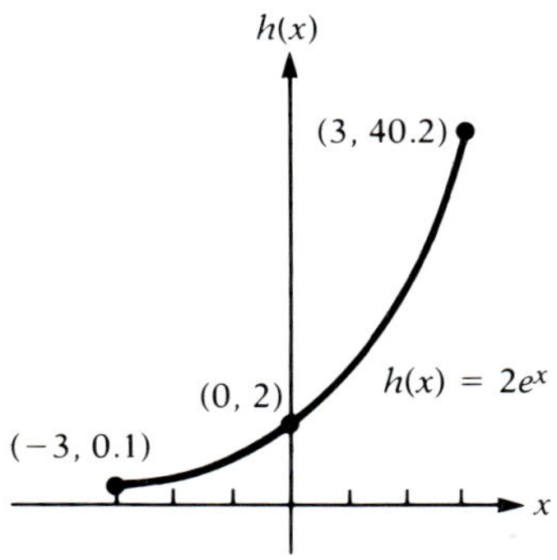

13.

x	-3	0	3
$f(x)$	4034	10	0.02

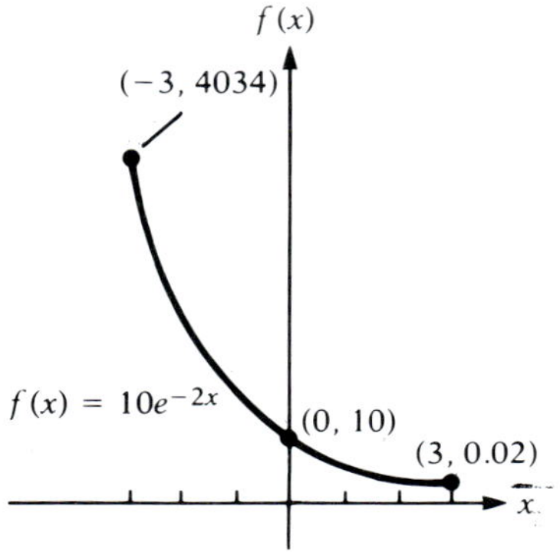

15.

x	-3	0	3
$g(x)$	-1909	0	95

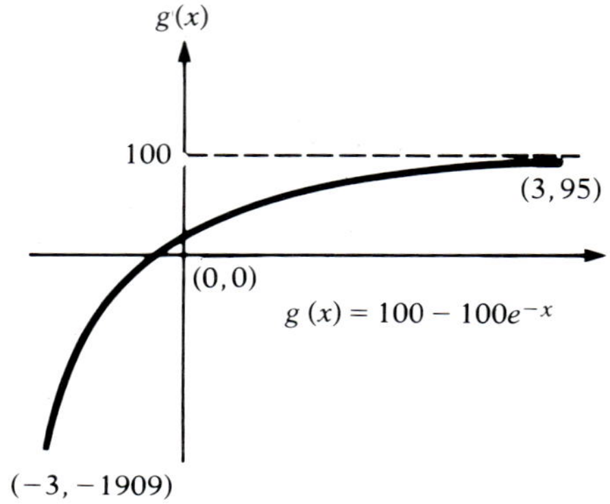

17. a) 8.2664×10^{10} (Today, 1986) b) 3.4443×10^{9} (Today, 1986)

19. a) 10 lb, b) 0.0000002 c) 2.3195×10^{-15}

Exercise Set 9.2, page 369

1. 2 **3.** -2 **5.** 2 **7.** 3 **9.** 3 **11.** 2 **13.** 343 **15.** 0.2 **17.** 1
19. 27 **21.** 17670 **23.** 2346 **25.** 0.8814 **27.** 362403.99 **29.** 742.542 **31.** 2.6667
33. 54 **35.** 12.8 **37.** 0.23385 **39.** 135,438.95 **41.** 1.7378×10^{8} times base intensity I_0
43. 0.00001 times more intense **45.** 8.7, 9.301 **47.** 16.496 more intense **49.** 28

Chapter 9 Review Problems, page 371

1. a)

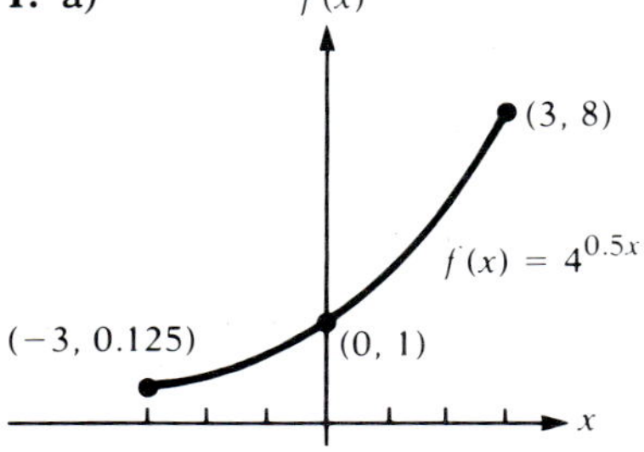

b)

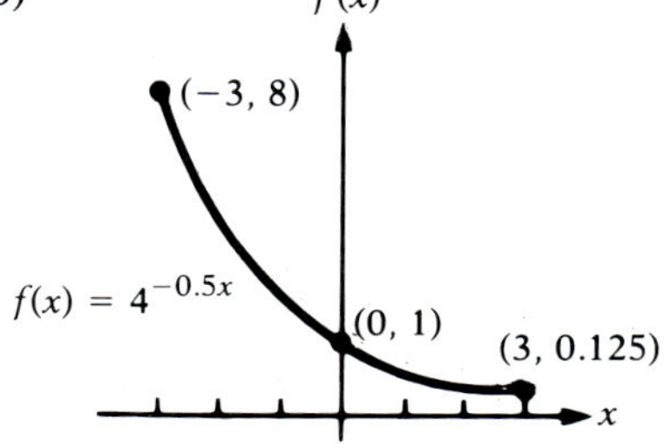

3. $x = 18.52$ **5.** $b = 32{,}768$ **7.** 5

9.

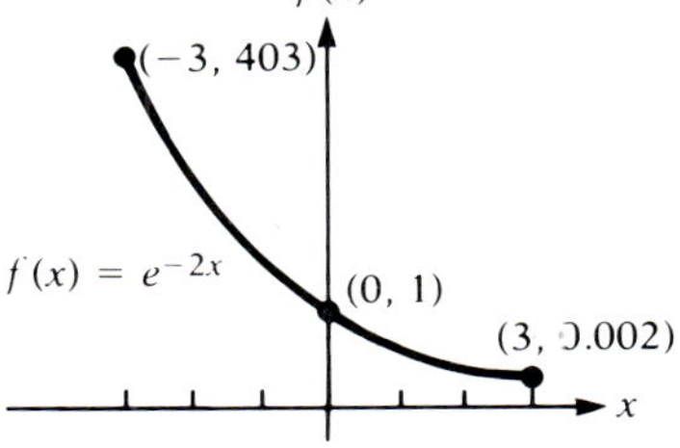

11.

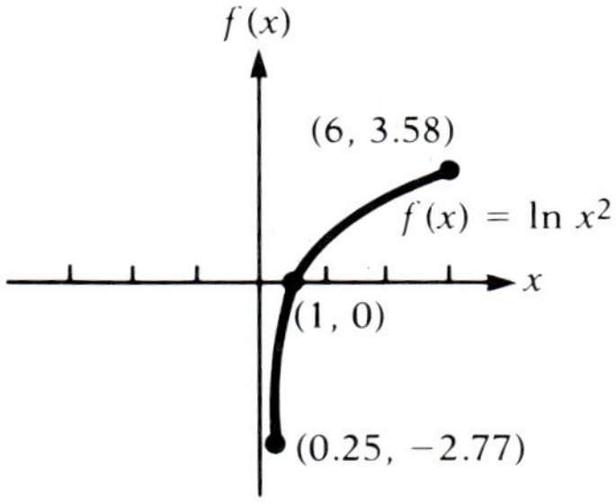

13. Learning rate = 0.96697, 500th unit = 3.6999

CHAPTER 10

Exercise Set 10.1, page 384

1. $I = 240$, $A = \$1040$ **3.** $t = 30$ days, $A = \$8100$ **5.** $r = 0.05$, $A = \$165$ **7.** $P = \$214.28$, $I = 5.71$ **9.** $P = \$7835.46$, $I = 164.54$ **11.** $P = \$48$ **13.** $I = \$287{,}500$ **15.** \$102.375 interest **17.** a) \$1275 b) \$225 c) 17.65% **19.** \$1185.1852 **21.** \$2307.69 **23.** a) \$5916 b) \$1116 **25.** a) \$7249.19 b) 14.24

Exercise Set 10.2, page 394

1. $A = \$1{,}797{,}481.9$ **3.** $r = 0.00447$ **5.** $P = \$1362.29$ **7.** $t = 22.426$ **9.** $A = \$1552.92$ **11.** $A = \$1{,}602{,}485.8$ **13.** $P = \$23{,}728.38$ **15.** $P = \$16{,}495.44$ **17.** 0.0618 **19.** 0.0618 **21.** $A = \$2615.07$ **23.** $P = \$226.21$ **25.** a) $t = 34.66$ b) 54.93 **27.** $r = 0.3466$ **29.** a) $r = 0.000124$ b) $t = 11200$

Exercise Set 10.3, page 405

1. $S = \$2312.37$ **3.** $S = \$417{,}172.14$ **5.** $R = \$323.09$ **7.** $R = \$293.86$ **9.** $R = \$1834.70$ **11.** $R = \$1{,}005.78$ **13.** $A = \$2{,}318.04$ **15.** $A = \$34{,}850.26$ **17.** $S = \$698{,}992.78$ **19.** $R = \$424.43$ **21.** $R = \$547.12$ **23.** Should pay cash, payments equal \$8,233.77 **25.** $R = \$38.25$ **27.** $R = \$175.34$

Exercise Set 10.4, page 412

1. $S = \$2347.05$ **3.** $S = \$120{,}681.94$ **5.** $R = \$322.02$ **7.** $R = \$290.23$ **9.** $R = \$1{,}798.73$ **11.** $R = \$998.29$ **13.** $A = \$239.21$ **15.** $A = \$35{,}198.76$ **17.** $S = \$37{,}351.75$ **19.** $R = \$188.13$ **21.** $S = \$776{,}341.74$ **23.** $R = \$121.42$

Exercise Set 10.5, page 418

1. $S = \$1{,}657.13$, $A = \$1{,}271.68$ **3.** $S = \$4{,}012.36$, $A = \$1{,}653.04$ **5.** $R = \$80.26$ **7.** $R = \$172.88$ **9.** $R = \$216.14$ **11.** a) $R = \$212.78$ b) \$210.67 **13.** $R = \$85.15$

Chapter 10 Review Problems, page 421

1. a) $I = \$202.50$ b) $A = \$2702.50$ **3.** \$3430 interest
5. a) 0.0867 b) 0.0873 c) 0.0876 d) 0.0876 **7.** $R = \$1{,}580.91$ **9.** $R = \$321.60$
11. $R = \$309.49$

CHAPTER 11

Exercise Set 11.1, page 434

1. $S = \{00, 0, 1, 2, 3, 4, 5, 6, 7, 8, 9, 10, 11, 12, 13, 14, 15, 16, 17, 18, 19, 20, 21, 22, 23, 24, 25, 26, 27, 28, 29, 30, 31, 32, 33, 34, 35, 36\}$

3. 1/38 = 0.026 **5.** 18/38 = 0.474 **7.** Simple, compound, compound, compound **9.** 16

11. 15/16 = 0.9375 **13.** b) probability cannot be negative c) probabilities do not add to 1

15. 75/105 = 0.714

17. a) P (4 heads) = 30/500 = 0.06
P (3 heads) = 125/500 = 0.25
P (2 heads) = 175/500 = 0.35
P (1 head) = 135/500 = 0.27
P (0 head) = 35/500 = 0.07

b) P (4 heads) = 1/16 = 0.0625
P (3 heads) = 4/16 = 0.25
P (2 heads) = 6/16 = 0.375
P (1 head) = 4/16 = 0.25
P (0 head) = 1/16 = 0.0625

19. a) 30/240 = 0.125 b) 80/240 = 0.333 c) 20/240 = 0.083

Exercise Set 11.2, page 440

1. P(failure) = 0.02

3. N = {Maine, N.H., Vermont, Mass., R.I., Conn.} C = {Conn., Colorado, Calif.} a) 8/50 = 0.16
M = {Montana, Miss., Missouri, Minn., Michigan, Maine, Mass.} b) 11/50 = 0.22

5. 5/16 = 0.3125, 11/16 = 0.6875 **7.** 8/52 = 0.154 **9.** 13/52 + 13/52 = 26/52 = 0.5

11. 13/52 + 12/52 − 4/52 = 21/52 = 0.404 **13.** $1 - P(S \cup H) = 1 - 21/52 = 31/52 = 0.596$

15. P (even or double) = 18/36 + 6/36 − 6/36 = 18/36 = 0.5 **17.** 2/36 = 0.056

19. 12/36 + 18/36 − 6/36 = 24/36 = 0.667

Exercise Set 11.3, page 448

1. $P(A') = 1/2$

3. $P(A \mid B) = \dfrac{P(A \cap B)}{P(B)} = \dfrac{1/10}{1/5} = 1/2$

5. $P(B \mid A) = P(B)$, are independent, 1/5 = 1/5

7. $P(A \cup B) = P(A) + P(B) - P(A \cap B) = 1/2 + 1/3 - 1/4 = 7/12$

9. $P(A \mid B) = \dfrac{P(A \cap B)}{P(B)} = \dfrac{1/4}{1/3} = 3/4$

11. a) 0.15 b) 3/7 = 0.429 c) 0.50

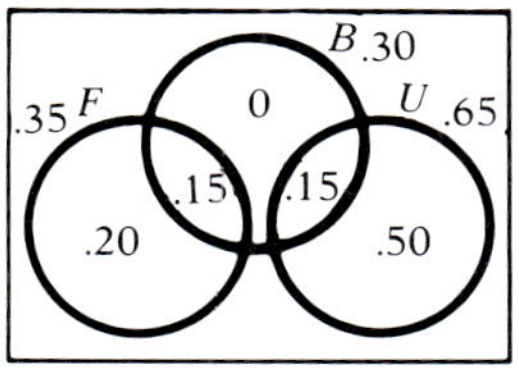

13. $P(L' \mid F) = \dfrac{P(L' \cap F)}{P(F)} = \dfrac{80/1200}{640/1200} = 0.125$

15. $P(L \mid \text{over } 25) = \dfrac{P(L \cap \text{over } 25)}{P\,(\text{over } 25)} = \dfrac{325/1200}{470/1200} = 0.691$

17. $P(L' \mid F) = 0.125$
$P(L') = 150/1200 = 0.125$, independent

19. a) $P(R' \cap F) = 0.20$ b) $P(F \mid R) = \dfrac{P(F \cap R)}{P(R)} = \dfrac{0.30}{0.45} = 0.667$ c) $P(R \mid F') = \dfrac{P(R \cap F')}{P(F')} = \dfrac{0.15}{0.50} = 0.3$
$P(R) = 0.45$ $P(R \mid F') \neq P(R)$, dependent

21. a) $P(B, G) = 0.16$ b) $P(B, B) = 0.04$

Exercise Set 11.4, page 455

1. $P(A_1 \mid B) = 0.348$ **3.** $P(A_3 \mid B) = 0.261$ **5.** $P(A_2 \mid B') = 0.352$
7. $P(Q \mid A) = 0.8$, $P(Q' \mid A) = 0.2$
9. a) $P(D \mid y) = 0.721$ b) $P(R \mid y) = 0.230$
11. a) $P(D \mid I) = 0.718$ b) $P(D \mid I') = 0.0001295$
13. $P(1 \mid E) = 0.213$
$P(2 \mid E) = 0.340$
$P(3 \mid E) = 0.128$
$P(4 \mid E) = 0.319$

Exercise Set 11.5, page 470

1. $P(8, 3) = 336$ **3.** $C(6, 2) = 15$ **5.** $C(10, 4) = 1260$ **7.** a) $P(6, 6) = 720$ b) $P(6, 4) = 360$
9. 386,216,000 committees **11.** a) 8.8186×10^9 b) 0.002641 **13.** $P = 0.0002966$
15. $P(\text{zero women}) = 0.05652$

Exercise Set 11.6, page 476

1. Yes **3.** Yes **5.** Yes **7.** Yes **9.** Yes **11.** No

13. a) $[0.6 \quad 0.4]\begin{bmatrix} 0.95 & 0.05 \\ 0.10 & 0.90 \end{bmatrix}$ b) $[0.61 \quad 0.39]$ c) $[2/3 \quad 1/3]$

Chapter 11 Review Problems, page 480

1. a) 0.15 b) No. $P(G \cap A) \neq P(G) \cdot P(A)$ c) 4/7
3. a) $P(S \mid IS) = 0.7826$ b) $P(S \mid IS') = 0.1176$
5. a) 0.1279 b) 0.2762
7. a) 80% will take dessert b) 80% will take dessert, 20% will not take dessert

CHAPTER 12

Exercise Set 12.1, page 490

1. a)

Interval	Frequency	Cumulative Frequency
0.95–0.982	2	2
0.983–1.015	4	6
1.016–1.048	1	7
1.049–1.081	10	17
1.082–1.114	6	23
1.115–1.147	1	24

b), c)

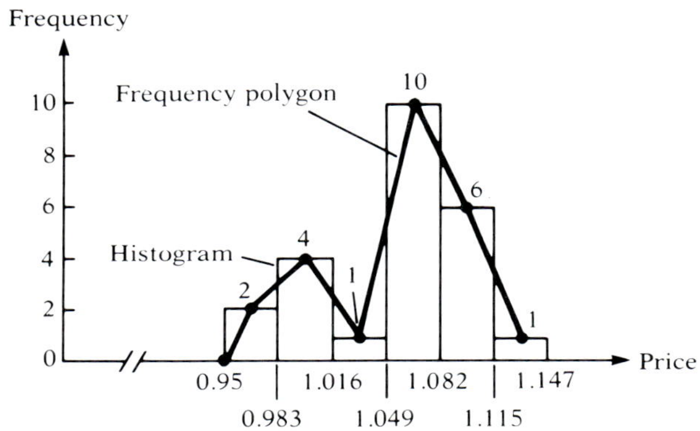

d)

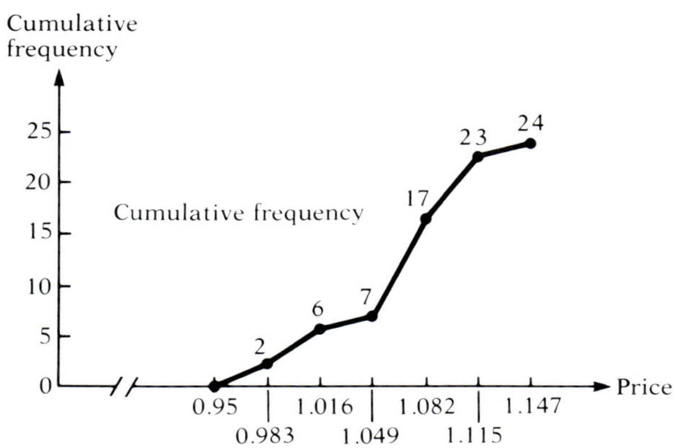

3. a)

Interval	Frequency	Cumulative Frequency
600–725	2	2
726–851	5	7
852–977	12	19
978–1103	21	40
1104–1229	18	58
1230–1355	10	68
1356–1481	7	75
1482–1607	5	80

b), c)

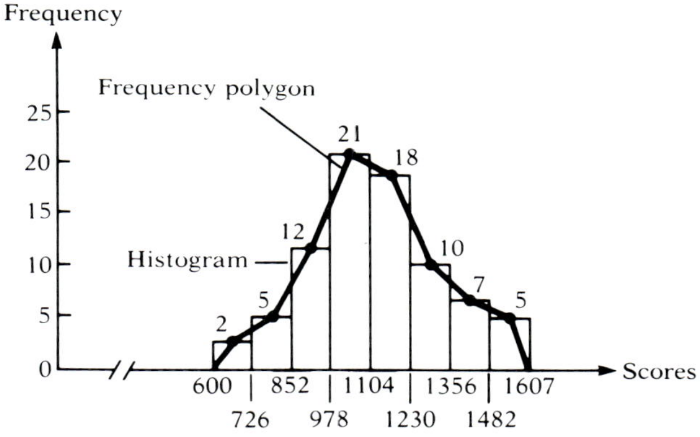

d)

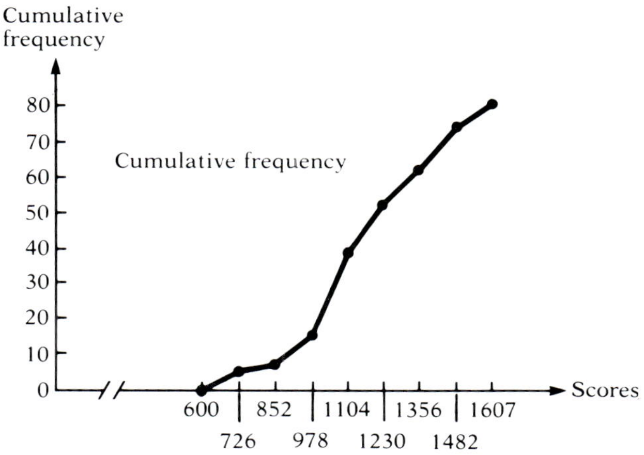

Exercise Set 12.2, page 496

1. Numerical **3.** Categorical **5.** Categorical **7.** Categorical

9. Mean = 41,583.33, median = 43,750, mode = 47,500

Exercise Set 12.3, page 503

1. Range = 6, standard deviation = 2.61, coefficient of variation = 48.29%
3. Range = 30, standard deviation = 9.54, coefficient of variation = 9.20%
5. a) Coefficient of variation (weight) = 21.43%, coefficient of variation (height) = 7.77% b) Weights
7. Standard time = 11.5, $10.87 \le x \le 12.13$, three workers were within the acceptable range

Exercise Set 12.4, page 509

1. a)

x	$P(x)$
0	0.132
1	0.329
2	0.329
3	0.165
4	0.041
5	0.004

b) 5/3 c) 1.054

3. a)

x	$P(x)$
0	0.95
1	0.048
2	0.00097
3	0.0000098
4	0.0
5	0.0

b) 4.95 c) 0.22

5. a)

x	$P(x)$
0	0.001
1	0.016
2	0.082
3	0.219
4	0.329
5	0.263
6	0.088

b) 1 c) 0.913

7. Mean = 2500, standard deviation = 35.355

Exercise Set 12.5, page 514

1. 0.1606 **3.** 0.0025 **5.** 0.0803 **7.** a) 0.99 b) 100%

Exercise Set 12.6, page 526

1. 0.50 **3.** 0.683 **5.** 0.309 **7.** 0.092 **9.** 11,920 **11.** a) 0.933 b) 0.067 **13.** 1
15. 0.873 **17.** 0.154

Exercise Set 12.7, page 529

1. 0.135 **3.** 0.393 **5.** 0.223 **7.** 0.000 **9.** 0.018

Chapter 12 Review Problems, page 533

1. a)

Interval	Frequency	Cumulative Frequency
0–0.999	5	5
1–1.999	9	14
2–2.999	13	27
3–3.999	12	39
4–4.999	6	45
5–5.999	5	50

b), c)

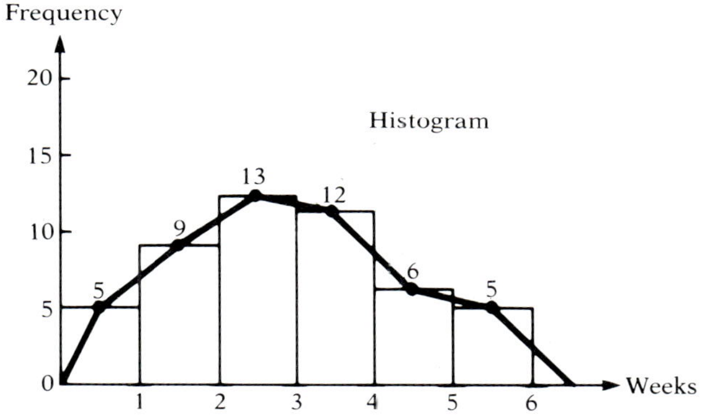

d)

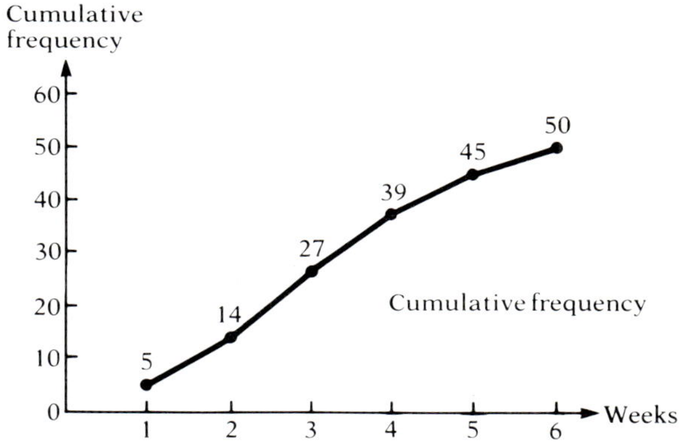

3. Mean = 206.875
median = 205
mode = 190
range = 40
standard deviation = 15.59
coefficient of variation = 7.526%

5. 0.96 **7.** 4 **9.** a) 0.65 b) $26,250 **11.** 0.0892 **13.** 0.0136 **15.** 0.2424 **17.** 0.866
19. 6827 **21.** 0.94295 **23.** 0.2636 **25.** 0.811

CHAPTER 13

Exercise Set 13.1, page 544

1. *EV* = 2 **3.** *EV* = 57 **5.** *EV* = 5 **7.** a) *EV* (red) = 1.5 b) *EV* (spade) = 0.7
c) *EV* (king) = 0.2 **9.** a) *EV* (gamble) = −0.05 b) *EV* (gamble) = −0.027
11. a) Expected daily demand = 275 b) *EV* = $46.70 **13.** *EV* = −0.18 **15.** $56,250

Exercise Set 13.2, page 551

1. *EV* (strategy 1) = $375, *EV* (strategy 2) = $475, *EV* (strategy 3) = $500

3. *EV* (strategy 1) = $390, *EV* (strategy 2) = $430, *EV* (strategy 3) = $550

5. a)

	State of Nature		
Strategies	**Upturn**	**Unchanged**	**Downturn**
Mutual funds	23,600	20,800	18,000
Corporate bonds	22,000	22,000	22,000

b) *EV* (mutual funds) = $21,500
EV (corporate bonds) = $22,000
Best investment = corporate bonds

7. a) *EV* (new design) = $10,000
EV (modify design) = $14,000
EV (current design) = 0
Best = modify current design

b) *EV* (new design) = $40,000
EV (modify design) = $26,000
EV (current design) = 0
Best = new design

9. EV (strategy 100) = \$100
EV (strategy 110) = \$102.50
EV (strategy 120) = \$100
EV (strategy 130) = \$ 90
Order 110 packages

Exercise Set 13.3, page 563

1. EV (fund) = \$20,000, Protean Products should fund.

3. a) EV (buy) = \$700, Sam should buy the skis and accessories

b) P (heavy | cold) = 0.571429
P (moderate | cold) = 0.214286
P (light | cold) = 0.214286
EV (buy | cold) = \$2500
Should buy if September is unseasonably cold.

P (heavy | not cold) = 0.3076923
P (moderate | not cold) = 0.3461538
P (light | not cold) = 0.3461538
EV (buy | not cold) = \$269.22
Sam should not buy the skis and accessories.

Exercise Set 13.4, page 575

1. Player A strategy 1, 0; player B strategy 1, 0; 0
3. Player A strategy 2/3, 1/3; player B strategy 1/3, 2/3; 2/3
5. Player A strategy 2/3, 1/3; Player B strategy 2/3, 1/3; −2/3
7. Player A strategy 28/67, 39/67; player B strategy 34/67, 33/67; 0.075
9. Player A strategy 8/13, 5/13; player B strategy 3/13, 10/13; 0.077
11. Player A strategy 1/4, 3/4; player B strategy 1/2, 1/2; 19/4
13. a)

$$\begin{array}{cc} & B \\ & \begin{array}{cc} 1 & 2 \end{array} \\ A \begin{array}{c} 1 \\ 2 \end{array} & \begin{bmatrix} -2 & 3 \\ 3 & -4 \end{bmatrix} \end{array}$$

b) Player A strategy 7/12, 5/12; player B strategy 7/12, 5/12; 1/12

15. Ace strategy 10/27; 17/27; 1/9
17. Precision should buy 150 tons raw materials at \$20 a ton.

Chapter 13 Review Problems, page 579

1. EV = 8 **3.** a) EV(number of hearts) = 0.5 b) EV(number of face cards) = 0.462

5. a) EV(one-month supply) = −300
EV(three-month supply) = −450
Buy one-month supply

b) EV(one-month suypply) = −1500
EV(three-month supply) = −250
Buy three-month supply

7. EV(8/4) = 7.5
EV(4/8) = 6
EV(6/6) = 13.75
She should study 6 hours for each

9. Player A strategy 3/7, 4/7; player B strategy 5/7, 2/7; 29/7
11. Saddle point at −10, −10

INDEX